U.S.-JAPAN WORKSHOP ON ION TEMPERATURE GRADIENT-DRIVEN TURBULENT TRANSPORT

AIP CONFERENCE PROCEEDINGS 284

U.S.-JAPAN WORKSHOP ON ION TEMPERATURE GRADIENT-DRIVEN TURBULENT TRANSPORT

AUSTIN, TX 1993

EDITORS: **W. HORTON**
A. WOOTTON
UNIVERSITY OF TEXAS
AUSTIN, TX

M. WAKATANI
KYOTO UNIVERSITY
KYOTO, JAPAN

American Institute of Physics   New York

Authorization to photocopy items for internal or personal use, beyond the free copying permitted under the 1978 U.S. Copyright Law (see statement below), is granted by the American Institute of Physics for users registered with the Copyright Clearance Center (CCC) Transactional Reporting Service, provided that the base fee of $2.00 per copy is paid directly to CCC, 27 Congress St., Salem, MA 01970. For those organizations that have been granted a photocopy license by CCC, a separate system of payment has been arranged. The fee code for users of the Transactional Reporting Service is: 0094-243X/87 $2.00.

© 1994 American Institute of Physics.

Individual readers of this volume and nonprofit libraries, acting for them, are permitted to make fair use of the material in it, such as copying an article for use in teaching or research. Permission is granted to quote from this volume in scientific work with the customary acknowledgment of the source. To reprint a figure, table, or other excerpt requires the consent of one of the original authors and notification to AIP. Republication or systematic or multiple reproduction of any material in this volume is permitted only under license from AIP. Address inquiries to Series Editor, AIP Conference Proceedings, AIP Press, American Institute of Physics, 500 Sunnyside Boulevard, Woodbury, NY 11797-2999.

L.C. Catalog Card No. 93-72460
ISBN 1-56396-221-7
DOE CONF-9301107

Printed in the United States of America.

CONTENTS

Preface ix

1. INTRODUCTION

1.1. Introduction 3
1.2. Neoclassical Heat Conduction 4
1.3. Mechanism of Ion Temperature Gradient (ITG)-Driven
     Turbulent Transport 6

2. ION THERMAL TRANSPORT IN TOKAMAKS

2.1. Parametric Variations of Ion Transport in TFTR 19
   S. D. Scott, C. W. Barnes, D. Ernst, J. Schivell, E. J. Synakowski,
   M. G. Bell, R. E. Bell, C. E. Bush, E. D. Fredrickson, B. Grek, K. W. Hill,
   A. Janos, D. L. Jassby, D. Johnson, D. K. Mansfield, D. K. Owens, H. Park,
   A. T. Ramsey, B. C. Stratton, M. Thompson, and M. C. Zarnstorff
2.2. Ion Temperature Profile Simulation of JT-60 Plasmas with Ion Temperature
     Gradient Mode Transport Models 60
   H. Shirai, T. Hirayama, Y. Koide, and M. Azumi
2.3. Ion Transport Analysis Based on η_i Mode Turbulence in JT-60 ... 87
   M. Yagi and M. Azumi
2.4. A Comparison Between the Measured Thermal Diffusivity χ_i^{Exp} in TEXT and
     the Predicted $\chi_i^{\eta_i}$ 116
   A. Ouroua, A. J. Wootton, and the TEXT Group

3. BASIC GRAD-T_i PLASMA EXPERIMENTS

3.1. ITG Instabilities and Transport Studies in the Columbia Linear Machine .. 135
   A. K. Sen, J. Chen, B. Song, and R. G. Greaves
3.2. Another Look At Experimental Evidence for Ion Temperature
     Gradient-Driven Turbulence in Tokamaks 165
   D. L. Brower
3.3. Plasma Transport Analysis for the Compact Helical System and the Large
     Helical Device 188
   K. Y. Watanabe, K. Yamazaki, H. Yamada, and T. Amano

4. PARTICLE SIMULATIONS

4.1. Kinetic Simulation of Microinstabilities in Tokamak Plasmas 213
    W. W. Lee, S. E. Parker, R. A. Santoro, and J. C. Cummings
4.2. 3D Gyrokinetic Particle Simulations of Ion Temperature Gradient-Driven Turbulence and Transport 224
    R. D. Sydora
4.3. Transport in the Self-Organized Relaxed State of Ion Temperature Gradient Instability 255
    T. Tajima, Y. Kishimoto, M. J. LeBrun, M. G. Gray, J. Y. Kim, W. Horton, V. Wong, and M. Kotschenreuther
4.4. The Marginal Stability of Experimental Profiles 276
    M. Kotschenreuther

5. FLUID AND GYROFLUID SIMULATIONS

5.1. Reactive Drift Wave Model for Tokamak Transport 295
    H. Nordman and J. Weiland
5.2. Toroidal Turbulence Simulations with Gyro-Landau Fluid Models in a Nonlinear Ballooning Mode Representation 310
    R. E. Waltz and G. D. Kerbel
5.3. Nonlinear Gyrofluid Model of ITG Turbulence 344
    W. Dorland, G. W. Hammett, T. S. Hahm, and M. A. Beer

6. STABILITY AND TRANSPORT THEORY

6.1. Kinetic Toroidal Analysis of Transport Trends in TFTR Plasmas 371
    G. Rewoldt and W. M. Tang
6.2. Progress in the Understanding of the Ion Temperature Gradient-Driven Transport 391
    F. Romanelli
6.3. Modeling of Anomalous Ion Thermal Transport in Hot Plasma 403
    J. Y. Kim and W. Horton
6.4. Cross-Field Energy Flux Due to Ion Temperature Gradient Mode in a Tokamak 428
    T. Yamagishi
6.5. Anomalous Particle and Electron Energy Pinches 444
    R. R. Dominguez
6.6. Ion Temperature Gradient-Driven Impurity Modes 469
    S. Migliuolo
6.7. Impurity Effect on Kinetic η_i-Mode in Tokamak Plasma 486
    J. Q. Dong, W. Horton, and X. N. Su

7. TURBULENCE THEORY

7.1. Transport Analysis Based on K-ϵ Anomalous Transport Model 509
  H. Sugama, M. Okamoto, and M. Wakatani
7.2. Resistive Interchange Instabilities in Heliotron E 526
  M. Wakatani
7.3. Point Vortex Dynamics in a Magnetized Plasma 559
  M. Kono, H. Shibahara, and K. Yabuki
7.4. Comments on Single-Helicity Versus Multi-Helicity Simulations 579
  H. Sugama, A. D. Beklemishev, and W. Horton
Subject Index 587
Author Index 591

PREFACE

The subject of this book is the plasma physics problem of determining the thermal transport rate of the ion kinetic energy across a confining magnetic field. In high-temperature, fully ionized gases confined by magnetic fields, principally in toroidal confinement geometries, the plasma transport rates of particles and thermal energy greatly exceed the transport rate calculated from Coulomb collisions. Finding the mechanism and the appropriate formulae describing the actual transport rates is of fundamental importance both for the intellectual understanding of basic plasma physics and for the design of controlled thermonuclear fusion reactors. The theoretical and experimental understanding of the plasma transport is based on statistical physics, the physics of turbulence, and knowledge of the long-range collective interactions between the charged particles that form the natural oscillations of the plasma system.

The current understanding of the transport mechanism is developed in this book by leading plasma physicists. The experimental and theoretical investigations presented here show that the long-range collective oscillations of the drift-wave type driven by the temperature gradient give rise to small-scale turbulent excursions of ions and electrons across the magnetic field. Chapter 1 reviews the basic mechanisms giving rise to the fluctuations. On the macroscopic scale of the confinement systems, the resulting transport is typically expressed in terms of an anomalous ion thermal diffusivity χ_i. Chapters 2 and 3 explore our experimental knowledge of thermal flux from ion temperature gradients of up to 50 keV/m or 5.8×10^8 °C/m, sufficient for a fusion reactor, and the associated effective diffusivities of χ_i which typically range from 0.1 m^2/s to 5 m^2/s, which is a few times larger than desirable for an efficient fusion reactor.

Chapters 4 through 7 develop the underlying understanding of the thermal transport mechanisms in terms of the basic plasma physics. The methods used in describing the nonlinear dynamics and transport are particle simulation, fluid and gyro-fluid simulation, and basic theory.

The editors give many thanks to the staff of the Institute for Fusion Studies who have worked diligently on this book. In particular, we owe much to Ms. Suzy Mitchell for technical word processing, to Mr. Ed Bailey and Dr. Mikhail Isichenko for working on the electronic form of the book, and to Ms. Joan Gillette for managing the documents and correspondence with the publisher. Finally, we note that the book grew out of a workshop sponsored by the US-Japan Joint Institute for Fusion Theory (JIFT), that was held in January of 1993 at The University of Texas at Austin.

W. Horton
M. Wakatani
A. Wootton

Chapter 1
INTRODUCTION

1. INTRODUCTION

One of the central issues in the magnetic confinement approach to controlled thermonuclear fusion is to determine the process and the rate at which ion thermal energy is transported across the magnetic field. The intrinsic, irreducible rate of thermal transport arises from the direct Coulomb collisions between the ions. The relevant formulas for this collisional transport are briefly reviewed in Sec. 1.2. In the 1970's when the plasma temperatures were typically in the few kilovolt range these collisional formulas were thought to provide a basis for understanding the observed ion thermal loss rates. As the plasma temperatures have increased to tens of kilovolts and the role of the turbulent fluctuations has become understood as a primary mechanism for the transport of both plasma particles and thermal energy across the magnetic field, the scientific view has changed to recognize that some form of drift wave turbulence driven by the ion temperature gradient is the dominant mechanism for the transport of thermal energy. The basic mechanism for the drift wave turbulent transport is outlined in Sec. 1.3 under the topic of Ion Temperature Gradient (ITG) Driven Turbulence which is the main issue of this book. A historical perspective on the change in the understanding of the thermal transport mechanism is one theme in the comprehensive chapter 2.1 by Scott *et al.* on the parametric variations of the ion transport.

In this volume leading research scientists from the large U.S. and Japanese tokamak experiments, computer simulation scientists, experimentalists from the basic plasma property experiments and leading theorists present their latest results on the mechanisms for the turbulent transport of the ion thermal energy in magnetically confined plasma. Plasmas with central ion temperatures in the range of 30 keV or 3×10^8 °K are routinely produced in the TFTR experiment at the Princeton Plasma Physics Laboratory. In JT-60U of the Japanese Atomic Energy Research Institute at Naka, which is the recent upgrade of JT-60 to increase the plasma current and the plasma volume, the central ion temperature has increased up to 38 keV at the average density of 4.3×10^{19} m^{-3} in the high poloidal beta operations. Some of those high ion temperature experimental results, which are especially relevant to the ITG mode, are contained in this volume. At the same time, the latest theories and simulations attempting to analyze those experimental results with ITG turbulence model are presented.

Guides to the contents of this volume are given in Tables 1.1 and 1.2. Table 1.1 identifies the principal experimental studies contained in the volume, and in Table 1.2 the material is catalogued by issues. The second grouping by issue is not unique since there is interplay between various physical effects, and the table is meant to be a tour guide for the newcomer to this complex array of plasma/fusion research.

2. NEOCLASSICAL HEAT CONDUCTION

In the absence of fluctuations the ion orbits in a tokamak divide into those with pitch angles smaller than the value required for mirroring from the nonuniform magnetic field B variation ($B_{max} - B_{min})/B_0$ ($\simeq 2r/R$), which are passing particles and those particles with larger pitch angles which are trapped. The trapped ions have the larger radial orbit width controlled by $\rho_{i\theta} = c(m_i T_i)^{1/2}/eB_\theta$ — the ion gyroradius in the poloidal magnetic field $B_\theta \ll B$, and the larger effective collision frequency $\nu_{\text{eff}} = \nu_i/\epsilon$ where $\epsilon = r/R \ll 1$. For the large banana orbits to be formed the ion-ion collisional frequency ($\nu_i = 230 \hat{n}_{13}/T_i^{3/2}$(keV)/ sec) must be sufficiently low that $\nu_{*i} = (R/r)^{3/2}(qR\nu_i/v_i) < 1$ which is well satisfied in today's magnetic confinement experiments. In this regime the collisional component of the tokamak thermal heat flux is given by (Chang and Hinton, 1986)

$$q_i = -n_i \left(\frac{r}{R}\right)^{1/2} \nu_i \rho_{i\theta}^2 \left(0.66 + 1.88 \left(\frac{r}{R}\right)^{1/2}\right) \frac{dT_i}{dr}$$

$$\equiv -n_i \chi_i^{\text{neo}} \frac{dT}{dr} \ . \tag{1}$$

In Eq. (1) the first term (0.66) arises from the pitch angle scattering and the second term from energy scattering (Bolton and Ware, 1983). Formula (1) is the base rate of thermal conduction in a pure hydrogen plasma in the absence of significant neutral beam heating. Under practical laboratory conditions the collisional thermal flux may exceed the rate given in Eq. (1) due to two effects: (1) impurities and (2) neutral beam heating. In the presence of an ion particle flux Γ_i the total thermal flux Q_i becomes $Q_i = q_i + (3/2)\Gamma_i T_i$.

In particular, the influx of impurity ions (Z) allows a larger outflow of fast hydrodynamic ions while still maintaining a charge neutral, or ambipolar particle flux $\Gamma_i + Z\Gamma_z = -\Gamma_e \simeq 0$ where Γ_e is the small electron particle flux controlled by the electron gyroradius in the poloidal magnetic field. The collisional Γ_e is in fact negligible: the turbulent $\langle \tilde{n} \tilde{\mathbf{E}} \times \mathbf{B} \rangle$ flux produces the relevant ambipolar particle flux. Secondly, strong neutral beam heating (NBI) elevates the population of ions with kinetic energies greater than $\frac{3}{2} k_B T_i$ resulting in an enhanced thermal flux. In certain cases these enhancements to the neoclassical thermal flux q_{NC} in Eq. (1) can be rather large. As an example, Ware (1988) works out that these enhancements give up to a factor of six for the PDX experiment which was a strongly beam heated plasma with $T_i = 1.6\,\text{keV}$, $T_e = 0.6\,\text{keV}$, $Z_{\text{eff}} \simeq 3$. In particular the beam injection effect gives an enhancement of $(6/Z_{\text{eff}})q_{NC}$ and the impurity influx effect gives up to $4q_{NC}$. The power balance studies in that experiment require an anomaly factor of the order of 6 and according to Ware (1989) enhancements to the collisional transport by such a factor appear possible for this experiment. Generally, however, the use of enhancements to the basic neoclassical ion thermal flux (1) has not been able to explain the persistent "anomalies" in the ion thermal flux. Chapter 2.1, by Scott et al., gives a historical account of the problems associated with the use of the collisional transport formulas to explain the measured losses.

Recents confinement experiments in this same lower temperature range of a few kilovolts reported in this volume are the two TEXT tokamak experiments and the Heliotron/Torsatron experiment in the Heliotron E and CHS (Chapter 3). In each case there is substantial evidence for anomalous ion thermal conductivity. In the case of the TEXT experiment there are detailed measurements of the fluctuation spectrum in the interior and the edge plasma (Brower et al., Chapter 3.2).

In the case of the JT-60 and TFTR plasmas the ion temperature is sufficiently high ($T_i(r) \geq 10\,\text{keV}$ in the transport zone) that the power loss inferred from the thermal flux is over an order of magnetic greater than the flux given by q_i^{NC} from Eq. (1).

Perturbative transport studies of the type described by Zarnstorff *et al.* (1990) yield insight into the physical mechanism controlling the transport. A typical, widely studied perturbative experiment is the TFTR profile modification discharge that was performed by Zarnstorff, *et al.* for the purpose of studying the ion temperature gradient turbulent transport dependence on the density profile. Detailed studies of the stability of this discharge are reported in the joint IFS-PPPL work (Horton *et al.*, 1992), by Kotschenreuther, *et al.* (1992), and by Nordman, *et al.* (1992). The results of these transport studies show that, while the issues of the role of the ionized carbon component remains somewhat uncertain, the discharge power balance in both the steep density gradient and flatter density profile regime appears consistent with the ion temperature gradient driven thermal transport.

3. MECHANISM OF ION TEMPERATURE GRADIENT (ITG) DRIVEN TURBULENT TRANSPORT

The mechanism for the turbulent ion thermal transport may be characterized as the magnetized plasma analog of the Rayleigh-Bernard convection in neutral fluids. In the nonuniform magnetized plasma there are collective oscillations from the long-range part of the Coulomb interaction that gives rise to drift wave-ion acoustic oscillations $(\omega, \mathbf{k})$ with self-consistency determining the dispersion relation where all the fluctuations are taken to vary as the real part of $\exp(\mathbf{k} \cdot \mathbf{x} - i\omega t)$. These oscillations with angular frequency ω [100 KHz to 1 MHz] and wavenumbers $\mathbf{k}$ [1cm^{-1} to 50 cm^{-1}] are the natural modes of the particles interacting through the collective electric forces. The dispersion connecting ω and $\mathbf{k}$ in simplest approximation is

$$(1 + k_\perp^2 \rho_s^2)\omega^2 - \omega\omega_{*e}\left(1 - 2\varepsilon_n - k_\perp^2 \rho_i^2(1 + \eta_i)\right) + \omega_{*pi}\omega_{Di} = 0 \quad (2)$$

where the characteristic drift frequencies ω_{*e}, ω_{*pi}, ω_{Di} are now derived and discussed. Note that in Eq. (2) ω is the angular frequency in the ion rest frame (the plasmas are often in motion with respect to the laboratory frame and thus the observed frequency can show large Doppler shifts that are non-uniform across the plasma). Recent measurements of the correlation lengths and the frequency spectra in TFTR are given by Mazzucato and Nazikian (1993), with a comparison between the NBI heated plasmas with the injected power of 14×10^6 Watts and the Ohmic heated discharges. The injected plasma shows a much higher level and broad band of turbulent fluctuations than the Ohmic discharge.

The inhomogeneous, magnetized plasma contains low frequency normal modes that result from the long-range part of the Coulomb interactions between charged particles. The ion temperature gradient instability arises out of the drift wave-ion acoustic wave branch of these collective modes (Horton, 1984). The basic plasma experiments confirming the properties of the drift wave-ion acoustic waves were performed by Hendel *et al.* (1968). In these oscillations the electrons establish a near-Boltzmann response

$$n_e = N_0 \exp\left(\frac{e\Phi}{T_e}\right) = n_0 \left(1 + \frac{e\delta\phi}{T_e}\right) \quad (3)$$

to the fluctuating part of the electrostatic potential $\Phi = \Phi_0(r) + \delta\phi(\mathbf{x}, t)$ and $n_0 = N_0 \exp(e\Phi_0(r)/T_e)$. The ions respond in the long wavelength limit with oscillations in the ion density δn_i, the parallel velocity $\delta v_\parallel$ and the pressure δp_i by the linearized fluid equations

$$-i\omega \delta n_i + v_{Ex} \frac{\partial n_0}{\partial x} + n_0 \nabla \cdot \mathbf{v}_E + ik_\parallel n_0 \delta v_\parallel = 0 \quad (4)$$

$$-i\omega \delta v_\parallel = \frac{e_i}{m_i}\left(\tilde{E}_\parallel - \frac{1}{e_i n_i} ik_\parallel \delta p_i\right) \quad (5)$$

$$-i\omega \delta p_i + v_{Ex} \frac{\partial p_i}{\partial x} + \gamma_\perp p_i \nabla \cdot \mathbf{v}_E + \gamma_\parallel p_i ik_\parallel \delta v_\parallel = 0 \ . \quad (6)$$

These three hydrodynamic equations, while approximations to the full gyrokinetic equations, describe the essential features of the ion temperature gradient driven instability. We briefly explore the plasma flows described by Eqs. (3)–(6) using the diagrams in Figs. 1 and 2. The full gyrokinetic equations and the intermediate level description of gyrofluid are developed in Chapters 3 and 4.

For the benefit of the new researcher in this field, it may be useful to write out the dispersion relation, along with the conventional definitions of the characteristic frequencies, that results from the linear ion field oscillations $(\widehat{\delta n}_i, \widehat{\delta v}_\parallel, \widehat{\delta p}_i)\exp(ik_x x + ik_y y + ik_\parallel z - i\omega t)$ and the condition

8 Introduction

of quasineutrality

$$\delta n_e = n_0 \frac{e\delta\phi}{T_e} = \delta n_i \tag{7}$$

when applied to Eqs. (4)–(6). After some algebra one derives the local dispersion $D(\omega, \mathbf{k}) = 0$ that gives the collective modes of the system, where

$$D(\omega, \mathbf{k}) = 1 - \frac{\omega_{*e}}{\omega} - \frac{k_\parallel^2 c_s^2}{\omega^2}\left(1 - \frac{\omega_{*pi}}{\omega}\right) + \left(k_\perp^2 \rho_s^2 - \frac{\omega_{Di}}{\omega}\right)\left(1 - \frac{\omega_{*pi}}{\omega}\right). \tag{8}$$

Here we define the following important characteristic frequencies of the system:

| electron drift frequency | $\omega_{*e} = -\dfrac{k_y c T_e}{eBn}\dfrac{dn}{dx}$ | (9) |

| ion acoustic frequency | $\omega_s = k_\parallel c_s = k_\parallel \left(\dfrac{T_e}{m_i}\right)^{1/2}$ | (10) |

| ion inertial scale | $\rho_s = \dfrac{c(m_i T_e)^{1/2}}{e_i B}$ | (11) |

| ion pressure gradient drift frequency | $\omega_{*pi} = \dfrac{k_y c}{e_i B n_i}\dfrac{dp_i}{dx}$ | (12) |

| ion toroidal drift frequencies | $\omega_{Di} = \dfrac{2 k_y c T_i}{e_i B R}$ | (13) |

Again, we note that the derivation of Eq. (8) is in the ion fluid rest frame and that in the laboratory frame there can be large, nonuniform Doppler shifts to $\omega \to \omega + \mathbf{k} \cdot \mathbf{v}_i$ (for a recent analysis of the stability effects from sheared ion mass flow $\rho \mathbf{v}_i$, see Dong and Horton (1993)).

The conventional definition for the parameters of a confinement system is as follows:

$$\frac{1}{L_n} = -\frac{1}{n_0}\frac{dn_0}{dx} \tag{14}$$

$$\frac{1}{L_{pi}} = -\frac{1}{p_i}\frac{dp_i}{dx} = \frac{1}{L_n}(1+\eta_i) \qquad (15)$$

and for the dimensionless parameters

$$\eta_i = \frac{d\ell n\, T_i}{d\ell n\, n_0} \qquad (16)$$

$$\varepsilon_n = \frac{L_n}{R}. \qquad (17)$$

While these symbols (9)–(17) have generally become the standard choices, the reader must be aware of occasional variations that occur with different researchers (such as $L_n \to r_n$ and $\varepsilon_n \to 2\varepsilon_n$, or $G = 2\varepsilon_n$, etc.).

The driving mechanisms for instabilities are clearest in two limiting cases: (1) The slab limit in which the toroidal drift frequency $\omega_{Di} = -2\varepsilon_n(\omega_{*e}T_i/T_e) \to 0$ in Eq. (13). In this cylindrical limit, the instability is called the slab ion temperature gradient mode with the parallel flows $\delta v_\parallel$ and the ion pressure convection giving $(k_\parallel^2 c_s^2/\omega^2)(1-\omega_{*pi}/\omega) < 0$ for $0 < \omega < \omega_{*pi}$ and the exponential $e^{\gamma t}$ growth rate is $\gamma_{\max} \sim (k_\parallel^2 c_s^2 \omega_{*pi})^{1/3}$. The dynamics of the slab instability are indicated in Fig. 1.1 and shown experimentally in the Columbia Linear Machine in Chapter 3. (2) The second regime is the toroidal limit where $X = |k_\parallel c_s/\omega_{Di}| \sim \frac{\varepsilon_n}{q}\frac{1}{k_y \rho_i} < 1$ so that the toroidal drift frequency ω_{Di} drives the instability with $\omega^2 \cong -\omega_{*pi}\omega_{Di}/(1+k_\perp^2 \rho_s^2)$ giving the maximum growth rate $\gamma_{\max} = |k_y|\rho_s v_i/(L_{pi} R)^{1/2}$. The convective plasma motions for the toroidal ion temperature gradient mode are shown in Fig. 1.2 in the form that the instability is investigated in tokamaks in Chapter 2.

In the slab regime the parallel mass flows give rise to the unstable growth of the local hot spots in the $m = 2$ convective cells as shown in Fig. 1.1. On the other hand, in the toroidal regime the mechanism is the interchange of the higher and lower pressure cells as shown in Fig. 1.2. Due to the so-called "unfavorable" toroidal curvature $\omega_{*pi}\omega_{Di} > 0$ the unstable convection occurs on the outside of the torus shown in Fig. 1.2. This cells grow exponentially in time while propagating in the either ion or electron diamagnetic direction in θ. In Fig. 1.2 it was necessary to exaggerate the size and coherence of the convective motion which in both experiments in Chapter 3 and simulations in Chapter 4, occur as small scale fluctuations. The basic plasma physics studies of the toroidal temperature gradient mode are given in Chapter 3.

These brief introductory remarks may help the reader understand the nature of the critical issues discussed and analyzed by leading fusion physicsists in the following chapters.

REFERENCES

1. C. Bolton and A.A. Ware, Phys. Fluids **26**, 459 (1983).

2. C.S.Chang and F.L. Hinton, Phys. Fluids **29**, 3314 (1986).

3. H.W. Hendel, T.K. Chu, and P.A. Politzer, Phys. Fluids **11**, 2426 (1968)

4. W. Horton, D-I. Choi, and W.M. Tang, Phys. Fluids **24**, 1077 (1981).

5. W. Horton, in *Handbook of Plasma Physics II*, edited by M.N. Rosenbluth and R.Z. Sagdeev (North-Holland, Amsterdam, 1984), p.383.

6. W. Horton, D. Lindberg, J.Y. Kim, J.Q. Dong, G.W. Hammett, S.D. Scott, M. Zarnstorff, and S. Hamaguchi, Phys. Fluids B **4**, 953 (1992).

7. J.Q. Dong and W. Horton, Phys. Fluids B **5**, 1581 (1993).

8. M. Kotschenreuther, H.L. Berk, M. LeBrun, J.Q. Dong, W. Horton, J-Y. Kim, D.W. Ross, T. Tajima, P.M. Valanju, H.V. Wong, B. Miner, D.C. Barnes, J.U. Brackbill, K.M. Ling, R.A. Nebel, W.D. Nystrom, J.A. Byers, B.I. Cohen, R.H. Cohen, A.M. Dimits, L.L. Lodestro, N. Mattor, G.R. Smith, T.J. Williams, G.D. Kerbel, J.M. Dawson, R.D. Sydora, B.A. Carreras, N. Dominguez, C.L. Hedrick, J-N. Leboeuf, H. Naitou, and T. Kamimura in *Plasma Physics and Controlled Nuclear Fusion Research*, 1992, Proceedings of the 14th International conference, Wurzburg (IAEA, Vienna, 1993).

9. E. Mazzucato and R. Nazikian, "The radial scale length of turbulent fluctuations in the main core of TFTR plasma," preprint, (1993).

10. H. Nordman and J. Weiland, Nucl. Fusion, **9**, 1653 (1992)

11. A.A. Ware, in *Plasma Physics and Controlled Nuclear Fusion Research*, 1988, Proceedings of the 12th International conference, Nice (IAEA, Vienna, 1989), Vol. 2, p. 106.

12. M.C. Zarnstorff, C.W. Barnes, P.C. Efthimion, G.W. Hammett, W. Horton, R.A. Hulse, D.K. Mansfield, E.S. Marmar, K. McGuire, G. Rewoldt, B.C. Stratton, E.J. Synakowski, W.M. Tang, J. Terry, X.Q. Xu, M.G. Bell, M. Bitter, N.L. Bretz, R. Budny, C.E. Bush, R.J. Fonck, E.D. Fredrickson, H.P. Furth, R.J. Goldston, B. Grek, R.J. Hawryluk, K.W. Hill, H. Hsuan, D.W. Johnson, D.C. McCune, D.M. Meade, S.S. Medley, D. Mueller, D.K. Owens, H.K. Park, A.T. Ramsey, M.N. Rosenbluth, J. Schivell, G.L. Schmidt, S.D. Scott, G. Taylor, and R.M. Wieland, in *Plasma Physics and Controlled Nuclear Fusion Research*, 1990, Proceedings of the 13th International conference, Washington (IAEA, Vienna, 1991), Vol. 1, p. 109.

Experimental Studies

- Identification of the ITG Mode
  3.1 Sen, 3.2 Brower, 3.3 Watanabe
- Parametric Dependence of Steady State Transport
  2.1 Scott, 2.2 Shirai, 2.3 Yagi, 2.4 Ouroua
- Perturbative Experiments
  density modification and parameter variations
  2.4 Ouroua, 3.2 Brower

Table 1.1

Ion Temperature Gradient Transport Issues

- Spatial Structures
   radial correlation scale (4.3 Tajima, 6.2 Romanelli)
   vortex versus wave turbulence (7.3 Kono)
   radial variation of turbulence level and diffusivity (5.2 Waltz, 6.3 Kim, 7.4 Sugama)
- Transport Mechanism and Rates
   mixing length transport (6.2 Romanelli, 7.1 Sugama, 7.2 Wakatani)
   quasilinear transport, heat pinch (5.1 Nordman, 6.1 Rewoldt, 6.4 Yamagish, 6.5 Dominguez)
   critical gradient models (4.3 Tajima)
   vortex transport (7.3 Kono)
- Linear Stability Properties
   impurity effects and impurity modes (4.4 Kotschenreuther, 6.3 Kim, 6.6 Migliuolo, 6.7 Dong)
   beam and anisotropy effects (2.1 Scott, 4.4 Kotschenreuther, 6.3 Kim)
   electromagnetic effects (6.3 Kim)
- Fluid and Gyrofluid Simulations (5.1 Nordman, 5.2 Waltz, 5.3 Dorland)
- Particle Simulations (4.1 Lee, 4.2 Sydora, 4.3 Tajima, 4.4 Kotschenreuther)

Table 1.2

14 Introduction

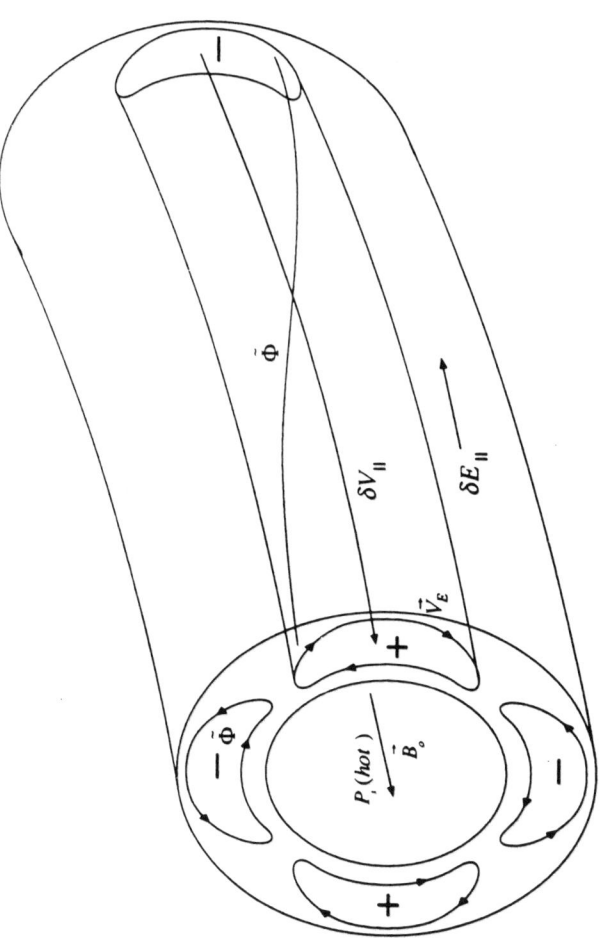

Figure 1. Slab ∇T_i Instability Mechanism

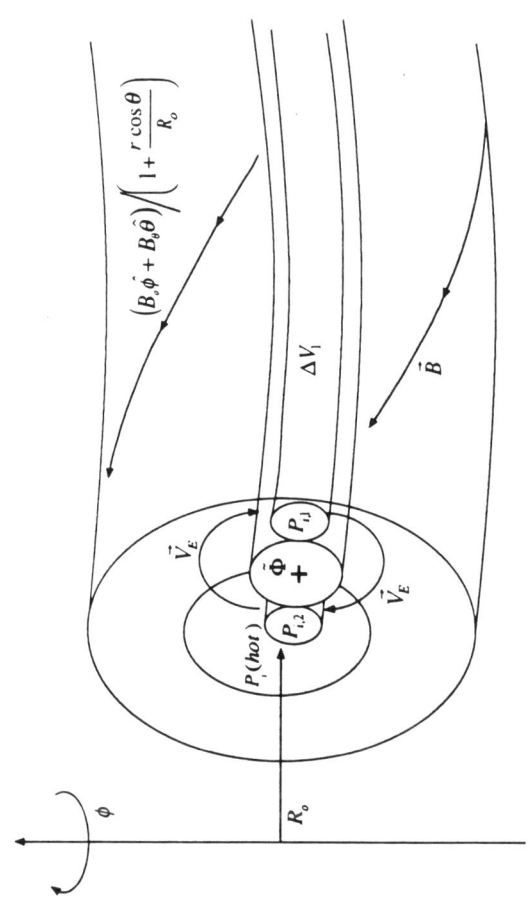

Figure 2. Toroidal ∇T_i Instability Mechanism

Chapter 2

ION THERMAL TRANSPORT IN TOKAMAKS

2.1 PARAMETRIC VARIATIONS OF ION TRANSPORT IN TFTR

S.D. Scott, Cris W. Barnes[1], D. Ernst[2], J. Schivell, E.J. Synakowski, M.G. Bell, R.E. Bell, C.E. Bush[3], E.D. Fredrickson, B. Grek, K.W. Hill, A. Janos, D.L. Jassby, D. Johnson, D.K. Mansfield, D.K. Owens, H. Park, A.T. Ramsey, B.C. Stratton, M. Thompson, and M.C. Zarnstorff
Princeton Plasma Physics Laboratory, Princeton, New Jersey 08543

2.1.1 INTRODUCTION

This chapter is divided into three roughly independent sections. The first is a historical review of the twenty year history of experimental ion heat transport measurements from many tokamaks. The second is a study of ion heat transport in Ohmic TFTR plasmas which shows that $\chi_i \sim \chi_e \approx 15\chi_i^{\text{neo}}$. Thus, ion heat transport is demonstrated to be strongly anomalous even in the absence of auxiliary heating. The third section describes the variation of χ_i with local ion temperature in TFTR during auxiliary heating, with emphasis on characterizing the differences between transport in the L-mode and supershot regimes. The results are consistent with the conjecture that improved ion energy confinement in supershot plasmas is caused by a high ratio of T_i/T_e.

2.1.2 REVIEW OF χ_i MEASUREMENTS

We begin with a review of measurements of ion heat transport in Ohmic and beam-heated plasmas. Our objective here is principally to provide an annotated bibliography of various experimental χ_i studies, rather than a complete review of the literature. Mostly for historical perspective, the early χ_i measurements will be compared to the predictions of then-current neoclassical theory. This approach does not provide

---

[1] Physics Division, Los Alamos National Laboratory, Los Alamos, New Mexico
[2] Massachusetts Institute of Technology, Cambridge, Massachusetts
[3] Oak Ridge National Laboratory, Oak Ridge, Tennessee

a time-independent benchmark because the neoclassical theory itself evolved during the 1970's and early 1980's. Also, it is slightly deficient in that χ_i is found to be so large a 'multiple' of χ_e^{neo} in current, large tokamaks that the relevance of neoclassical theory to ion heat transport is questionable. But it remains the most convenient approach because the original measurements themselves were typically compared to neoclassical theory. More recent χ_i measurements will be considered also in comparison to χ_e and to qualitative predictions of ion temperature gradient turbulence (ITGDT) theory.

Interestingly, transport studies in about a dozen different tokamaks worldwide found χ_i to be roughly consistent with neoclassical theory in both Ohmic and auxiliary-heated plasmas for more than ten years, from 1970 to the early to mid-1980's. Due to limited diagnostic capability and strong ion-electron coupling, initial studies of ion transport in small tokamaks (T-3,TM-3,T-4) could establish only that χ_i was roughly the correct order of magnitude as expected from neoclassical theory (Table 2.1). Subsequent Ohmic studies in JFT, ATC, TFR, Alcator, T-10, PLT and ISX-A, including some analyses based on ion temperature profile measurements, claimed more quantitative agreement, implying $\chi_i \leq 2\chi_i^{neo}$. These observations naturally led to a prolonged focus on electron transport as the dominant anomalous loss mechanism in tokamak plasmas. Since the mechanisms underlying ion versus electron heat transport were thought to be different, only a few of the early studies compared the relative magnitude of χ_i to χ_e. TFR reported that the scaling of *central* temperature could be reproduced by assuming that the $\tau_{Ei} = \tau_{Ee}$ [1], while profiles of $T_i(r)$ measured by charge-exchange could be adequately modelled using $\chi_i \leq 2\chi_i^{neo}$ [2]. Measurements in T-4, although roughly consistent with neoclassical χ_i, were also comparable to the core χ_e [3]. $\tau_{ei} \approx 3\tau_{Ee}$ was observed in some regimes in Ohmic PLT, however here the electrons were dominated by radiation.

One early experimental anomaly was the failure of τ_E to follow its linear scaling with $\bar{n}_e$ at high density in Ohmic ISX-A plasmas. Murakami [4] observed that ion heat transport could dominate the total power flow even if $\chi_i \approx \chi_i^{neo}$, which explained the high-density saturation of τ_E without the need to postulate an additional deterioration of χ_e. Global τ_E measurements at high density in ISX-A and Alcator were found to be consistent with ion heat transport at roughly the neoclassical level. This provided further support for the supposition that ion heat transport could be adequately understood in a neoclassical framework, with no need to invoke anomalous ion losses driven by microturbulence or MHD.

Of course, inferences of χ_i in Ohmic plasmas required accurate temperature measurements, since the power flow to the ions depends on $T_e - T_i$. Most of the Ohmic studies relied on neutron emission or analysis of the energy spectrum of escaping neutrals to measure the ion temperature [5]. Global neutron emission provides a reasonably accurate measurement of $T_i(0)$, provided that the deuterium depletion (Z_{eff}) is measured with modest accuracy. However, analysis of χ_i in Ohmic plasmas based only on $T_i(0)$ is relatively insensitive to the magnitude of ion transport beyond $r = a/2$, and so agreement with neoclassical predictions of $T_i(0)$ does not provide much assurance that χ_i is close to neoclassical throughout the confinement region. Multi-chord neutral particle analysis (NPA) can measure a $T_i(r)$ profile; however, without a doping neutral beam the signal is chord-integrated. The measurement requires numerical modelling of the neutral density profile, whose accuracy is difficult to evaluate, to deduce the $T_i(r)$ profile.

Therefore it was with some relief that early studies of beam-heated plasmas in the late 1970's, some extending into the 1980's, also found $\chi_i \approx \chi_i^{\text{neo}}$. TFR roughly doubled $T_i(0)$ to 1.9 keV with 650 kW of neutral injection [6] in 1976. True, the inferred χ_i was independent of collision frequency, contrary to the neoclassical prediction; hence the "multiplier" on χ_e^{neo} was ~ 1 at $\nu/\epsilon^{\frac{3}{2}} > 4$ and $\sim 3-4$ for $\nu/\epsilon^{\frac{3}{2}} < 1$ (here $\nu = \nu_i q R/v_i$ and $\epsilon = r/R$). However, given diagnostic and modelling uncertainties of the power source, they could not conclude that the ion transport was more than a factor two greater than χ_i^{neo}. The magnitude of χ_i^{neo} was still evolving as the neoclassical theory was refined, so a factor of two discrepancy with experiment was regarded as corroborative evidence for the basic neoclassical mechanism of ion heat transport. ORMAK [7] also concluded that χ_i was neoclassical during beam injection, based mostly on comparisons of measured and calculated $T_i(0)$. Interestingly, their published data shows that neoclassical theory overpredicted the $T_i(0)$ outside the error bars both at high power and low density. Furthermore, the χ_i inferred from some $T_i(r)$ measurements agreed with χ_e^{neo} in the plasma center, but $\chi_i/\chi_i^{\text{neo}}$ increased to a factor

~ 3 at $r = a/2$, and to a factor ~ 7 close to the edge.

Ion collisionality was pushed down to $\nu^* \le 0.01$ and $T_i(0)$ up to 7 keV in the first high-power heating experiments in PLT [8, 9], where trapped-particle modes were predicted to enhance energy transport. Due to the keen interest in documenting the high temperatures and the non-existence of catastrophic transport driven by the trapped-particle modes, the central ion temperature was measured with three independent diagnostics (see Table 2.2). It was found to be in good agreement with the temperature predicted from χ_i^{neo}. However, the authors were careful to point out that the temperature measurements could not exclude enhancements of χ_i over χ_e^{neo} by as much as a factor 3-5. Also, the toroidal rotation speed induced by the beam torque was much less than expected from neoclassical momentum transport. Subsequently, neutral-beam heating studies in mostly L-mode plasmas extending through 1984 on ISX-B, T-11, ASDEX, and PDX all found χ_i to be described by χ_e^{neo} to within a factor of three.

The earliest claims of anomalous ion heat transport originated from studies of high density Ohmic plasmas. Unlike the linear increase of τ_E with $\bar{n}_e$ observed in Alcator-A, τ_E started to saturate above $\bar{n}_{e_{e19}} = 25$ in Alcator-C. Due to the strong ion-electron coupling at high density, it was not possible to unambiguously attribute the τ_E saturation to anomalous ion transport. In 1980, Fairfax *et al.* could demonstrate only that χ_e^{neo} would have to be multiplied by 2-4 to reproduce the observed τ_E if ions alone were blamed for the entire transport [10]. In 1982, Blackwell *et al.* [11] reported that local transport analysis in Alcator C implied a neoclassical multiplier of 2-4 in the core (which dominates the neutron emission), but they could not unambiguously exclude the alternate possibility of increased core electron transport. The discovery that $\tau_E \propto \bar{n}_e$ was recovered at high density if pellets were injected [12], thereby forming a peaked density profile, stimulated further experiments and analysis. Global energy confinement times with pellets were up to 70% higher than gas-fuelled shots at the highest densities. Subsequent studies of transport in moderate-density Ohmic plasmas led to a firmer conclusion that χ_i was several times the neoclassical value in broad-density, gas-fuelled plasmas, which was reduced to roughly the neoclassical value in pellet-fuelled plasmas with peaked density profiles [13, 14, 15, 16]. The Alcator group observed that the correlation of improved χ_i with peaked density profiles suggested a possible role for ion-temperature-gradient driven turbulence in driving "anomalous" ion transport in broad-density, gas-fuelled plasmas [13]. This work motivated a number of studies in other tokamaks of confinement differences between peaked-density versus broad-density plasmas, with peaked density profiles created by pellet injection or other means.

Increased global τ_E was observed in peaked-density, pellet-fuelled TFTR Ohmic plasmas compared to gas-fuelled plasmas at the same $\bar{n}_e$ [17]. However, it was not possible to disentangle the ion and electron power flows at the plasma center and thus to determine whether χ_i dropped to χ_e^{neo} following pellet injection. The total heat flow (ions+electrons) following pellet injection was $\sim 2 - 3$ times the value expected from χ_e^{neo} alone, but the comparable multiplier before pellet injection was not evaluated. Similarly, ASDEX observed a reduction of χ_i down to neoclassical levels in a variety of Ohmic regimes that involve peaked density profiles, including pellet fuelling [18], the Improved Ohmic Confinement regime (IOC) obtained by reducing gas flow [19], and the sawtooth-free Ohmic regime [20]. The enhanced core χ_i in the IOC regime was comparable to that expected simply from sawteeth, making a comparison with ITGDT theory problematic. As with most Ohmic studies, these conclusions derived from relatively small differences in measured $T_e - T_i$, but the latter two papers were based on profiles of $T_i(r)$ from neutral-particle analysis with an active doping beam which improved the spatial localization of the measurement.

Throughout the 1980's, additional studies in gas-fuelled Ohmic plasmas were conducted in the larger tokamaks then coming on line (DIII, TFTR, JET). Fairly large anomaly factors ($\chi_i/\chi_i^{\text{neo}} = 2 - 7$) were reported as early as 1982 by Ejima et al. in DIII [31]. They noted that their results were not qualitatively different from previous comparisons with neoclassical theory (which obtained smaller anomaly factors) because the expressions for χ_e^{neo} were also evolving. For example, they pointed out that the TFR results would require a multiplier of 5 on the Hazeltine-Hinton expression for χ_i^{neo}. They also presented some indirect evidence from scans at $a = 23$, 32, and 44 cm, based not on T_i measurements but on the assumption that $n_e \chi_e$ remained constant, that the neoclassical anomaly factor increases with a or a/R. This trend was confirmed by later size-scaling studies in TFTR [32] which found roughly

Parametric Variations of Ion Transport

Year	Tok	χ_i/χ_i^{neo}	χ_i/χ_e	a	I_p	$\bar{n}_{e,19}$	$T_i^m(0)$	Diag	neo	Reference
70	T-3	$\sim\sim 1$	-	0.15	0.06-0.11	1.5-3.8	0.5	cx	$T_i(0)$ gs	Artsimovich [21]
70	TM-3	$\sim\sim 1$	-	0.08	0.02-0.04	1.0-3.0	0.2	cx	$T_i(0)$ gs	Artsimovich [21]
73	T-4	$\sim\sim 1$	-	0.20?	~ 0.15		0.6	cx,n	$T_i(0)$ -	Gorbunov [22]
74	JFT	~ 1	-	0.25	0.08	0.5-1.0	0.2	cx	$T_i(0)$ -	Fujisawa [23]
74	ORMAK	~ 1	-	0.23	0.06-.16	0.5-2.5	0.4	cx	$T_i(0)$ hr	Berry [24, 25]
75	ATC	~ 1	-	0.17	0.08	0.6-2.0	0.2	cx	$T_i(0)$ hh	Stott [26]
76	TFR	≤ 2	-	0.20	0.30	4.5	1.0	cx	$T_i(r)$ hr	TFR [2, 1]
78	T-11	~ 1	~ 1	0.20	0.1	0.5-5.0	0.3	cx	$T_i(0)$ -	Vlasenkov [3]
78	ALC	~ 1	-	0.10	0.13-0.24	10-30	0.6	cx,n	$T_i(0)$ hh	Gondhalekar [27]
78	ALC	≤ 1.5	-	0.10	0.13-0.24	60	0.6	e	e hh	Gondhalekar [27] [28]
78	T10	~ 1	-	0.35	.4-.56	3-7	1.1	cx,n	$T_i(0)$ -	Bagdasarov [29]
78	PLT	~ 1	-	0.40	0.4	$\sim 2-4$	1.2	cx,$K\alpha$	$T_i(r)$ hr	Brusati [30]
79	ISX-A	1-2	-	0.26	0.12	1- 5.5	0.5	cx	$T_i(0)$ hh	Murakami [4]
80	ALC-C	2-4	-	0.16	0.25-0.54	20-60	-	cx,n	e -	Fairfax [10]
82	ALC-C	2-4	-	0.16	0.10-0.75	20-70	1.0	cx,n	$T_i(0)$ hh	Blackwell [11]
82	DIII	2-7	-	0.44	0.46	2.0-4.3	0.7	cx	$T_i(0)$ hh	Ejima [31]
84	TFTR	1-3	>1	0.41	0.9-3.2	1-3	-	n	$T_i(0)$ ch1	Efthimion [32]
84	TFTR	0-8	-	0.83	0.6-1.4	2.7	-	cx,n,$K\alpha$	$T_i(0)$ ch1	Efthimion [32]
84	JET	~ 1-3	-	1.1	1.2 - 3.5	0.7-3.4	-	cx	$T_i(0)$ ch1	Cordey [33]
88	JET	$\sim 4-8$	-	0.8-1.2	1.0-4.0	~ 1	~ 3	cx,n,ns	$T_i(0)$ ch2	Bartlett [34]
88	JET	$\sim 1-3$	-	0.8-1.2	1.0-4.0	~ 3	~ 2.5	cx,n,ns	$T_i(0)$ ch2	Bartlett [34]
92	TFTR	15	1-2	0.80	2.0	2.4	2.4	cxrs	$T_i(r)$ ch2	Figure 2.1
84	ALC-C PEL	$5 \to 1$	-	0.60	0.4-0.8	30-60	1.5	n	$T_i(0)$ ch1	Greenwald [13]
86	TFTR PEL	≤ 3	-	0.70	1.6	≤ 14	-	n	$T_i(0)$ -	Schmidt [17]
86	ALC-C PEL	$5 \to 1$	-	0.16	0.25-0.55	10-30	1.5	n	$T_i(0)$ ch1	Wolfe [15] [14, 16]
88	ASDEX PEL	$4 \to 1$	-	0.40	0.38	5	~ 0.5	cx,n	$T_i(0)$ ch2	Gruber [18]
89	ASDEX IOC	$20 \to 1$	-	0.4	0.38	2.5-4.7	1.0	acx	$T_i(r)$ ch2	Stroth($q < 1$) [19]
91	ASDEX SFO	~ 1	-	0.40	0.25-0.32	4-6	0.8	acx	$T_i(r)$ ch2	Stroth [20]

Table 2.1: Historical survey of χ_i measurements in tokamaks. Diagnostics: cx=passive neutral-particle analysis of charge-exchange efflux; acx = active neutral-particle analysis with diagnostic beam; n=total neutron emission; ns = neutron spectroscopy; e=total energy content; $K\alpha$= measurement of Doppler-broadened x-ray radiation; cxrs=charge-exchange recombination spectroscopy. Neoclassical models: gs = Galeev-Sagdeev [35]; hr = Hinton-Rosenbluth [36];hh =Hinton-Hazeltine [37, 38], ch1=Chang-Hinton [39];ch2 = Chang-Hinton [40]. Regimes: PEL=peaked-density following pellet injection; IOC=peaked density improved-Ohmic regime; SFO=peaked-density sawtooth-free Ohmic. $T_i^m(0)$ = maximum central ion temperature in dataset.

neoclassical transport ($\chi_i/\chi_e^{neo} \leq 3$) in small plasmas ($a = 0.41$ m), but anomaly factors up to 8 in larger plasmas ($a = 0.80$ m).

Year	Tok	Regime	χ_i/χ_i^{neo}	χ_i/χ_e	a	I_p	$\bar{n}_{e_{e19}}$	P_b	$T_i^m(0)$	Diag	Prof	neo	ref
75	ATC		$\sim\sim 1$	-	0.17	0.08	0.6-2.0	0.1	0.3	cx	$T_i(0)$	hh	Stott [26]
76	TFR		$\sim 1\ (\nu^* > 4)$	-	0.20	0.2-0.4	3.5-7.5	0.65	1.9	cx	$T_i(r)$	hh	TFR [6, 1]
76	TFR		$3-4\ (\nu^* < 1)$	-									
76	ORMAK		~ 1		0.23	0.16-0.18	2-6	0.4	1.8	cx,n	$T_i(0)$	hr,hh	Berry [7]
79	PLT		$\sim 1(l5)$		0.40	0.43	3.0	2.9	6.3	cx,n,$K\alpha$	$T_i(0)$	-	Eubank [8] [9]
81	ISX-B	lmode	1-3		0.26	.1-.2	3-7	2.5	1.3	cx,n	$T_i(0)$	hh	Swain [41]
82	T-11		~ 1		0.20	0.1	5-6	0.6	0.5	cx	$T_i(0)$	-	Barsukov [42]
82	ASDEX	lmode	1-3		0.40	0.3-0.4	1.5-6.5	2.2	4.0	cx	$T_i(r)$	hh	Becker [43, 44]
84	ISX-B	lmode	~ 1		0.26	0.14	-	2.5	1.1	cx	$T_i(0)$	ch1	Murakami [45]
84	PDX	lmode	0.5-3.0	-	0.4	0.2-0.5	2.5-4.2	7.2	6.5	cx,$K\alpha$	$T_i(0)$	ch1	Kaye [46]
86	DIII	lmode	$5-15(a/2)$	2-5	0.40	.35-.9	3.7-7.9	6.1	3.8	cxrs	$T_i(r)$	ch1	Groebner [47]
86	DIII	lmode	$\sim 1(r=0)$	>1									
86	PBX	hmode	~ 1	~ 1	0.30	0.38	-	2.1	2.5	cxrs	$T_i(r)$	ch1	Okabayashi [48]
88	TFTR	super	$\sim 25(a/2)$	1-2	0.80	1.4	3.9	22	27	cxrs	$T_i(r)$	ch2	Zarnstorff [49]
88	TFTR	super	$\sim\sim 1(r=0)$	-									
88	TFTR	lmode	$\gg 1$	~ 4	0.80	1.4	3.9	22	4	cxrs	$T_i(r)$	ch2	Zarnstorff [49]
89	TFTR	super	$\sim 10(a/2)$	~ 2	0.8	0.9	2.5	20	30	cxrs	$T_i(r)$	ch1	Fonck [50]
90	TFTR	lmode	$\gg 1$	1.3-5	0.9	0.9-1.8	2.9-4.4	12	6	cxrs	$T_i(r)$	ch2	Scott [51]
92	JT60	lmode	10-50	3-6	0.6-0.9	1.0-2.7	1.2-6.5	17	6	cxrs	$T_i(r)$	ch2	Hirayama [52]
92	JET	lmode	$\sim 15-30\ (a/2)$	5-10	1.16	3.1	2.4-3.1	18	10	cxrs	$T_i(r)$	ch2	Balet [53, 54]
92	JET	lmode	$\sim 2-6\ (r=.1)$	1-2									
92	JET	hihm	$\sim 10\ (a/2)$	~ 1	1.08	4.0	~ 4	18	22	cxrs	$T_i(r)$	ch2	Balet [53]
92	JET	hihm	$\sim 1\ (r=0.2)$	~ 0.15							cxrs	$T_i(r)$	ch2	Balet [53]

Table 2.2: Historical Survey of χ_i measurements in tokamaks in beam-heated plasmas. Regimes: super = supershot, hihm = hot-ion H-mode.

Early analysis (1984) of JET Ohmic ion transport by Cordey [33] obtained $\chi_i = 1 - 3\chi_i^{neo}$. A later study with improved diagnostics based on $T_i(0)$ measurements from charge-exchange, supplemented by neutron emission and neutron spectroscopy [34], implied neoclassical multipliers of 4-8 at low density ($\bar{n}_{e_{19}} \approx 1$), decreasing to 1-3 at moderate density ($\bar{n}_{e_{19}} \approx 3$). However, Bartlett et al. concluded that although the JET data suggested a large anomaly over χ_i^{neo}, they could not exclude the possibility that $\chi_i \approx \chi_i^{neo}$ without measurements of the $T_i(r)$ profile. A prior JET study [55] using four NPA diagnostics to measure a coarse $T_i(r)$ profile concluded that $\chi_i \approx 5\chi_i^{neo}$.

Charge-exchange recombination spectroscopy (CXRS) was developed and implemented on several tokamaks in the 1980's, and represented a major advance in the study of ion heat and momentum transport. Its advantages over prior techniques are that it is a local measurement, fairly accurate, and straightforward to construct as a diagnostic to measure the full radial $T_i(r)$ profile. Groebner et al. used CXRS in their pioneering work on DIII to demonstrate clearly that during neutral injection $\chi_i(r)$ had a radial profile quite different from that of χ_i^{neo}, with the ratio χ_i/χ_i^{neo} increasing strongly with radius [47]. The observed χ_i was within error bars of χ_e^{neo} near the plasma center, but increased to 5-15 times χ_e^{neo} at the half-radius. Ion heat transport, rather than electron heat transport, was found to be the dominant heat loss mechanism in auxiliary-heated plasmas. This conclusion was subsequently confirmed by CXRS measurements in TFTR, JET, and JT60 L-mode plasmas (see Table 2.2), for which the ratio χ_i/χ_i^{neo} was as high as 50 at the half-radius and was rarely less than 10. Even in enhanced-confinement regimes, such as the H-mode and supershot, χ_i was more than a factor 10 larger than χ_i^{neo} at the half-radius. Only near the center of the supershot and hot-ion H-mode plasmas (characterized by peaked density profiles, $T_i > 10$ and $T_i > T_e$), where heat transport due to particle convection becomes a significant or dominant term in the ion power balance, could χ_i still be within error bars of the neoclassical prediction.

Tokamaks comparable in size to DIII but which used NPA diagnostics to measure ion temperature, including PDX [46] and ASDEX [43, 56], reported smaller anomaly factors relative to χ_e^{neo} during auxiliary heating. For example, Becker [43] found χ_i to be within a factor of three of neoclassical over a wide range of ASDEX L-mode co-injected plasmas, including ion temperatures of $0.5 - 3.9$ keV and electron densities $1.5 - 6.5$. Subsequently, Gruber [56] observed that χ_i was reduced to neoclassical levels during counter injection (with peaked density) on the basis that the total plasma energy would be underestimated if the neoclassical multiplier observed during co-injection (3-4) was used, even in the limit of no heat transport by electrons.

The observed large anomaly of χ_i over neoclassical at the half-radius discredited the use of χ_i^{neo} as a benchmark for reporting χ_i measurements, and it became more common to compare the magnitude of χ_i

with χ_e or with the predictions of ITG theory. There is consistent evidence from the larger tokamaks that $\chi_i > \chi_e$ in L-mode plasmas, outside the error bars, at $r \approx a/2$; ratios ranging from 1.3 to 10 have been published from TFTR, JET, and JT60, with values of 3-4 being typical. The ratio χ_i/χ_e at $r = a/2$ decreases to 1-2 in the supershot and hot-ion H-mode, reflecting the general trend that ion transport improves more strongly than electron transport in these regimes. One issue that remains unclear from an experimental perspective is the relationship between χ_i and χ_ϕ; TFTR consistently observes $\chi_i \sim \chi_\phi$ across the confinement region in L-mode and supershot plasmas [51, 57], whereas JET observes that χ_i/χ_ϕ ranges up to 10 at the half-radius in L-mode plasmas. Furthermore JET and DIII-D [58] find that in the hot-ion H-mode regime (whose core resembles the TFTR supershot regime), the radial profile of χ_ϕ more closely resembles that of χ_e rather than χ_i.

2.1.3 DIAGNOSTICS

This section describes the TFTR diagnostic set that was used for the transport studies described in Secs. 2.1.5 and 2.1.6. The ion temperature and toroidal rotation profiles were measured by a charge-exchange recombination spectroscopic (CXRS) diagnostic [59] with viewing sightlines every 5-10 cm. The electron density profile was measured by a 10-chord interferometer [60, 61]. Four electron temperature diagnostics were used: Thompson scattering [62], an absolutely calibrated Michelson interferometer [63], an absolutely calibrated ECE radiometer [64, 65, 66, 67], and a grating polychromator [68] normalized to Thomson scattering. Table 2.3 lists the T_e diagnostic used for particular scans. Total neutron emission was measured with a calibrated set of fission detectors [69, 70, 71] with an uncertainty of ±15 [72]. The neutron emission profile was also measured by ten detectors in a multi-channel neutron collimator [73, 74, 75]. Being a line-integrated measurement, the measured profile does not put strong constraints on the ion temperature profile. Preliminary results using the neutron collimator data are consistent with the results of this paper. The total stored plasma energy is determined from magnetic analysis of the diamagnetic flux and equilibrium magnetic field [76, 77], with an estimated uncertainty of 70 kJ. The plasma Z_{eff}, assumed radially constant, was deduced from a single-channel visible Bremsstrahlung measurement in the horizontal midplane [78]. In Ohmic plasmas this measurement was checked against pulse-height analysis (PHA) of the soft x-ray spectrum [79] and from the surface voltage, assuming neoclassical resistivity [80]. The magnitude of neutral sources due to deuterium efflux recycled on the inner bumper limiter was inferred from measurements of the $H\alpha$ emission along five poloidal viewing sightlines [81, 82]. The profile of flux-surface-averaged radiated power density was deduced by Abel-inverting [83] measurements of two bolometer arrays [84, 85] which viewed the plasma poloidally.

2.1.4 TRANSPORT ANALYSIS

Local heat and momentum transport coefficients are inferred from the measured profiles using the equilibrium transport code SNAP [86, 87, 88]. The measured profiles are first smoothed with a triangular weighting function with a full width at half-maximum of approximately 7-10 cm. The calculated beam deposition in the SNAP code includes ionization and charge-exchange processes by thermal ions and electrons, as well as by the slowing-down beam ion population. The beam-ion slowing down population is simulated as a separate species not subject to anomalous transport. The beam-ion distribution function is calculated from a solution of the Fokker-Planck equation in the rotating plasma frame [89], neglecting radial diffusion. Beam ions are treated as joining the background thermal ion species when their energy falls below $\frac{3}{2} T_i$. Electron-ion energy exchange is assumed to be classical. The energy transport is analyzed assuming that the radial transport mechanisms are cross field diffusion, represented by anomalous diffusivities χ_i and χ_e, radial convection due to particle flows, and charge-exchange losses, which are important for the ion power balance near the edge;

$$\begin{aligned} Q_i &\equiv -\chi_i \left(\sum_j n_j\right) \nabla T_i + \tfrac{3}{2} \Gamma_i T_i \equiv -\chi_i^{\text{tot}} \left(\sum_j n_j\right) \nabla T_i \\ Q_e &\equiv -\chi_e n_e \nabla T_e + \tfrac{3}{2} \Gamma_e T_e \equiv -\chi_e^{\text{tot}} n_e \nabla T_e \end{aligned} \quad (1)$$

where Q_i and Q_e are the total ion and electron heat flux, Γ_i and Γ_e are the ion and electron particle flux (Γ_i is assumed hydrogenic with average mass $\langle m_h \rangle$), $\sum_j$ represents a sum over all ion species including impurities, Γ is the radial flux of toroidal momentum, and χ_i^{tot} and χ_e^{tot} are "total" diffusivities which include both the convective and conductive fluxes. Use of such total diffusivities remove the uncertainty involved in the choice of 3/2 or 5/2 as the convective multiplier [90, 91]. Since steady-state is assumed, the particle fluxes Γ_i and Γ_e are simply the integrated volumetric particle source terms (wall and beam) and divided by the flux-surface area. This analysis neglects pinch effects and off-diagonal terms, as well as local damping due to toroidal field ripple. SNAP also calculates the total stored energy from the kinetic measurements (including the beam stored energy) and the total neutron emission (including beam-beam interactions), which provide consistency checks when compared to the corresponding measured quantities [72].

2.1.5 ION TRANSPORT IN OHMIC PLASMAS

This section discusses measurements of χ_i in Ohmic TFTR plasmas using two techniques: active measurements of the ion temperature by charge exchange recombination spectroscopy, and measurements of the neutron emission. The active $T_i(R)$ measurement was performed during the first 20 ms of beam injection, before the ions could receive significant heating power from the beam ion population (neglecting radial heat transport, the beam power deposited to ions in this period would increase $T_i(0)$ by 0.24 keV or 10%, corresponding to an average increase over the measurement period of 5%). Both ion temperature techniques indicate that χ_i is of order $\sim 10 - 20$ times larger than neoclassical theory. The plasma for the study utilizing CXRS measurements of $T_i(R)$ was a 2.0 MA deuterium Ohmic discharge with other parameters as described in Scan OH1 in Table 2.3. Sawteeth with period ~ 75 ms and inversion radius $r_{inv} = 0.23$ m modulated the central electron temperature by $\Delta T_e = 0.5$ keV. Carbon was the dominant impurity ($Z_{eff}^{metals} = 0.03$). Other data including the profile measurements of $n_e(R)$ and $T_e(R)$ were averaged over a period of 100 ms and thus represent sawtooth averages. There was good agreement amongst the measurements of Z_{eff}: $Z_{eff}^V = 2.4$, $Z_{eff}^{PHA} = 2.0$, and, from a measurement of the surface voltage, $Z_{\text{eff}}^{\text{neo}} = 1.7 - 1.8$. The neutron emission calculated from the measured $T_i(r)$ and $n_e(r)$ profiles, using the $Z_{\text{eff}}^{\text{VB}}$ for ion depletion, is only about 30% lower than the measured value. Since $\langle \sigma v \rangle_{dd}$ increases as $T_i^{\sim 4}$ for $T_i \sim 2$ keV, the neutron emission would imply a central ion temperature about 9% higher than CXRS, assuming the same profile shape.

Figure 2.1a and b show the measured profiles mapped to a minor radius grid. The ion-electron temperature difference is large in this plasma, which allows the ion power balance to be evaluated with reasonable accuracy. As shown in Figure 2.1c, electron heat conduction is the dominant loss channel in the plasma core, but it exceeds the ion loss channel by only $\sim 50\%$ at $r/a = 0.5$. Radiation and convection become the dominant loss channels in the outer 20 cm. The profiles of χ_i and χ_e inferred from the power balance are illustrated in Fig. 2.1d. Throughout the confinement region we observe that $\chi_i \approx \chi_e \gg \chi_i^{\text{neo}}$. The error bars illustrate the effect of moving the measured $T_i(r)$ profile to the top and bottom of its error bars, or moving the measured $T_e(r)$ profile (by the Michelson interferometer) up and down by 7%. Clearly, χ_i is much greater than χ_i^{neo} throughout the confinement region, well outside the experimental uncertainties. It is interesting to note that the χ_e is relatively "flat" across the minor radius, and does not increase significantly at the outside of the plasma. Such a flat χ_e profile is typical of medium-to-high density TFTR Ohmic plasmas, though the robustness of the result is clouded by the increasing effects of convection and radiation at the outside.

Alternately, a neoclassical model of χ_i can be adopted to *calculate* the ion temperature profile, using an global multiplier adjusted to match the measured neutron emission. As shown in Fig. 2.1c, a multiplier of 15.3 is required to reproduce the measured neutron emission. The width of the calculated $T_i(r)$ is slightly broader than the measured $T_i(r)$ profile. The required neoclassical multiplier varies from 13-17 as the T_e profile is varied up or down 7% and from 12-19 as Z_{eff} is varied up or down 15%. Thus, measurements of both the ion temperature profile and the neutron emission indicate that $\chi_i \gg \chi_i^{\text{neo}}$ in this Ohmic plasma.

To determine the behavior of χ_i over a wider parameter range, we consider the 2.0 MA deuterium Ohmic density scan OH2 described in

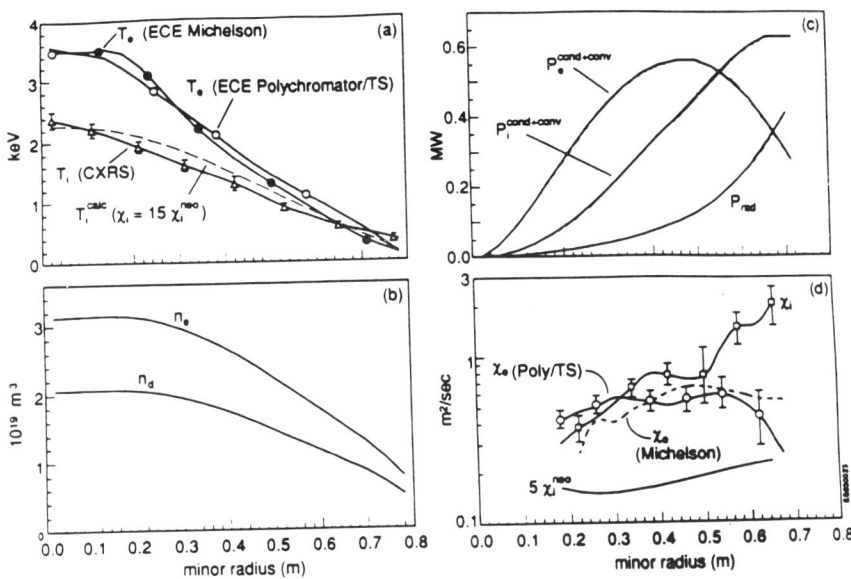

Figure 2.1: (a) Measured temperature profiles for Ohmic pulse 52499. The dashed line represents an ion temperature profile calculated using $\chi_i = 15.3\chi_i^{neo}$ which matches the neutron emission. from Z_{eff}. (c) Integrated radiated power, and power flows through ions and electrons. The shaded region represents the contribution of convection ($q_{\text{conv}} \equiv \frac{3}{2}\Gamma T$) to the power flow. (d) Local heat diffusivities χ_i and χ_e inferred from power balance analysis. The two lines for χ_e represent analysis based on the ECE grating polychromator (normalized to Thomson scattering) and Michelson measurements of T_e.

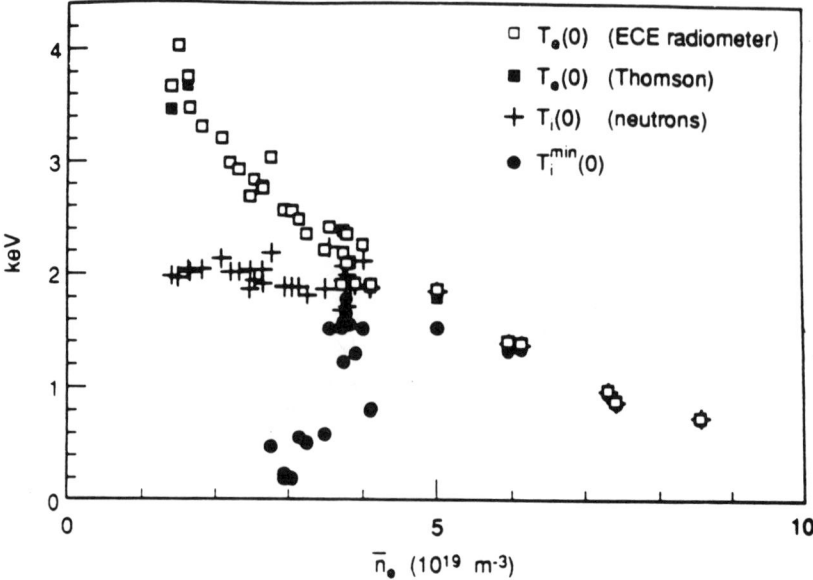

Figure 2.2: Variation of central temperatures in the Ohmic density scan. T_i^{min} is the minimum possible central ion temperature consistent with ion-electron power coupling, assuming no radial transport of heat by electrons.

Table 2.3.

For these plasmas, the $T_i(r)$ profile is calculated assuming some multiplier on neoclassical transport from the measured neutron emission. Figure 2.2 shows the variation of central T_e and T_i with density. Gas fuelling only was used for most plasmas with $\bar{n}_{e_{e19}} \leq 4$, while higher densities were achieved only with deuterium pellet fuelling. The pellet data represent measurements 300-1500 ms after pellet injection, when the time rate of change of stored energy was small($W/(\partial W/\partial t > 1.$ sec). The gas-only discharges had the customary broad density profiles ($P_{ne} < 2.0$), while quite peaked density profiles ($P_{ne} \leq 4.5$) were obtained with pellet injection. As reported previously in TFTR Ohmic plasmas [32, 94], Z_{eff} decreases systematically with in-

Scan	Sym	Gas	R	a	B_t	I_p	P_b	$\bar{n}_{e_{e19}}$	S	P_{ne}	T_{io}	T_{eo}	R_n	R_W
OH1		D	2.45	0.80	4.8	2.0	0.0	2.4	-	1.6	2.4	3.5	0.70	0.92(-33kJ)
OH2		D	2.58	0.93	4.5	2.0	0.0	1.0-3.0	-	2.0-1.8	2.0-2.0	4.0-2.4	≡1	±50kJ
L1	▲	D	2.58	0.93	3.8	1.2	2.4-17	3.0-3.6	1.24-1.13	[1.48-1.54]	2.1-4.2	2.6-3.6	0.7-1.3	1.12-1.04
L2	■	D	2.58	0.93	3.8	1.8	2.2-20	[3.0-3.2]	[1.2-1.3]	1.66-1.91	3.3-8.2	2.7-5.0	[0.98-1.24]	1.06-1.01
L3	●	D+He	2.45	0.80	4.0	1.3	4.3-22	3.2-3.6	1.16-1.47	1.95-1.73	3.5-9.5	3.1-5.2	[0.84-0.90]	[1.00-1.04]
S1	△	D	2.45	0.79	4.8,5.1	1.4	12-22	2.5-3.9	1.7-2.8	1.8-3.1	14-27	5.7-8.0	[1.00-1.30]	[0.92-1.04]
S2	+	D	2.45	0.80	4.8	1.6	10-24	2.3-3.8	1.5-2.4	1.8-2.9	13-28	6.0-8.6	0.95-0.73	[0.88-1.04]
S3	○	D	2.45	0.80	4.8	1.3	4.3-21	1.6-3.2	1.4-2.5	2.0-2.4	9.6-27	5.8-8.6	[0.66-0.76]	[0.94-1.04]
S4	⊠	D	2.45	0.80	4.8	1.1	2.3-14	1.2-2.6	1.4-2.4	2.08-2.36	8.1-19.4	4.0-6.3	[0.90-1.00]	1.09-1.00†

Scan	τ_E	Z_{eff}	$V_{\phi o}$	W_b/W	W_i/W	W_e/W	Diag	Year	Shots	Ref
OH1	270	2.4	-	-	0.35	0.65	YG,YM	1990	52499	Figure 2.1
OH2	200-350	3-1.5	-	-	0.33-0.46	0.67-0.56	YS,TS	1990	45195-46501	Figure 2.3
L1	170-73	[1.8,2.6]	[-0.4-1.0]	0.12-0.40	0.38-0.23	0.49-0.37	YS,TS	1990	45620-46588	Scott [51]
L2	230-100	1.8-4.0	[-0.3-1.3]	0.09-0.52	0.42-0.16	0.49-0.31	YS,TS	1990	45585-45621	Scott [51]
L3	128-81	[2.2-2.4]	2.4-1.5	0.23-0.55	0.28-0.16	0.48-0.28	YS	1990	51068-51169	Scott [92]
S1	132-166?	3.3-2.9	[-0.2,1.3]	[0.50-0.60]	[0.20-0.26]	[0.21-0.23]	TS,YS	1988	35782-37266	Zarnstorff [49]
S2	[130,160]	[2.3,3.5]	[1,4]	[0.44-0.50]	[0.24-0.32]	[0.22-0.30]	YM	1990	47613-49539	Strachan [93]
S3	163-134	3.9-3.3	[1.6,3.4]	[0.48-0.56]	[0.18-0.26]	[0.22-0.28]	YS	1990	51023-51073	Scott [92]
S4	[139,172]	[2.1-2.5]	[-2.7,2.0]	0.41-0.54	0.33-0.24	0.27-0.22	YG	1992	65684-65703	-

Table 2.3: Summary of TFTR. Units and definitions: Sym = symbol identifying scan shots in Figures 2.6–2.9; R(m); a(m); B_t(Tesla); I_p(MA); $\bar{n}_{e_{e19}}(10^{19}\text{m}^{-3})$; P_b(MW); $S(W_{tot}/W_{L-mode})$; $F_{ne} = n_e(0)/\langle n_e \rangle$; T_{io}, T_{eo}(keV); R_n=Neutron emission[kinetic]/Neutron emission[measured]; $R_W = W_{tot}^{kin}/W_{tot}^{dia}$; τ_E = global τ_E including beam-ion contribution; $V_{\phi o}$ is the central toroidal rotation speed (10^5m/s); W_b/W, W_i/W, and W_e/W are the ratios of beam, thermal ion, and electron energy to total energy. Where a range of values is quoted it represents a systematic variation with increasing density or beam power. Values in brackets [ ] indicates that the parameter's value varied during the scan over the indicated, but without a systematic trend with power. Electron temperature diagnostics: TS (Thompson scattering); YG (ECE grating polychromator normalized to TS); YS (absolutely calibrated ECE radiometer); YM (absolutely calibrated Michelson interferometer). The supershot values exclude one exceptional shot (35782) in scan S1 with $Z_{\text{eff}} = 2.5$, $W_b/W = 0.39$, $W_i/W = 0.38$, $W_e/W = 0.23$, and one exceptional shot (49426) in scan S2 at $P_b = 22.6$ MW which attained $P_{ne} = 3.29$ and $T_{eo} = 9.8$. † Kinetic measurements of W_{tot} agree with diamagnetic measurements within 0.07 MJ across the entire scan. The ratio R_W is 1.18 at $P_b = 2.3$ MW because the total stored energy is small (0.35 MJ).

creasing $\bar{n}_e$, in this scan dropping from 3.0 to ~ 1. Figure 2.3 shows the global energy confinement time in these plasmas, evaluated from the kinetic data. The confinement times for gas-fuelled discharges are similar to those obtained in previous density scans [32], however pellets provided little or no increase in τ_E in constrast to earlier results in slightly smaller plasmas, which obtained $\tau_E = 500$ ms [17]. The reason for the lack of improvement at high density is not understood.

The transport analysis will focus on the lower density plasmas ($\bar{n}_{e_{e19}} \leq 3$) where the ion-electron coupling is not too strong. $T_i(r)$ profiles were calculated for all the plasmas in Fig. 2.1, using transport models of the form $\chi_i(r) = c_e \chi_e(r)$ or $\chi_i = c_n \chi_i^{neo}$ with c_e and c_n iteratively chosen to reproduce the global neutron emission. Of course, although the models calculate the radial profile of χ_i as part of the convergence algorithm, analysis based on measurements of the global neutron emission cannot provide any information regarding the radial shape of χ_i because the global neutron emission is mostly determined by the central plasma conditions. SNAP calculates that more than 70% of the neutron emission originates from within $r/a < 0.33$. Thus, the "multiplier" on χ_e or χ_e^{neo} required to match the global neutron emission represents a complicated radial average (somewhat centrally weighted because $T_i \approx T_e$ is enforced at the edge by strong ion-electron equilibration, irrespective of χ_i/χ_i^{neo}) of the local ratio of χ_i to χ_e^{neo} or χ_e.

The dominant errors which affect our inferences of χ_i in these Ohmic plasmas are the measurement uncertainties in T_e, n_e, Z_{eff}, and neutron emission. Since the metallic contribution to Z_{eff} is small ($Z_{\text{eff}}^{\text{metal}} \leq 0.3$, implying $n_m Z_m/n_e \leq 0.01$), it is adequate to consider a single impurity, carbon.

Then the deuterium concentration is simply related to Z_{eff} by $n_d/n_e = (6 - Z_{\text{eff}})/5$. Redundant measurements of Z_{eff} by PHA, by resistive loop voltage measurements (assuming neoclassical resistivity), and by a radially-viewing VB monitor complement the standard Z_{eff} measurement by the tangential VB monitor. These measurements agree within ± 0.5 units and correspond to an uncertainty in central T_i of 7%. The calibration uncertainty in neutron emission ($\pm 15\%$) corresponds to an error in T_{io} of 3.5%. The uncertainty in density becomes significant only in highly peaked density profiles following pellet injection, when the density scale length becomes less than the spacing between in-

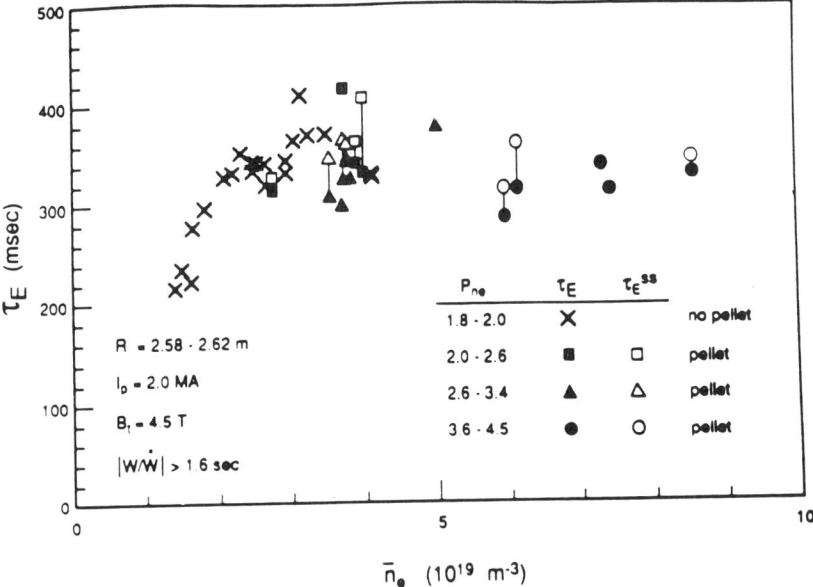

Figure 2.3: Global energy confinement time as a function of line-average density in the Ohmic density scan. τ_E^{ss} is the steady-state global energy confinement time evaluated from the kinetic data. τ_E is the same quantity but corrected for the time-dependence of stored energy, $\tau_E \equiv \tau_E^{ss}/(1-W/\dot{W})$, and is shown only where the correction exceeds 10%.

terferometer chords (10-20 cm). However, it is difficult to deduce a meaningful X_i in high density Ohmic plasmas anyway, due to strong ion-electron coupling.

The local ion-electron heat transfer is given by $q_{ei} = c_{ei}(T_e-T_i)Z_{eff}\, n_e^2/T_e^{3/2}$ where c_{ei} is a constant. Thus in regions of high electron density and/or low electron temperature, small ion-electron temperature differences drive large power flows between the two species. A lower bound on T_i can be calculated by assuming no radial transport of heat by electrons, i.e. that all of the Ohmic input power q_{OH} is *locally* transferred to the ions. This yields

$$T_i^{min} \equiv T_e - \frac{q_{OH}\, T_e^{\frac{3}{2}}}{c_{ei}\, Z_{eff}\, n_e^2}. \qquad (2)$$

At high density, the T_i is constrained to be very close to T_e irrespective of the magnitude of X_i (i.e. $T_i^{min} \approx T_e$). Thus measurements of ion temperature can provide meaningful inferences of X_i only when T_i^{min} is considerably less than T_e.

Figure 2.4 plots the inferred neoclassical multiplier as a function of $\bar{n}_e$. Throughout the density range where the temperature difference is large enough to permit accurate inferences of X_i, the neoclassical multiplier is approximately 15, with no apparent trend with density. A fixed neoclassical multiplier of 3 in the analysis yields T_i profiles whose neutron emission exceeds the measured emission by a factor 2-3.5 for $\bar{n}_{e_{e19}} \leq 3$, which is enormously outside the error bars. The multiplier on X_e required to match the neutron emission increases from 0.8–2.0 over the density range $\bar{n}_{e_{e19}} = 1$–3. Analysis of transport in reduced-bore Ohmic plasmas ($a = 0.50$ and 0.61 m) indicates considerably smaller neoclassical multipliers, $X_i = 3 - 6X_i^{neo}$, which is consistent with the observations in DIII size-scaling experiments [31].

The pellet-fuelled, high-density shots in Fig. 4 are potentially interesting because η_i at the half-radius is decreased from its value of $\sim 1.5 - 2.5$ at moderate density ($\bar{n}_{e_{e19}} = 2-3$) to ~ 0.5 at $\bar{n}_{e_{e19}} = 7.5$ and to ~ 0.2 at $\bar{n}_{e_{e19}} = 8.6$ ($P_{ne} = 4.1$). Thus if ITGDT mechanisms control the ion heat transport, one might expect to observe improved X_i in these plasmas. Due to the strong ion-electron power exchange, it is not possible to deduce an accurate X_i in these plasmas from neutron emission and measured $T_e(r)$, but one might expect to observe an im-

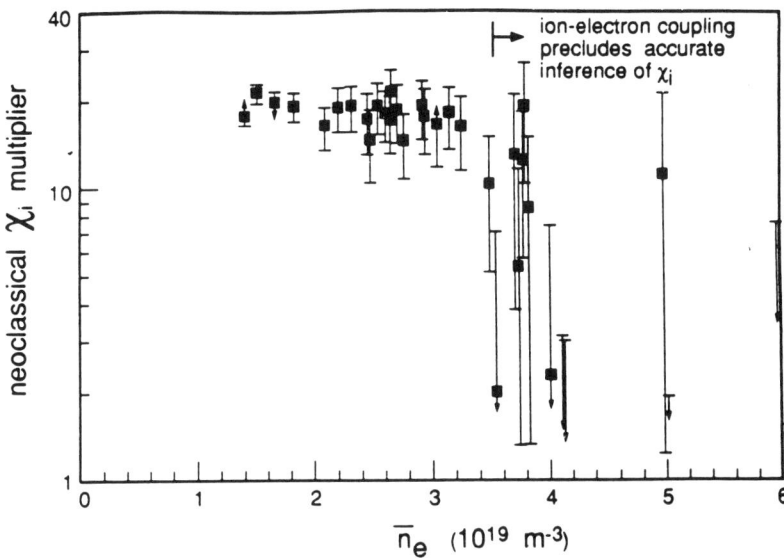

Figure 2.4: Multiplier on neoclassical χ_i required to match the neutron emission in the Ohmic density scan. Error bars represent the variation of χ_i as the T_e profile measured by ECE is increased or decreased 7%.

proved global τ_E to the extent that ion heat transport is a significant loss mechanism.

In contrast to Ohmic pellet-fuelling experiments conducted in slightly smaller plasmas ($a = 0.7$) in 1986, which produced a higher τ_E with pellets [17], there is no indication of improved energy confinement with peaked density in this data set. Both the gas-fuelled and pellet-fuelled τ_E saturate at approximately 300 ms. Unfortunately, for $\bar{n}_{e_{e19}} \geq 6$ the energy balance is complicated by a large fraction of core radiated power, which ranges from 80% to 150% of the Ohmic input power (with considerable uncertainty) at $r/a = 0.4$ at these high densities. Correspondingly, the $T_e(r)$ profile is flat or inverted inside this radius at high density. We will not discuss the highest-density plasmas further except to note that if radiation is neglected, and all heat transport is attributed to the ions, the implied χ_i is again more than ten times larger than neoclassical at the half radius.

In the intermediate density region ($\bar{n}_{e_{e19}} = 5$, $P_{ne} = 2.6$)) where core radiation is small (30% at $r = a/2$, 50% at $r = a$) and where the $T_e(r)$ profile remains peaked, there is tantalizing evidence that χ_i is indeed reduced. The multiplier on neoclassical χ_i required to match the neutron emission for one discharge (45271) is less than two even when the $T_e(r)$ profile measured by ECE or Thomson scattering is taken at the top of the $1 - \sigma$ error bars. However a detailed transport and error analysis of this discharge has not been completed.

2.1.6 VARIATION OF χ_i WITH TEMPERATURE

The variation of heat transport coefficients with temperature in tokamaks is not well understood. Theoretical models based on ITGDT turbulence are gyroBohm in character and so have an intrinsic temperature dependence $\chi \propto T_i^\alpha T_e^\beta / B^2$ with $\alpha + \beta = \frac{3}{2}$, although this is sometimes modified by the temperature dependence implicit in other dimensionless factors which enter the theoretical expressions for χ. The relative exponents on T_i and T_e vary considerably from theory to theory (see Connor et al. [95]). Experimentally, both regression analysis of τ_E across a wide database and analysis of dedicated heating power scans find that the global energy confinement time τ_E decreases with increasing power in

L-mode plasmas. This behavior is often interpreted as a consequence of χ_i and χ_e increasing with local temperature.

By adjusting the gas puff rate or recycling state of the limiter, it is possible to maintain constant density as the power is increased. Measurements of local transport in such constant-density power scans on TFTR found that both particle transport (measured with perturbative techniques) and heat transport increase as $\propto T^m$ with $m = 1.5 - 2.5$ across the plasma radius [96]. Preliminary analysis of complementary experiments which vary the density at constant temperature suggest there is little change in χ as the density increases by a factor ~ 2 (Fig. 2.5). The lack of increase in χ in the density scan at constant temperature would imply that the increased χ observed in the temperature scan is really associated with temperature rather than pressure.

However, it is well known that several features of the TFTR transport measurements are contrary to the behavior expected from ITGDT theory, which has stimulated further study both theoretically and experimentally. First, as observed on most tokamaks, the measured χ_i *increases* with minor radius, despite the monotonically decreasing temperature. Second, nondimensional scans which vary $\rho^* = \rho_s/a$ at approximately constant β and ν^* find that the heat flux is better described by Bohm rather than gyroBohm scaling [98]. Finally, in supershot plasmas the central ion temperature increases by a factor 3-5 compared to L-mode plasmas with comparable density and heating power, driven by a *reduction* of χ_i of about the same magnitude. Within experimental error, the supershot temperature and density profiles yield values of η_i that are close to or below the threshold for exciting ITG turbulence [99, 57, 100], which in principle could reduce the ion heat transport dramatically. However, no increase in local χ_i was observed when a supershot density profile was perturbed with gas puffing or pellets to drive $\eta_i \gg \eta_i^c$ [101, 100], indicating that the supershot profiles are *not* marginally stable with respect to a virulent transport mechanism. These studies are described more fully in chapter ??? of these proceedings (Kotschenruther).

The energy confinement realized in supershot plasmas correlates with a number of parameters including edge ion temperature [102], $P_{ne} = n_e(0)/\langle n_e \rangle$, T_i, T_i/T_e, and β [99]. Empirically, the strongest correlation of τ_E in supershots is with low rates of hydrogenic and carbon

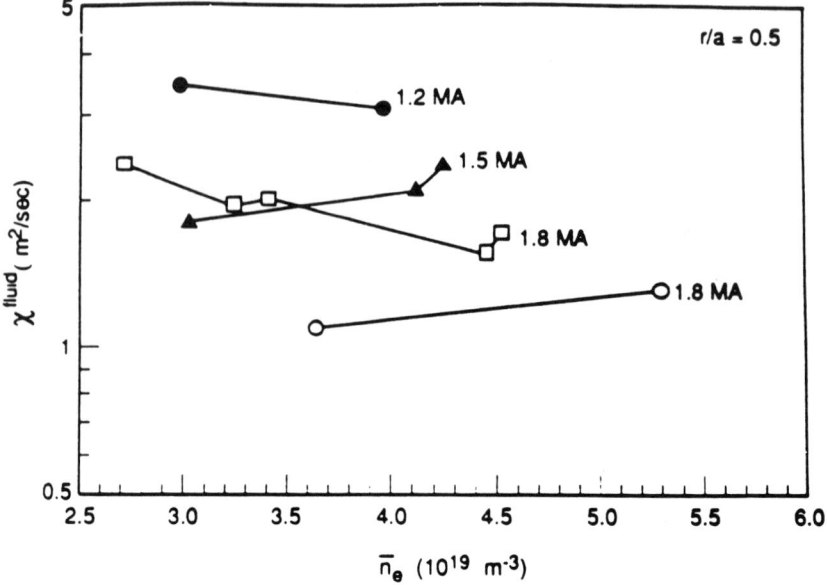

Figure 2.5: Variation of single-fluid thermal diffusivity at $r = a/2$ with density in L-mode scans at constant T_e. Conditions: (a) $I_p = 1.2$ MA, $B_t = 3.8$ T, $R = 2.58$ m, $T_{eo} = 3.5$ KeV; (b) $I_p = 1.5$ MA, $B_t = 3.8$ T, $R = 2.58$ m, $T_{eo} = 3.4$ keV; (c) (squares) $I_p = 1.8$ MA, $B_t = 3.8$ T, $R = 2.58$ m, $T_{eo} = 4.0$ keV; (d) (open circles) [97] $I_p = 1.8$ MA, $B_t = 4.8$ T, $R = 2.45$ m, $T_{eo} = 5.0$ keV.

recycling at the limiter, but it remains unclear how edge recycling affects χ_i in the plasma interior. This section focuses on the role played by T_i/T_e, which is ~ 1 in L-mode plasmas, and ≥ 2 in supershots across most of the plasma. Our fundamental observation is that the reduction of χ_i between L-mode and supershot plasmas correlates more closely with T_i/T_e than it does with η_i. This trend is clearest in the region $r \geq a/2$ and in low-power supershots, where the density scale lengths in L-mode and supershot plasmas are more similar than they are at high power near the plasma center.

This study used the set of 7 beam power scans described in Table 2.3, comprising 4 scans with low-recycling limiters (yielding supershots) and 3 scans with high-recycling limiters (yielding L-mode). All scans were conducted with the plasma resting on the carbon-carbon composite inner bumper limiter. Each scan was conducted at fixed R, B_t, and I_p. As shown in Table 2.3, there is good agreement between the total stored energy and neutron emission evaluated from kinetic measurements, and direct measurements by magnetic and neutron diagnostics. The ratio of kinetic values (neutrons, W_{tot}) to measured values changes only slightly across each scan, indicating no systematic bias of the kinetic measurements with heating power or temperature. Scans L1 and L2 are standard L-mode beam power scans at $I_p = 1.2$ and 1.8 MA, obtained with the limiter saturated with deuterium. Gas puffing and limiter conditioning were adjusted to maintain the density approximately constant as the power was increased. Scan L3 (4-21 MW) was taken with a low-recycling limiter, however helium gas puffing in the Ohmic target plasma provided high-recycling edge conditions. Its performance can be compared directly with supershot scan S3, which was taken under the same conditions but without the helium. Scans L3 and S3 have corresponding symbols (closed and open circles) to facilitate comparison in the figures. Beam injection was partially unbalanced in scans L3 and S3 (2 or 3 more beam sources directed co- than counter-) to permit accurate measurements of momentum transport. The other scans had nominally balanced injection.

Scans S1 and S2 are standard supershot power scans which span a more limited range in beam power (10-24 MW). At the high-power end, these scans include some of the best supershots obtained in TFTR to date. These scans exhibit the customary supershot correlation among beam power, density peakedness, and energy confinement. By contrast, the density profile peakedness in the lower-power supershot scans (S3 and S4) increases only weakly with power and at high power is only marginally greater than in the L-mode. These scans provide an opportunity to evaluate the possible role of parameters other than the density gradient scale length in reducing transport from L-mode to supershot values. As shown in Table 2.3, unthermalized beam ions represent about 50% of the total plasma energy at high power for *both* L-mode and supershot plasmas. The beam-ion density in the center of high-power discharges $\sim$ 30% of n_e in supershots and $\sim$ 20% in L-modes.

Figure 2.6 plots several measures of plasma performance as a function of heating power for these scans. The supershot plasmas attain a factor $\sim$3 higher central T_{io} and modestly higher n_{io} than comparable L-modes at high power, yielding an overall improvement of 4–5 in the central product $n_{io}T_{io}$ and 2.5 in volume-integrated thermal ion energy. The beam heating profile is slightly more peaked in supershot scans owing to the increased peakedness of the electron density profile, but the heating effectiveness parameter defined by Callen and coworkers [103] is almost the same. Thus the changes in the heat deposition profile are responsible for only a small fraction of the improved central ion heating in supershots, as was reported previously [99]; the dominant cause is a reduction in local heat transport coefficients.

Standard steady-state transport analysis was carried out for each of the discharges in the scans using the models described in Sec. 2.1.4. Figure 2.7 plots the total ion thermal diffusivity χ_i^{tot} as a function of heating power at $r/a = 0.32$ and $r/a = 0.72$ for each of the scans. In the L-mode, χ_i^{tot} increases strongly with power at both radii. By contrast, χ_i^{tot} increases weakly with power in the supershot at $r/a = 0.72$ and is a weak or *decreasing* function of power at $r/a = 0.32$ [104]. At low power χ_i^{tot} is comparable in the L-mode and supershot, but owing to the differing variations with power, at high power χ_i^{tot} is smaller in the supershot by a factor $\sim$4 ($r/a = 0.32$) and $\sim$ 2 ($r/a = 0.72$). This improvement is more striking when viewed from the perspective of the expected *increase* in χ_i^{tot} associated with the higher supershot ion temperatures. Figures 2.7c and 2.7d normalize the measured χ_i^{tot} by

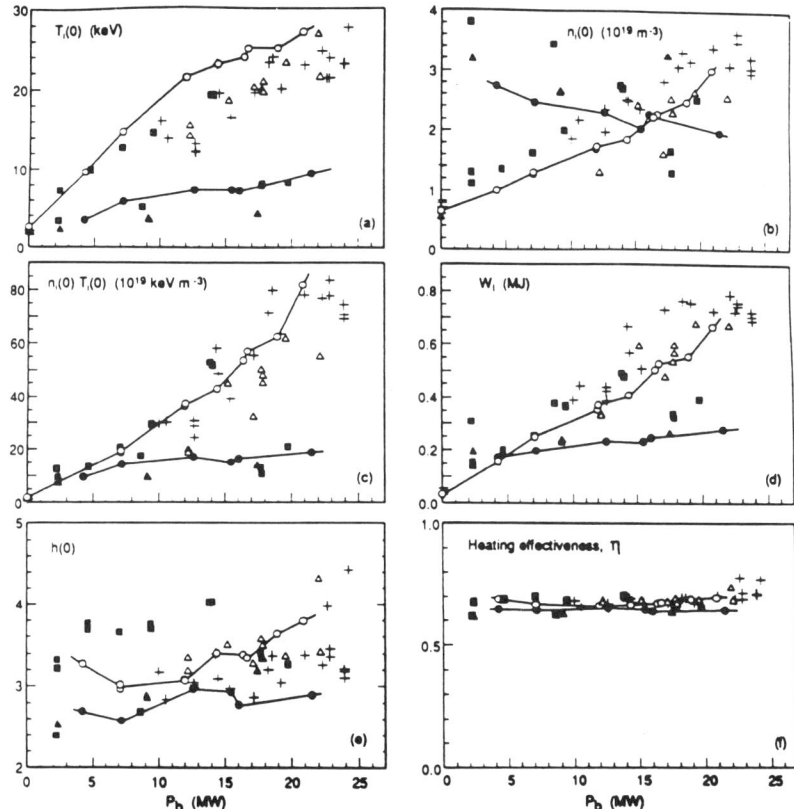

Figure 2.6: Heating parameters of beam power scans in low and high-recycling plasmas. For graphical clarity, data points from three exceptional supershot plasmas (with $n_i \leq 5.9 \times 10^{19}\text{m}^{-3}$, $n_i T_i \leq 160$ keV 10^{19}m^{-3}, and $W_i \leq 1.3$ MJ) have been omitted. (a) Central ion temperature. (b) Central thermal ion density. (c) Product of central $n_i T_i$. (d) Total thermal ion energy. (e) Peakedness of beam power deposition profile $h(0) \equiv P_b(r = 0)/(P_b^{tot}/\text{Volume})$. (f) Calculated heating effectiveness parameter, assuming a fixed radial profile shape for $\chi \propto \exp(1.6r/a)$. Symbols defined in Table 2.3.

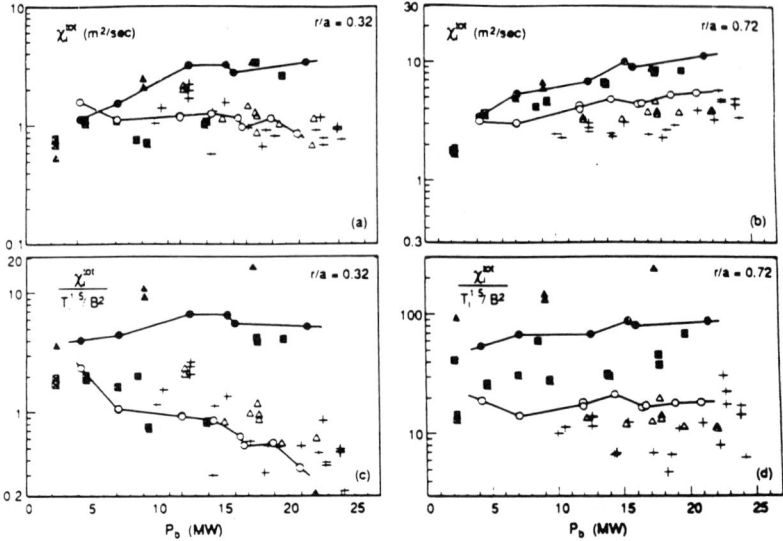

Figure 2.7: Measured χ_i^{tot} versus heating power at fixed radius. (a,b) χ_i^{tot} versus beam power at $r/a = 0.32$ and $r/a = 0.72$. (c,d) χ_i^{tot} normalized to a gyroBohm scaling ($T_i^{1.5}/B^2$). Symbols defined in Table 2.3.

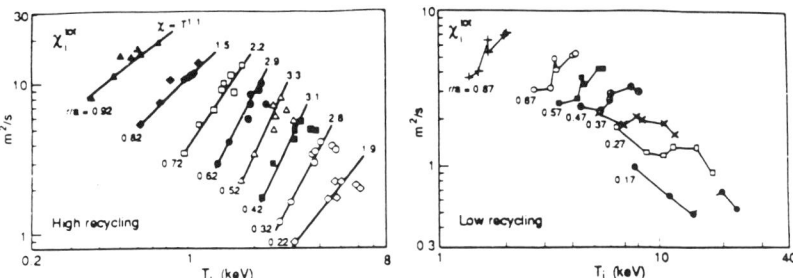

Figure 2.8: Total ion heat diffusivity as a function of local ion temperature in the L-mode and supershot power scans L3 and S3.

a factor $T_i^{3/2}/B^2$ which represents a simple gyroBohm scaling of local transport. The gyroBohm-normalized transport improves by a factor 4–5 at the edge and by a factor ~15 at the plasma center between the most comparable L-mode and supershot scans, L3 and S3.

Figure 2.8 addresses the local temperature dependence of χ_i^{tot} directly by plotting it as a function of local T_i, rather than power, at several radii. Power-law fits to the L-mode data are shown for guidance only and should not be regarded as quantitative measurements of the temperature scaling of local χ_i^{tot}. The important conclusion is that χ_i^{tot} increases throughout the entire plasma cross section in L-mode plasmas, roughly as $T_i^{2\pm1}$. The temperature dependence of χ_i^{tot} is also positive in the edge of supershots, becomes flat at about the half-radius, and then *reverses* near the plasma center. A robust theoretical understanding of the reduction in transport between supershot and L-mode plasmas must address these two fundamental observations: (1) that χ_i^{tot} is reduced *throughout* the plasma, not just in the region of steep electron density gradients, and (2) that χ_i^{tot} improves more in the plasma center than in the edge.

Very close to the plasma edge ($\approx$ 10 cm inside the last closed flux surface), the electron temperature scales roughly linearly with total beam power per electron, showing little or no difference between high and low recycling regimes. However, the edge *ion* temperature is strongly correlated with recycling, being almost an order of magnitude higher in supershots than in L-modes (2.0 versus 0.2 keV) for the same power per particle [102]. Initial studies of the edge power balance could not identify a simple atomic physics mechanism, such as charge-exchange, ion-electron energy exchange, or particle convection sufficiently large to account for the low edge ion temperatures in high-recycling plasmas. Nevertheless, some mechanism associated with reduced particle recycling plasmas apparently allows $T_i > T_e$ in the edge of supershot plasmas. Transport may be reduced in regimes where $T_i > T_e$ because ions with finite Larmor radius and drift surface size effectively average the field fluctuations during the course of their gyro- or transit period [105, 106]. Orbit averaging may also reduce the effect of particles on the fluctuations, thereby modifying the stability and spectrum of modes responsible for turbulent transport [107]. The observation of small rates of fast-ion diffusion [92, 108, 109, 110, 111], and decreased χ_i^{tot} and D_e in the core of supershots (where $T_i/T_e = 2 - 3$) is qualitatively consistent with the expected orbit-averaging effects.

It is difficult to ascertain whether high T_i/T_e or some other parameter such as η_i controls the reduction of χ_i^{tot} in high-power supershots, due to the usual correlation amongst the density profile shape, beam power, and T_i/T_e, and χ_i^{tot}. To address this issue, Fig. 2.9 plots the temperature ratio T_i/T_e, the gradient scale lengths L_{n_e} and L_{T_i}, and $\eta_i^{tot} \equiv L_{n_e}/L_{T_i}$ as a function of power at $r/a = 0.32$ and $r/a = 0.72$. At both radii, there is a distinct difference in T_i/T_e between the L-mode and supershot scans, with $T_i/T_e = 1.0 - 1.4$ in L-mode and $T_i/T_e \geq 1.8$ in the supershot.[4] As expected, there is a considerable difference in L_{n_e} at the plasma center between supershots and L-modes. However, further out in radius ($r/a = 0.72$) there is little difference between the supershot and L-mode values of L_{n_e}, L_{T_i}, or η_i^{tot}, particularly for the paired scans L3 and S3. Recall that at this radius the gyroBohm-normalized χ_i^{tot} is lower by a factor 4-5 in the supershot compared to the L-mode. Thus, at least beyond $r/a \geq 1/2$, these plasmas exhibit sub-

---

[4]An exception is the low-power supershot scan S4 at $r/a = 0.72$, which has T_i/T_e only 1.4-1.6. Interestingly, this scan has higher χ_i^{tot} at $r/a = 0.72$ than the other supershot scans (Fig. 2.7d).

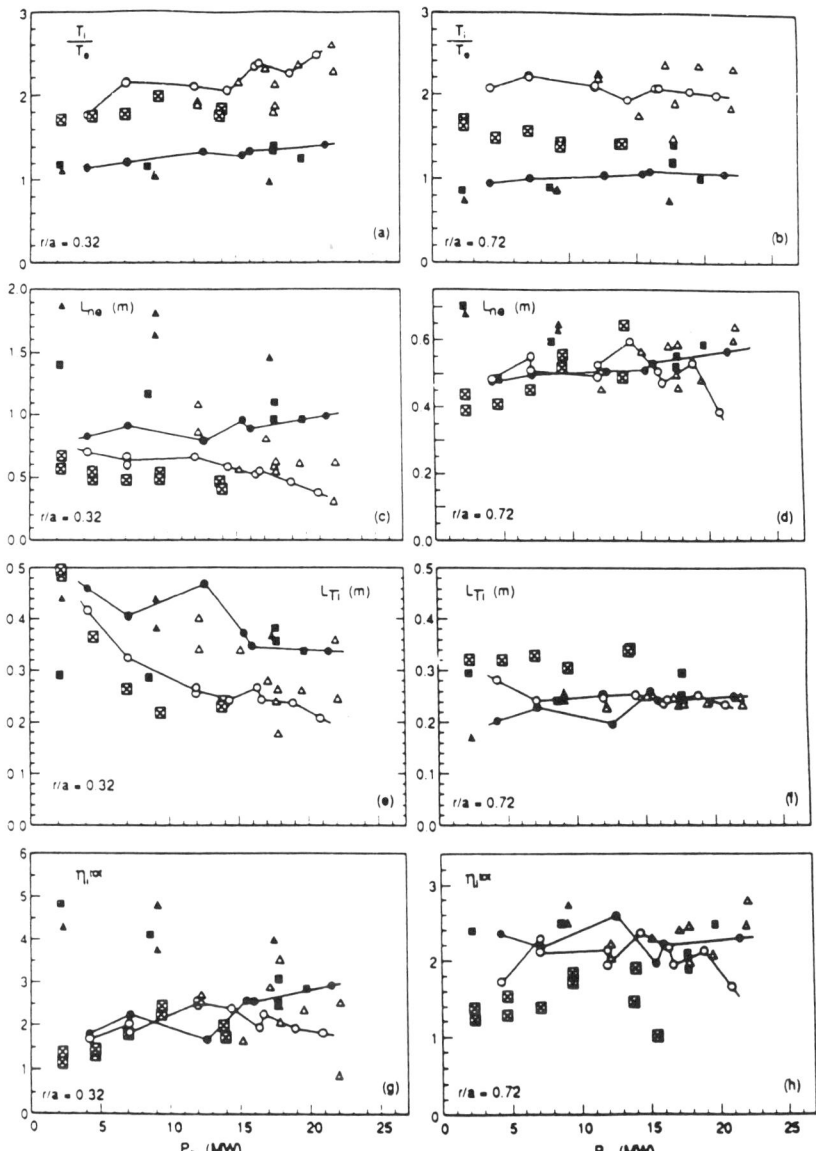

Figure 2.9: Variation of T_i/T_e, L_{ne}, L_{Ti}, and η_i^{tot} with heating power at $r/a = 0.32$ and $r/a = 0.72$. Scan S24 has been omitted for clarity. Symbols defined in Table 2.3.

stantially different transport properties despite having similar values of parameters (η_i, L_{T_i}) which are expected to govern ITG turbulence, yet they do differ in the ratio of T_i/T_e.

The crucial variable is not the value of η_i *per se* but its magnitude relative to the threshold value for exciting ITGDT turbulence. Various theoretical expressions for η_i^c such as the recent work by Guo et al. ($\eta_i^{cGuo} = \frac{4}{3}(1 + \frac{T_i}{T_e})(1 + 2\frac{s}{q})\frac{L_n}{R}$) [112] explicitly involve the temperature ratio, which in principle allows the L-mode plasmas to exceed the threshold condition while the supershots do not, despite having comparable values of η_i. Indeed, at $r/a = 0.72$ the supershot discharges in Fig. 2.9 have $\eta_i \approx \eta_i^{cGuo}$, whereas η_i for the L-mode discharges lie about 0.5-1.0 units above η_i^{cGuo}. Ascertaining whether a difference of this magnitude could be sufficient to cause a factor ~ 4 change in the gyroBohm-scaled χ_i^{tot} in these plasmas will require comprehensive kinetic microinstability calculations of the type developed by Rewoldt and Tang [113, 114, 115]. Although not implausible, this explanation appears somewhat unlikely, given that the collisionless trapped electron mode is calculated to drive significant transport even when η_i falls below η_i^c [116], and in view of the experimental results of Zarnstorff et al. showing little increase in χ_i^{tot} even when the threshold criterion for exciting ITGDT turbulence is strongly violated.

Finally, we consider the conjecture that the reduced χ_i^{tot} in supershot plasmas is governed *primarily* by the T_i/T_e becoming large, either through an orbit-averaging effect or through some other mechanism. The data presented in Figs. 2.7-2.9 represents mostly a consistency check, in that the χ_i^{tot} always appears to be reduced relative to L-mode values when T_i/T_e substantially exceeds unity. It is also true that the improvement in supershot ion heat confinement is greatest at the plasma center, where the T_i/T_e ratio is the largest. Perhaps the most attractive feature of the T_i/T_e conjecture is that provides a natural mechanism for edge recycling to affect χ_i^{tot} across the plasma cross-section, through its effect on the transport boundary condition, i.e. the edge ion temperature and edge T_i/T_e. The conjecture can be tested in TFTR by examining the variation of χ_i^{tot} in supershots heated with a mixture of neutral beams and ICRF power. Since most of the ICRF power is delivered to electrons, it should be possible to vary the T_i/T_e ratio by varying the mix of ICRF and beam power, and observe the consequent variations in χ_i^{tot}.

2.1.7 CONCLUSION

The ion thermal diffusivity in large Ohmic TFTR plasmas, inferred from both CXRS measurements of $T_i(R)$ and global neutron emission, has been found to be comparable to the electron thermal diffusivity and more than an order of magnitude greater than expected from neoclassical theory. A smaller neoclassical multiplier (3-6) is obtained for reduced-bore Ohmic plasmas. The implied size scaling of the neoclassical 'multiplier' is qualitatively consistent with results from DIII [31] and with a multitude of measurements on small tokamaks, which typically report agreement with neoclassical theory within a factor of three. The large discrepancy between measured χ_i and neoclassical theory in TFTR suggests that ion heat transport is fundamentally anomalous even in ohmic plasmas; it does not appear to be a phenomenon associated exclusively with auxiliary heating. The observation that $\chi_i \sim \chi_e$ in ohmic TFTR plasmas (although there may be differences in the radial profile) also parallels the observation in beam-heated plasmas.

In L-mode power scans at constant density, χ_i increases with temperature throughout the entire confinement region, qualitatively in accord microturbulence theory. A similar but weaker temperature dependence is observed in the periphery ($r/a \sim 0.7$) of supershot plasmas, so that at high power, χ_i is lower in supershots compared to L-mode plasmas despite having higher temperature. As one approaches the center of supershot plasmas the apparent temperature dependence of χ_i becomes progressively weaker, and actually reverses in some scans, resulting in a χ_i that decreases with increasing temperature. Normalized to a gyroBohm scaling ($\chi_i \propto T_i^{\frac{3}{2}}/B^2$), the χ_i is a factor of ~ 4 lower in high-power supershots than in L-mode plasmas at the periphery, and more than an order of magnitude at the center. It is difficult to associate the improvement in peripheral confinement with variations in the density profile shape, which changes little between the L-mode and supershot conditions. A more attractive conjecture is that the reduced transport is a direct consequence of the local ratio of T_i/T_e, which remains substantially larger in the supershot regime throughout the plasma cross-section.

ACKNOWLEDGMENT

We wish to thank J. Strachan for analysis of neutron collimator data in the ohmic density scan and G. Rewoldt for valuable discussions.

Bibliography

[1] Equipe TFR, Nucl. Fusion **16**, 279 (1976).

[2] Equipe TFR, in *Plasma Physics and Controlled Nuclear Fusion Research 1976 (Proc 6th Int Conf., Berchtesgaden, 1976)* (IAEA,Vienna, ADDRESS, 1977), Vol. 1, pp. 35–47.

[3] V. Vlasenkov *et al.*, in *Plasma Physics and Controlled Nuclear Fusion Research 1978, Proceedings of the Seventh International Conference, Innsbruck* (IAEA,Vienna, ADDRESS, 1979), Vol. 1, pp. 211–229.

[4] M. Murakami *et al.*, Phys. Rev. Lett. **42**, 655 (1979).

[5] J. D. Strachan, *Diagnostics for Fusion Reactor Conditions* (PUBLISHER, Varenna, 1982), Vol. I, pp. 383–400, eUR 8351-I-EN.

[6] Equipe TFR, in *Plasma Physics and Controlled Nuclear Fusion Research 1976 (Proc 6th Int Conf., Berchtesgaden, 1976)* (IAEA,Vienna, ADDRESS, 1977), Vol. 1, pp. 69–84.

[7] L. A. Berry *et al.*, in *Plasma Physics and Controlled Nuclear Fusion Research 1976 (Proc 6th Int Conf., Berchtesgaden, 1976)* (IAEA,Vienna, ADDRESS, 1977), Vol. 1, pp. 49–68.

[8] H. Eubank *et al.*, Phys. Rev. Lett. **43**, 270 (1979).

[9] W. Stodiek *et al.*, in *Plasma Physics and Controlled Nuclear Fusion Research 1980, Proceedings of the Eighth International Conference, Brussels* (IAEA,Vienna, ADDRESS, 1981), Vol. 1, pp. 9–22.

[10] S. Fairfax et al., in *Plasma Physics and Controlled Nuclear Fusion Research 1980, Proceedings of the Eighth International Conference, Brussels* (IAEA,Vienna, ADDRESS, 1981), Vol. 1, pp. 439-451.

[11] B. Blackwell et al., in *Plasma Physics and Controlled Nuclear Fusion Research 1982 (Proc 9th Int Conf., Baltimore, 1982)* (IAEA,Vienna, ADDRESS, 1983), Vol. 2, pp. 27-40.

[12] M. Greenwald et al., Phys. Rev. Lett. **53**, 352 (1984).

[13] M. Greenwald et al., in *Plasma Physics and Controlled Nuclear Fusion Research 1984* (IAEA, Vienna, ADDRESS, 1985), Vol. 1, pp. 45-55, proc. 10th Int. Conf. London, 1984.

[14] R. R. Parker et al., Nucl. Fusion **25**, 1127 (1985).

[15] S. M. Wolfe et al., Nucl. Fusion **26**, 329 (1986).

[16] M. Greenwald et al., in *Plasma Physics and Controlled Nuclear Fusion Research 1986* (IAEA,Vienna, ADDRESS, 1987), Vol. 1, pp. 139-149.

[17] G. Schmidt et al., in *Plasma Physics and Controlled Nuclear Fusion Research 1986* (IAEA,Vienna, ADDRESS, 1987), Vol. 1, pp. 171-178.

[18] O. Gruber et al., in *Proceedings of the 15th European Conference on Controlled Fusion and Plasma Physics, Dubrovnik, 1988* (European Physical Society, Petit-Lancy, Switzerland, ADDRESS, 1988), Vol. 1, pp. 27-30.

[19] U. Stroth, H.-U. Fahrbach, W. Herrmann, and H.-M. Mayer, Nucl. Fusion **29**, 761 (1989).

[20] U. Stroth et al., Nucl. Fusion **31**, 2291 (1991).

[21] L. Artsimovich, A. Glukhov, and M. Petrov, JETP lett. **11**, 304 (1970).

[22] E. Gorbunov, V. Zaverjaev, and M. Petrov, in *Controlled Fusion and Plasma Physics (Proc. 6th European Conf. Moscow, 1973)* (Moscow, ADDRESS, 1973), Vol. 1, pp. 1–4.

[23] N. Fujisawa et al., in *Plasma Physics and Controlled Nuclear Fusion Research 1974, Proceedings of the Fifth International Conference, Tokyo* (IAEA,Vienna, ADDRESS, 1975), Vol. 1, pp. 3–16.

[24] L. A. Berry et al., in *Plasma Physics and Controlled Nuclear Fusion Research 1974 (Proc 5th Int Conf., Tokyo, 1974)* (IAEA,Vienna, ADDRESS, 1975), Vol. 1, pp. 101–112.

[25] L. A. Berry, J. Clarke, and J. Hogan, Phys. Rev. Lett. **32**, 362 (1974).

[26] P. E. Stott, Plasma Physics **18**, 251 (1975).

[27] A. Gondhalekar et al., in *Plasma Physics and Controlled Nuclear Fusion Research 1978* (IAEA,Vienna, ADDRESS, 1979), Vol. 1, pp. 199–209.

[28] M. Gaudreau et al., Phys. Rev. Lett. **39**, 1266 (1977).

[29] A. Bagdasarov et al., in *Plasma Physics and Controlled Nuclear Fusion Research 1978, Proceedings of the Seventh International Conference, Innsbruck* (IAEA,Vienna, ADDRESS, 1979), Vol. 1, pp. 35–49.

[30] M. Brusati et al., Nucl. Fusion **18**, 1205 (1978).

[31] S. Ejima et al., Nucl. Fusion **22**, 1627 (1982).

[32] P. Efthimion et al., in *Plasma Physics and Controlled Nuclear Fusion Research 1984* (IAEA, Vienna, ADDRESS, 1985), Vol. 1, p. 29, proc. 10th Int. Conf. London, 1984.

[33] J. G. Cordey et al., in *Plasma Physics and Controlled Nuclear Fusion Research 1984* (IAEA, Vienna, ADDRESS, 1985), Vol. 1, pp. 167–179, proc. 10th Int. Conf. London, 1984.

[34] D. V. Bartlett et al., Nucl. Fusion **28**, 73 (1988).

[35] A. A. Galeev and R. Z. Sagdeev, Sov. Phys.-JETP **26**, 233 (1968).

[36] F. L. Hinton and M. N. Rosenbluth, Phys. Fluids **16**, 836 (1973).

[37] F. L. Hinton and R. D. Hazeltine, Rev. Mod. Phys. 239 (1976).

[38] R. D. Hazeltine and F. L. Hinton, Phys. Fluids **16**, 1883 (1973).

[39] C. S. Chang and F. L. Hinton, Phys. Fluids **25**, 1493 (1982).

[40] C. S. Chang and F. L. Hinton, Phys. Fluids **29**, 3314 (1986).

[41] D. Swain et al., Nucl. Fusion **21**, 1409 (1981).

[42] A. G. Barsukov et al., in *Plasma Physics and Controlled Nuclear Fusion Research 1982 (Proc 9th Int Conf., Baltimore, 1982)* (IAEA,Vienna, ADDRESS, 1983), Vol. 1, pp. 83–94.

[43] G. Becker, the ASDEX team, and Neutral Injection Team, Nucl. Fusion **22**, 1589 (1982).

[44] G. Becker et al., Nucl. Fusion **23**, 1293 (1983).

[45] M. Murakami et al., in *Plasma Physics and Controlled Nuclear Fusion Research 1984* (IAEA, Vienna, ADDRESS, 1985), Vol. 1, p. 87.

[46] S. Kaye et al., Nucl. Fusion **24**, 1303 (1984).

[47] R. Groebner et al., Nucl. Fusion **26**, 543 (1986).

[48] M. Okabayashi et al., in *Plasma Physics and Controlled Nuclear Fusion Research 1986* (IAEA,Vienna, ADDRESS, 1987), Vol. 1, pp. 275–290.

[49] M. Zarnstorff et al., in *Plasma Physics and Controlled Nuclear Fusion Research 1988* (IAEA,Vienna, ADDRESS, 1989), Vol. 1, pp. 183–191.

[50] R. Fonck et al., Phys. Rev. Lett. **63**, 520 (1989).

[51] S. Scott et al., Phys. Fluids B **2**, 1300 (1990).

[52] T. Hirayama et al., Nucl. Fusion **32**, 89 (1992).

[53] B. Balet, F. Cordey, and P. Stubberfield, Plasma Phys. Controlled Fusion **34**, 3 (1992).

[54] B.Balet et al., Nucl. Fusion **30**, 2029 (1990).

[55] S. Corti et al., in *Proceedings of the 13th European Conference on Controlled Fusion and Plasma Physics, Schliersee, 1986* (European Physical Society, Petit-Lancy, Switzerland, ADDRESS, 1987), Vol. 1, pp. 318–321.

[56] O. Gruber et al., in *Proceedings of the 15th European Conference on Controlled Fusion and Plasma Physics, Dubrovnik, 1988* (European Physical Society, Petit-Lancy, Switzerland, ADDRESS, 1988), Vol. 1, pp. 23–26.

[57] S. Scott et al., Phys. Rev. Lett. **64**, 531 (1990).

[58] K. Burrell et al., Transport in Auxiliary-heated, hot-ion H-mode and L-mode discharges in the DIII-D tokamak, paper IAEA-CN-53/A-2-3, presented at the Thirteenth International Conference on Plasma Physics and Controlled Fusion Research, Washington, D.C. 1-6 October 1990.

[59] B. C. Stratton et al., in *Proceedings of the IAEA Technical Committee Meeting on Time Resolved Two- and Three-Dimensional Plasma Diagnostics, (Nagoya, Japan, Novermber 1990)* (IAEA, Vienna, 1990), pp. 78–87.

[60] D. K. Mansfield et al., Appl. Opt. **26**, 4469 (1987).

[61] H. K. Park, Rev. Sci. Instrum. **61**, 2879 (1990).

[62] D. Johnson et al., Rev. Sci. Instrum. **56**, 1015 (1985).

[63] F. J. Stauffer et al., Rev. Sci. Instrum. **59**, 2139 (1988).

[64] G. Taylor et al., Rev. Sci. Instrum. **55**, 1739 (1984).

[65] G. Taylor et al., Rev. Sci. Instrum. **56**, 929 (1985).

[66] G. Taylor et al., Nucl. Fusion **26**, 339 (1986).

[67] G. Taylor et al., Rev. Sci. Instrum. **57**, 1974 (1986).

[68] A. Cavallo, R. C. Cutler, and M. P. McCarthy, Rev. Sci. Instrum. **59**, 889 (1988).

[69] H. W. Hendel, IEEE Trans. Nucl. Sci. **33**, 670 (1986).

[70] H. W. Hendel et al., Rev. Sci. Instrum. **59**, 1682 (1988).

[71] H. W. Hendel et al., Rev. Sci. Instrum. **61**, 1900 (1990).

[72] C. W. Barnes et al., Rev. Sci. Instrum. **61**, 3151 (1990).

[73] J. D. Strachan et al., *Proceedings of the 17th European Conference on Controlled Fusion and Plasma Heating, Amsterdam, 1990*, Vol. 14B of *Europhysics Conference Abstracts* (EPS, Petit-Lancy, 1990), pp. 1548–1551, part IV.

[74] A. L. Roquemore et al., Rev. Sci. Instrum. **61**, 3163 (1990).

[75] L. C. Johnson, Rev. Sci. Instrum. **63**, 4517 (1992), proceedings of the 9th Topical Conference on High Temperature Plasma Diagnostics (Santa Fe, NM, March 1992).

[76] J. Coonrod et al., Rev. Sci. Instrum. **56**, 941 (1985).

[77] M. G. Bell et al., Plasma Phys. Controlled Fusion **28**, 1329 (1986).

[78] A. T. Ramsey and S. L. Turner, Rev. Sci. Instrum. **58**, 1211 (1987).

[79] K. W. Hill et al., Rev. Sci. Instrum. **56**, 840 (1985).

[80] M. C. Zarnstorff et al., Phys. Fluids B **2**, 1852 (1990).

[81] D. Heifetz et al., J. Vac. Sci. Technol. A **6**, 2564 (1988).

[82] R. V. Budny and the TFTR Group, J. Nucl. Matl. **176-177**, 427 (1990).

[83] J. Schivell, Rev. Sci. Instrum. **58**, 12 (1987).

[84] J. Schivell, G. Renda, J. Lowrance, and H. Hsuan, Rev. Sci. Instrum. **53**, 1527 (1982).

[85] J. F. Schivell, Rev. Sci. Instrum. **56**, 972 (1985).

[86] H. Towner *et al.*, Rev. Sci. Instrum. **63**, (1992).

[87] J. A. Murphy, S. D. Scott, and H. H. Towner, Rev. Sci. Instrum. **63**, 4750 (1992).

[88] J. A. Murphy, S. D. Scott, and H. H. Towner, Technical Report No. PPPL-TM393, Princeton Plasma Physics Laboratory (unpublished).

[89] R. Goldston, in *Basic Physical Processes of Toroidal Fusion Plasmas (Proceedings of Course and Workshop, Varenna)* (Office for Official Publications of the European Communities, Luxemborg, ADDRESS, 1985), Vol. 1, pp. 165–186.

[90] M. C. Zarnstorff *et al.*, in *Proceedings of the 15th European Conference on Controlled Fusion and Plasma Heating (Dubrovnik, Yugoslavia, 1988)*, Vol. 12B of *Europhysics Conference Abstracts* (EPS, Petit-Lancy, 1988), pp. 95–98, part I.

[91] M. C. Zarnstorff *et al.*, *Plasma Physics and Controlled Nuclear Fusion Research, 1988* (IAEA, Vienna, 1989), Vol. 1, p. 183, (Nice Conference).

[92] S. Scott *et al.*, in *Plasma Physics and Controlled Nuclear Fusion Research 1990* (IAEA, Vienna, ADDRESS, 1991), Vol. 1, pp. 235–259.

[93] J. D. Strachan *et al.*, Technical Report No. PPPL-2858, Princeton Plasma Physics Laboratory (unpublished).

[94] K. W. Hill *et al.*, *Plasma Physics and Controlled Nuclear Fusion Research, 1986* (IAEA, Vienna, 1987), Vol. 1, pp. 207–216, (Kyoto Conference).

[95] J. W. Connor *et al.*, Technical Report No. JET-P(92)63, JET Joint Undertaking, (unpublished), submitted to Plasma Physics and Controlled Fusion.

[96] P. Efthimion *et al.*, Phys. Rev. Lett. **66**, 421 (1991).

[97] E.-S. Ghanem, J. Kinsey, C. Singer, and G. Bateman, in *Proceedings of the 19th European Conference on Controlled Fusion and Plasma Physics*, Vol. 16C of *Europhysics Conference Abstracts* (EPS, Petit-Lancy, 1992), pp. 38–41, part I.

[98] F. W. Perkins *et al.*, Phys. Fluids B **5**, (1992), (in press).

[99] M. Zarnstorff *et al.*, in *Proceedings of the 16th European Conference on Controlled Fusion and Plasma Heating, Venice, 1989* (European Physical Society, Petit-Lancy, Switzerland, ADDRESS, 1989), Vol. 1, pp. 35–38.

[100] M. Zarnstorff *et al.*, Advances in Transport Understanding using Perturbative techniques in TFTR, paper IAEA-CN-53/A-II-2, presented at the Thirteenth International Conference on Plasma Physics and Controlled Fusion Research, Washington, D.C. 1-6 October 1990.

[101] M. Zarnstorff *et al.*, in *Proceedings of the 17th European Conference on Controlled Fusion and Plasma Heating, Amsterdam, 1990* (European Physical Society, Petit-Lancy, Switzerland, ADDRESS, 1990), Vol. 1, pp. 42–45.

[102] S. D. Scott *et al.*, in *Proceedings of the 19th European Conference on Controlled Fusion and Plasma Physics*, Vol. 16C of *Europhysics Conference Abstracts* (EPS, Petit-Lancy, 1992), part I.

[103] J. Callen *et al.*, Nucl. Fusion **27**, 1857 (1987).

[104] S. D. Scott *et al.*, *Plasma Physics and Controlled Nuclear Fusion Research, 1990* (IAEA, Vienna, 1991), Vol. 1, pp. 235–260, (Washington D. C. Conference).

[105] H. E. Mynick and J. A. Krommes, Phys. Rev. Lett. **43**, 1506 (1979).

[106] H. Mynick and J. Krommes, Phys. Fluids **23**, 1229 (1980).

[107] H. E. Mynick and S. J. Zweben, Nucl. Fusion **32**, 518 (1992).

[108] R. Radeztsky *et al.*, in *Proc. 15th Eur. Conf. Controlled Fusion and Plasma Physics, Dubrovnik (1988)* (European Physical Society, ADDRESS, 1988), pp. 79–82.

[109] C. W. Barnes *et al.*, in *Proceedings of the 15th European Conference on Controlled Fusion and Plasma Physics, Dubrovnik, 1988* (European Physical Society, Petit-Lancy, Switzerland, ADDRESS, 1988), Vol. 1, pp. 87–90.

[110] R. Hawryluk *et al.*, Overview of TFTR transport studies, invited paper presented at the 18th European Conference on Controlled Fusion and Plasma Physics, Berlin, 1991. To be published in Plasma Physics and Controlled Fusion.

[111] S. J. Zweben *et al.*, Nucl. Fusion **31**, 2219 (1991), preprint in PPPL-2770.

[112] S. C. Guo, F. Romanelli, G. Rewoldt, and W. M. Tang, in *Proceedings of the 19th European Conference on Controlled Fusion and Plasma Physics, Innsbruck (1992)* (European Physical Society, ADDRESS, 1992), Vol. 2, pp. 1437–1440.

[113] G. Rewoldt, W. Tang, and M. S. Chance, Phys. Fluids **25**, 480 (1982).

[114] W. Tang, G. Rewoldt, and L. Chen, Phys. Fluids **29**, 3715 (1986).

[115] G. Rewoldt, W. Tang, and R. Hastie, Phys. Fluids **30**, 807 (1987).

[116] G. Rewoldt and W. Tang, Phys. Fluids B **2**, 318 (1990).

2.2 Ion Temperature Profile Simulation of JT-60 Plasmas with Ion Temperature Gradient Mode Transport Models

H. Shirai , T. Hirayama , Y. Koide and M. Azumi
Japan Atomic Energy Research Institute
Naka-machi, Ibaraki-ken, 311-01 Japan

Abstract. *Ion temperature profiles of neutral beam heated plasmas in JT-60 have been simulated by using models of ion thermal diffusivity, χ_i, based on ion temperature gradient mode (η_i mode) turbulence and drift wave turbulence (trapped electron mode and circulating electron mode). Three different χ_i models proposed by Dominguez & Waltz, Lee & Diamond, and Romanelli are adopted and the predicted T_i profiles are compared with those measured by CXRS (charge exchange recombination spectroscopy). All three χ_i models can fit the T_i profile over wide range of plasma parameters of JT-60 L-mode plasmas except for 1 MA limiter cases. However, in the high ion temperature (high T_i) plasmas, the predicted T_i profiles are much broader than that of experiment.*

2.2.1 INTRODUCTION

Extensive transport analyses of tokamak plasmas have been conducted worldwide from small size machines to large tokamaks. Various kinds of transport models have been proposed and adopted in the transport analyses in order to reproduce the measured density or temperature profiles [1, 2, 3, 4, 5, 6, 7]. Among these models, the η_i mode turbulence [1, 8, 9, 10, 11, 12, 13, 14, 15, 16, 17, 18, 19, 20, 21] has been regarded as a candidate for ion energy transport, because the improvement of energy confinement time has been often observed in the peaked density plasmas, in which η_i ($\equiv d \ln T_i / d \ln n$) is small and η_i mode is suppressed. These models are adopted to explain the measured density and temperature profiles in many devices.

Some recent analyses of heat transport in Alcator C [22] and AS-DEX [23] showed agreement with the transport by η_i mode turbulence, while TEXT [24], JET [25] and TFTR [26] showed disagreement. One point of objection to the η_i mode turbulence model is that the the profile of χ_i^{pre} is often different from that of χ_i^{exp}, where χ_i^{pre} is the predicted χ_i which is calculated by substituting measured plasma parameters in the χ_i formula based on the η_i mode turbulence theory, and χ_i^{exp} is the experimentally evaluated χ_i by the ion energy balance equation with all plasma parameters such as n_e, T_e and T_i are fixed at theie measured values. The χ_i^{exp} usually increases from the plasma center toward the plasma edge, whereas χ_i^{pre} often decreases from around the half of minor radius toward the plasma surface. However, because the conduction loss is not the main energy loss channel in the plasma edge region, the different behavoir between χ_i^{pre} and χ_i^{exp} in the plasma edge region does not immediately lead to a definite conclusion that the η_i mode turbulence is not the candidate of ion heat transport.

Another point of objection to the η_i mode turbulence model is that a large discrepancy between η_i^{exp} and η_{ic}, the threshold values, is seen in some discharges [25, 26] and in such cases the calculated χ_i is expected to deviate from χ_i^{exp}. Early theoretical analysis predicted very large heat transport when η_i exceeds η_{ic}, thus theoretically T_i profile could not significantly exceed the "marginally stable" value. It seems that this hypothesis is justified for the strong transport processes such as ideal MHD instabilities. However, the transport by η_i mode turbulence may not be sufficiently virulent. Our perspective is that the transport increases continuously, not abruptly, as η_i increases when η_i crosses over the η_{ic}, thus η_i^{exp} may exceed η_{ic} is less important. The significant test is whether we can reproduce the T_i profile when it is far from marginal stability. Recent η_i mode turbulence theory [2], including a self-consistent linear spectrum of $k_\parallel$ which is determined by the ballooning stability analysis, well explains the pellet injected TFTR plasma, in which χ_i^{pre} shows no increase after the pellet injection although η_i becomes far above η_{ic}.

We adopt the χ_i model based on η_i mode turbulence and drift wave turbulence. The η_i mode has two branches according to the destabilization mechanism. In the slab branch, the driving force is the destabilized ion acoustic wave by the ion temperature gradient. In the toroidal branch, interchange effect destabilize eigenmode localizes at the bad curvature region of tokamak like ballooning mode.

We adopted three different χ_i formula of η_i mode turbulence proposed by Dominguez & Waltz [1], Lee & Diamond [10] and Romanelli [11]. We will express these models as D&W model, L&D model and R model for simplicity in the following. In the D&W model, the diffusion by random walk process; $D \propto \gamma/k_\perp^2$ and the dispersion relation of toroidal η_i mode is used obtaining a χ_i formula is obtained by assuming mixing length theory. In L&D model, the nonlinear saturation level of χ_i is estimated from the wavenumber spectrum in a sheared slab geometry by using two point renormalization theory without using a mixing length estimate. In R model, the equation of the electrostatic potential which is obtained from the equation of quasineutrality and the drift kinetic equation is used to evaluate the ion energy flux by quasilinear theory. Then χ_i is obtained from mixing length theory. As is shown in the following sections, D&W model and R model are rather similar. Though L&D model alone does not adopt the mixing length estimation, its χ_i formula resembles the others except that it has a strong η_i dependence, while the other two models have a weak η_i dependence.

We calculate T_i profiles using the above three χ_i models and compare the results in order to study the validity of each model. We determine the overall magnitude of each χ_i model by matching the predicted and measured temperatures in a few typical JT-60 L-mode discharges. We use same factor in JT-60 plasmas and do not adjust that factor shot by shot.

We restrict our attention to beam heated plasmas because neutral beam is necessary to measure the T_i with the CXRS diagnostics [28, 29, 30, 31, 32, 33, 34]. We calculate the T_i profiles of JT-60 plasmas over a wide range of plasma parameters in L-mode plasmas and high ion temperature (high T_i) plasmas by using χ_i models and solving ion energy balance equation using the one dimensional (1-D) steady state transport analysis code SCOOP [35]. Plasma parameters other than T_i such as n_e, T_e, Z_{eff} and P_{rad} are fixed at their measured value. The profile of toroidal plasma current, J_ϕ, and the neutral beam heating of ions, P_{NBI}^i, are calculated and used in the transport analysis code.

In the next section, we present the χ_i formula based on η_i mode turbulence and drift wave turbulence, which are used in this paper. The method of analysis is also shown. Section III shows results of T_i profile analysis in L-mode plasmas. Section IV shows the results of high T_i plasmas in JT-60. The effects of η_{ic} on the simulation results are discussed in Sec. V. In Sec. VI, the results are summarized.

2.2.2 SIMULATION METHOD

Ion Thermal Diffusivities

We define the formula of ion thermal diffusivity shown as follows;

$$\chi_i = \chi_i^{\eta_i} + \chi_i^{\text{TE/CE}} + \chi_i^{\text{NC}} \qquad (1)$$

where $\chi_i^{\eta_i}$ is the diffusivity based on the η_i mode turbulence. $\chi_i^{\text{TE/CE}}$ represents the diffusivity based on the drift wave turbulence: the trapped electron mode and the circulating electron mode. χ_i^{NC} is the neoclassical diffusivity proposed by Chang & Hinton [36]. The term $\chi_i^{\eta_i}$ and $\chi_i^{\text{TE/CE}}$ are explained in detail in the following subsections.

ION THERMAL DIFFUSIVITIES DUE TO η_i MODE

We adopted three different χ_i models based on the η_i mode turbulence:
(a) D&W model [1]

$$\chi_i^{\eta_i} = C^{\eta_i} 2.5 \frac{\omega_{*e}}{k_\theta^2} \left(\frac{2T_i L_n \eta_i}{T_e R} \right)^{1/2} f(\eta_i) \qquad (2)$$

(b) L&D model [10]

$$\chi_i^{\eta_i} = C^{\eta_i} 0.4 \left\{ \frac{\pi}{2} \frac{T_i}{T_e} (1+\eta_i) \ln(1+\eta_i) \right\}^2 \frac{\rho_s^2 C_S}{L_S} f(\eta_i) \qquad (3)$$

(c) R model [11]

$$\chi_i^{\eta_i} = C^{\eta_i} 3 \frac{V_i \rho_i^2}{L_n} \varepsilon_n^{1/2} g(\eta_i) \qquad (4)$$

where

$$\eta_i \equiv \frac{L_n}{L_{T_i}} = \frac{d\ln T_i}{dr} \Big/ \frac{d\ln n_e}{dr}, \quad L_n = -n_e \Big/ \frac{dn_e}{dr}, \quad L_{T_i} = -T_i \Big/ \frac{dT_i}{dr},$$

$$L_S = Rq^2/r \frac{dq}{dr}, \quad \rho_i = \frac{V_i}{\Omega_i}, \quad \rho_S = \frac{C_S}{\Omega_i}, \quad V_i = \left(\frac{T_i}{m_i}\right)^{1/2},$$

$$C_S = \left(\frac{T_e}{m_i}\right)^{1/2}, \quad \Omega_i = \frac{eB_t}{m_i}, \quad \omega_{*e} = \frac{C_s k_\theta \rho_S}{L_n}, \quad \varepsilon_n = L_n/R$$

and

$$f(\eta_i) = \frac{1}{1 + \exp(-6(\eta_i - \eta_{ic}))}, \tag{5}$$

$$g(\eta_i) = \begin{cases} 0 & (\eta_i \leq \eta_i - 0.4) \\ 1.398\,(\eta_i - \eta_{ic} + 0.4)^2 & (\eta_{ic} - 0.4 \leq \eta_i \leq \eta_{ic}) \\ 1.118\,(\eta_i - \eta_{ic}) + 0.224 & (\eta_{ic} \leq \eta_i \leq \eta_{ic} + 0.2) \\ (\eta_i - \eta_{ic})^{0.5} & (\eta_i \geq \eta_{ic} + 0.2) \end{cases} \tag{6}$$

In the above formula, ρ_i is the ion Larmor radius, ρ_S is the Larmor radius measured by electron temperature, V_i is the ion thermal velocity, C_S is the sound velocity, Ω_i is the ion cyclotron frequency and k_θ is the wave number. We assume $k_\theta \rho_S = 0.3$. We set coefficients $C^{\eta_i} = 0.5$ and 1 for D&W model and R model, respectively. These coefficients are determined by adjusting the calculated χ_i in a few typical L-mode plasmas. In Lee & Diamond's paper, the absolute value of χ_i is determined without the coefficient C^{η_i}. However with $C^{\eta_i} = 1$, the predicted T_i becomes much higher than that of experimental data. Therefore we set $C^{\eta_i} = 2.5$ for L&D model in this paper. The plasma parameter dependence of each $\chi_i^{\eta_i}$ are

$$\chi_i^{\eta_i} \propto T_e\, T_i^{0.5}\, B_t^{-2}\, (R\, L_{T_i})^{-0.5}\, f(\eta_i) \quad \text{(D\&W model)}$$
$$\chi_i^{\eta_i} \propto T_e^{-0.5}\, T_i^2\, B_t^{-2} L_S^{-1} \{(1+\eta_i)\ln(1+\eta_i)\}^2\, f(\eta_i) \quad \text{(L\&D model)}$$
$$\chi_i^{\eta_i} \propto T_i^{1.5}\, B_t^{-2}\, (RL_n)^{-0.5}\, g(\eta_i) \quad \text{(R model)}$$

They have $T^{1.5} B_t^{-2}$ dependence in common. The η_i dependence of $\chi_i^{\eta_i}$ is similar between D&W model and R model, while L&D model is entirely different.

Theoretically, when η_i becomes larger than the threshold value, η_{ic}, the η_i mode suddenly becomes unstable. This causes χ_i discontinuous at $\eta_i = \eta_{ic}$ like step function. The function $f(\eta_i)$ is introduced in order to provide a smooth transition from stable to unstable state of η_i mode around $\eta_i = \eta_{ic}$. This function avoids the numerical conversion problem of χ_i which would otherwise arise when the calucated η_i is near η_{ic}. It must be kept in mind that because of the characteristics of the function $f(\eta_i)$ $\chi_i^{\eta_i}$ is not zero even if η_i is lower than η_{ic}. The $\chi_i^{\eta_i}$ is 0.1 at $\eta_i - \eta_{ic} \approx -0.37$ and is 0.9 at $\eta_i - \eta_{ic} \approx 0.37$.

The number 6 which appears in the denominator of Eq. (5) denotes the degree to which the function $f(\eta_i)$ resembles the step function. If we increase this number, the function $f(\eta_i)$ becomes much more like the step function and the value of $f(\eta_i)$ changes from zero to unity more rapidly around $\eta_i = \eta_{ic}$. We have changed this number between 6 and 20. The results show little difference because the η_i mode is in the fully unstable region almost everywhere, that is $f(\eta_i) \approx 1$, with L-mode plasma parameters as are shown in the following section. Even in the high T_i plasmas in JT-60 in which the η_i mode is in the marginally stable region, i.e. $f(\eta_i) \approx 0.5$, the results show little difference.

In the R model, the original formula indicates $g(\eta_i) = \sqrt{\eta_i - \eta_{ic}}$ [11]. We modified this formula as Eq. (6) in order to aviod the numerical convergence problem of χ_i around $\eta_i = \eta_{ic}$, where $\partial \chi_i / \partial \eta_i$ becomes infinity by the original $g(\eta_i)$ formula. The modified $g(\eta_i)$ is continuous and differentiable at $\eta_i = \eta_{ic} - 0.4$, $\eta_i = \eta_{ic}$ and $\eta_i = \eta_{ic} + 0.2$. This formula means that η_i mode is unstable if $\eta_i \geq \eta_{ic} - 0.4$. When η_i is larger than $\eta_{ic} + 0.2$, Eq. (4) becomes just the same as the original χ_i formula. As is shown in the following section, the condition $\eta_i \geq \eta_{ic} + 0.2$ holds in almost all region in the L-mode plasmas in JT-60.

The value of η_{ic} is [9, 11];

$$\eta_{ic} = \begin{cases} 1 & (\varepsilon_n \leq 0.2) \\ 1 + 2.5(\varepsilon_n - 0.2) & (\varepsilon_n \geq 0.2) \end{cases} \quad (7)$$

In the Dominguez & Waltz' paper [1], $\eta_{ic} = 1$. Equation (7) adds the effect of the magnetic field line curvature. This formula indicates that η_{ic} becomes large where the density profile is flat, for example, near the plasma central region.

According to the recent theories [12, 13, 14], the η_{ic} depends on the ratio of ion temperature to the electron temperature: as T_i/T_e increases, η_{ic} increases, especially when ε_n is large. However, because T_i/T_e is usually between 0.7 and 1.3, (at most 1.6) in JT-60 L-mode plasmas, the correction of η_{ic} with T_i/T_e is practically unnecessary. Even in the high T_i plasmas in JT-60 in which T_i/T_e reaches 3, η_{ic} does not change much because the supershot plasmas are accompanied by a peaked density profile which makes ε_n small. Thus the modification of η_{ic} affects very little the simulation results. For this reason, we do not include the T_i/T_e modification of η_{ic} in this work.

MODEL OF DRIFT WAVE TURBULENCE

The term $\chi_i^{\text{TE/CE}}$ is made up of the contribution from the trapped electron mode, χ_i^{TE}, and the circulating electron mode, χ_i^{CE} [1], that is;

$$\chi_i^{\text{TE/CE}} = C^{\text{TE}}\chi_i^{\text{TE}} + C^{\text{CE}}\chi_i^{\text{CE}} \qquad (8)$$

Originally, $\chi_i^{\text{TE/CE}}$ was introduced in the transport analysis of JT-60 L-mode plasmas in order to compensate for the decrease of $\chi_i^{\eta_i}$ toward the plasma peripheral region, which is unaviodable with JT-60 plasma parameters. As is shown in Sec. III, $\chi_i^{\text{TE/CE}}$ increases from the plasma central region toward the peripheral region in JT-60 L-mode data. We set $C^{\text{TE}} = C^{\text{CE}} = 0.2$; these values are also adjusted to fit T_i profile of JT-60 L-mode plasmas. The χ_i^{TE} term includes contributions from the collisionless trapped electron mode ($\nu_{ei} < \omega_{*e}\varepsilon$) and the dissipative trapped electron mode ($\nu_{ei} > \omega_{*e}\varepsilon$), where $\varepsilon = r/R$;

$$\chi_i^{\text{TE}} = \frac{5}{2}\frac{\omega_{*e}}{k_\theta^2}\varepsilon^{1/2}\min\left(1, \frac{\omega_{*e}\varepsilon}{\nu_{ei}}\right) \qquad (9)$$

where ν_{ei} is the electron-ion collision frequency. Again we assume $k_\theta\rho_S = 0.3$.

The χ_i^{CE} term includes contributions from the collisionless circulating electron mode ($\nu_{ei} < \omega_{be}$) and the collisional circulating electron mode ($\nu_{ei} > \omega_{be}$);

$$\chi_i^{\text{CE}} = \frac{5}{2}\frac{\omega_{*e}^2}{k_\theta^2\omega_{be}}\max\left(1, \frac{\nu_{ei}}{\omega_{be}}\right) \qquad (10)$$

where the bounce frequency of trapped electron is $\omega_{be} = (T_e/m_e)^{1/2}/qR$. The χ_i^{CE} term is very small except for the strongly collisional region near the plasma surface.

METHODS OF CALCULATION

In this section, we describe calculation methods used to simulate JT-60 data by SCOOP code [35]. Diagnostic data such as n_e and T_e which are measured on different spatial grids are interpolated and put on the grid in the minor radius direction taking into account of Shafranov shift of magnetic surfaces due to the finite beta of plasma. We will use a volume averaged radius r in the following; that is $r = \sqrt{V/2\pi^2 R}$ where R and V are the major radius and the plasma volume within a magnetic surface, respectively. In the calculation, these data are rearranged to a 32 point evenly spaced grid in SCOOP using a fitting function, $f(r) = (f(0) - f(a))\{1 - r^2 + \alpha r^2 (1 - r) + \beta r^2 (1 - r^2)\} + f(a)$, where the coefficient α and β are determined by the least square method. The function $f(r)$ gives not only smooth n_e and T_e profile, but also smooth η_{ic} profile.

In the following sections, we compare the profile of calculated χ_i with χ_i^{exp}, which is defined as follows;

$$\chi_i^{exp}(r) = \frac{\frac{3}{2}T_i\Gamma_i - \frac{1}{r}\int_0^r \left(P_{NBI}^i - P_{eq} - P_{CX}\right)r\,dr}{n_i \left\langle |\nabla r|^2 \right\rangle \frac{\partial T_i}{\partial r}} \quad (11)$$

where Γ_i, P_{CX} and P_{eq} represent the ion particle flux, the charge exchange loss and the equi-partition energy transfer between electrons and ions, respectively. Bracket is the average over a magnetic surface. In the cylindrical configuration, $\langle |\nabla r|^2 \rangle = 1$. Γ_i is defined as follows.

$$\Gamma_i(r) = \frac{1}{r}\int_0^r (S_{NBI} + S_n)\,r\,dr \quad (12)$$

where S_n and S_{NBI} are the profile of particle source due to the recycling at the first wall and the fast ion thermalization profile of neutral beam, respectively.

The total absorbed power, P_{abs}, is made up of ohmic heating power, P_{OH}, and the neutral beam heating power, P_{NBI}. P_{NBI} is the injected beam power subtracted by shine through, orbit loss and beam charge exchange loss. P_{NBI} is divided into the absorbed power by electrons and ions; we identify them P_{NBI}^e and P_{NBI}^i, respectively.

These calculations are carried out by Orbit Following Monte Carlo code (OFMC code) [35] in SCOOP.

In JT-60 plasmas, the particle confinement time, τ_p, is evaluated from the empirical scaling of ; τ_p (sec)$=0.05/\bar{n}_e$ $(10^{20}m^{-3})/\sqrt{P_{abs}(MW)}$ [37, 38]. The τ_p of typical shots of JT-60 are listed on Table 2. The effective charge number, Z_{eff}, which is inferred from the visible Bremsstrahlung intensity is assumed to be spatially constant. We calculate plasma current profile assuming neoclassical resistivity.

The ion temperature profile calculated by the method described above is compared experimental data measured by CXRS at up to eight points [33].

2.2.3 RESULTS OF CALCULATION IN L-MODE PLASMAS

The parameter ranges of JT-60 plasmas which we simulated in this paper are shown in Table 1. We select shot E10737 as the typical lower X-point divertor shot in JT-60. Plasma parameter profiles such as n_e, T_e, P_{NBI}^e and P_{NBI}^i are shown in Fig. 1. Other plasma parameters are shown in Table 2.

Figure 2 shows the simulation results of this shot with three different $\chi_i^{\eta_i}$ model. The η_i profile in Fig. 2 is estimated by the predicted T_i profile and the measured n_e profile. The η_{ic} profile is estimated by Eq. (7) using the measured n_e profile. Small density gradient around $r = 0.4$ m causes a peak in η_i there. While the calculated η_i is larger than η_{ic}, $\eta_i - \eta_{ic}$ is not so large, at most 1.5 in D&W and R model. It is smaller in L&D model because of strong η_i dependence of χ_i. In Fig. 2, we can see that $\chi_i^{\eta_i}$ is the dominant conduction loss term in $r < 2a/3$ for three models. From the plasma center to $r \sim a/2$, the calculated χ_i increases due to the dependence of $L_{T_i}^{-0.5}$ (D&W model), dq/dr (L&D model) and $L_n^{-0.5}$ (R model). It gradually decreases from $r \sim a/2$ toward the plasma edge region due to the $T^{1.5}$ temperature dependence. In L&D model, small increment in η_i around r=0.4 m results in large $\chi_i^{\eta_i}$.

The $\chi_i^{TE/CE}$ term also has positive temperature dependence, which makes χ_i decrease from the plasma central region toward the plasma edge region. On the other hand, it also has L_n^{-1} and L_n^{-2} dependence, which makes χ_i increase toward the plasma edge. With JT-60 L-mode plasma parameter, the latter becomes the dominant dependency. Thus $\chi_i^{TE/CE}$ increases with minor radius and compensates for the decrease in $\chi_i^{\eta_i}$. The dominant contribution to $\chi_i^{TE/CE}$ near the plasma surface comes from the circulating electron mode, χ_i^{CE}. Slightly inside the plasma surface, χ_i^{CE} decreases rapidly and the contribution from the trapped electron mode, χ_i^{TE}, dominates (although χ_i^{TE} becomes negligibly small in the plasma central region). The neoclassical diffusivity, χ_i^{NC}, is at least ten times smaller than $\chi_i^{\eta_i}$ or $\chi_i^{TE/CE}$ throughout most of the plasma; it does not play a significant role in the conduction loss. It becomes comparable to $\chi_i^{\eta_i}$ only very near the magnetic axis where χ_i^{NC} increases as $\varepsilon^{-1.5}$.

Figure 3(a) is the comparison of the calculated χ_i with χ_i^{exp}. In most of JT-60 plasmas, the χ_i^{exp} profile is almost spatially constant or increases slightly from the plasma central region toward the plasma edge region. The calculated χ_i from the D&W model and R model agrees with χ_i^{exp} and the predicted T_i profile agrees with experimental data measured by CXRS (Fig. 3(b)). With L&D model, the calculated χ_i becomes large around r=a/2 because of strong η_i dependence of $\chi_i^{\eta_i}$ and the predicted T_i profile becomes smaller than the measured temperature. However, the discrepancy is not very large.

In the plasma edge region, where $\chi_i^{TE/CE}$ overwhelms $\chi_i^{\eta_i}$, the magnitude of convection and equi-partition energy transfer become comparable to or even larger than that of conduction. Consequently conduction plays a relatively weak role in the energy loss near the plasma surface.

We compare the predicted T_i profiles with and without $\chi_i^{TE/CE}$ for three models (Fig. 4). C^{η_i} is 0.6, 3 and 1.2 for D&W model, L&D model and R model, respectively, without $\chi_i^{TE/CE}$ case in order to suppress the increase of central T_i by eliminating $\chi_i^{TE/CE}$. The predicted T_i near the plasma surface becomes slightly higher than that of experiment. However, the difference of T_i profile with and without $\chi_i^{TE/CE}$ is not so large. The inclusion of $\chi_i^{TE/CE}$ in the calculation improves the agreement with the experimental data in the peripheral region but is not essential.

With fixed heating power of 11 MW, we compare the predicted and measured T_i profiles in a current scan by using three χ_i models as shown in Fig. 5. We also examined a power scan with fixed plasma current of 1 MA divertor plasmas (Fig. 6), 1.5 MA divertor plasmas (Fig. 7) and 2 MA limiter plasmas (Fig. 8). Generally agreement between the predicted and the measured T_i in the wide range of plasma parameters can be seen in these figures.

For these shots, it must be kept in mind that the line averaged electron density is not always in the same range because an increase of the plasma current or an increase of the neutral beam heating power accompanies the increase of the electron density, especially in limiter discharges.

In some of these shots, the calculated χ_i profile differs from χ_i^{exp}. However, since the radially averaged χ_i^{cal} is comparable to χ_i^{exp}, the predicted T_i lies near measurements. In other words, the predicted T_i profiles are fairly similar for the three $\chi_i^{\eta_i}$ models (although the predicted T_i profile by L&D model tends to resemble the n_e profile), and it is difficult to single out a 'best' χ_i model.

Besides the above shots, there are two 1 MA limiter shots available for the transport analysis in JT-60. Figure 9 shows the comparison of predicted and the measured T_i in a power scan of 1 MA limiter plasmas. The predicted T_i profiles are much larger than measured ones for all $\chi_i^{\eta_i}$ models. At the present time, there are only two 1 MA limiter shots and we cannot make clear the reason of disagreement.

2.2.4 RESULTS OF CALCULATION IN HIGH T_i PLASMAS

In this section, we simulate JT-60 high ion temperature ($T_i \geq 10\,\text{keV}$) plasmas. They are obtained with high power neutral beam heating, $P_{abs} \geq 15\,\text{MW}$, and low plasma current, $I_p \leq 0.6\,\text{MA}$ [39]. The analyzed plasma parameter ranges are shown in Table I. We select shot E10300 as the typical high T_i plasma whose plasma parameters are shown in Table II. The profiles of n_e, T_e, P_{NBI}^e and P_{NBI}^i of this shot are shown in Fig. 10. Most of the neutral beam heating power is absorbed by ions.

The characteristics of high T_i plasmas are that profiles of n_e, T_e and T_i are altogether highly central peaked ones with a pedestal. In shot E10300, T_i is about 2 keV at r=a/2 whereas the central T_i is about 10 keV. It is as if there are two plasmas of different character, the central plasma with very high temperature and peaked profile and the edge plasma with low temperature, adjoining each other.

Figure 11 shows the calculation results with three $\chi_i^{\eta_i}$ models. In the plasma central region, the η_i is about 0.5. The characteristics of function $f(\eta_i)$ and $g(\eta_i)$ (Equations (5) and (6)) imply that the η_i mode is practically stable. The calculated η_i and χ_i reach their maximum values at $r = 0.5$ m where the density gradient is small. This is especially noticeable with the L&D model. In contrast to the L-mode plasmas, $\chi_i^{TE/CE}$ becomes large in the plasma central region because the electron temperature is high; the dominant term there is the collisionless trapped electron mode. Because $\chi_i^{TE/CE}$ depends on T_e not T_i, the large $\chi_i^{TE/CE}$ suppresses both the central T_i and its gradient thereby stabilizing the η_i mode in the central region.

Both the profile shape and the magnitude of the calculated χ_i differ considerably from those of χ_i^{exp} (Fig. 12(a)). The calculated χ_i is almost spatially constant at about 4 m^2 sec^{-1}. On the other hand, χ_i^{exp} is very low in the plasma central region and is large at the peripheral region. A similar profile of χ_i^{exp} is seen in other high T_i shots. This discrepancy between the calculated χ_i and χ_i^{exp} results in predicted T_i profiles which are too broad (Fig. 12(b)).

In order to reduce the discrepancy in the plasma central region due to the large $\chi_i^{TE/CE}$, we carried out the simulation without $\chi_i^{TE/CE}$ to investigate whether only $\chi_i^{\eta_i}$ and χ_i^{NC} could reproduce the experimental T_i (Fig. 13). If we compare Fig. 13 with Fig. 12, it is found that η_i mode becomes more unstable to some extent in the central region because $\chi_i^{\eta_i}$ is not suppressed by $\chi_i^{TE/CE}$ any longer. However, the total calculated χ_i is lower and the predicted central T_i increases up to 9 $\sim$ 10 keV, which is comparable to the measured value. It must be kept in mind that T_i is too high for 0.2 m $\leq$ r $\leq$ 0.6 m, as before. In other words, if χ_i were adjusted to the level of χ_i^{exp} in this region, high central T_i would not be realized. The predicted T_i profile obtained without $\chi_i^{TE/CE}$ is still less peaked than measured one. As we have pointed out at the beginning of this section, good thermal confinement in the central region and poor confinement of high T_i plasmas in the outer region should be treated separately.

Apart from the plasma edge region, we would like to clarify whether the present $\chi_i^{\eta_i}$ model can explain the T_i profile of high T_i plasmas, at least, in the central region. In order to reproduce the measured T_i in the edge region, we will use χ_i^{exp} instead of $\chi_i^{TE/CE}$. However, since we intend to investigate the validity of $\chi_i^{\eta_i}$ model in the plasma central region, we must eliminate the influence of χ_i^{exp} in the plasma central region. To satisfy these conditions, we have carried out the simulation by using the following χ_i formula;

$$\chi_i = \chi_i^{\text{model}} + \left(\chi_i^{\text{exp}} - \chi_i^{\text{model}}\right) F(r) \qquad (13)$$

where

$$\chi_i^{\text{model}} = \chi_i^{\eta_i} + \chi_i^{NC} \qquad (14)$$

$$F(r) = \frac{1 + \tanh\{(r - r_b)/\Delta\}}{2} \qquad (15)$$

we set $r_b = 0.45\,\text{m}$ and $\Delta = 0.1\,\text{m}$. In the region of $r > r_b$, $F(r)$ rapidly approaches unity and makes $\chi_i \sim \chi_i^{\text{exp}}$ there. In this region, the predicted T_i is almost the same as measurement. In the region of $r < r_b$, on the other hand, F(r) rapidly approaches zero and and makes $\chi_i \sim \chi_i^{\text{model}}$ there. This method imitates setting up boundary condition at $r = r_b$. It enable me to concentrate the heat transport in the plasma core region.

Figure 14 shows the predicted and measured T_i profiles. The predicted central temperature is smaller than the experimental data. It was found in the further calculations that the reduction of C^{η_i} by a factor 5 is necessary in order to reproduce the central T_i.

After all, the T_i profile calculated by $\chi_i^{\eta_i}$ models based on the η_i mode turbulence does not agree with the pedestal and center peaked T_i profile shapes of high T_i shots. Other transport mechanism, for example inward heat pinch in the plasma central region, must be considered to explain the experimental data.

2.2.5 DISCUSSION

In this paper, we adopt the η_{ic} model proposed by Romanelli [11]. Although η_{ic} is the variable of $O(1)$, it changes a little bit by authors. In this section, we investigate the effect of η_{ic} value on the calculation results.

We compare the four cases of spatially constant η_{ic}; $-1, 1, 1.5$ and 2. The case of $\eta_{ic} = -1$ means that η_i mode is almost always destabilized. The results of Romanelli's η_{ic} model presented in the Secs. III and IV is almost the same with the $\eta_{ic} = 1$ case with JT-60 plasma parameters. We set C^{η_i} as 0.5, 0.5, 0.8 and 3 for $\eta_{ic} = -1, 1, 1.5$ and 2, respectively in order to adjust the central T_i to the measured one. In the following calculation, we neglect $\chi_i^{TE/CE}$.

Figure 15(a) is the comparison of η_i profile in shot E10737 with the D&W model. In the $\eta_{ic} = -1$ case, the η_i becomes maximum at $r \sim 0.4$ m where L_n takes maximal value. Apart from $r \sim 0.4$ m, η_i deceases and is almost one at the plasma center and the edge region. As η_{ic} increases, η_i of plasma center and edge region increases whereas it deceases at $r \sim 0.4$ m; thus the the η_i profile becomes flatter. Figure 15(b) shows the comparison of $f(\eta_i)$ profiles. As η_{ic} increases, $f(\eta_i)$ becomes smaller because the η_i mode is easier to be stabilized with high η_{ic}.

Figure 15(c) shows the calculated χ_i profiles. In spite of the different value and different shape of $f(\eta_i)$ for each η_{ic}, the obtained χ_i profile becomes similar because of the different C^{η_i}. As a result, the predicted T_i profiles (Fig. 15(d)) are almost the same and agree well with experimental data, although the T_i profile of $\eta_{ic} = 2$ case somewhat resembles the n_e profile (see Fig. 1(a)).

The reason why η_i at $r \sim 0.4$ m decreases as the increase of η_{ic} is as follows: As the η_{ic} increases, the η_i mode stabilized first at the plasma center and the edge region where originally η_i is small. This results in the decrease of χ_i in this area. Since the central T_i is unchanged due to the adjustment of C^{η_i}, χ_i around $r \sim 0.4$ m should become large instead of plasma center and the edge region; thus the gradient T_i around $r \sim 0.4$ m is compelled to decrease. Then η_i at $r \sim 0.4$ m decreases. In short, the increase of both η_{ic} and C^{η_i} makes flatter η_i profile.

We also tried η_{ic} scan for high T_i shot (E10300) with the D&W model. C^{η_i} for each η_{ic} is set as same as L-mode case. As η_{ic} increases, the η_i mode is stabilized in the first place in the plasma central region where η_i is essentially small comparing with plasma edge region (Fig. 16(a)). It gives rise to the reduction of χ_i in the plasma central region as the increase of η_{ic}. Different from the L-mode case, the predicted T_i profile drastically changes as the increase of η_{ic}. We can see better agreement between the predicted and the measured T_i with the higher η_{ic}. Originally n_e profile and T_i profile of high T_i plasma is similar; central peaked profile with pedestal (see Fig. 10(a)). Since higher η_{ic} causes flat η_i profile, It is not supprising that higher η_{ic} makes better agreement. The same tendency is also obtained for the L&D model.

We will not discuss the validity of η_{ic} model in this paper. Only we should keep in mind that η_{ic} value may affect the simulation results when transport analysis by η_i mode turbulence is carried out.

2.2.6 CONCLUSION

The ion temperature profiles of JT-60 neutral beam heated plasmas both in L-mode phase and high T_i phase, have been simulated in a wide range of plasma parameters. T_i profiles predicted with three different $\chi_i^{\eta_i}$ models based on the η_i mode turbulence and drift wave turbulence have been compared with experimental data measured by CXRS.

With fixed coefficients predicted T_i profiles in JT-60 L-mode plasmas by three χ_i models show good agreement with experimental data in the wide range of plasma parameters: $I_p = 1.0 \sim 1.8\,\text{MA}$, $P_{abs} = 1.3 \sim 16.7$ MW, $\bar{n}_e = 1.2 \sim 5.0 \times 10^{19}\,\text{m}^{-3}$ for the lower X-point divertor plasmas and $I_p = 2.0 \sim 2.7$ MA, $P_{abs} = 3.0 \sim 17.4$ MW, $\bar{n}_e = 1.5 \sim 6.5 \times 10^{19}$ m^{-3} for the limiter plasmas. Only two 1 MA limiter shots show disagreement between the predicted and the measured T_i profiles.

In these calculations, the dominant conduction loss in the bulk plasma is caused by the η_i mode turbulence. The decrease of $\chi_i^{\eta_i}$ toward the plasma edge is compensated by the rising $\chi_i^{TE/CE}$ and improves the agreement with the measured T_i profile in the edge region. However, the inclusion of $\chi_i^{TE/CE}$ is not essential because of the relatively weak role of conduction in the energy balance of plasma periphery.

In the high T_i plasmas, the predicted T_i profiles are below the measured profiles which are highly centrally peaked with a pedestal. Even if we eliminate $\chi_i^{TE/CE}$, which causes the suppression of the core ion temperature, we cannot obtain the observed high central ion temperature of 10 keV in the simulation. Other transport mechanism, for example inward heat pinch, must be considered in the plasma central region in order to explain the peaked T_i profiles seen in the experimental data.

2.2.7 ACKNOWLEDGMENTS

The continuous encouragement of Dr. S. Tamura is gratefully acknowledged. The authors acknowledge Drs. M. Yagi, M. Kikuchi and the JT-60 team.

Table 1. Analyzed Plasma Parameter Region

Type	L-mode	L-mode	High T_i
Divertor/Limiter	D	L	D
R (m)	2.90 ~ 2.95	3.00 ~ 3.03	2.92
a (m)	0.71 ~ 0.74	0.87 ~ 0.89	0.69 ~ 0.72
I_p (MA)	1.0 ~ 1.8	1.0 ~ 2.7	0.37 ~ 0.57
B_t (T)	4.50 ~ 4.67	4.45 ~ 4.52	4.58 ~ 4.64
P_{abs} (MW)	1.3 ~ 16.7	2.4 ~ 17.4	15.0 ~ 17.9
$\bar{n}_e$ ($\times 10^{19} m^{-3}$)	1.2 ~ 5.0	1.2 ~ 6.5	2.6 ~ 3.9

Table 2. Plasma Parameter of Selected Shot

Type	L-mode	High T_i
Shot Number	10737	10300
Divertor/Limiter	D	D
R (m)	2.92	2.92
a (m)	0.72	0.71
I_p (MA)	1.5	0.57
B_t (T)	4.63	4.62
P_{abs} (MW)	11.1	15.0
$\bar{n}_e$ ($\times 10^{19} m^{-3}$)	2.9	3.5
τ_p (msec)	52	37
Z_{eff}	3.5	2.8

Bibliography

[1] R.R. Dominguez, R. E. Waltz, Nucl.Fusion **27**, (1987) 65.

[2] M.H. Redi, W.M. Tang, P.C. Efthimion, D.R. Mikkelsen, G.L. Schmidt, Nucl. Fusion **27**, (1987) 2001.

[3] H. Shirai, T. Hirayama, K. Shimizu, T. Takizuka, M. Azumi, Nucl. Fusion **29**, (1989) 805.

[4] R.R. Dominguez, R.E. Waltz, Nucl. Fusion **29**, (1989) 885.

[5] J. Shefield, Nucl. Fusion **29**, (1989) 1347.

[6] C.E. Singer, E.S. Ghanem, G. Bateman, D.P. Stotler, Nucl. Fusion **30**, (1990) 1595.

[7] M.H. Redi, G. Bateman, Nucl. Fusion **31**, (1991) 547.

[8] B. Coppi, M.N. Rosenbluth, Phys. Fluids **10**, (1967) 582.

[9] P. Terry, W. Anderson, W. Horton, Nucl. Fusion **22**, (1982) 487.

[10] G.S. Lee, P.H. Diamond, Phys. Fluids **29**, (1986) 3291.

[11] F. Romanelli, Phys. Fluids B **1**, (1989) 1018.

[12] F. Romanelli, Plasma Phys. and Controlled Fusion **31**, (1989) 1535.

[13] T.S. Hahm, W.M. Tang, Phys. Fluids B **1**, (1989) 1185.

[14] F. Romanelli, S. Briguglio, Phys. Fluids **32**, (1990) 754.

[15] H. Biglari, P.H. Diamond, M.N. Rosenbluth, Phys. Fluids B **1**, (1989) 109.

[16] B. Hong, W. Horton, Phys. Fluids B **2**, (1990) 978.

[17] J. Weiland, A. Jarmén, H. Nordman, Nucl. Fusion **29**, (1989) 1810.

[18] H. Nordman, J. Weiland, A. Jarmén Nucl. Fusion **30**, (1990) 983.

[19] A. Paccagnella, F. Romanelli, S. Briguglio, Nucl. Fusion **30**, (1990) 545.

[20] S.C. Guo, L. Chen, S.T. Tsai, P.N. Guzdar, Plasma Phys. and Controlled Fusion **31**, (1989) 423.

[21] S. Hamaguchi, W. Horton, Phys. Fluids B **2**, (1990) 1833.

[22] S.M. Wolfe, M. Greenwald, R. Gandy, R. Granetz, C. Gomez, D. Gwinn, B. Lipschultz, S. McCool, E. Marfe, J. Becker, R.R. Parker, J. Rice, Nucl. Fusion **26**, (1986) 326.

[23] F.X. Söldtner, E.R. Müller, F. Wagner, H.S. Bosch, A. Eberhagen, H.U. Fahrbach, G. Fussmann, O. Gehre, *et al.*, Phys. Rev. Lett. **61**, (1988) 1105.

[24] W.L. Rowan, R.V. Bravanec, J.C. Wiley, R.D. Bengtson, R.D. Durst, *et al.*, Nucl. Fusion **30**, (1990) 903.

[25] B. Balet, D.A. Boyd, D.J. Campbell, C.D. Challis, J.P. Christiansen, Nucl. Fusion **30**, (1990) 2029.

[26] M.C. Zarnstorff, *et al.*, Advances in Transport Understanding Using Perturbative Techniques in TFTR, in Plasma Physics and Controlled Nuclear Fusion Research 1990 (Proc. 13th Int. Conf. Washington D.C., 1990), Vol. 1, IAEA, Vienna (1991) 109.

[27] W. Horton, D. Lindberg, J.Y. Kim, J.Q. Dong, G.W. Hammett, S.D. Scott, M.C. Zarnstorff, S. Hamaguchi, Phys. Fluids B **4**, (1992) 953.

[28] R.C. Isler, L.E. Murray, S. Kasai, J.L. Dunlap, S.C. Bates, P. H. Edmonds, *et al.*, Phys. Rev. A **24**, (1981) 2701.

[29] J. Fonck, R. J. Goldston, R. Kaita, D. Post, Appl. Phys. Lett. **42**, (1983) 239.

[30] R.J. Jaehing, R.J. Fonck, K. Ida, E.T. Powell, Rev. Sci. Instrum. **56**, (1985) 865.

[31] R.P. Seraydarian, K.H. Burrell, N.H. Brooks, R.J. Groebner, C. Kahn, Rev. Sci. Instrum. **57**, (1986) 155.

[32] H. Weisen, M.G. von Hellermann, A. Boileau, L.D. Horton, W. Mandl, H.P. Summers, Nucl. Fusion **29**, (1989) 2187.

[33] Y. Koide, S. IShida, A. Sakasai, H. Shirai, T. Hirayama, H. Kubo, T. Sugie, A. Funahashi, Nucl. Fusion **33**, (1993) 251.

[34] R.J. Fonck, R. Howell, K. Jaehing, L. Roquemore, G. Schilling, S. D. Scott, M. C. Zarnstorff, *et al.*, Phys. Rev. Lett. **63**, (1989) 520.

[35] T. Hirayama, K. Shimizu, K. Tani, H. Shirai, M. Kikuchi, Experimental Transport Analysis Code System in JT-60, Rep. JAERI-M 88-043, Japan Atomic Energy Research Institute (1988).

[36] C.S. Chang, F.L. Hinton, Phys. Fluids **29**, (1986) 3314.

[37] S. Tsuji and JT-60 TEAM, Energy and Particle Confinements of Combined Heating Discharges in JT-60, in Controlled Fusion and Plasma Physics (Proc. 14th Eur. Conf.on Controlled Fusion and Plasma Phys., Madrid , 1986) Part I, Eur. Phys. Soc. (1987) 57.

[38] K. Yamada, S. Tsuji, K. Shimizu, T. Nishitani, K. Nagashima and JT-60 Team Nucl. Fusion **27**, (1987) 1203.

[39] S. Ishida, M. Kikuchi, T. Hirayama, *et al.*, High Poloidal Beta Experiments with a Hot Ion Enhanced Confinement Regime in the JT-60 Tokamak, in Plasma Physics and Controlled Nuclear Fusion Research 1990 (Proc. 13th Int. Conf. Washington D.C., 1990), Vol. 1, IAEA, Vienna (1991) 195.

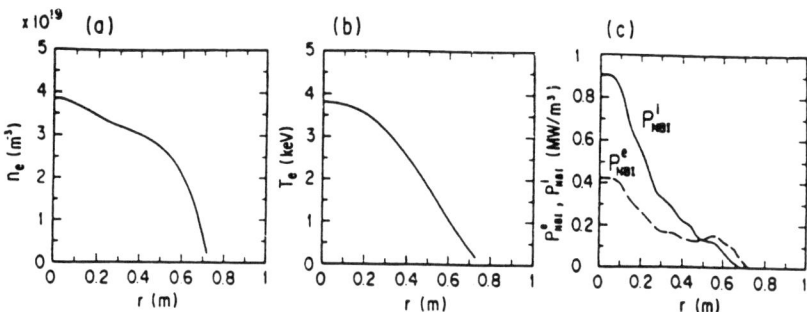

1. The profiles of (a) n_e, (b) T_e, (c) calculated P_{NBI}^e and P_{NBI}^i in the 1.5 MA lower X-point divertor shot (E10737).

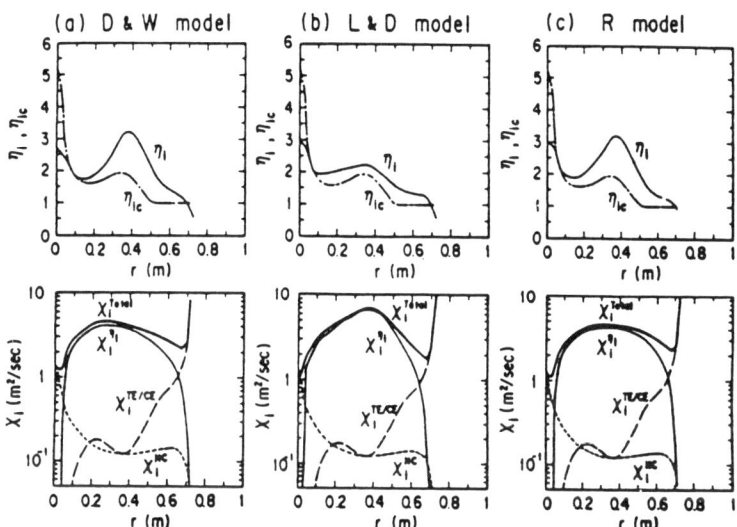

2. The profiles of η_i and η_{ic}, calculated χ_i and its components in shot E10737 with three $\chi_i^{\eta_i}$ models; (a) D&W model, (b) L&D model and (c) R model.

80 Ion Temperature Profile Simulation

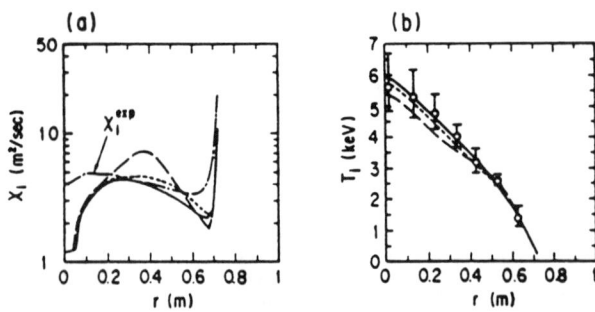

3. Comparison of (a) calculated χ_i profiles with χ_i^{exp} and (b) predicted T_i profiles with experimental data in shot E10737 with D&W model (solid line), L&D model (broken line) and R model (dotted line).

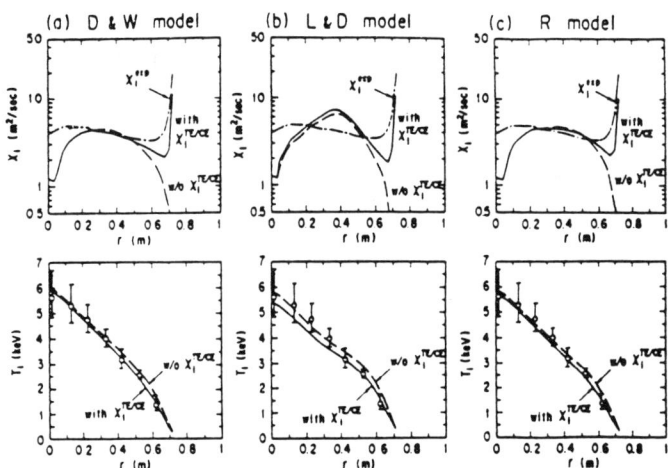

4. Comparison of calculated χ_i and χ_i^{exp}, predicted and measured T_i with and without $\chi_i^{TE/CE}$ in shot E10737 with (a) D&W model, (b) L&D model and (c) R model.

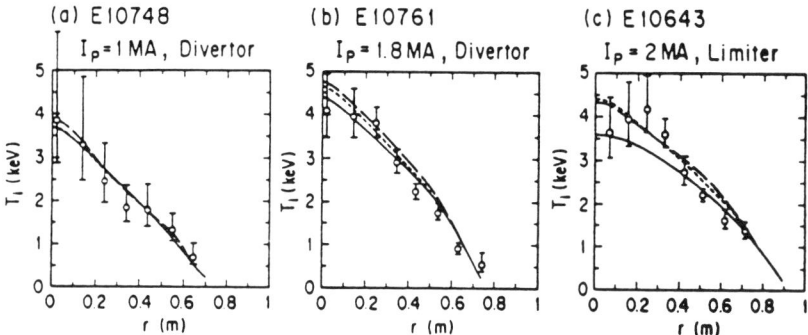

5. Comparison of predicted and measured T_i profiles in a plasma current scan with D&W model (solid line), L&D model (broken line) and R model (dotted line). The heating power is about 11 MW.

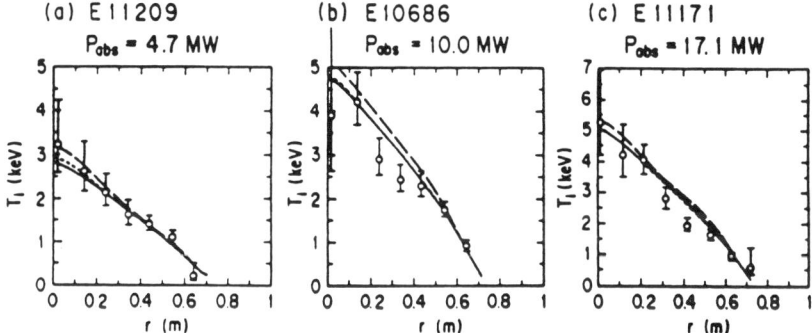

6. Comparison of predicted and measured T_i profiles in 1 MA lower X-point divertor shots in a heating power scan with D&W model (solid line), L&D model (broken line) and R model (dotted line).

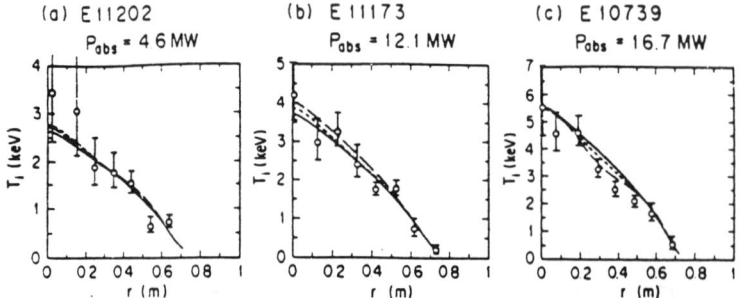

7. Comparison of predicted and measured T_i profiles in 1.5 MA lower X-point divertor shots in a heating power scan with D&W model (solid line), L&D model (broken line) and R model (dotted line).

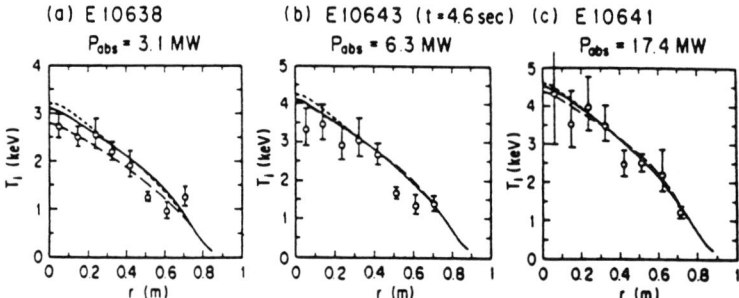

8. Comparison of predicted and measured T_i profiles in 2 MA limiter shots in a heating power scan with D&W model (solid line), L&D model (broken line) and R model (dotted line).

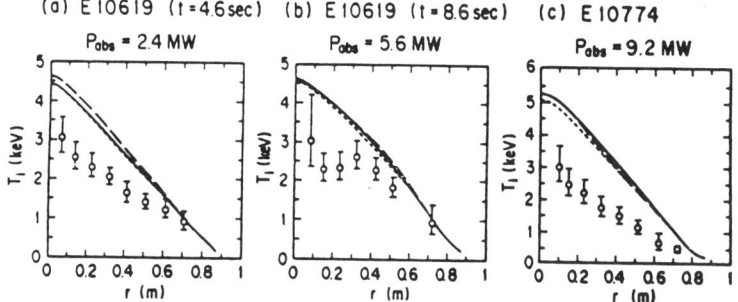

9. Comparison of predicted and measured T_i profiles in 1 MA limiter shots in a heating power scan with D&W model (solid line), L&D model (broken line) and R model (dotted line).

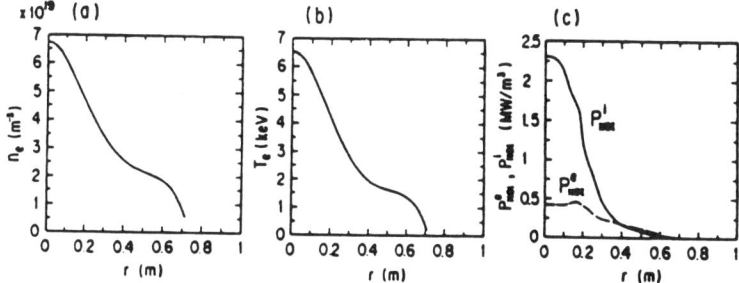

10. The profiles of (a) n_e, (b) T_e, (c) calculated $P_{\rm NBI}^e$ and $P_{\rm NBI}^i$ in the 0.57 MA high T_i shot (E10300).

84 Ion Temperature Profile Simulation

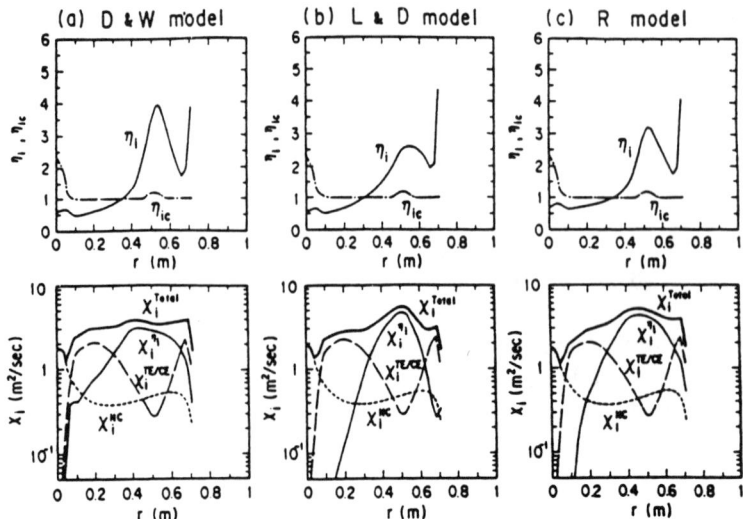

11. The profiles of η_i and η_{ic}, calculated χ_i and its components in shot E10300 with (a) D&W model, (b) L&D model and (c) R model.

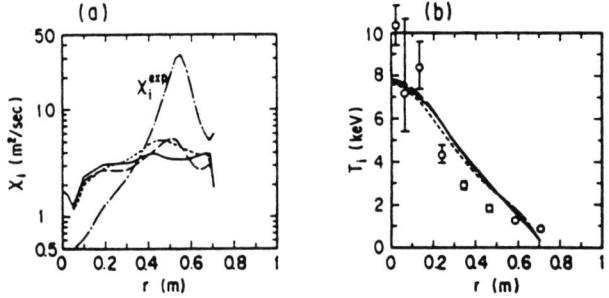

12. Comparison of (a) calculated χ_i profiles with χ_i^{exp} and (b) predicted and measured T_i profiles in shot E10300 with D&W model (solid line), L&D model (broken line) and R model (dotted line).

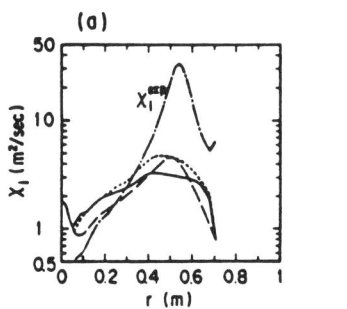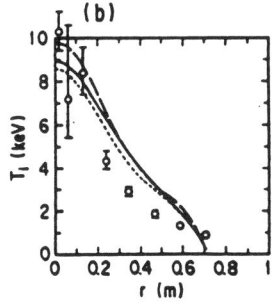

13. Comparison of (a) calculated χ_i profiles with χ_i^{exp} and (b) predicted and measured T_i profiles in shot E10300 without $\chi_i^{TE/CE}$. Results of D&W model (solid line), L&D model (broken line) and R model (dotted line) are presented.

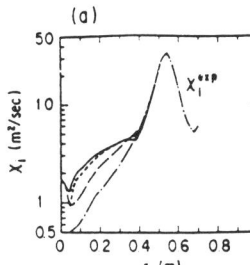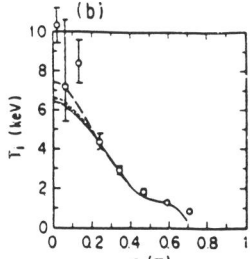

14. Comparison of (a) calculated χ_i profiles with χ_i^{exp} and (b) predicted and measured T_i profiles in shot E10300. χ_i^{exp} is adopted in stead of $\chi_i^{TE/CE}$. Results of D&W model (solid line), L&D model (broken line) and R model (dotted line) are presented.

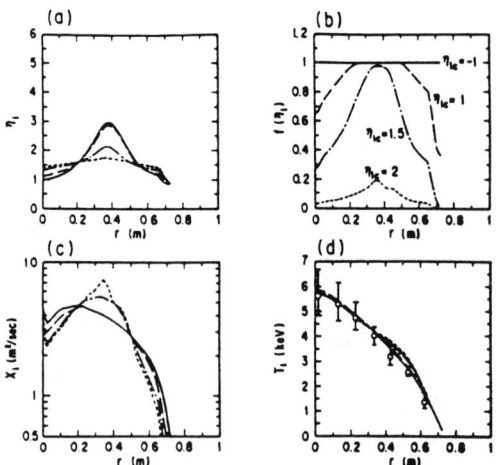

15. Comparison of (a) η_i, (b) $f(\eta_i)$, (c) χ_i and (d) T_i with different η_{ic} of -1, 1, 1.5 and 2 in shot E10737. D&W model is adopted. $C^{TE/CE}$ is set 0. C^{η_i} are 0.5, 0.5, 0.8 and 3, respectively.

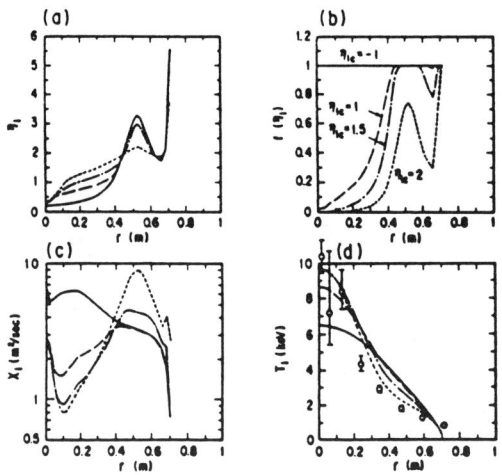

16. Comparison of (a) η_i, (b) $f(\eta_i)$, (c) χ_i and (d) T_i with different η_{ic} of -1, 1, 1.5 and 2 in shot E10300. D&W model is adopted. $C^{TE/CE}$ is set 0. C^{η_i} are 0.5, 0.5, 0.8 and 3, respectively.

2.3 ION TRANSPORT ANALYSIS BASED ON η_i MODE TURBULENCE IN JT-60

M. Yagi and M. Azumi
Japan Atomic Energy Research Institute, Naka, Ibaraki, Japan

Abstract. Validity of the ion temperature gradient mode (η_i mode) driven turbulent transport has been studied by comparing the nonlinear simulation results based on fluid-type moment equations and profile data observed in JT-60 experiments. Toroidal η_i modes give the reasonable value of thermal diffusivity in the half-way region of the plasma, while both toroidal and slab modes give the diffusivity strongly decreasing near the plasma periphery. Both the trapped electron effect and the neoclassical one have been also studied and tested in the edge region, but these effects are found to be not enough to explain experimental results.

2.3.1 INTRODUCTION

The η_i modes are considered to be the strong candidate of the anomalous ion energy transport in a tokamak and the linear stability and nonlinear state of these instabilities have been extensively studied from the theoretical points of view. The validity of these theories is investigated by the comparison of the theoretically evaluated thermal diffusivity itself with experimental results at some specific radial points or by the transport code simulation based on this diffusivity. For example, Shirai *et al.* have tested the mixing length theory of η_i mode turbulence on JT-60 plasmas, by performing transport simulations for discharges with different plasma parameters, and they have concluded that the η_i mode turbulence can show the fairly good agreement of calculated ion temperature profiles with experimental ones in L-mode discharges, except central and peripheral regions [1]. In these analyses, the absolute value of ion thermal diffusivity is adjusted such that the simulation result for some specific discharge is similar to experimental result. However, it is determined by the saturation amplitude of the mode and it may not be the constant value independent of plasma parameters. To test the validity of η_i mode turbulent transport in more quantitative sense, in the first part of this work, we perform the nonlinear simulation of η_i mode turbulence by using Hong and Horton model and evaluate the absolute value of the thermal conductivity χ_i to compare with experimentally analyzed one. In the peripheral region of the plasma, the pure fluid like theory gives the radially decreasing diffusivity on contrast with experimental observations. Then, in the second part, we study the trapped electron effect and neoclassical effect on η_i modes as the possible candidates and check whether these effects can increase the ion thermal diffusivity in this region.

2.3.2 ION TRANSPORT ANALYSIS BASED ON THE NONLINEAR SIMULATION OF η_i MODE TURBULENCE

We perform the nonlinear simulations of η_i modes both in the 2 dimension and the 3 dimension, based on the following model equations [2];

$$(1-\nabla_\perp^2)\frac{\partial \phi}{\partial t} = -\left(1 - G + \frac{1+\eta_i}{\tau}\nabla_\perp^2\right)\frac{\partial \phi}{\partial y} + G\frac{\partial p}{\partial y}$$

$$-\nabla_\parallel v + [\phi, \nabla_\perp^2 \phi] - \mu_{\perp,1}\nabla_\perp^4 \phi \, , \tag{1}$$

$$\frac{\partial v}{\partial t} = -\nabla_\parallel(\phi + p) - [\phi, v] + \mu_{\perp,2}\nabla_\perp^2 v + \mu_{\parallel,2}\nabla_\parallel^2 v \, , \tag{2}$$

$$\frac{\partial p}{\partial t} = -\frac{1+\eta_i}{\tau}\frac{\partial \phi}{\partial y} + \frac{\Gamma}{\tau}G\left(1 - \frac{1}{\tau}\right)\frac{\partial \phi}{\partial y}$$

$$-\frac{\Gamma}{\tau}\frac{1+\eta_i}{\tau}\nabla_\perp^2\frac{\partial \phi}{\partial y} + \frac{2\Gamma}{\tau}G\frac{\partial p}{\partial y} - \frac{\Gamma}{\tau}\nabla_\parallel v$$

$$-[\phi,p] + \chi_\perp \nabla_\perp^2 p + \chi_\parallel \nabla_\parallel^2 p \, , \tag{3}$$

where $\Gamma = 5/3$, $G = 2L_n/R$, $\eta_i = L_n/L_{T_i}$, $\tau = T_e/T_i$, $\nabla_\parallel = \partial/\partial z + Sx\partial/\partial y$, $S = L_n/L_s$, $\nabla_\perp^2 = \partial^2/\partial x^2 + \partial^2/\partial y^2$ and $[\ ,\ ]$ is the Poisson bracket. μ and χ are the viscosity and thermal conductivity, respectively. The coordinates (x, y, z) are taken such that x is in the direction of density and temperature gradient, z in that of the magnetic field at $x = 0$, and y in that of perpendicular to x and z, respectively, and correspond to the cylindrical coordinates $(r - r_0, \theta, z - q(r_0)\theta)$. The energy conservation of these equations is written as follows;

$$\frac{dE_T}{dt} = \left(\frac{1+\eta_i}{\Gamma} + \frac{G}{\tau}\right)\left\langle \phi \frac{\partial p}{\partial y}\right\rangle - \frac{1+\eta_i}{\tau}\left\langle p\nabla_\perp^2 \frac{\partial \phi}{\partial y}\right\rangle$$

$$-\mu_{\perp,1}\left\langle (\nabla_\perp^2 \phi)^2\right\rangle - \mu_{\perp,2}\left\langle (\nabla_\perp v)^2\right\rangle - \mu_{\parallel,2}\left\langle (\nabla_\parallel v)^2\right\rangle$$

$$-\chi_\perp \left\langle (\nabla_\perp p)^2\right\rangle - \chi_\parallel \left\langle (\nabla_\parallel p)^2\right\rangle , \qquad (4)$$

and

$$E_T = \frac{1}{2}\left\langle \phi^2 + (\nabla\phi)^2\right\rangle + \frac{1}{2}\left\langle v^2\right\rangle + \frac{1}{2}\frac{\tau}{\Gamma}\left\langle p^2\right\rangle . \qquad (5)$$

According to Hamaguchi and Horton [3], we evaluate the ion thermal diffusivity by using the following expression,

$$\chi_i = -\frac{c_s \rho_s^2}{L_n}\frac{\tau}{1+\eta_i}\overline{\left\langle p\frac{\partial \phi}{\partial y}\right\rangle} , \qquad (6)$$

$$\overline{g(t)} = \lim_{T\to\infty} \frac{1}{T}\int_0^T g(t)dt , \qquad (7)$$

$$\langle f(x,y)\rangle = \frac{1}{\Delta_x L_y}\int_0^{L_x} dx \int_0^{L_y} dy\, f(x,y) . \qquad (8)$$

For the 3 D case $\Delta_x = L_x$ and for the 2 D case,

$$\Delta_x = \int_0^{L_x} \theta(f(x) - 0.1 f_{\max})dx , \qquad (9)$$

where θ is the Heaviside step function. The error bar is obtained from

$$\Delta \chi_i = \left\{\overline{(\chi_i(t) - \chi_i)^2}\right\}^{1/2} . \qquad (10)$$

When we set $G = 0$ and neglect the term $\Gamma(1+\eta_i)/\tau^2 \nabla_\perp^2 \partial\phi/\partial y$ in the ion pressure evolution, these equations reduce to Hamaguchi and Horton model which are used in the analysis of the slab η_i modes. Then,

90   Ion Transport Analysis

before we evaluate the ion thermal diffusivity of η_i modes by using parameters of JT-60 L-mode discharges, we investigate some characteristics of the slab η_i modes based on the Hamaguchi and Horton model equations.

Quasilinear effect on slab η_i modes

In order to study the quasilinear effect on the mode saturation, as the first step, we have done the 2 mode simulations with $m = 0$ and $m = 1$, where m is the poloidal mode number. In this case, nonlinear terms are schematically written as

$$[\phi, \nabla_\perp^2 \phi] \rightarrow \frac{d\phi_0}{dx} \frac{\partial}{\partial y} \nabla_\perp^2 \phi - \frac{d}{dx} \nabla_\perp^2 \phi_0 \frac{\partial \phi}{\partial y},$$

$$[\phi, v] \rightarrow \frac{d\phi_0}{dx} \frac{\partial v}{\partial y} - \frac{dv_0}{dx} \frac{\partial \phi}{\partial y},$$

$$[\phi, p] \rightarrow \frac{\partial \phi_0}{\partial x} \frac{\partial p}{\partial y} - \frac{dp_0}{dx} \frac{\partial \phi}{\partial y}. \tag{11}$$

Figure II.A.1-a shows the time evolutions of total, $m = 0$, and $m = 1$ mode energies in the case of all quasilinear terms retained. The parameters are chosen as 300 equidistant grids in the x direction with $L_x = 80\rho_s$, $k_y\rho_s = 0.8$, $K \equiv (1 + \eta_i)/\tau = 2$, $S = 0.1$, $\Gamma = 2$, $\mu_\perp = \chi_\perp = 0.01$, and $\mu_\parallel = \chi_\parallel = 0.1$. After the first phase, in which the m=1 mode energy $E^{m=1}$ shows the rapid linear growth ($t < 100$) and the slow decrease to the quasisteady state, $E^{m=1}$ shows again the increase at $t = 800$ and then settles down at the same level as the value at $t < 700$, while the total energy E^{total} and the $m = 0$ energy $E^{m=0}$ increase considerably during this second phase.

In Fig. II.A.1-b, the mode structures of $p^{m=0}$ and $\phi^{m=1}$ are plotted at the time marked by $((A), (B), (C),$ and (D)) in Fig. II.A.1-a. In the quasisteady phase ($t = 300 - 800$), although there is a weak source of the pressure gradient, d^2v_0/dx^2 is almost constant around the rational surface. This means the quasisteady state sustains the pressure gradient destabilization and parallel velocity shear stabilization. It is clearly shown that the radial mode number l of ϕ is changed from $l = 1$

at $t = 600(A)$ to $l = 2$ at $t = 3000(D)$. This transition of the mode structure makes the change of $E^{m=0}$ energy. In this phase d^2v_0/dx^2 strongly varies around the rational surface and the stabilization effect of the parallel velocity shear becomes weak and $d\phi_0/dx$ or electric filed is peaking around the rational surface. It seems that electric field tries to compensate the stabilization effect.

Figure II.A.1-c shows the linear growth rate evaluated by taking into account the quasilinear modification of $m = 0$ mode at each time in Fig. II.A.1-a. The symbols of circle, triangle and square represent the cases with all quasi-linear ($m = 0$) terms retained, the cases with $d\phi_0/dx = 0$ and the cases with $dv_0/dx = d\phi_0/dx = 0$, respectively. We can see the parallel velocity shear dv_0/dx generated by the quasilinear effect stabilizes the mode, while the radial electric field $d\phi_0/dx$ and the vorticity shear $d\nabla_\perp^2 \phi_0/dx$ have little effect on the stability.

Figure II.A.1-d shows the radial structures of $-K + dp_0/dx$, dv_0/dx, $d\phi_0/dx$, and $d\nabla_\perp^2 \phi_0/dx$ at $t = 1600$ in Fig. II.A.1-a. The quantity $-K + dp_0/dx$ corresponds to the total pressure gradient and this figure shows that there still remains a driving source of slab η_i mode in the resonance region, in spite of the saturation of the $m = 0$ component of the pressure. On the other hand, the parallel velocity shear is sharply changed in this region and the negative electric field is also formed in the same region.

Next, we have performed the simulation for the same parameter set, neglecting the quasilinear modification of the pressure profile ($dp_0/dx = 0$).

Figure II.A.2-a shows the temporal evolution of the total energy and $m = 0$ and $m = 1$ energies in this case. The saturation level of energy is lower than that in the case of Fig. II.A.1-a. In contrast to the case of Fig. II.A.1-a, the radial mode structure is not changed during the evolution.

Figure II.A.2-b shows the linear growth rate evaluated for parameters at $t = 3000, 3200, 3400, 3600$ by using quasilinear terms as Fig. II.A.1-c case. The symbols of square, circle and triangle correspond to the case where all terms of $dv_0/dx \neq 0$, $d\phi_0/dx \neq 0$ and $d\nabla_\perp^2 \phi_0 \neq 0$ are retained, the case with $d\nabla_\perp^2 \phi_0/dx = 0$, and the case with $d\phi_0/dx = 0$ and $d\nabla_\perp^2 \phi_0/dx = 0$, respectively. From this figure, it can be seen that the radial electric field due to the quasilinear effect

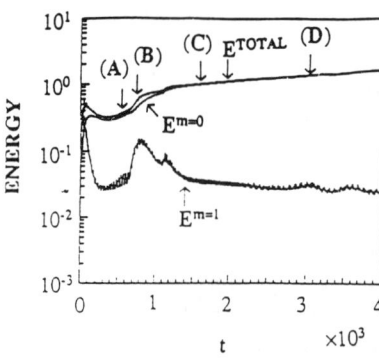
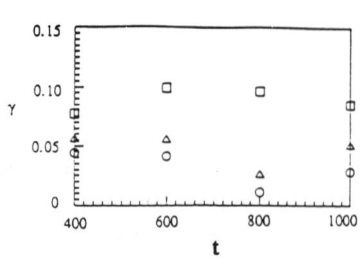

Fig.II.A.1-a The time evolution of the total energy and $m=0$ and $m=1$ energies for the two mode simulation with all quasi-linear terms retained.

Fig.II.A.1-c The linear growth rate at each time for Fig.II.A.1-a.
$\square,\triangle,\circ$ denote the case with $dv_0/dx = 0$, $d\phi_0/dx = 0$, the case with $d\phi_0/dx = 0$, and the case with all quasi-linear terms, respectively.

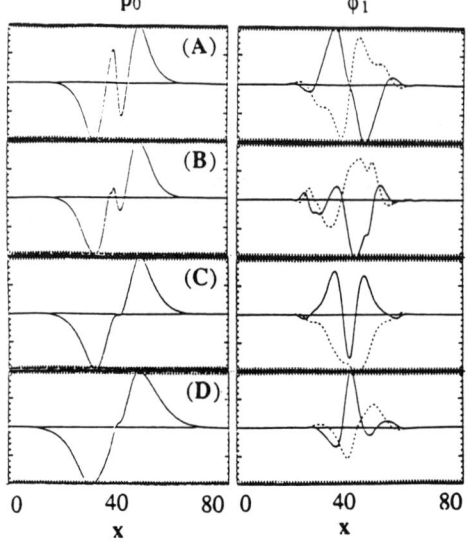
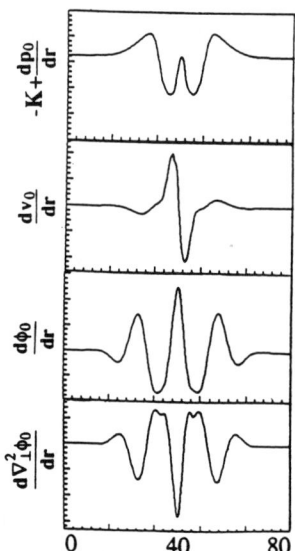

Fig.II.A.1-b The mode structure of $p^{m=0}$ and $\phi^{m=1}$ at the time marked in Fig.II.A.1-a.

Fig.II.A.1-d The radial structures of $-K+dp_0/dx$, dv_0/dx, $d\phi_0/dx$, and $d\nabla_\perp^2\phi_0/dx$ at $t=1600$ in Fig.II.A.1-a.

stabilizes the mode.

As the third case, we have done the simulation with $dp_0/dx = 0$, and $dv_0/dx = 0$. The result is shown in Fig. II.A.3, where the $m = 1$ energy shows the strongly oscillatory behavior and the quasilinear stabilization effect is weak. From these three simulations, we can see that, for the slab η_i mode, the quasilinear radial electric field is correlated with the ion pressure modification and, also, the parallel velocity shear plays important role for the mode stabilization through the ion pressure modification.

Next we compare the results of quasilinear (QL) simulation with nonquasilinear (NQL) simulation, in which all quasilinear terms are dropped.

Figure II.A.4 shows the time evolution of the total energy in three dimensional QL and NQL simulations. The parameters are chosen as $L_x = 80\rho_s$, $L_y = 10\pi\rho_s$, $L_z = 7.5\pi L_n$, 200 grid in the x direction, $k_{y0}\rho_s = 0.2$, $k_{z0}L_n = 0.267$, $0 \leq m \leq 9$, $-6 \leq n \leq 6$, $S = 0.1$, $K = 3$, $\Gamma = 2$, $\mu_\perp = \chi_\perp = 0.1$, and $\mu_\parallel = \chi_\parallel = 1$. For comparison, we plot the result of the two dimensional NQL simulation, in which parameters are chosen as $L_x = 80\rho_s$, $k_{ymax}\rho_s = 2.4$, and poloidal mode number, $9(k_{ymin}\rho_s = 2.4/9)$, and $32(k_{ymin}\rho_s = 2.4/32)$, other parameters are the same as those in the 3 D case. Energy saturation levels of 3-D-QL, 3-D-NQL and 2-D-NQL simulations are almost the same. In the 3-D cases, the stabilization due to the quasilinear effect is not essential for the energy saturation and boundary conditions strongly affects the saturation. This is because the local pressure flattening at some place induces the local pressure steeping at another place due to the constraint on the ion pressure at the calculation boundary.

Figure II.A.5 shows the time evolution of the ion thermal diffusivity χ_i evaluated from the 2-D and 3-D NQL simulations. The parameters are $L_x = 80\rho_s$, $L_y = 53.333\pi\rho_s$, 300 grids in x direction, poloidal mode number 64, $k_{ymin}\rho_s = 0.0375$, and $k_{max}\rho_s = 2.4$, $S = 0.1$, $K = 2.0$, $\Gamma = 2.0$, $\mu_\perp = \chi_\perp = 0.1$, $\mu_\parallel = \chi_\parallel = 1.0$ for the 2-D case, and, for the 3-D case, $L_x = 80\rho_s$, $L_y = 10\pi\rho_s$, $L_z = 7\pi L_n$, 400 grid in x direction, poloidal mode number 9, and toroidal mode number 15, $k_{ymin}\rho_s = 0.2$, $k_{ymax}\rho_s = 1.8$, $k_{zmin}L_n = 0.2666$, $k_{zmax}L_n = 4.0$, other parameters are same as the 2-D case. The thermal diffusivity reaches the level evaluated from the mixing length theory at $t = 100$ in the case of 2-D simulation and at $t = 60$ in the case of 3-D one, and then it settles down to the lower level. At this saturation state, the 2-D NQL simulation gives the χ_i at the same level as the 3-D NQL one.

Figure II.A.6 shows the radial profiles of the thermal flux defined by $Q(x) = (-1/L_y) \int_0^{L_y} p \partial \phi / \partial y \, dy$ for the linear calculation (top), the 2-D NQL simulation (middle), and the 3-D NQL simulation (bottom), respectively. The parameters are same as those in Fig. II.A.5. The thermal flux for the 2-D simulation is localized around the rational surface, therefore the thermal diffusivity depends on the system size if we use L_x for Δ (Eq. (8)). For this reason, we use Eq. (9) for 2-D calculation to get rid of this effect.

The dependences of thermal diffusivity on η_i and the shear S are summarized in Fig. II.A.7 and 8. The parameters are the same as Fig. II.A.5. The symbols of square, triangle and circle denote the mixing length estimation, the 3-D NQL calculation, and the 2-D NQL calculation, respectively, in these figures. For the mixing length estimation, we use $\chi_i = \gamma \delta^2$, where γ is the linear growth rate (in this case, the maximum growth rate is given at $k_y \rho_s = 0.6$), and δ is the radial width of the linear eigenfunction obtained from numerical simulations. For 2-D and 3-D cases, we have plotted the thermal diffusivity averaged in the time interval of $t = 150 - 300$ (Fig. II.A.5). The 2-D NQL calculation gives a good approximation of 3-D NQL calculation for both η_i and S dependence, although the 2-D case gives the slight stronger dependence on η_i and the weaker dependence on S. The mixing length estimation in Fig. II.A.7 is one order higher than others, while it's dependence on η_i is similar to those of nonlinear simulations. The difference of the absolute value of mixing length estimation in Fig. II.A.7 comes from the shear dependence as seen in Fig. II.A.8. For high shear limit, the mixing length estimate gives the thermal diffusivity of order comparable with those for 2-D and 3-D calculations. But in the low shear limit, the difference is large. The main reason of this difference is that, in the weak shear limit, the mode structure is broad so that $\gamma \delta^2$ is larger even though the growth rate is reduced with shear. From analytic estimation, we obtain $\chi_i \propto K^{3/2} S$ for $KS \ll 1$ and $\chi_i \propto K^{3/4} S^{1/4}$ for $KS \gg 1$ but the shear dependence has clearly the opposite tendency compared with the numerical result. The mixing length estimation obtained from the numerical result is only valid in

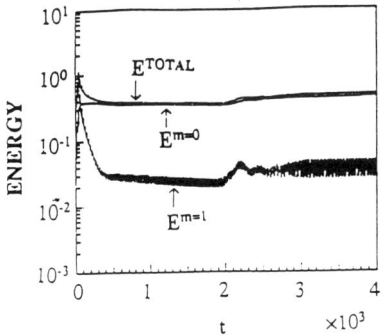

Fig.II.A.2-a The time evolution of the total energy and $m = 0$ and $m = 1$ energies for the two mode simulation with $dp_0/dx = 0$.

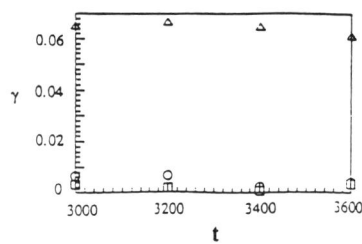

Fig.II.A.2-b The linear growth rate at each time for Fig.II.A.2-a.
$\triangle, \circ, \square$ denote the case with $d\phi_0/dx = 0$, $d\nabla_\perp^2 \phi_0/dx = 0$, the case with $d\nabla_\perp^2 \phi_0/dx = 0$, and the case with $dv_0/dx \neq 0, d\phi_0/dx \neq 0, d\nabla_\perp^2 \phi_0 \neq 0$, respectivelty.

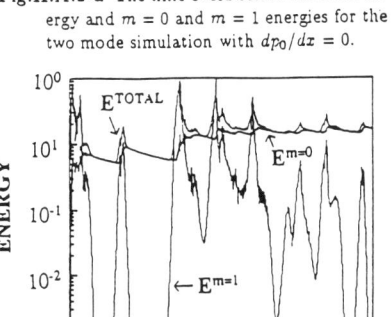

Fig.II.A.3-a The time evolution of the total energy and $m = 0$ and $m = 1$ energies for the two mode simulation with $dp_0/dx = 0$, and $dv_0/dx = 0$.

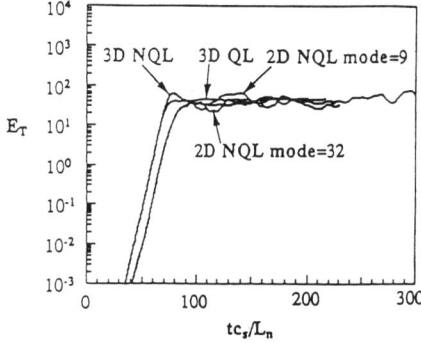

Fig.II.A.4 The time evolution of the total energy in 2D-NQL, 3D-QL and 3D-NQL simulations.

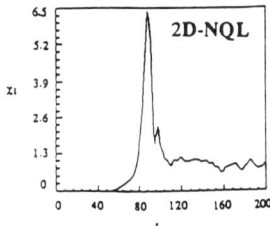 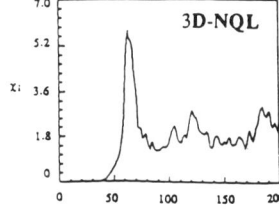

Fig.II.A.5 The time evolution of the ion thermal diffusivity evaluated from 2D and 3D NQL simulation.

the high shear limit.

In the above simulations, the artificial dissipations, $\chi_\perp$, $\chi_\|$, $\mu_\perp$ and $\mu_\|$, are set to be some proper values. In order to check the validity of these values, the $\chi_\perp$ and $\chi_\|$ dependences on χ_i for 2-D NQL simulations are shown in Fig. II.A.9. The parameters are same as Fig. II.A.5, where we assume $\chi_\perp = \mu_\perp$ for the $\chi_\perp$ scan in Fig. II.A.9-a and $\chi_\| = \mu_\|$ for the $\chi_\|$ scan in Fig. II.A.9-b. It is easily seen that, for $\chi_\perp < 0.1$ and $\chi_\| < 1$, the thermal diffusivity is almost independent on $\chi_\perp$ and $\chi_\|$, otherwise it considerably reduces. Therefore, we can confirm that results in Fig. II.A.7 and 8 do not depend on these artificial dissipations describing subgrid scale dissipation and mode coupling.

Finally, we study the thermal diffusivity in the flat density limit of $L_n \to \infty$. For this, we renormalize the basic equations, Eqs. (1)–(3) by using L_{T_i} instead of L_n.

Figure II.A.10 shows the η_i^{-1} dependence of χ_i evaluated from 2-D NQL simulations. The parameters are chosen as $S^* = 0.1$, $\Gamma = 2$, $\chi_\perp^* = \mu_\perp^* = 0.1$, and $\chi_\|^* = \mu_\|^* = 1.0$, where the suffix "*" represents the value normalized by L_{T_i}, and results are plotted by circles with error bar. After reducing almost linearly to reducing η_i^{-1}, the thermal diffusivity normalized by L_{T_i} tends to settle down at the constant value. For comparison, we have also calculated the case of $S^* = 0.1/5$, $\chi_\perp^* = 0.1/5$ and $\chi_\|^* = 1.0/5$ which correspond to $S = 0.1$, $\chi_\perp = 0.1$, and $\chi_\| = 1.0$, and the result is plotted by the square in the same figure. The result normalized by L_n gives $\chi_i/(c_s \rho_s^2/L_n) \sim 6.0$ and it agrees with the result at $\eta_i = 5$ in Fig. II.A.7. In the range of $2 \leq \eta_i \leq 5$, we obtain $\chi_i^{2D} \propto \eta_i^{0.617}$ from the calculation with L_{T_i} normalization. This scaling is somewhat weaker than the scaling of $\chi_i^{2D} \propto \eta_i^{1.07}$ in the same η_i range in Fig. II.A.7. This difference comes from the effectively larger values of S, $\chi_\perp$ and $\chi_\|$ in these simulations; that is, constant values of $S^* = 0.1$, $\chi_\perp^* = 0.1$ and $\chi_\|^* = 1.0$ correspond to $S = 0.2 - 0.5$, $\chi_\perp = 0.2 - 0.5$ and $\chi_\| = 2 - 5$. These values affect the thermal diffusivity, as seen in Fig.II.A.9.

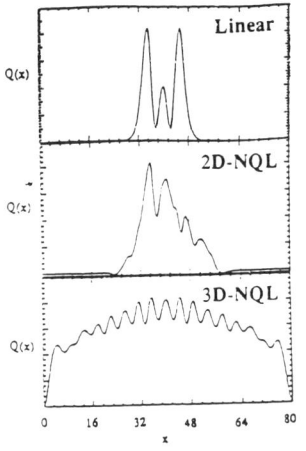

Fig.II.A.6 The radial profiles of the thermal flux for linear, 2D-NQL, 3D-NQL simulation results, respectively.

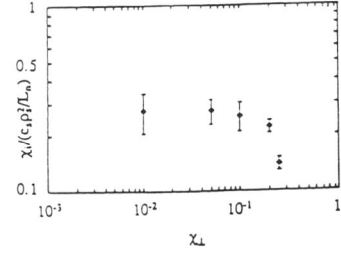

Fig.II.A.9-a The thermal diffusivity versus $\chi_\perp$.

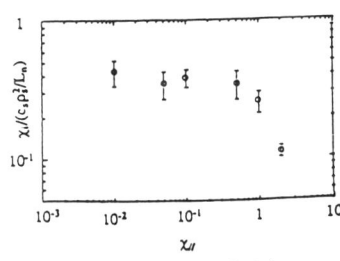

Fig.II.A.9-b The thermal diffusivity versus $\chi_\parallel$.

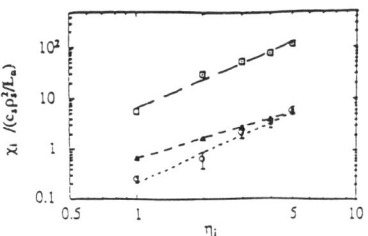

Fig.II.A.7 The thermal diffusivity versus η_i. $\square, \triangle, \circ$ denote linear, 3D-NQL, 2D-NQL simulation results.

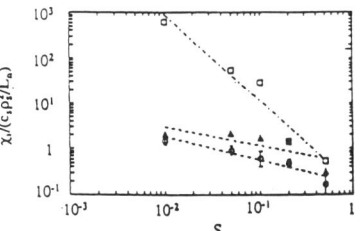

Fig.II.A.8 The thermal diffusivity versus S. $\square, \triangle, \circ$ denote linear, 3D-NQL, 2D-NQL simulation results.

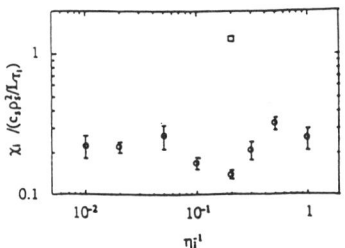

Fig.II.A.10 The thermal diffusivity versus η_i^{-1}. $\circ$ denotes the results from flat density normalization. $\square$ denotes the result from normal density normalization ($\eta_i = 5$).

JT-60 L-mode ion transport analysis

In this subsection, we evaluate the thermal diffusivity from the nonlinear simulation of η_i modes by using the plasma parameters in the JT-60 L-mode discharge. Simulations have been performed in two dimensional geometry and all quasilinear terms in Eqs. (1)–(3) have been omitted.

Figure II.B.1 shows radial profiles of plasma parameters and parameters related to the η_i mode stability for the typical L-mode discharge: shot number $\#E11171$ with $I_p = 1\,\text{MA}$, $B_T = 4.5T$, $\bar{n}_e = 3.46 \times 10^{19}\text{m}^{-3}$ and $P_{\text{abs}} = 10\,\text{MA}$ in the divertor configuration and the comparison of ion thermal diffusivities for this typical discharge. The shaded region shows the experimentally estimated diffusivities with experimental ambiguity. The closed circle and closed triangle correspond to the diffusivity determined from the nonlinear simulation and the value evaluated from the mixing length model by using the result of linear stability analysis, respectively. The upper figure shows results for the toroidal η_i mode, which is the unstable drift wave driven by the unfavorable magnetic curvature and the ion pressure gradient, while the lower figure corresponds to the slab η_i mode, which is the unstable ion acoustic wave driven by the ion pressure gradient. For comparison, the profile of thermal diffusivity of the Domingues-Waltz model employed in the transport analysis is plotted by the solid line [1]. The anomalous diffusivity due to the trapped electron modes in addition to the η_i mode is included in this model to enhance the diffusivity near the plasma periphery and the absolute value of the total diffusivity is adjusted such that the evaluated ion temperature becomes the same as the experimental one.

From these figures, we can see the following points. Firstly, for the slab η_i mode turbulence, the absolute values of the mixing length formula and the nonlinear result are crucially different. This is because the shear parameter of JT-60 plasma is very small ($S < 0.1$ in the whole plasma region, as seen in Fig. II.B.1) and the linear mode width of the slab η_i mode is broad. In this weak shear limit, the mixing length formula overestimates the diffusivity as seen in Fig. II.A.8. Also, the weaker shear makes the saturation level of slab η_i mode lower because the shear plays the key role in the linear stability of the slab mode through the parallel compressibility. Then the resultant diffusivity is fairly low in the low shear like in JT-60 discharges and we can omit the possibility of slab η_i mode for JT-60 L-modes. For the toroidal η_i mode turbulence, the mixing length estimate and nonlinear result are in the same order and both are fairly good in the half way region of the plasma but they don't explain the anomaly in the edge region. We note that the magnetic shear is relatively unimportant for the toroidal mode with $G > S$, and the linear mode is localized near the resonance region even in the low shear limit. Both in toroidal and slab modes, the diffusivities radially decrease in the peripheral plasma region due to the decreasing value of η_i.

The simulation results of toroidal η_i mode for discharges with different parameters are summarized in Fig. II.B.2, where the dashed line denotes the profile of thermal diffusivity deduced from experimental data, while the circle with error bar denotes the value evaluated from the simulation.

Figure II.B.2-a shows the power scan of divertor discharges (#E11209A with 1 MA, $\bar{n}_e = 3.01 \times 10^{19}$ m^{-3}, $P_{abs} = 4.69$ MW and #E10686A with 1 MA, $\bar{n}_e = 3.16 \times 10^{19}$ m^{-3}, $P_{abs} = 10$ MW). For the high power case with $P_{abs} = 10$ MW, the toroidal η_i mode predicts the diffusivity somewhat larger than the experimental value, while, in the region of $r > 0.5m$, the mode is linearly marginal due to the η_i profile. On the other hand, for the low power case, the calculated thermal diffusivity is somewhat smaller than the experimental value.

Results of the I_p scan are shown in Fig. II.B.2-b for limiter discharges, and Fig. II.B.2-c for divertor discharges. shots (#E10774G with 1 MA, $\bar{n}_e = 2.54 \times 10^{19}$ m^{-3}, $P_{abs} = 9.25$ MW and #E10643C with 2 MA, $\bar{n}_e = 3.93 \times 10^{19}$ m^{-3}, $P_{abs} = 10.3$ MW) and Divertor shots (#E10748D with 1 MA, $\bar{n}_e = 3.97 \times 10^{19}$ m^{-3}, $P_{abs} = 9.6$ MW and #E10761A, with 1.8 MA, $\bar{n}_e = 3.54 \times 10^{19}$ m^{-3}, $P_{abs} = 9.39$ MW). For limiter discharges, it seems that the calculated thermal diffusivities are fairly good in the halfway region. On the other hand, for divertor discharges, absolute values of the diffusivity in the low current case are lower than experimental values In this low current divertor discharge, the η_i parameter is near the marginal value of linear stability and then the diffusivity may be sensitive to the theoretical model of the critical η_i value, η_{ic}.

In Fig. II.B.2-d, results are plotted for the case of high T_i divertor discharge (#E10300A with 0.57A, $\bar{n}_e = 3.49 \times 10^{19} m^{-3}$, and $P_{abs} = 15\,\mathrm{MW}$). The experimentally deduced diffusivity shows the strong improvement toward the plasma center, where the η_i value is large, while the calculated diffusivity does not change so much in the radial direction in this region.

From the calculations in Fig. II.B.2, we may conclude that, except low current divertor discharges and high T_i discharges, the toroidal η_i mode gives the diffusivity of the same order as experimental value in the halfway region of the plasma and seems to be a good candidate for ion anomalous transport, although additional effect may be needed in the peripheral region.

2.3.3 ADDITIONAL EFFECTS ON THE η_i MODE STABILITY

In the previous section, we found that the fluid η_i mode can explain the anomalous ion transport in the halfway region of the plasma but, in the peripheral region, it gives the large discrepancy with experimental observation. This means that we need have some additional effects on η_i mode stability in this region. The reason of the reduction of thermal diffusivity is that the mode becomes marginal and the linear growth rate considerably reduced in this region. The candidates to increase the free energy source of η_i are the dissipative effects Then, in this section, we consider the trapped electron effect and the neoclassical effect on the η_i mode stability and study whether these effects can enhance the ion anomalous transport in the peripheral region or not.

Trapped electron effect on η_i modes

Microinstability based on η_i mode and trapped electron mode are briefly investigated by many authors. Romanelli has shown that, when trapped electron dynamics is taken into account, the η_i mode is modified at several points compared with the fluid limit [5]. One is the threshold value of η_i stability; that is, the threshold disappears when the mode frequency is in the electron diamagnetic direction, while, when the frequency is in the ion diamagnetic direction, threshold still exits. That is due to the competition of trapped electron destabilization and ion magnetic resonance stabilization. Recently Nordman et al. showed the nonlinear simulation results of the η_i mode using the fluid model equations [6]. They treated the trapped electron as the fluid (Kadomtsev model [7]). Fluid model is advantageous to simulate nonlinear phenomena because of its relative simplicity. But we should carefully treat the results obtained from fluid simulations. Fluid model is at least valid in the limit $k_y\rho_s \to 0$, $\omega_{Di}/\omega \to 0$. To obtain the steady state spectrum, we usually use the artificial viscosity to drop off the short wavenumber modes. At this point, we loose the information on the short wave length modes ($k_\perp \rho_s \gtrsim 1$). From this viewpoint, firstly we clarify the difference of fluid and kinetic model equations in order to understand the limitation of fluid model equations. Then we test the analytic formula of thermal diffusivity obtained by Romanelli, by using experimental data in JT-60 L-mode discharges.

Comparison of kinetic and fluid model equations for η_i modes with trapped electron effect We use the kinetic model equations used by Romanelli and the fluid model equations used by Nordman. We will compare three model equations; that is, the kinetic ion and kinetic electron model (kinetic model), the kinetic ion and fluid electron model (hybrid model), the fluid ion and fluid electron model (fluid model). Using the normalization of $\tilde{n}/n_0 \to \tilde{n}$ and $e\tilde{\phi}/T_e \to \tilde{\phi}$, the kinetic ion density response is written as

$$\tilde{n}_i = -\tau \left[1 - \frac{\eta_i}{2\epsilon_n}\Gamma_0(b) - \left\{ z\left(1 - \frac{\eta_i}{2\epsilon_n}\right) + \frac{1}{\epsilon_n}\left(1 - \frac{3}{2}\eta_i\right) \right\} F_1 + \frac{\eta_i}{4\epsilon_n} F_3 \right] \tilde{\phi} \quad (12)$$

where

$$F_n = \frac{2}{\sqrt{\pi}} \int_0^{+\infty} dv_\perp v_\perp^n \exp(-v_\perp^2) J_0^2(\sqrt{2b}v_\perp) \int_{-\infty}^{+\infty} dv_\| \exp(-v_\|^2) \frac{1}{z + v_\perp^2/2 + v_\|^2} \quad (13)$$

and $b = k_y^2 \rho_i^2/2 = k_y^2 \rho_s^2/(2\tau)$, $z = \tau\omega/(2\epsilon_n k_y \rho_s \omega_{*el})$, $\omega_{*el} = c_s/L_n$, $\epsilon_n = L_n/R$ and ω_{*el} is the electron diamagnetic frequency with $k_y \rho_s =$

1. For the kinetic electron density response, we have

$$\tilde{n}_e = \left[1 - \sqrt{2\epsilon}\frac{\eta_e}{\alpha\epsilon_n} - \sqrt{2\epsilon}\left\{z\left(1 - \frac{\eta_e}{\alpha\epsilon_n}\right) - \frac{\tau}{2\epsilon_n}\left(1 - \frac{3}{2}\eta_e\right)\right\}H\right]\tilde{\phi} \quad (14)$$

where

$$H = \frac{2}{\sqrt{\pi}}\int_0^\infty dy\, y^{1/2}\exp(-y)\frac{1}{z - \tau\alpha y/2} \quad (15)$$

and $\alpha = 4/3$. For the fluid ion density response,

$$\tilde{n}_i = -\tau\left[1 - \frac{\eta_i}{2\epsilon_n} - \left\{z\left(1 - \frac{\eta_i}{2\epsilon_n}\right) + \frac{1}{2\epsilon_n}(1 - \eta_i a_2) - \frac{k_y^2\rho_s^2}{\tau a_1}\left(z + \frac{1+\eta_i}{2\epsilon_n}\right)\right\}\frac{a_1}{za_1 + 1}\right]\tilde{\phi} \quad (16)$$

and $a_1 = 1 + 2/(3z + 5)$, $a_2 = 1 + 1/(3z + 7)$. And finally, the electron fluid density response is expressed as

$$\tilde{n}_e = \left[1 - \sqrt{\epsilon}\frac{\eta_e}{2\epsilon_n} - \sqrt{\epsilon}\left\{z\left(1 - \frac{\eta_e}{2\epsilon_n}\right) - \frac{\tau}{2\epsilon_n}(1 - \eta_e a_4)\right\}\frac{a_3}{za_3 - \tau}\right]\tilde{\phi} \quad (17)$$

and $a_3 = 1 - 2/(3z - 5\tau)$, $a_4 = 1 - \tau/(3z - 7\tau)$.

Firstly, we check the kinetic η_i mode with an adiabatic electron response. Figure III.A.1-a shows the wavenumber dependence on the real frequency (ω_r) and growth rate (ω_i). We have chosen the parameters as $\omega_g = 0.3$, $\eta_i = 2$, $\tau = 1$ where ω_g is the ion magnetic frequency normalized by the ion diamagnetic frequency ($\omega_g = 2\epsilon_n$). The diamond and circle correspond to the result of the full kinetic calculation and the result without FLR effect ($k_y\rho_i = 0$), respectively. For comparison, we plot the $\eta_i = 1$ case with the same other parameters by the square symbol. The negative value of ω_r means that the mode rotates in the ion diamagnetic direction. Without FLR effect (circles), the real frequency is almost constant and the growth rate increases linearly to $k_y\rho_s$. With FLR effect (diamonds), the mode is stabilized at $k_y\rho_s \sim 3$ for $\eta_i = 2$. The growth rate has the tail in the region of $k_y\rho_s \sim 2 - 3$, where the dependence of real frequency on $k_y\rho_s$ is very weak. In the limit of $\omega_r \gg \omega_i$, the wave-particle resonance is occurred at $v_\parallel^2 = -z_r - v_\perp^2 \geq 0$, or $0 \leq v_\perp/2 \leq -2z_r$. So roughly speaking, $-2z_r$ is the resonance width in the velocity space. In the range of $0 \leq k_y\rho_s \leq 1.5$, $-2z_r$ is linearly increasing and then it decreases slowly in the region of $1.5 \leq k_y\rho_s$.

This means the resonance stabilization becomes weaker in the short wave length range and the tail with small growth rate is formed in this region. On the other hand, in the case of $\eta_i = 1$ (squares), the mode is near the marginal point, and it is easily stabilized in the region of small $k_y \rho_s$ where the real frequency still increases almost linearly to $k_y \rho_s$.

Secondly, we study the trapped particle response on the η_i mode stability. Figure III.A.1-b shows the wavenumber dependence on frequency and growth rate. The parameters are $\epsilon_n = 0.3$, $\omega_g = 0.3$, $\eta_i = \eta_e = 2$, $\tau = 1$, and symbols of diamond, circle and square denote the kinetic, hybrid and fluid model, respectively. For $k_y \rho_s \leq 0.4$, the fluid model (square) gives almost the same value as that of the hybrid and kinetic models, but for $k_y \rho_s \geq 0.4$, the mode from the fluid model is stabilized.

Figure III.A.1-c shows the η_i dependence on the real frequency and growth rate. Parameters used are $\epsilon_n = 0.3$, $\omega_g = 0.3$, $k_y \rho_s = 0.6$, $\eta_e = 2$, $\tau = 1$. Symbols have the same meaning as in Fig. III.A.1-b. For the fluid model (square) there exits the threshold η_i value for stability, while the kinetic model (diamond) and the hybrid model (circle) give unstable modes in the case with electron magnetic resonance. The frequency is in the electron diamagnetic direction so that the mode does not have a stabilizing effect due to the ion magnetic resonance. In the fluid model, $\omega_{Di}/\omega \sim 1$ near the threshold, and this means that the fluid model is inappropriate for the linear stability analysis, while, for η_i greater than the threshold, the fluid model gives the similar tendency as the hybrid and kinetic models.

The dependences on η_e, τ and ω_g are shown in Fig. III.A.1-d $\sim$ 1-f, respectively. The η_e parameter weakly destabilizes the η_i mode for the kinetic model (diamond) but the fluid model (square) and hybrid model (circle) do not depend on this parameter as shown in Fig. III.A.1-d. For $\tau \leq 0.5$, the η_i mode is strongly stabilized for all models of kinetic (diamond), hybrid (circle) and fluid (square). This stabilization due to small τ may be related with super shots or hot ion mode discharges. The mode is also stabilized at large ω_g as shown in Fig. III.A.1-f. This corresponds to the flat density limit or H mode case. The threshold η_i value of kinetic (diamond) and hybrid (circle) model is smaller than the fluid model (square). The stabilization is due to the kinetic effect of ion magnetic resonance.

Ion Transport Analysis

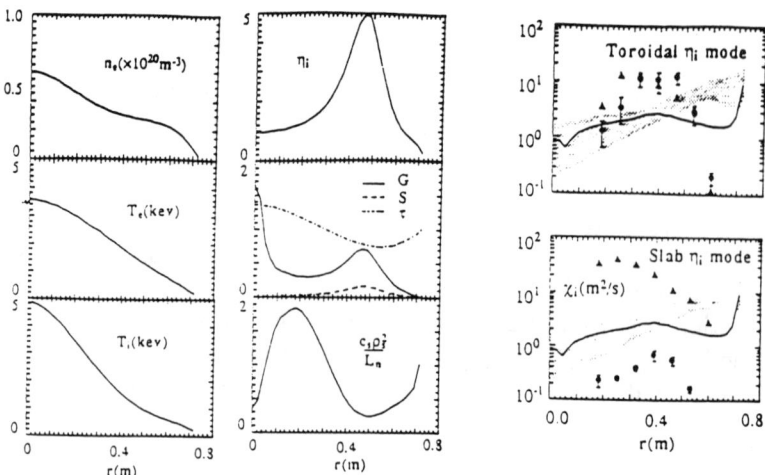

Fig.II.B.1 The radial profiles of plasma parameters in typical L-mode discharge #E11171 in JT-60 and the ion thermal diffusivities for this shot.
The shaded region shows the experimentally estimated diffusivities, the closed circle, and closed triangle correspond to the diffusivity determined from nonlinear simulation and the mixing length estimation, respectively. Domingues-Waltz model is plotted by the solid line.

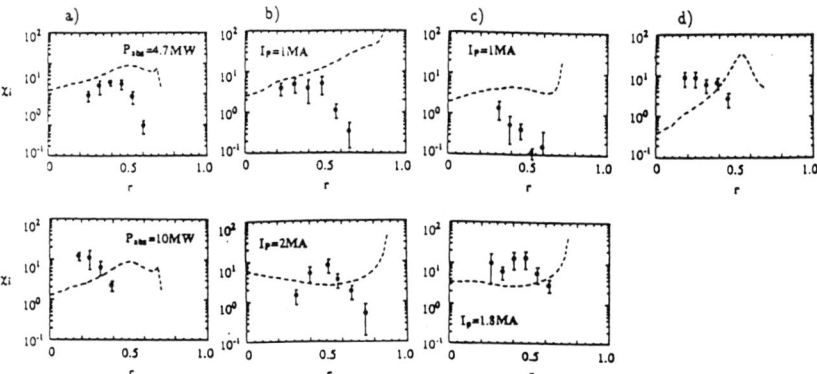

Fig.II.B.2 The comparision of thermal diffusivity evaluated from 2D-NQL simulation and the experimentally deduced one.
  a) Power scan of divertor discharges.
  b) I_p scan of limiter discharges.
  c) I_p scan of divertor discharges.
  d) High T_i shot.

Test of η_i and trapped electron model for JT-60 L-mode discharges In this subsection, we test the validity of the analytic expression of ion thermal diffusivity derived by Romanelli for JT-60 L-mode plasmas. To derive the thermal diffusivity expression, FLR effect is neglected for simplicity and $k_y \rho_i$ is set to be 0.3. From Fig. III.A.1-a in the previous subsection, this approximation seems to be justified. The main difference from the Dominguez and Waltz model of thermal diffusivity is that this model contains only one numerical factor in contrast to that the Dominguez and Waltz model contains two numerical coefficients, which come from their derivation in which they separately treat the η_i mode turbulence and trapped/circulating electron turbulence. The ion thermal diffusivity is given by

$$\chi_i = 7C \frac{v_{thi}\rho_i^2}{R}\tau(q_i^{\eta_i} + q_i^{TE}) + \chi_i^{NC} + 100\theta(a/q_a - r) \quad (18)$$

where the second term corresponds to the neoclassical contribution and the third the sawtooth effect. The parameters are given by [5]

$$q_i^{\eta_i} = \frac{0.7}{\sqrt{\tau\epsilon_n}}\sqrt{\eta_i - \eta_{ic}}\theta(\eta_i - \eta_{ic})\theta(\eta_i - \eta_0\epsilon_{n1}),$$

$$\frac{q_i^{TE}}{q_0^{TE}} = (1 - \epsilon_n/\epsilon_{n1})^{2.5}\theta(\eta_i - \eta_0\epsilon_{n1})\theta(\epsilon_{n1} - \epsilon_n)$$

$$+ (1 - \epsilon_n/\epsilon_{nc})^{2.5}\theta(\eta_0\epsilon_{n1} - \eta_i)\theta(\epsilon_{nc} - \epsilon_n)$$

$$q_0^{TE} = 0.75\sqrt{2\epsilon}\frac{\eta_e + 2.5}{0.045\nu + 1.5}$$

$$\eta_0 = \frac{4}{3}\left\{1 + \frac{1}{\tau}\left(1 - \frac{3}{4}\frac{\sqrt{2\epsilon}}{1+2.5\nu}\right)\right\}, \quad \epsilon_{n1} = \frac{1 + \frac{1-\eta_e}{\tau}\frac{\sqrt{2\epsilon}}{1+2.5\nu}}{2\left(1+\frac{1}{\tau}\right) - \frac{1}{\tau}\frac{\sqrt{2\epsilon}}{1+2.5\nu}},$$

$$\eta_{ic} = \eta_0 \max(\epsilon_n, \epsilon_{n1}), \quad \nu = 50\tau\frac{\nu_{eff}}{\omega_{be}}, \quad \tau = \frac{T_e}{T_i},$$

$$\epsilon_{nc} = \frac{\eta_i}{\eta_0}\theta(\eta_i - \eta_0\epsilon_{n1}) + 1.5\frac{1 - \eta_i + \frac{\sqrt{2\epsilon}(1-\eta_e)}{(1+2.5\nu)\tau}}{1 + \frac{1}{\tau}}\theta(\eta_0\epsilon_{n1} - \eta_i)$$

(19)

By using this expression of X_i and plasma parameters measured in experiments, we solve the ion energy transport equation and compare the resultant ion temperature profile with the experimental one. In this calculation, we impose the boundary condition of T_i as $T_i = T_i^{EXP}$ at $r = 0.8a$.

Figure III.A.2.1 shows the radial profiles of X_i and other parameters, which are evaluated from the calculation result for the shot #E11171A. The X_i is flat in the halfway region. Although the trapped electron term q_i^{TE} in X_i is finite in the region where the η_i term $q_i^{\eta_i}$ decreases, it is quite small and have little contribution on the total X_i. As a result, the total X_i is almost determined by the $q_i^{\eta_i}$ term. Another trapped particle effect comes from the modification of the threshold value of η_i. The calculated threshold value of η_i is very flat in the almost whole plasma region, while the calculation result of η_i has the strong peak at the half radius. This corresponds to that the calculated ion temperature has the steep gradient here in spite that η_i largely exceeds η_{ic}. For comparison, we plot the η_{ic} without trapped electron correction [8], η_{ic}^{OLD}. The trapped particle correction reduces the η_{ic}, but the reduction is not so large compared with the difference between η_i and η_{ic}. This means that the trapped particle modification of η_{ic} also has a small effect on X_i.

The same calculation has been performed for discharges with three different absorbed power ($P_{abs} = 5.58$, 9.6 and 17.1 MW) and results are shown in Fig. III.A.2.2. The plasma current and average plasma density are nearly the same in these discharges: $I_p = 1$ MA and $\bar{n}_e = (3.0-4.0) \times 10^{19}$ m^{-3}. In all cases, the calculated values X_i^{sim} of X_i have the flat profiles and the same values as the experimentally deduced one X_i^{exp} in the halfway radius of the plasma. However, in the peripheral region of plasma, X_i considerably decreases in each case as shown before. This means that the trapped electron modification of X_i does not work in the peripheral region in any case. The calculated T_i profiles show good agreements with experimental ones (circles in the figure) within the experimental values, in spite of the large discrepancy of X_i^{sim} and X_i^{exp} in the peripheral region. This is because, in the L-mode discharge, X_i near the edge is not so important for the core plasma transport.

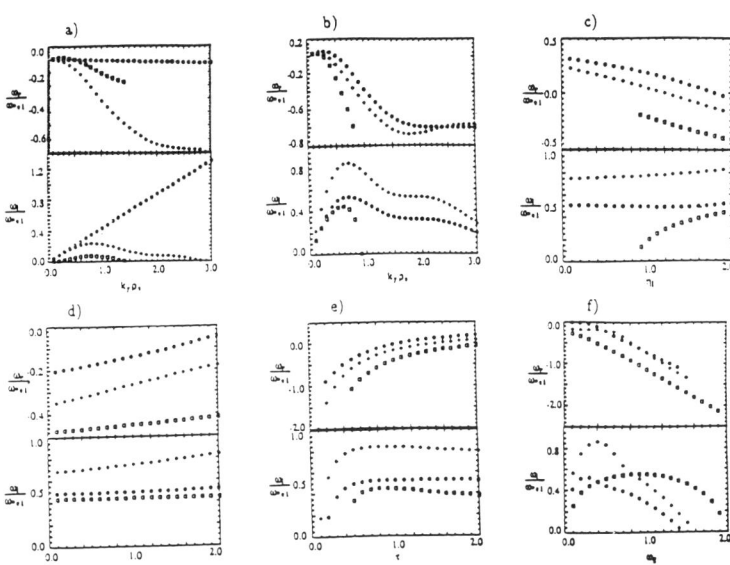

Fig.III.A.1 Comparision of parameter dependence on the growth rate based on kinetic(•), hybrid(∘), and fluid models(□).
a) ω_r and ω_i versus $k_y\rho_s$. The adiabatic electron response is assumed in this calculation. ∘ shows the case with $\eta_i = 2$, •, the case without FLR effect, and □, the case with $\eta_i = 1$.
b) ω_r and ω_i versus $k_y\rho_s$. The trapped electron responce is taken account of. c) ω_r and ω_i versus η_i. d) ω_r and ω_i versus η_e. e) ω_r and ω_i versus τ. f) ω_r and ω_i versus ω_g.

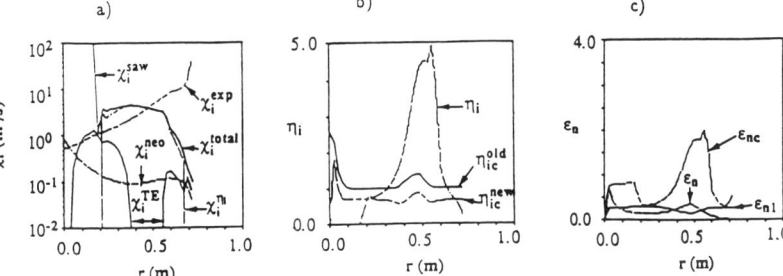

Fig.III.A.2.1 The numerical result based on the Romanelli's model and typical parameters deduced from experimental data.
a) Thermal diffusivity versus the minor radius. The dashed line denotes the experimentally deduced thermal diffusivity. χ^{TE} corresponds trapped electron contribution on η_i mode, χ^{SAW}, sawthooth effect, and χ^{neo}, neoclassical thermal transport.
b) η_i versus the minor radius. η_{ic}^{OLD} means the threshold value without trapped electron effect, and η_{ic}^{new}, with trapped electron effect.
c) ϵ_n versus the minor radius.

Neoclassical effect on η_i mode

In the peripheral region, collisional effects may play important effects on the η_i mode stability. The extension of fluid equations to the low collisionality regime relevant to the fusion plasma has been first studied by Callen and Shaing by introducing the neoclassical viscosity terms [9]. Based on the new moment equations, they found that neoclassical effect destabilizes resistive ballooning modes in the banana-plateau regime. These model equations have been further revised by Sundaram and Callen, who reevaluated the neoclassical viscosity term and investigated neoclassical ballooning modes and they have shown the resistive ballooning modes are controlled by effects associated with the perturbed bootstrap current, the enhanced ion polarization and pinch-type currents [10]. These neoclassical fluid equations have been applied to the analysis of η_i mode stability by Kim and his coworkers and they found that the neoclassical polarization destabilized the slab η_i mode in the banana-plateau regime [11]. However, they only considered $\nabla \cdot \Pi$ effects. They used Hirshman and Sigmar relation to close the neoclassical viscosity, which shows the thermal viscosity $\nabla \cdot \Theta$ makes also important contributions [12]. Therefore it is important to check not only $\nabla \cdot \Pi$ but also $\nabla \cdot \Theta$ effect on the η_i mode stability. Using the conventional normalization; $r/\rho_s \to r$, $c_s t/L_n \to t$, $z/L_n \to z$, we obtain the model equations in the following dimensionless form [13]

$$\frac{\partial v_\|}{\partial t} + [\phi, v_\|] = -\nabla_\| F + 4\mu \nabla_\perp^2 v_\| - \Sigma_1 , \qquad (20)$$

$$\frac{3}{2}\left(\frac{\partial p}{\partial t} + [\phi, p] + (1+\eta_i)\frac{1}{r}\frac{\partial \phi}{\partial \theta}\right) = \frac{5}{2}\left(\frac{\partial \phi}{\partial t} + \frac{1}{r}\frac{\partial \phi}{\partial \theta}\right) - \nabla_\| q_\|$$

$$+\frac{5}{2}\frac{G}{\tau}\left(\sin\theta \frac{\partial T}{\partial r} + \frac{\cos\theta}{r}\frac{\partial T}{\partial \theta}\right) + \chi_\perp \nabla_\perp^2 T + \Lambda_2 , \qquad (21)$$

$$\frac{\partial \phi}{\partial t} + \frac{1}{r}\frac{\partial \phi}{\partial \theta} = G\left(\sin\theta \frac{\partial F}{\partial r} + \frac{\cos\theta}{r}\frac{\partial F}{\partial \theta}\right)$$

$$-\nabla_\| v_\| + \left(\frac{\partial}{\partial t}\nabla_\perp^2 F + [\phi, \nabla_\perp^2 F] - [\nabla_\perp \frac{p}{\tau}; \nabla_\perp F]\right)$$

$$-\mu \nabla_\perp^4 F - \Lambda_1 , \tag{22}$$

$$\frac{\partial q_\|}{\partial t} + [\phi, q_\|] = -\frac{5}{2}\frac{1}{\tau}\nabla_\| T + \Sigma_2 - \frac{2}{5}\nu q_\| , \tag{23}$$

where $F = \phi + p/\tau$, $T = p - \phi$, $G = 2L_n/R_0$, $\nu = \sqrt{2}(L_n \nu_{ii}/c_s)$, $\tau = T_e/T_i$ and $[A; B] = \sum_k [A_k, B_k]$. The subscripts 'i' denoting ion species are dropped of for convenience. Σ_1 and Σ_2 correspond to the parallel component of particle and heat flow viscous forces, respectively. Taking the flux surface averaged part of neoclassical viscosity, these terms can be expressed as

$$\Sigma_1 = \frac{L_n \nu_{ii}}{c_s}[\mu_1 v_* - \mu_2 q_*], \quad \Sigma_2 = \frac{L_n \nu_{ii}}{c_s}[\mu_2 v_* - \mu_3 q_*] \tag{24}$$

where $v_* = v_\| + (q/\epsilon)\partial F/\partial r$ and $q_* = (2/5)q_\| + (q/\epsilon\tau)\partial T/\partial r$ correspond to the poloidal component of particle and heat flow, respectively. On the other hand, Λ corresponds to the divergence of the perpendicular component of neoclassical viscous force. By retaining the derivative perpendicular to the flux surface and neglecting the variation of $p_\| - p_\perp$ along the field line, the following expressions is obtained [9]:

$$\Lambda_1^{C/S} = -\frac{L_n \nu_{ii}}{c_s}\frac{1}{\epsilon}\left[\mu_1 \frac{\partial}{\partial r} q v_* - \mu_2 \frac{\partial}{\partial r} q q_*\right] ,$$

$$\Lambda_2^{C/S} = -\frac{L_n \nu_{ii}}{c_s}\frac{1}{\epsilon}\left[\mu_2 \frac{\partial}{\partial r} q v_* - \mu_3 \frac{\partial}{\partial r} q q_*\right] \tag{25}$$

The reduced set of neoclassical MHD equations Eqs. (20)-(23) gives the following energy conservation law:

$$\frac{dE_T}{dt} = \frac{3}{2}\frac{\eta_i}{\tau}\left\langle \phi \frac{\partial p}{\partial y}\right\rangle$$

$$-4\mu \left\langle (\nabla_\perp v_\|)^2\right\rangle - \mu \left\langle (\nabla_\perp^2 F)^2\right\rangle - \frac{4}{25}\nu \left\langle q_\|^2 \right\rangle$$

$$-\langle F \Lambda_1 \rangle + \left\langle \frac{T}{\tau}\Lambda_2 \right\rangle - \langle v_\| \Sigma_1 \rangle + \frac{2}{5}\langle q_\| \Sigma_2 \rangle \tag{26}$$

where the total energy is given by

$$E_T = \left\langle \frac{v_\parallel^2}{2} \right\rangle + \left(1 + \frac{1}{\tau}\right)\left\langle \frac{\phi^2}{2} \right\rangle + \left\langle \frac{(\nabla_\perp F)^2}{2} \right\rangle + \frac{3}{2}\left\langle \frac{T^2}{2\tau} \right\rangle + \frac{2}{5}\left\langle \frac{q_\parallel^2}{2} \right\rangle$$

and the neoclassical dissipation terms are positive definite as follows

$$\langle F\Lambda_1 \rangle - \left\langle \frac{T}{\tau}\Lambda_2 \right\rangle + \langle v_\parallel \Sigma_1 \rangle - \frac{2}{5}\langle q_\parallel \Sigma_2 \rangle =$$

$$\frac{L_n}{c_s}\nu_{ii}\left[\frac{\mu_1\mu_3 - \mu_2^2}{\mu_1}\left(\frac{2}{5}q_\parallel + \frac{q}{\epsilon}\frac{1}{\tau}\frac{\partial T}{\partial r}\right)^2\right.$$

$$\left. + \frac{1}{\mu_1}\left\{\mu_1\left(v_\parallel + \frac{q}{\epsilon}\frac{\partial F}{\partial r}\right) - \mu_2\left(\frac{2}{5}q_\parallel + \frac{q}{\epsilon}\frac{1}{\tau}\frac{\partial T}{\partial r}\right)\right\}^2\right] \quad (28)$$

where $\mu_1\mu_3 - \mu_2^2 > 0$.

Figure III.B.1 shows the normalized growth rate versus collision frequency for different model equations. The parameters are chosen as $\eta_i = 3$, $S = 0.2$, $G = 0.2$, $k_y \rho_s = 0.4$, $\tau = 1$, $l = 0$, $q/\epsilon = 8.0$, where l is the radial mode number. The symbols B, P and P.S. denote Banana, Plateau and Pfirsch-Schlüter regimes, respectively. We retain G terms finite since the slab η_i mode is completely stabilized for these parameters due to the $\nabla \cdot \Theta$ effect in plateau regime. In the figure, the symbol "6" is for the four field model $\{\phi, v_\parallel, p, q_\parallel\}$ with neoclassical viscosities. For comparison, results for other models are also plotted. The numbers "1" and "2" are for the three field models with $q_\parallel = 0$ ("1": the dissipationless fluid model), and with $q_\parallel = \chi_\parallel \nabla_\parallel^2 T$ ("2": the conductive fluid model), respectively. The numbers "3-6" are for the four field models: "3" for the case with $\chi_\perp$, $\nabla \cdot \Pi$ and $\nabla \cdot \Theta = 0$, "4" for the case with $\chi_\perp \neq 0$, and "5" for the case with $\chi_\perp$, $\nabla \cdot \Pi \neq 0$, respectively. From the comparison of "1" and others, we can see that the growth rate of η_i mode is largely reduced by the effect of $\nabla_\parallel q_\parallel$ in the whole collisional regimes. In the plateau regime, terms of $\nabla \cdot \Pi$ contribute the destabilization of η_i mode ("5"), while those of $\nabla \cdot \Theta$ stabilize the mode, and, as a total, the growth rate of η_i mode decreases with the collision frequency ("6"). The strong stabilization in the P-S regime comes from the increase of classical thermal diffusivity. These results imply that the neoclassical effect can not be the candidate of the edge anomaly of η_i mode driven transport.

Finally, we briefly comment on the Sundaram and Callen's model of neoclassical viscosity. They employ the neoclassical viscosity with retaining the variation of $p_\parallel - p_\perp$ along the field line as follows

$$\nabla_\perp \cdot \frac{\mathbf{B} \times \nabla \cdot \mathbf{\Pi}}{B^2} \simeq \frac{\partial}{\partial \Psi} \frac{I}{B^2} \left((p_\parallel - p_\perp) \mathbf{B} \cdot \nabla B - \frac{2}{3} \mathbf{B} \cdot \nabla (p_\parallel - p_\perp) \right) \tag{29}$$

while the viscous energy dissipation can be written as

$$\int \mathbf{V} \cdot \nabla \cdot \mathbf{\Pi} d^3 x = \int \nabla \cdot (\mathbf{V} \cdot \mathbf{\Pi}) - \int \mathbf{\Pi} : \nabla \mathbf{V}$$

$$= -\int (p_\parallel - p_\perp)(\mathbf{BB} : \nabla \mathbf{V} - \frac{1}{3} \nabla \cdot \mathbf{V}) d^3 x$$

$$\simeq -\int (p_\parallel - p_\perp) \left(\mathbf{V} \cdot \nabla \ln B + \frac{2}{3} \nabla \cdot \mathbf{V} \right) , \tag{30}$$

From those two equations, it is clear that retaining the last term contradicts the usage of the Hirshman and Sigmar relation to close the neoclassical viscosity, because the Hirshman and Sigmar relation retains only the magnetic pumping term and it does not ensure the positive definiteness of dissipation energy. In fact, the sign of Λ is changed when the last term is retained, and we have the fictitious instability even for the $k_y \to 0$ limit. In order to ensure the positive definiteness of the viscous energy dissipation in this case, we have to leave the Hirshman-Sigmar relation and have to take into account the compressibility. We have found that the compressibility effect on the neoclassical viscosity strongly reduces the magnetic pumping effect (by factor 1/9). And, if we take account of the compressibility consistently, not only the neoclassical η_i mode but also other neoclassical instabilities become unimportant in the plateau regime.

2.3.4 CONCLUSION AND DISCUSSION

To summarize our results, firstly, we have studied the nonlinear behaviors of η_i modes by using the Hong/Horton model equations and have compared the numerically evaluated ion thermal diffusivity with experimental data in JT-60 L-mode discharges. By checking the role of each quasilinear term in the two mode coupling simulation, we have found that the parallel velocity shear generated by the quasilinear effect plays an important role for the mode stabilization through the ion pressure modification in the slab η_i mode. By comparison of 3-D quasilinear, 3-D nonquasilinear and 2-D nonquasilinear simulations, it was found that (1) the quasilinear stabilization is not important in the 3-D simulation, (2) the mixing length theory overestimates the thermal diffusivity in the low shear region, and (3) the 2-D simulation gives a similar parameter dependence of the thermal diffusivity with the 3-D simulation. Based on these observations, the thermal diffusivities evaluated from the 2-D nonquasilinear simulation results have been compared with experimentally deduced ones in JT-60 L-mode discharges, and it was found that (1) the simulation results of the toroidal η_i mode are in fairly good agreement with the experimental results as for the magnitude at the halfway of the plasma column but (2) the thermal diffusivities have the large discrepancy near the peripheral region, where η_i modes are marginal or stable in the present model. As the possible candidates for the destabilization of η_i modes, the effects of trapped electrons and neoclassical viscosity have been studied. As for the trapped electron effect, the role of kinetic response on the linear stability has been studied, by comparing the kinetic, hybrid and fluid model equations. We show that the kinetic effects extend the unstable regime to large $k_y \rho_s$ but the mode has the maximum growth rate at $k_y \rho_s < 1$, where the fluid model gives the similar tendency with the hybrid or kinetic one, if $\eta_i > \eta_{ic}$. Romanelli's model of thermal diffusivity with trapped electron effect has been tested with JT-60 L-mode discharges, by solving the ion energy balance equation. The absorbed power dependence of the calculated thermal diffusivity is in good agreement with experimental one in the halfway region of the plasma, as in the case of nonlinear simulations, while both the trapped particle effect on the diffusivity and the η_{ic} threshold formula have little effect on the enhancement in the peripheral region. However, the calculated ion temperature profiles agree with experimental ones, because the edge transport is not so important in L-mode discharges. We have also checked the neoclassical effects on η_i modes and found that the Hirshman-Sigmar type of neoclassical viscosity stabilizes η_i modes in the plateau regime. However, we need a more consistent treatment of compressibility to derive the neoclassical viscosity in the perturbed fields.

As the other possibility enhancing the edge diffusivity near the peripheral region, the coupling with current diffusive ballooning mode (λ mode) may destabilize the toroidal η_i mode. The analysis of this destabilization mechanism is now going on.

ACKNOWLEDGMENTS

The continuous encouragement of Dr. S. Tamura is gratefully acknowledged. Author acknowledge Drs. Shirai, Hirayama, Kikuchi, and JT-60 team.

114   Ion Transport Analysis

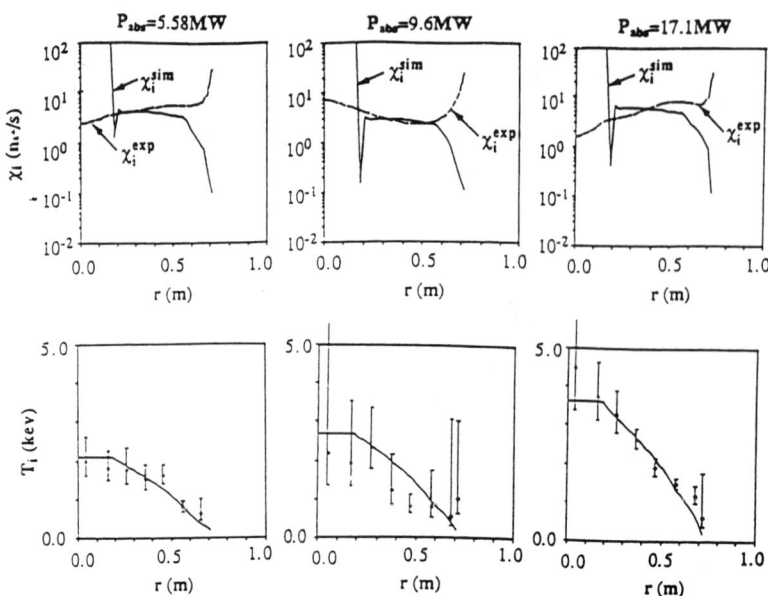

Fig.III.A.2.2 Comparision of simulation results based on Romanelli's model (solid line) and experimental data in JT-60 L-mode discharges.

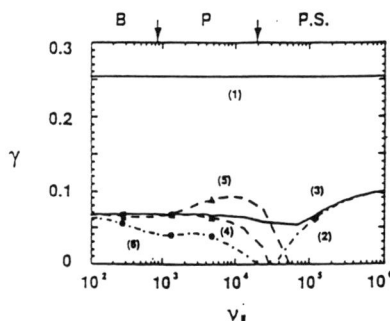

Fig.III.B.1 The growth rate versus collision frequency.
(1): the dissipationless fluid model.
(2): the conductive fluid model.
(3): four field model ($\chi_\perp$, $\nabla \cdot \Pi$, and $\nabla \cdot \Theta = 0$).
(4): four field model ($\chi_\perp \neq 0$).
(5): four field model ($\chi_\perp, \nabla \cdot \Pi \neq 0$).
(6): four field model ($\chi_\perp, \nabla \cdot \Pi$, and $\nabla \cdot \Theta \neq 0$).

Bibliography

[1] H. Shirai, T. Hirayama, M. Azumi, JAERI-M 91-018 (1991).

[2] B.G. Hong and W. Horton, Phys. Fluids B **2**, 978 (1990).

[3] S. Hamaguchi and W. Horton, Phys. Fluids B **2**, 1833 (1990).

[4] R.R. Dominguez, and R.E. Waltz, Nucl. Fusion **27**, 65 (1987).

[5] F. Romanelli, and S. Briguglio, Phys. Fluids B **2**, 754 (1990).

[6] H. Nordman, J. Weiland, and A. Jarmen, Nucl. Fusion **17**, 983 (1990).

[7] in *Review of Plasma Physics*, edited by M.A. Leontovich (Consultants Bureau, New York, 1965), Vol. 5, p. 249.

[8] F. Romanelli, Phys. Fluids B **1**, 109 (1989).

[9] J.D. Callen and K.C. Shaing, Phys. Fluids **28**, 1845 (1985).

[10] A.K. Sundaram and J.D. Callen, Phys. Fluids B **3**, 336 (1991).

[11] Y.B. Kim, P.H. Diamond, H. Biglari and J.D. Callen, Phys. Fluids B, 384 (1991).

[12] S.P. Hirshman and D.J. Sigmar, Nucl. Fusion **21**, 1079 (1981).

[13] M. Yagi, J.P. Wang, Y.B. Kim, and M. Azumi, accepted to Phys. Fluids B.

2.4 A COMPARISON BETWEEN THE MEASURED THERMAL DIFFUSIVITY χ_i^{Exp} IN TEXT AND THE PREDICTED $\chi_i^{\eta_i}$

A. Ouroua[5] A.J. Wootton, and the TEXT Group
Fusion Research Center, The University of Texas at Austin, Austin, Texas

Abstract. Measured profiles of the ion thermal diffusion coefficients $\chi_i^{\text{Exp}}(r)$ are compared with neoclassical predictions as well as with several predicted theoretical expressions that describe transport driven by the ion temperature gradient instability. While the multitude of η_i-models presently available makes a conclusive comparison between theory and experiment difficult, the overall magnitude of the predicted quantities, for some of the models, are reasonable. Steepening of the density profile by pellet injection did not reduce the anomaly in χ_i^{Exp}.

2.4.1 INTRODUCTION

The ion thermal transport in tokamaks has not always been known with confidence as compared to the well established anomalous character of electron thermal transport. In an early study the TFR group[1] reported an anomalous ion behavior, while in PLT[2, 3] the ions were found to be within neoclassical predictions. In DIII-D[3], χ_i^{Exp} was found anomalous in magnitude and shape, and in T-10[5], χ_i^{Exp} was 1.5 to 4 times larger than the neoclassical values. In TEXT[18], the anomaly factor was between 2 and 6, and in TFTR[7], ASDEX[8], and JT-60[9] the experimentally determined χ_i^{Exp} was also found to be larger than the predicted neoclassical values[10]. Generally, except for the case of large machines with auxiliary high power heating, such as in the TFTR supershot regime where the anomaly factor was as large as 40[7, 6], the inferred χ_i^{Exp} is found to be similar to or larger than the neoclassical values by a factor of 1 to 10. The most prominent cause

---

[5]Presently at ECSE Dept., RPI.

responsible for the anomaly in χ_i is the temperature drift instability[12, 13, 14, 9, 16, 17]. This is an electrostatic instability driven by the ion temperature gradient and triggered when the parameter $\eta_i = \frac{d\ln T_i}{d\ln n}$ is larger than a critical value η_i^{crit}. Here we present a comparison between the measured anomaly ($\chi_i^{\text{Exp}} - \chi_i^{\text{Neo}}$) and $\chi_i^{\eta_i}$ predicted by various η_i-models[12, 13, 14, 9, 16, 17]. A comparison between χ_i^{Exp} and the neoclassical predictions is presented first.

2.4.2 χ_i^{Exp} COMPARED WITH $\chi_i^{\text{Neoclassical}}$

Power balance analysis is used to determine the ion thermal diffusion coefficient

$$\chi_i^{\text{Exp}}(r) = -\frac{q_i(r)}{n(r)\frac{\partial T_i(r)}{\partial r}}$$

where q_i is the ion heat conduction flux. The plasma ion temperature profiles are measured using a diagnostic neutral beam and a scanning multichannel neutral particle analyzer. The heat flux $q_i(r)$ is determined from the steady-state power balance analysis. The details about the transport equations and diagnostics used to determine χ_i^{Exp} as well as the error analysis method used in this study can be found in Refs. [6,18]. Measured χ_i profiles for three different discharges are shown in Fig. 1 along with the predicted neoclassical profiles and the electron thermal diffusion coefficient χ_e^{Exp} for comparison. In the confinement region ($\rho = 0.4$), we found χ_i^{Exp} larger than the neoclassical χ_i^{Neo} by a factor of 2 to 6.

Pellet injection experiments have been carried out on TEXT in order to study the improved confinement and transport characteristics of pellet fuelled discharges[19]. Under controlled conditions plasma density profiles become more peaked after pellet injection, reducing the parameter η_i below a threshold η_{ic}, which would prevent the occurrence of the predicted ion temperature gradient instability. In this respect, the observed χ_i anomaly in gas fuelled discharges may be due to the presence of η_i modes which would be switched off after pellet injection by the steep density gradient. In order to check this possibility, ion power balance analyses were performed for two different discharges after pellet injection. The results are shown in Fig. 2. From this experiment we found no clear evidence of a reduction of ion thermal transport in pellet fuelled discharges.

2.4.3 EXCESSIVE THERMAL DIFFUSIVITY COMPARED WITH $\chi_i^{\eta_i}$

As indicated earlier, one of the possible causes of the measured enhanced transport is the ion temperature gradient (ITG) instability.

Several expressions describing transport induced by the (ITG) instability are available for comparison with experimental results. However, such a variety of $\chi_i^{\eta_i}$ formulae makes this comparison difficult. This is because the expressions are different from one another and, consequently, give different results. In fact, an important observation we would like to point out in this paper is the lack of a single $\chi_i^{\eta_i}$ expression with which all experimental results can be compared, such as in the case of the neoclassical χ_i^{Neo}. To illustrate this point, we chose six different expressions of $\chi_i^{\eta_i}$[12, 13, 14, 9, 16, 17], and compared their values with the excess thermal diffusivity $\chi_i^{Anom} = \chi_i^{Exp} - \chi_i^{Neo}$. The anomalous diffusivity formulas chosen are:

$$\chi_{i1}^{\eta_i} = \frac{1}{40} \frac{cT_i}{eB} \frac{\rho_i L_s}{L_{T_i}^2} \quad [Ref. \; [12]]$$

$$\chi_{i2}^{\eta_i} = \frac{cT_i}{eB} \rho_i \left(\frac{1}{qR}\right)^{1/4} \left(\frac{T_e}{T_i L_{T_i}}\right)^{3/4} \quad [Ref. \; [13]]$$

$$\chi_{i3}^{\eta_i} = 0.4 \left(\frac{\pi}{2} \ln(1+\eta_i)\right)^2 \left(\frac{1+\eta_i}{\tau}\right)^2 \frac{\rho_s^2 c_s}{L_s} \quad [Ref. \; [14]]$$

$$\chi_{i4}^{\eta_i} = (K_{\theta i}) \frac{cT_i}{eB} \frac{L_s}{R} \frac{\rho_s^2}{L_n \rho_i}(1+\eta_i) \quad [Ref. \; [15]]$$

$$\chi_{i5}^{\eta_i} = 5 \left(\frac{2}{\tau}\right)^{1/2} c_s \frac{\rho_i^2}{L_n} \varepsilon_n^{1/2} [\eta_i - \eta_{ic}(\varepsilon_n)]^{1/2} \quad [Ref. \; [16]]$$

$$\chi_{i6}^{\eta_i} = \left[K_\theta \frac{2\varepsilon_n}{\tau S}\left[(1+\eta_i)\left(\eta_i - \frac{2}{3}\right)\right]^{1/2} - \frac{K_\theta}{2}\left[2\varepsilon_n \frac{\eta_i - \frac{2}{3}}{\tau}\right]^{1/2}\right] \frac{\rho_s}{L_n} \frac{cT_e}{eB} \cdot \quad [Ref. \; [17]]$$

In formulas[9] and[17] the parameter K_θ describes the mean poloidal wavenumber $K_\theta \equiv k_\theta \rho_s$ of the fluctuations responsible for the anomalous transport. These expressions are evaluated using the nominal experimental profiles. In the figures we omitted the error bars for clarity. However, if experimental errors are systematically included in the evaluation of the $\chi_i^{\eta_i}$ coefficients as well as in χ_i^{Anom}, some of the compared coefficients would overlap within error limits. Results for the gas fuelled discharges are shown in Figs. 3, 4, and 5; and in Figs. 6 and 7 for the pellet fuelled discharges. Profiles of η_i and η_e for the gas fuelled and pellet fuelled discharges are shown, respectively, in Figs. 8 and 9.

2.4.4 DISCUSSION AND SUMMARY

In the confinement region ($r/a \sim 0.4$) of Ohmic gas fuelled discharges, as well as in pellet fuelled discharges, the ion thermal diffusivity is between two and six times the neoclassical value. Thus, the results in TEXT indicate that processes responsible for anomalous ion heat transport are present not only in gas fuelled discharges but also in pellet fuelled discharges. While we found no clear evidence of a reduction of ion thermal transport in pellet fuelled discharges, an ion mode observed by laser scattering[20, 21] in the microturbulence spectra in high density gas fuelled discharges is clearly suppressed under conditions similar to those of one of the pellet discharges studied here. The suppression of this ion mode was interpreted as a turn-off of the (ITG) instability by the peaked density profiles after pellet injection. The discrepancy between the conclusions from the transport analysis and the fluctuation analysis may be due to several possible reasons. First, the discharges studied may have been different. This is unlikely because most gross features of those discharges were similar. Second, our error bars may be too large for conclusive results for the transport. Finally, other mechanisms responsible for the enhanced transport may be present in both gas and pellet fuelled discharges. A possible mechanism we suspect to play an important role in ion transport in TEXT is the convective transport driven by the toroidal magnetic field ripple[22] (the ripple in the edge of TEXT is $\approx 3\%$). However, analysis by Gentle and Hazeltine[23] indicate that this is not relevant with the real electric field found in experiments. They argued that the radial electric field would drive poloidal motion even for the ripple-trapped ions and suggested that ripple effects should not seriously degrade thermal ion transport. This ripple transport mechanism will be investigated in more detail in TEXT-U with a larger ripple.

In summary, the nominal experimental profiles have been used to evaluate $\chi_i^{\eta_i}$ as predicted by six different η_i models. Generally, the predicted thermal diffusivities are different in shape and magnitude and do not consistently reproduce the nominal profiles of χ_i^{Anom} in all discharges analyzed, making a final conclusion concerning the presence of the (ITG) transport mechanism difficult. However, some thermal diffusivities are reasonably close to the experimental results, and those formulas (Romanelli[18], and Hong and Horton[17]) with an explicit dependence on $\eta_i^{critical}$ seem to give the better results.

Bibliography

[1] Equipe TFR, Nucl. Fusion **18**, 1271 (1978).

[2] Brusati, M., Davis, S.L., Hosea, J.C., *et al.*, Nucl. Fusion **18**, 1205 (1978).

[3] Heidbrink, W.W., Loveberg, J., Strachan, J.D., *et al.*, Nucl. Fusion **27**, 129 (1987).

[4] Groebner, R.J, Pfeiffer, W., Blau, F.P., *et al.*, Nucl. Fusion **26**, 543 (1986).

[5] Berezovskij, E.L., Dnestroskij, Yu.N., Efremov, S.L., *et al.*, Nucl. Fusion **27**, 2019 (1987) 2019.

[6] Ouroua, A., Wootton, A.J., Bravenec, R.V., *et al.*, Nucl. Fusion **30** (1990) 2585.

[7] Fonck, R.J., Bitter, M., Goldston, R.J., *et al.*, in Controlled Fusion and Plasma Heating (Proc. 15th Eur. Conf. Dubrovnik,1988), Vol. 12B, Part I, European Physical Society (1988) 83.

[8] Gruber, O., Kallenbach, A., Fahrbach., H.U., *et al.*, in Controlled Fusion and Plasma Physics (Proc. 16th Eur. Conf. Venice, 1989), Vol. 13B, Part I, European Physical Society (1989) 171.

[9] Hirayama, T., Shirai, H., Yagi, M., *et al.*, Nucl. Fusion **32** (1992) 89.

[10] Chang, C.S., Hinton, F.L., Phys. Fluids **29** (1986) 3314.

[11] Zarnstorff, M.C., Goldston, R.J., Bell, M.G., *et al.*, in Controlled Fusion and Plasma Physics (Proc. 16th Eur. Conf. Venice, 1989), Vol. 13B, Part I, European Physical Society (1989) 35.

[12] Kadomtsev, B.B., Pogutse, O.P., in *Reviews of Plasma Physics*, Vol. 5 (Leontovich, M.A., Ed.), (Consultant Bureau, New York, 1970) 249.

[13] Connor, J.W., Nucl. Fusion **26** (1986) 193.

[14] Lee, G.S., Diamond, P.H., Phys. Fluids **29** (1986) 3291.

[15] Biglari, H., Diamond, P.H., Rosenbluth, M.N., Phys. Fluids B **1** (1989) 109.

[16] Romanelli, F., Phys. Fluids B **1** (1989) 1018.

[17] Hong, B.G., Horton, W., Phys. Fluids B **2** (1990) 978.

[18] Ouroua, A., Ion thermal diffusion in TEXT, Rep. FRCR 343, Fusion Research Center, Univ. of Texas at Austin (1989).

[19] Rowan, W.L., Bravenec, R.V., Wiley, J., *et al.*, Nucl. Fusion **30** (1990) 903.

[20] Brower, D.L., Kim, S.K., Tang, W.M., *et al.*, in *Controlled Fusion and Plasma Heating* (Proc. 15th Eur. Conf. Dubrovnik, 1988), Vol. 12B, Part I, European Physical Society (1988) 183.

[21] Brower, D.L., Redi, M.H., Tang, W.M., *et al.*, Nucl. Fusion **29** (1989) 1247.

[22] Gurevich, A.V., Dimant, Ya.S., Nucl. Fusion **21** (1981) 159.

[23] Gentle, K.W., Hazeltine, R.D., Phys. Fluids B **3** (1991) 3198.

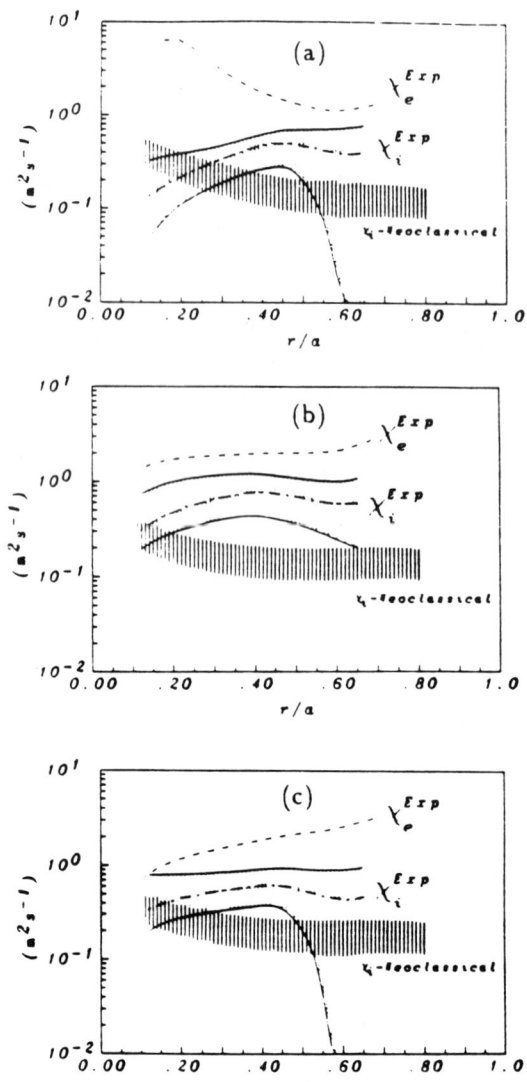

Figure 1. Inferred χ_i and χ_e profiles.

(a): $B_t = 2.8\,\text{T}$, $I_p = 300\,\text{kA}$, and $\bar{n}_e = 3 \times 10^{19}\,m^{-3}$
(b): $B_t = 2.8\,\text{T}$, $I_p = 300\,\text{kA}$, and $\bar{n}_e = 1.5 \times 10^{19}\,m^{-3}$
(c): $B_t = 2.0\,\text{T}$, $I_p = 200\,\text{kA}$, and $\bar{n}_e = 1.5 \times 10^{19}\,m^{-3}$

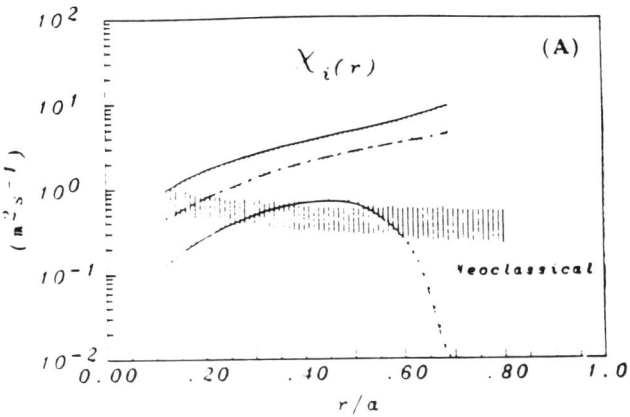

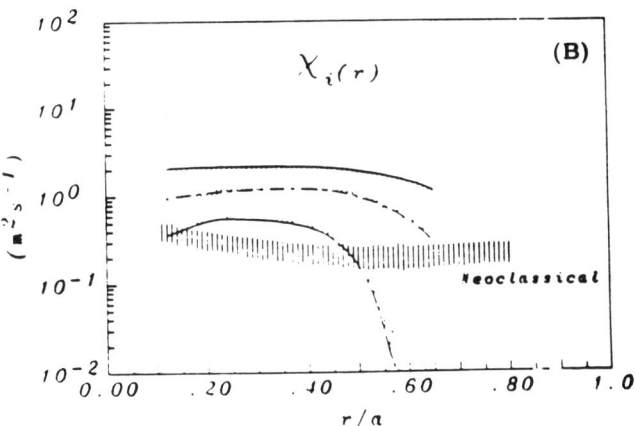

Figure 2. Inferred χ_i profiles after pellet injection (hydrogen pellets into hydrogen plasma).

(a): $B_t = 2.0\,\text{T}$, $I_p = 200\,\text{kA}$, and $\bar{n}_e = 3 \times 10^{19}\,m^{-3}$ + 1 Pellet $\longrightarrow 6 \times 10^{19}\,m^{-3}$

(b): $B_t = 2.8\,\text{T}$, $I_p = 250\,\text{kA}$, and $\bar{n}_e = 3 \times 10^{19}\,m^{-3}$ + 2 pellets $\longrightarrow 6.5 \times 10^{19}\,m^{-3}$

126 Measured Thermal Diffusivity χ_i^{Exp} in TEXT

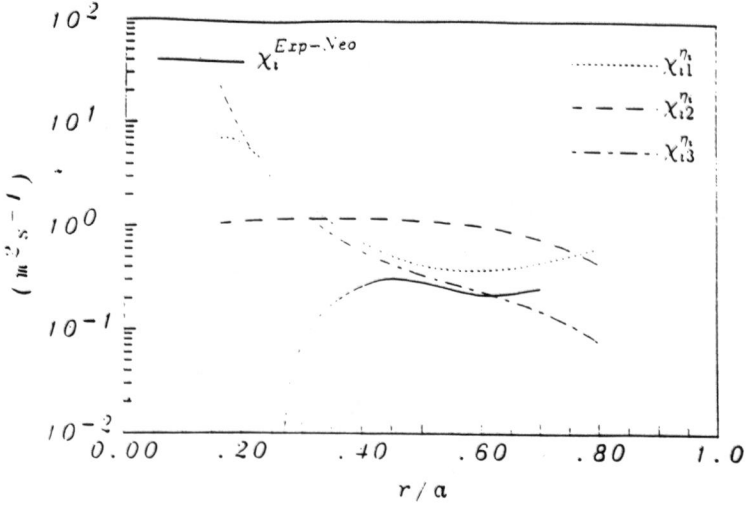

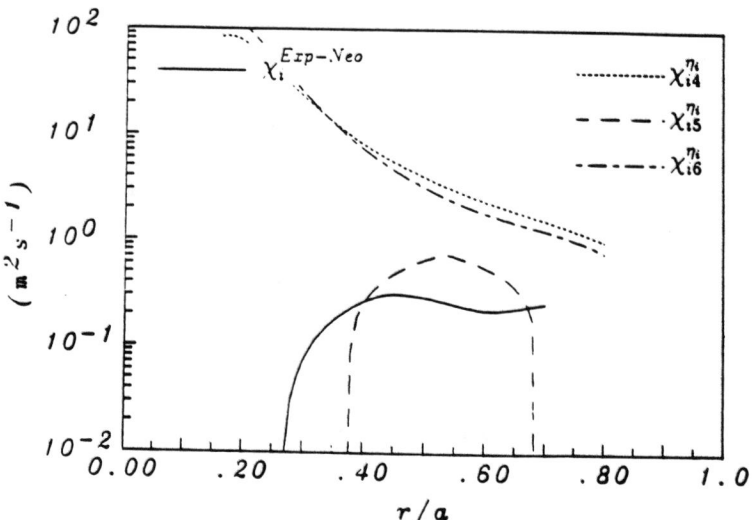

Figure 3. $(\chi_i^{Exp} - \chi_i^{Neoclassical})$ vs. $\chi_i^{\eta_i}$ for discharge (a).

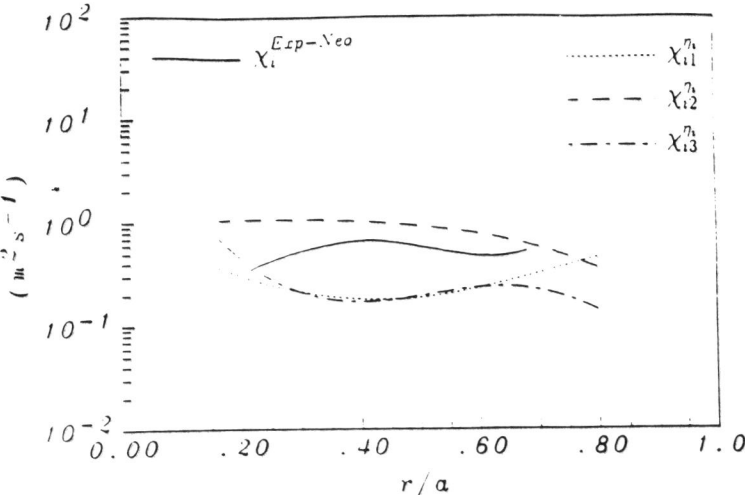

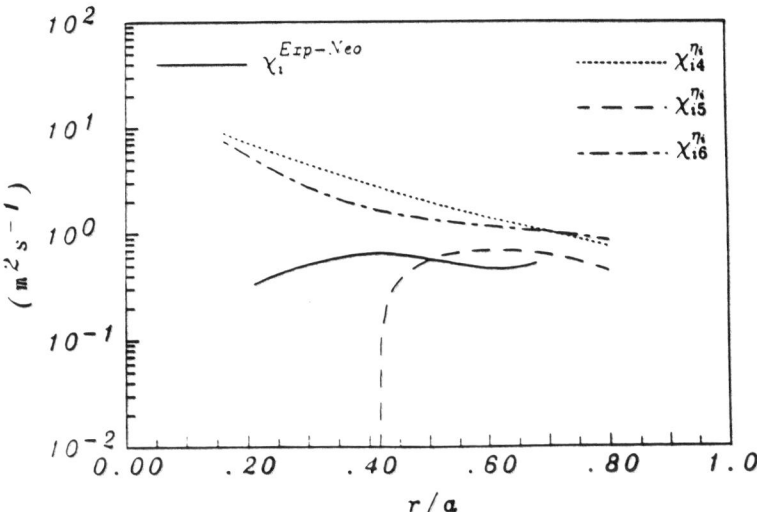

Figure 4. $(\chi_i^{\text{Exp}} - \chi_i^{\text{Neoclassical}})$ vs. $\chi_i^{\eta_i}$ for discharge (b).

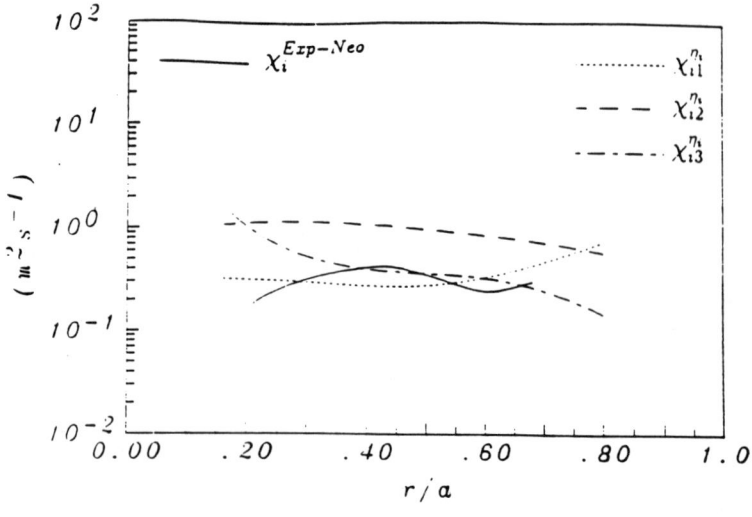
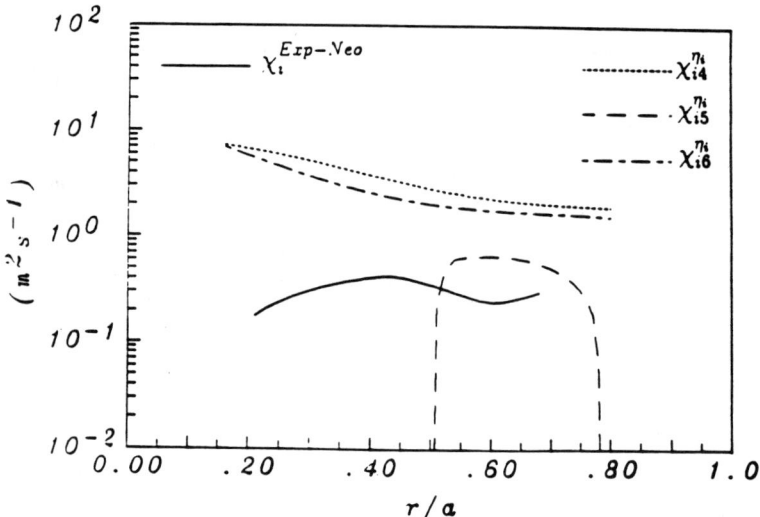

Figure 5. $(\chi_i^{\text{Exp}} - \chi_i^{\text{Neoclassical}})$ vs. $\chi_i^{\eta_i}$ for discharge (c).

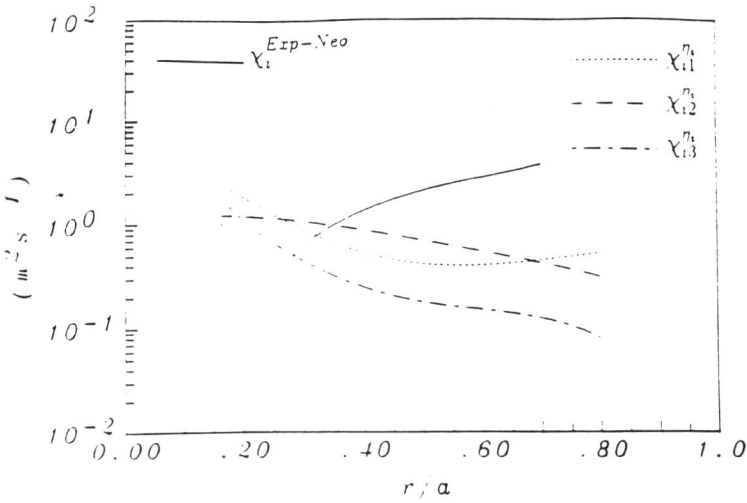

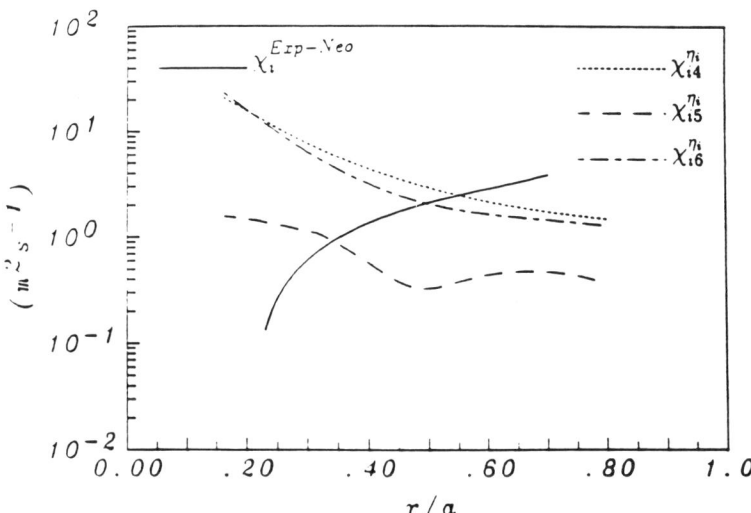

Figure 6. $(\chi_i^{\text{Exp}} - \chi_i^{\text{Neoclassical}})$ vs. $\chi_i^{\eta_i}$ for pellet fuelled discharge (a).

130 Measured Thermal Diffusivity χ_i^{Exp} in TEXT

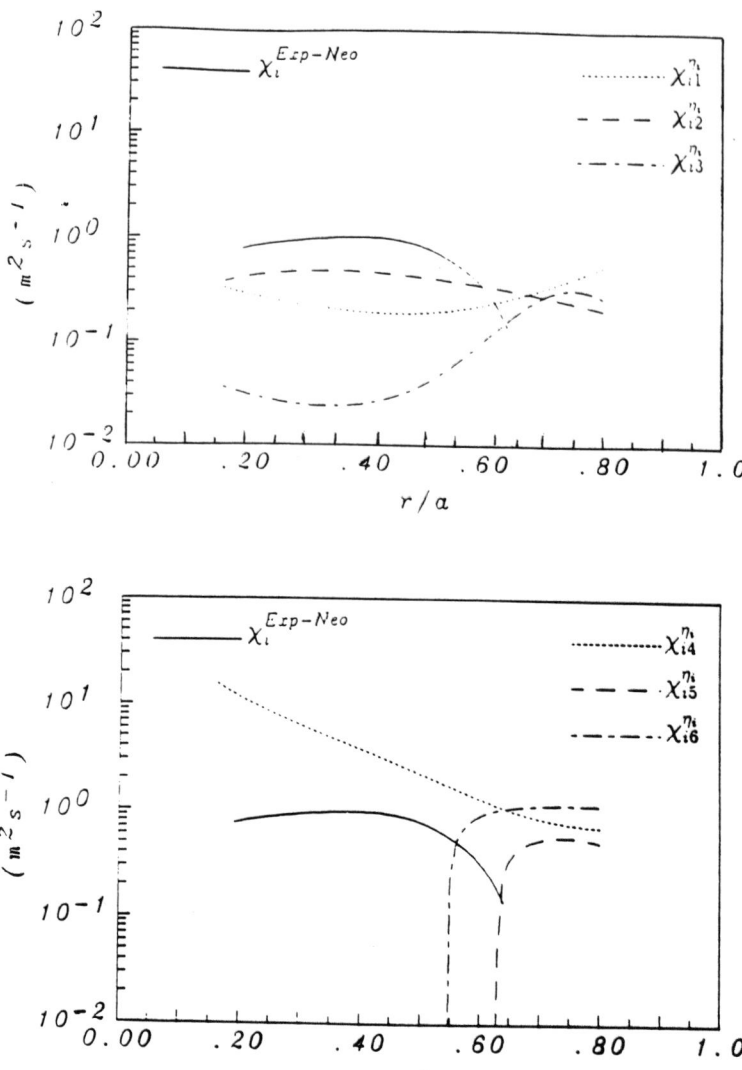

Figure 7. $(\chi_i^{Exp} - \chi_i^{Neoclassical})$ vs. $\chi_i^{\eta_i}$ for pellet fuelled discharge (b).

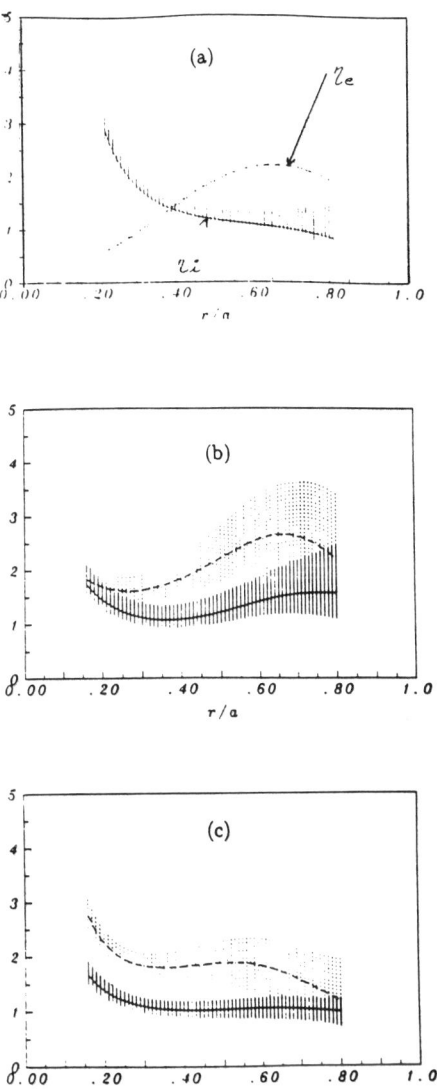

Figure 8. η_i and η_e profiles for discharges (a), (b), and (c).

132 Measured Thermal Diffusivity χ_i^{Exp} in TEXT

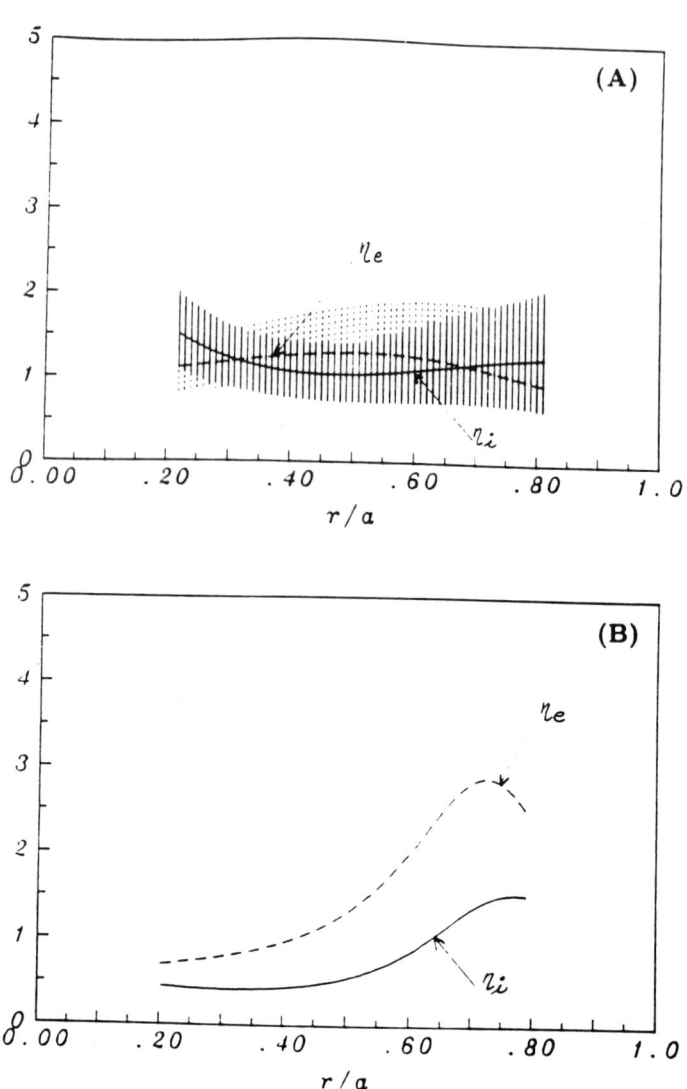

Figure 9. η_i and η_e profiles for pellet fuelled discharges (A) and (B).

Chapter 3
BASIC GRAD-T_i PLASMA EXPERIMENTS

3.1 ITG INSTABILITIES AND TRANSPORT STUDIES IN THE COLUMBIA LINEAR MACHINE

A.K. Sen, J. Chen, B. Song, and R.G. Greaves
Plasma Research Laboratory, Columbia University, New York, New York

Abstract. Both the slab and toroidal branches of the ion temperature gradient (ITG) driven instabilities have been studied in the Columbia Linear Machine (CLM). The ITG instabilities are observed as $m = 2$ modes when a sufficient parallel ion temperature gradient is reached by using either d.c. ion acceleration heating or rf-transit time heating methods. The slab branch, $m = 2$ mode, is identified as the ITG mode based on the four basic characteristics predicted by most theoretical work. The toroidal ITG mode is studied by starting with the slab ITG mode and gradually turning on the mirror current in CLM. The curvature effect is shown to further destabilize the mode and reduce the real frequency in the laboratory frame. The experimentally measured features in the presence of the ion temperature gradient and the mirror field also correspond closely to the theoretical predictions of a newly postulated hybrid ITG-dissipative trapped electron drift mode. The anomalous local ion thermal conductivities due to the slab mode has been measured via a temperature relaxation method. Its average value of 0.5 m^2/sec is about 3 orders of magnitude larger than the classical and 1 order of magnitude lower than the Bohm.

3.1.1 INTRODUCTION

One of the outstanding open physics issues for the present and future generation of high-temperature tokamaks is the anomalous ion thermal conductivity. In the past one important component of this transport was believed to be due to the ion-temperature-gradient instability (η_i mode) [1, 2, 3, 4]. In the last few years considerable amount of theoretical and experimental work (references in these proceedings) tend to substantiate this early conjecture, but not without some recent experimental reservations. However, it has been difficult, if not impossible, to identify this instability directly in tokamaks. Therefore, the production, identification and transport studies of this mode are presently being pursued in the simpler and experimentally convenient configuration of the Columbia Linear Machine (CLM) [5].

In the CLM a hydrogen plasma, with coincident radial temperature and density gradients, is produced by a $\mathbf{E} \| \mathbf{B}$ discharge source as shown in Fig. 1. The machine is approximately 3 m in overall length and has a uniform axial magnetic field. In the past, localized mirrors creating a mirror cell were used for the study of various trapped-particle instabilities as shown in Fig. 1. The typical parameters of the CLM before modifications are: $N \sim 6 \times 10^8 - 2 \times 10^9 \, \text{cm}^{-3}$, $T_e \sim T_i \sim 5 \, \text{eV}$, and $P_n \sim 7 \times 10^{-7} \, \text{Torr}$. The other parameters are $r_p \sim 2.7 \, \text{cm}$, $1/(r_p L_n) \sim 0.27 \, \text{cm}^{-2}$, $L_c \sim 50 - 150 \, \text{cm}$, $B_0 = 1 \, \text{kG}$, $\nu_e/\omega^* \sim 0.0015 - 0.2$, $\nu_{en}/\omega^* \sim 0.15 - 0.01$, ν_e/ω_{Be}, $\nu_e/\omega_{te} \sim \nu_i/\omega_{Bi}$, $\nu_i/\omega_{ti} \sim 3 \times 10^{-5} - 10^{-3}$, $\eta_e \sim 1.0$, $\eta_i \sim 0$, and $\rho_i \sim 0.2$. Therefore, the dimensionless parameters of CLM, relevant to electrostatic microinstabilities, can clearly range from those of TFTR, DIII, and JET to very collisionless plasmas of future reactor-type tokamaks. It is clear that for our present purpose the only inadequate parameter is η_i. This parameter has to be increased beyond the critical value $\eta_{i,c}$ for the instability, which is generally estimated to be of the order of 1. The methods and modifications to accomplish this are described below.

The parallel and transverse ion temperature were measured with gridded ion energy analyzers (IEA). Rectangular Langmuir probes (2 mm x 5 mm) were used to obtain electron temperatures, ion saturation current and to detect density fluctuations. The plasma potential was obtained from the floating potential of an emissive probe. These electrostatic probes are fixed at axial positions but could be moved radially to obtain profiles of the various plasma parameters.

3.1.2 PRODUCTION AND IDENTIFICATION OF THE SLAB BRANCH OF ITG MODE

$\nabla T_{i\parallel}$ production by D.C. ion heating

For the purpose of producing the slab branch of the ITG mode which does not depend on trapped particles or curvature drive, we do not energize the mirror coils in CLM. The parameter η_i can be increased either by flattening the density gradient or by increasing the ion temperature gradient (from zero). It must be emphasized that even for zero density gradient, one must have an appropriate ion temperature gradient to excite this instability, which is a temperature-gradient-driven mode. A "feathered" screen installed at the entrance of the experimental region can significantly lower the density gradient. Various electromagnetic ion-cyclotron-resonance-heating antennas have failed to produce appropriately peaked temperature profiles. Launching electrostatic ion cyclotron waves directly near the plasma axis sometimes succeeded in producing moderately peaked $T_{i\perp}$ profiles, but failed to excite any instability. This led to an examination of the consequences of only transverse heating and creation of

$$\eta_{i\perp} = \frac{d\ln T_{i\perp}}{d\ln N} \gg \eta_{i\parallel} = \frac{d\ln T_{i\parallel}}{d\ln N} \sim 0 \; .$$

A theory of ion-temperature-gradient instability with anisotropic η_i has been developed [6] which indicates the critical role of $\eta_{i\parallel}$ and a secondary role $\eta_{i\perp}$ for the slab branch. This motivated the development of a scheme for parallel heating of the ion.

The basic idea of the scheme is to accelerate the ions from the plasma source before they enter the experimental cell [7]. The acceleration parallel to the magnetic field is achieved via a 70% transparent tungsten mesh at the two ends of the plasma column, both biased from -30 to -50 V [7]. Lastly, the terminating end plate has a bias of $+4$ to $+8$ V to contain the heated ions, while the typical plasma potentials is -6 to -8 V. Thermalization in the high-neutral-pressure transition region, subsequent to acceleration, produces parallel ion heating roughly in the region covered by the accelerating meshes (or radii 1.0 cm and 0.6 cm). This leads to a peaked $T_{i\parallel}$ profile. Furthermore, the mesh in the transition region reduces the density in the central core and helps reduce the density gradient. Therefore the biased mesh can act as both a density flattening and parallel-temperature-peaking device and can produce high values of $\eta_{i\|}$.

Profiles of plasma parameters in the absence of parallel ion heating in Fig. 2(a) and those with ion heating are shown in Fig. 2(b). It is clearly seen that when the heating is off, the parallel ion temperature $T_{i\|}$ profile is almost flat over the entire plasma column, while with the heating on it develops a strong gradient in the vicinity of the accelerating mesh. The transverse ion temperature $T_{i\perp}$ remains flat in both cases, as there is no transverse heating in this scheme. This is a clear consequence of the fact that biased meshes can only accelerate ions in the parallel direction and yield an enhanced parallel ion temperature, while leaving the transverse ion temperature unchanged.

The frequency spectra of the density fluctuations corresponding to the profiles in Fig. 2 are shown in Fig. 3. The data taken from a Langmuir probe biased to collect ion saturation current is located at $r = 1.4$ cm, where the mode attains its maximum value. Without heating, in the case of the flat parallel ion temperature profile, a spectral feature is present at a frequency of 27 kHz [see Fig. 3(a)], which correlates with the equilibrium $\mathbf{E} \times \mathbf{B}$ rotation of the plasma. This feature is always present in our machine and is believed to be a centrifugally driven flute mode. The potential profiles responsible for the radial electric fields are shown in Fig. 2. The effect of the electric field will be discussed later. As the parallel heating is turned on, the profiles change, as shown in Fig. 2(b), and the corresponding spectrum is shown in Fig. 3(b). The lower frequency mode is still present at a slightly higher frequency of 35 kHz, since the equilibrium radial electrical field has increased, as shown in Fig. 2(b). However, a new mode is now clearly visible at a frequency of 62 kHz. Under a variety of plasma conditions, this new mode correlates with the existence of $\nabla T_{i\|}$ and is therefore considered as a candidate for the η_i mode. In order to identify whether the second peak in the spectrum in Fig. 3(b) represents a η_i mode, we first examine the character of this mode through the marginal stability criteria. For $\nabla N \sim 0$ and $\nabla T_{i\perp} \sim 0$ in the vicinity of $r \sim 1.0$ cm in Fig. 2(b), one needs to use the inverse-temperature-gradient scale length parameter $L_{T_{i\|}} = (dT_{i\|}/dr)/T_{i\|}$ rather than the η_i value. Using the results developed in Refs. 6 and 7, we calculate the critical temperature gradient scale length for the onset of the slab branch of the instability and compare with the experimental values as follows. The theoretical values are $(L_{i\|}) = 2.2\,\text{cm}$ ($m = 1$) and $4.0\,\text{cm}$ ($m = 2$), where m is the azimuthal number of the mode; while from the experimental data, we find $\langle L_{T_{i\|}} \rangle_{\text{exp}} \sim 2.4\,\text{cm}$. The experimental measurement of $L_{T_{i\|}}$ has been averaged over the radial mode width to yield $\langle L_{T_{i\|}} \rangle$. From the above numbers for theoretical and experimental ion-temperature-gradient scale lengths, we can expect the $m = 1$ mode to be marginally stable, but the $m = 2$ mode will be strongly unstable.

Next we investigate the azimuthal mode numbers for the modes shown in Fig. 3(b). The azimuthal mode numbers and the directions of propagation of the modes have been determined via cross-correlation of signals from two Langmuir probes separated azimuthally by 90°. The results show that the first mode, both with heating (at 35 kHz) and without heating (at 27 kHz) has a mode number $m = 1$, and travels in the direction of the equilibrium $\mathbf{E} \times \mathbf{B}$ rotation, which is the electron diamagnetic drift direction in the CLM. The second mode at 62 kHz has a mode number $m = 2$, and in accordance with the theoretical predictions, this could very possibly be an η_i mode.

The $m = 1$ mode, as we mentioned earlier, is always present in the CLM, and its amplitude peaks at the location where the equilibrium electric field is maximum. The observed frequency of the mode is correlated with the equilibrium electric field strength averaged over the mode width. The larger the electric field, the higher the mode frequency. This mode is believed to be a centrifugal flute instability driven by the equilibrium $\mathbf{E} \times \mathbf{B}$ rotation, which is the electron diamagnetic drift direction in the CLM.

Further identification of the mode is obtained from the measurements of parallel wavelength. The parallel wavelength is determined in a manner similar to that of the azimuthal wavenumber, but the signals are obtained from four Langmuir probes separated from each other axially by $15-32\,\text{cm}$. In order to get a reasonable estimate of the parallel wavelength in spite of shot-to-shot variations, several measurements from different pairs of probes were repeatedly carried out in different shots and on different days. The results show that parallel wavelengths are $\lambda_\| = 750 \pm 100\,\text{cm}$ ($m = 1$) and $\lambda_\| = 270 \pm 50\,\text{cm}$ ($m = 2$), where the machine length $L = 150\,\text{cm}$. This suggests that the $m = 1$ mode is a flute-like mode, whereas the $m = 2$ mode is a drift-like mode with $\lambda_\| \sim 2L$. From the physical picture and the analysis presented earlier, the η_i mode (the slab branch) can never be flute-like. Furthermore, the $m = 2$ mode cannot be simply the second harmonic of the $m = 1$ mode, because of their difference in parallel wavelengths, as well as the question of the real frequency of the mode discussed below. The measured wavelength of the $m = 2$ mode has a finite parallel wavelength, which is an indication of a drift-like mode in contrast to the $m = 1$ flute mode. Furthermore, in Sec. 5 the theoretical analysis of the slab η_i mode with this measured $\lambda_\|$ predicts a strongly unstable $m = 2$ mode.

We now discuss the real frequencies of the modes. The laboratory frame frequency of the $m = 1$ rotationally driven mode will be roughly the $\mathbf{E} \times \mathbf{B}$ rotation frequency. As the plasma-frame frequency of the ITG mode is a small negative value (propagating in the ion diamagnetic drift direction), we should then expect this mode at roughly the second harmonic of the rotational frequency. This is clearly seen in the spectrum of Fig. 3(b). Furthermore, we always see this mode at a frequency slightly lower $(6 - 10\,\text{kHz})$ than the second harmonic of the rotation frequency, which is consistent with the propagation of this mode in the ion diamagnetic direction rather than in the electron direction that would lead to an upshift in frequency.

$\nabla T_{i\|}$ production by transit time rf heating

We now discuss the production of the ITG mode by a very different ion heating scheme, namely the transit time rf heating. Unlike the dc heating scheme discussed above, this produces an excellent Maxwellian ion population leading to a more definitive measurement of the η_i number. The heating is achieved by applying an rf voltage of 30 to 60 V at a frequency of 1.3 MHz to a tungsten mesh in the plasma core, centered on the plasma axis [8]. An annular mesh covers the cross-sectional area of the plasma complementary to the center mesh and separated axially. The function of the annular mesh is simply to compensate for the density reduction in the center of the plasma column caused by the heating mesh which would otherwise produce an inverted density profile.

The results are similar to the case of dc heating discussed above we find that increasing the rf signal level steepens the ion temperature gradient and flattens the density gradient. Both of these effects act to increase η_i. In order to estimate the value of η_i, we plot the log $(T_{i\|})$ against the corresponding values of log (n). This is shown in Fig. 4(a). These data then yield profiles for $\eta_{i\|}(r)$ $(= \partial \ln T_{i\|}(r)/\partial \ln n(r))$. Unfortunately, the data in Fig. 4(a) are too noisy to obtain a meaningful local derivative. Instead we estimate the average $\langle \eta_{i\|} \rangle$, rather than its profile, by fitting a straight line to the data. The fitted curves in Fig. 4(a) are obtained from a linear regression on the data points between $r = 1.5\,\text{cm}$ and $r = 3.0\,\text{cm}$ where $\eta_{i\|}$ is enhanced by the rf heating. For the three cases shown, the regressions yield values for $\langle \eta_{i\|} \rangle$ of 0.3 ± 0.07, 1.5 ± 0.1 and 3.6 ± 0.8 respectively.

The spectra corresponding to the three curves in Fig. 4(a) are shown in Figs. 4(b)–4(d). For the case where no heating is applied, there is a coherent spectral feature at $f = 35\,kHz$ [Fig. 4(b)]. Azimuthal cross-correlation measurements show this to be a $m = 1$ mode propagating in the electron diamagnetic drift direction (which is also the direction of the $\mathbf{E} \times \mathbf{B}$ rotation in our experiment). Correlation measurements show no appreciable axial phase shift for the $m = 1$ mode, and this, together with the observation that the mode has a uniform amplitude along the plasma column, indicates that it is a flute mode. This mode, believed to be a centrifugally-driven Raleigh-Taylor type instability arising from the $\mathbf{E} \times \mathbf{B}$ rotation, is always present in the CLM. As the rf signal level is increased from zero, $\langle \eta_{i\|} \rangle$ increases. When $\langle \eta_{i\|} \rangle$ exceeds some critical value, a new feature in the frequency spectrum as shown in Fig. 4(c). Cross-correlation measurements show that the new mode has an azimuthal mode number $m = 2$. The increase in the electric field increases the frequency of the $m = 1$ peak as a consequence of the larger $\mathbf{E} \times \mathbf{B}$ velocity. As the rf signal level is further increased (with a corresponding increase in $\langle \eta_{i\|} \rangle$, the new $m = 2$ peak increases in amplitude and both peaks move to higher frequencies as shown in Fig. 4(d). We estimate the critical value of η_i that coincides with the onset of the new mode by noting that the $m = 2$ in Fig. 4(c) is close to marginal stability. We therefore estimate the critical value to be $\sim 1.3 - 1.5$ for this case. Using the results of a theory of ITG mode with anisotropic η_i [6], we find the critical $\eta_{i\parallel}$ to be

$$(\eta_{i\parallel})_{\text{crit}} = 1 - \eta_{i\perp} b G$$
$$+ \left\{ (1 - \eta_{i\perp} b G)^2 + 2[1 + \tau_\parallel + S_0(\tau_\parallel - \tau_\perp) + k_\perp^2 \lambda_D^2] - S_0/\xi^{*2} S_0^2 \right\}^{1/2}.$$

In the above equation $\xi \equiv \xi_i \equiv \omega/k_\parallel v_{thi}$, $\tau_\perp \equiv T_e/T_{i\perp}$, $b \equiv k_\perp^2 \rho_i^2$, where ρ_i is the ion Larmor radius, $S_0 \equiv e^{-b} I_0(b)$, $G \equiv 1 - I_1(b)/I_0(b)$, where the I's are the usual modified Bessel functions, λ_D is the electron Debye length, $\xi^* \equiv \omega_e^*/k_\parallel v_{thi}$, where ω_e^* is the electron diamagnetic frequency, and Z is the plasma dispersion function. For the experimental parameter, the above expression yields $(\eta_{i\parallel})_{\text{crit}} \sim 1.3$ which is quite close to the experimental estimate.

As before the $m = 1$ mode is a flute mode with real frequency $\omega_1 \approx \omega_E$, the $\mathbf{E} \times \mathbf{B}$ rotation frequency which will be the Doppler shift for the $m = 2$ mode. Therefore the $m = 2$ ITG mode real frequency ω_2 is given by $\omega_2 = -2\bar{\omega}_1 + \bar{\omega}_2$, where the overbar indicates the Dopler-shifted frequency and positive and negative signs of the angular frequency indicate propagation in the electron and ion diamagnetic drift directions, respectively. The $m = 2$ mode frequency is consistently downshifted from twice the $m = 1$ mode frequency over a wide range of plasma parameters and is typically in the range $|\omega_2|/(2\pi) \sim 6-18\,\text{kHz}$ in the ion diamagnetic drift direction.

A simple linear theory predicts that the mode frequency should depend on the ion temperature and on the value of η_i according to the equation [9]

$$\omega \sim k_\parallel v_{ti} \left(\frac{1 + \eta_i}{2} \right)^{1/2} \tag{1}$$

where $v_{ti} = (2T_i/m_i)^{1/2}$ is the ion thermal speed. Since we can vary both η_i and v_{ti} over a range of values, while maintaining the presence of the mode, we were able to investigate the dependence of the mode frequency on these parameters in order to test he dependence in equation (1). In Fig. 5, we plot the mode frequency (correct for the Doppler shift) against the product $v_{ti}[(1 + \eta_i)/2]^{1/2}$. The data follow the dependence that is expected for the η_i mode, i.e. they show an approximately linear relationship. From equation (1) we expect the slope of the curve to be roughly equal to the axial wavenumber of the mode. The slope of the curve in Fig. 5 is $0.015\,\text{cm}^{-1}$, which is typical of the axial wavenumber measured.

Comparison with theoretical predictions

A non-local dispersion relation (as a radial differential equation) of the slab ITG mode can be easily derived [8]. Using the experimental parameter profiles in the D.C. heating case, a shooting code solution of the radial differential equation has been performed [10]. The boundary conditions are (i) spatial decay at infinity and (ii) a match to the known solution at small radii. With $\lambda_\| \sim 270\,\text{cm}$ which is the experimental value, this analysis shows that the growth rate of the $m = 1$, η_i mode is slightly negative, but for the $m = 2$, η_i mode $\omega_r = 0.98 k_\| \, v_{ti}$, and the growth rate is positive ($\gamma = 0.17\, k_\| \, v_{ti}$). These are consistent with the experimental observation. Furthermore the computed radial mode structure peaks at $r \approx 1.0\,\text{cm}$ which is approximately the location of the maximum gradient of the parallel ion temperature.

Recently, a 3-D gyrokinetic particle simulation of ITG instability with the CLM parameters has been performed [11]. It shows $m = 2$ mode is dominant in a nonlinearly saturated state, which confirms our observation.

3.1.3 PRODUCTION AND IDENTIFICATION OF THE TOROIDAL ITG MODE

In order to investigate the transition of the ITG slab mode to the toroidal mode, we increase the effective magnetic curvature in the experimental cell by energizing the magnetic mirror coil. The magnetic curvature can then be adjusted either by changing the mirror cell length or by varying the mirror coil current. Unfortunately, this latter technique also has the effect of changing the fraction of trapped particles, which can introduce various trapped-particle effects.

In Fig. 6(a) we show the mode spectrum without any mirror. Figure 6(b) shows the effect on the mode amplitude at $R = 1.5\,\text{cm}$ by increasing the mirror ratio R_m from 1.0 to 2.5 at the fixed mirror cell length $\sim 50\,\text{cm}$. The radial mode structure remains basically the same as that without any mirror (14). As the mirror ratio is increased up to $R_m \sim 1.4$, the curvature drive further destabilizes the mode and its amplitude increases. Paradoxically, with further increase in the mirror ratio, the mode amplitude decreases. In addition to affecting the mode amplitude, energizing the mirror coil also affects the mode real frequency as shown in Fig. 3(c). As the mirror ratio increased from one, the mode frequency decreased monotonically.

In order to determine if the new drive results from changes in the plasma parameters, we obtained a comprehensive set of plasma profiles for various values of mirror ratio. We found that the shapes of the T_i and n profiles in the mirror cell were observed to remain roughly invariant with the mirror ratio. Therefore, the η_i drive was essentially unchanged. However, the increase of the mirror ratio substantially changed the trapped particle fraction and it could be up to 60% near the edge of the plasma column. In conclusion, the extra drive for the ITG mode could only arise either from curvature and ∇B drifts effects or trapped particle effects. Parallel wavelength, which is a crucial parameter for the η_i mode, was carefully measured for different mirror ratios and shown in Fig. 7. It was obtained by cross-correlating the signals from two Langmuir probes axially separated but located at the same azimuthal and radial positions. The parallel wavelength hardly changed for the low mirror ratios, but it increased dramatically for mirror ratios $R_m > 1.6$.

To investigate curvature effects, we obtained density fluctuation spectra sensed by a Langmuir probe for a fixed mirror ratio ($R_m = 1.8$) while increasing the length of the mirror cell. This has the effect of decreasing the bounce averaged magnetic curvature experienced by the trapped particles while leaving the trapped fraction essentially unchanged. The results of the mode amplitude and the real frequency vs. the mirror cell length are presented in Fig. 8. By increasing the mirror cell length and consequently reducing the effective curvature, we were able to reduce the amplitude but increase the frequency of the ITG mode. The measurements of the other plasma parameters indicated that the curvature was only the parameter which changed in this set of te experiments. Hence, the magnetic curvature provides an additional drive for the ITG mode and the reduction of the real frequency of the mode with increasing curvature is an indication of its increasing toroidal mode character as discussed later.

The effect of energizing the mirror coil on the ITG mode can be understood in terms of both magnetic curvature and particle trapping effects in the mirror cell. Although the effects of magnetic curvature have been discussed by many authors in the toroidal limit, their results are not entirely applicable to our present study of the transition from the slab to the toroidal branch. The main reason for this is the co-existence of both the slab and toroidal branches in our experiments as explained below.

The ITG mode amplitude decreases with further increase of the mirror ratio beyond $R_m \sim 1.4$ as shown in Fig. 6(b). This can be explained by a model of hybrid ITG-Dissipative-Trapped Electron mode for high mirror ratios and trapped fractions. In the CLM, the trapped particle fraction increases faster than the curvature with the high mirror ratio. The trapped fraction of both electrons and ions varies from 0 to 60% for R_m from 1 to 2.2 while ε_n from 0 to 0.025. With high trapped fractions and moderate collisionality of the CLM, one can expect a Dissipative Trapped Electron Mode dynamics. However, this non-adiabatic electron dynamics will be destabilizing for the ion modes and reduce the ITG mode amplitude as we observed. Furthermore, as the Dissipative trapped Electron mode propagates in the electron diamagnetic drift direction, this dynamics will tend to reduce the magnitude of the real frequency of the ion mode (propagating in the ion diamagnetic direction). This point is corroborated by the rapid decrease of the mode frequency with the high mirror ratio. Lastly, the parallel wavelength is seen to increase with increasing mirror ratio, which is a consequence of the curvature drive and trapped particles to produce a more 'flute'-like mode'. We develop local dispersion relation for this type of mode on the basis of: bounce average trapped ion kinetic response with curvature drive ω_c and ∇B drive ω_{GB}, slab-like transit ion response and bounce average dissipative trapped electron response. The dispersion relation is

$$1 + \tau_\perp + \left(1 - \frac{n_t}{n_0}\right) S_0 \left\{ \eta_\parallel - \tau_\perp - \frac{\eta_{i\parallel} \omega_{*i} \omega}{(k_\parallel v_{thi})^2} \right.$$

$$+ Z\left(\frac{\omega}{k_\| v_{thi}}\right)\left[\frac{\tau_\| \omega}{k_\| v_{thi}} - \frac{\omega_{*i}}{k_\| v_{thi}}\left(1 + \frac{\eta_{i\|}\omega^2}{k_\|^2 v_{thi}^2} - \frac{\eta_{i\|}}{2} - \eta_{i\perp}bG\right)\right]\Big\}$$

$$-\frac{n_t S_0}{n_0}\Big\{\frac{1}{1+i\nu_{\text{eff},i}/\omega}\left[\tau - \frac{\omega_{*i}}{\omega}\left(1 - bG\eta_{i\|}\right)\right]$$

$$+ \frac{\omega_c}{2\omega(1+i\nu_{\text{eff},i}/\omega)^2}\left[\tau - \frac{\omega_{*i}}{\omega}\left(1 + \eta_{i\|} - \eta_{i\perp}bG\right)\right]$$

$$+ \frac{2\omega_{GB}}{\omega(1+i\nu_{\text{eff},i}/\omega)^2}\left[(1-bG)\left(\tau_\perp - \frac{\omega_{*i}}{\omega}(1-\eta_{i\perp})\right)\right.$$

$$\left. + \frac{\eta_{i\|}\omega_{*i}}{\omega}(2 - b - 3bG + 2b^2 G)\right]\Big\}$$

$$-\frac{n_t}{n_0}\left[\frac{\omega+\omega_{*i}}{\omega+i\nu_{\text{eff},e}} - \frac{(\omega_{*i}(1+\eta_e)+\omega)(\tau_\|\omega_c/2 + 2\tau_\perp \omega_{GB})}{(\omega+i\nu_{\text{eff},e})^2}\right] = 0 \quad (2)$$

where n_t/n_0 is the fraction of trapped particles, $\nu_{\text{eff},e} = \nu_e/\varepsilon$ is the effective electron collision frequency, $\nu_{\text{eff},i} = \nu_i/\varepsilon$ is the effective ion collision frequency and other symbols have already been defined. Using the experimental parameters $\tau_\perp = 1$, $\tau_\| = 0.5$, $\eta_{i\perp} = 1.5$, $\eta_{i\|} = 3$, $L_n = 2$(cm), $b = 0.044$, $\varepsilon_n = 0 \sim 0.005$, $\nu_e/k_\| v_{ith} \sim 0.01$ and $k_\| L_n = 0.1 \sim 0.02$, we obtain the result shown in Fig. 9. It is in general agreement with the experimental observations.

3.1.4 MEASUREMENT OF ITG ION THERMAL TRANSPORT

We now discuss the measurement of the anomalous ion thermal conductivity due to the slab branch of the ITG instability discussed in Sec. 2.2. The typical parameters under normal operating conditions with a rf heating are: ion density $n_i \sim 5 \times 10^8 \text{ cm}^{-3}$, neutral pressure in cell region $P_c \approx 5 \times 10^{-7}$ torr, electron temperature $T_e \sim 10\,\text{eV}$, perpendicular ion temperatures $T_{i\perp} \sim 5\,\text{eV}$, parallel ion temperature $T_{i\|} \sim 15\,\text{eV}$, $\eta_{i\|} \sim 6$, $\eta_{i\perp} \leq 1$, magnetic field (experimental cell) $B \approx 1\,\text{kG}$, plasma cell length $L \sim 160\,\text{cm}$, plasma column radius $r_p \sim 3\,\text{cm}$. The ITG instability, produced in the transition region of the Columbia Linear Machine where a sharp temperature gradient in created, travels down the machine with the plasma flow as shown in Fig. 10. The mode is nonlinearly saturated and the ITG turbulence is fully developed by the time it reaches the experimental cell. The ions experience the ITG turbulence scattering as they flow down the machine and enhanced ion thermal transport may be produced. Two ion energy analyzers or Langmuir probes, calibrated against each other, were located at $z = 65\,\text{cm}$ (upstream) and $z = 125\,\text{cm}$ (downstream) (see Fig. 10). The position of the heating mesh was taken as $z = 0\,\text{cm}$. Therefore, the consequent ion particle or thermal transport can be measured by determining the ion density and ion temperature radial profile relaxation downstream.

The transverse particle diffusion (across **B** field lines), which can cause convective thermal transport, was carefully studied by measuring both upstream and downstream ion density profiles. Figure 11 shows these results with the presence of ITG mode. No significant density profile relaxation is observed. This is consistent with the presumption that the electron density response is approximately adiabatic for the ITG mode in CLM. The fluctuations of the density and the radial velocity are 90° out of phase in the adiabatic approximation and no particle transport can result.

The radial ion temperature profiles at $z = 65\,\text{cm}$ (upstream) and $z = 125\,\text{cm}$ (downstream) for the case without the ITG mode are displayed in Fig. 12. It is seen that both the upstream and downstream temperature profiles are roughly the same. With the ITG mode, the temperature relaxation shown in Fig. 13(a) is obvious. The profile relaxation is a result of heat flux directed from the plasma center to the edge, causing the plasma core ($r < 2.0\,\text{cm}$) to cool down and the plasma edge ($r > 2.0\,\text{cm}$) to heat up. Because of no particle transport observed, transverse thermal convection can be neglected. Therefore, the ion thermal transport in this experiment is dominated by the thermal conduction.

We now model the temperature profile relaxation and calculate the transverse ion thermal conductivity $\chi_\perp$. The ion energy transfer along the $\mathcal{B}$ field is basically determined by the plasma flow. As the transverse thermal convection can be neglected in accordance with the discussion above, the radial ion thermal flux is entirely due to the transverse thermal conduction, i.e. $q = -n\chi_\perp \nabla T_{i\|}$, where $q, \chi_\perp$, n, and $T_{i\|}$, are ion thermal flux across the B field line, transverse ion thermal conductivity, ion density, and parallel ion temperature respectively. Then in the steady state the transport equation governing the ion temperature relaxation can be written as follows:

$$nv_f \frac{\partial}{\partial z} T_{i\|} = \frac{1}{r} \frac{\partial}{\partial r} \left(rn\chi_\perp \frac{\partial}{\partial r} T_{i\|} \right).$$

Where v_f is the plasma flow velocity. It is clear from Eq. (3) that with experimental data on ion temperature and ion density profiles, there are two unknown parameters v_f and $\chi_\perp$, only one of which can be solved for. As we are able to measure the flow velocity $v_f \sim 1.5 \times 10^6$ cm/sec, then the solution of Eq. (3) yields $\chi_\perp$. (The measurement of the flow velocity was carried out by launching ion acoustic waves along the magnetic field B, i.e. parallel and antiparallel to the flow velocity.) Since no experimental data are available outside $r = 3.0$ cm, an exponentially decaying function is assumed in this region. The temperature profile at $z = 65$ cm (Fig. 13a) is chosen as the upstream boundary condition. Due to the negligible transverse particle transport, the density at $z = 65$ cm, fitted by an 8th order polynomial, is used in the equation. The thermal conductivity is modeled by a trial function with a number of adjustable parameters such that the calculated temperature profile at $z = 125$ cm is optimally fitted to the experimental downstream temperature profile. The calculated temperature profile is displayed in Fig. 13a (solid curve). The fit of the calculated profile to the experimental profile is fairly good. The corresponding ion thermal conductivity is shown in Fig. 13b. It is noted that the position of maximum thermal conductivity is around the location of maximum temperature gradient. More interestingly, it also corresponds to the peak of the ITG mode (Fig. 14). The radial profile of the ITG mode can be compared with the calculated thermal conductivity which indicated remarkable similarity. The conductivity is small at the center where the fluctuation levels are also low. Therefore it can be concluded that the measured thermal conductivity is due to the ITG mode.

We note that the average thermal conductivity ($\sim 0.5\,\mathrm{m^2/s}$) is much larger than the classical ($\sim 10^{-3}\,\mathrm{m^2/s}$) and less than the Bohm diffusion coefficient ($\sim 6\,\mathrm{m^2/s}$). For the CLM parameters, the quasilinear theory yields $\chi_\perp \sim 0.5\,\mathrm{m^2/s}$, the Kadomtsev strong turbulence estimate gives $\chi_\perp \sim 1\,\mathrm{m^2/s}$, and the Kadomtsev weak turbulence scaling yields $\chi_\perp \sim 0.3\,\mathrm{m^2/s}$.

3.1.5 CONCLUSIONS

In conclusion, both the slab and the toroidal branches of the ITG instabilities have been produced and definitively identified in the modified Columbia Linear Machine. These appear in parameter regimes and bear parametric signatures roughly in accordance with theories. As the relevant plasma parameters in tokamaks are similarly appropriate, one may expect these modes to be present there. The ion thermal transport associated with this mode has been measured and found to be highly anomalous in CLM.

The slab branch was produced by unique heating scheme with biased meshes to yield a peaked ion temperature and flat density profiles. Under these conditions, the parameter η_i exceeds the critical value and a strong instability has been observed. Very similar results have also been obtained by a transit-time rf technique to heat the core of the plasma column. Based on observations of the azimuthal and axial wavelengths, the real frequency, and increasing mode amplitude with increasing inverse-in-temperature-gradient scale length, it is identified as the η_i mode. Computational results from a nonlocal theory of the η_i mode corroborate the existence of an instability with the observed azimuthal mode number, axial wavelength, real frequency, and radial structure. Finally, a 3-D gyrokinetic particle simulation with the CLM parameters has confirmed the basic features of our observations.

The first production and identification of the toroidal ion temperature-gradient (ITG) mode was also achieved by first exciting the slab ITG mode and gradually raising the magnetic curvature in the Columbia Linear Machine. Modest levels of magnetic curvature destabilized and reduced the real frequency of the mode in accordance with the prediction from a simple theory. However, with further increase in mirror ratio and trapped particles, the observed mode amplitude decreased and the parallel wavelength greatly increased. These may be the signatures of a hybrid ITG-Dissipative Trapped Electron mode, as the theoretical predictions based on this mode agree fairly well with the experimental observations.

Lastly, the anomalous ion thermal transport due to the ITG mode has been measured in the CLM by the temperature profile relaxation method. The ITG mode does cause significant anomalous ion thermal transport across magnetic field lines primarily via thermal conduction. The measured local ion thermal conductivity $\chi_\perp(r)$, shows a strong correlation with the ITG mode radial profile, its magnitude is about an order of magnitude lower than Bohm and nearly three orders of magnitude larger than the classical.

Bibliography

[1] D.L. Brower, *et al.*, Phys. Rev. Lett. **59**, 48 (1987).

[2] S.M. Wolfe, *et al.*, Nucl. Fusion **26**, 329 (1986).

[3] R.J. Groebner, *et al.*, Nucl. Fusion **26**, 543 (1986).

[4] T. Antonsen, B. Coppi, and R. Englade, Nucl. Fusion **19**, 641 (1979).

[5] J. Chen, A.K. Sen, R.G. Greaves, and B. Song, "Production and Identification of the Toroidal Branch of the ITG Mode," accepted for publication in Phys. Rev. Lett. Jan. (1993).

[6] B. Song, J. Chen, and A.K. Sen, "Measurement of the Anomalous Ion Thermal Transport due to the Ion Temperature Gradient Driven Instability," submitted for publication to Phys. Rev. Lett., Jan. (1993).

[7] B. Song, A.K. Sen, and P. Tham, "Scaling Behavior of the ITG Ion Transport," submitted to Phys. Fluids B, Feb. (1993).

[8] R.G. Greaves, J. Chen, and A.K. Sen, Plasma Phys. Controlled Fusion **34**, 1253 (1992).

[9] H. Biglari, P.H. Diamond and M.N. Rosenbluth, Phys. Fluids B **1**, 109 (1989).

[10] J. Chen, A.K. Sen, and S. Migliuolo, Phys. Fluids B **4**, 512 (1992).

[11] S. Parker and W. Lee, *Proc. 14th Int. Conf. Numerical Simulation of Plasma*, Annapolis, Maryland, Sept. (1991).

152   ITG Instabilities

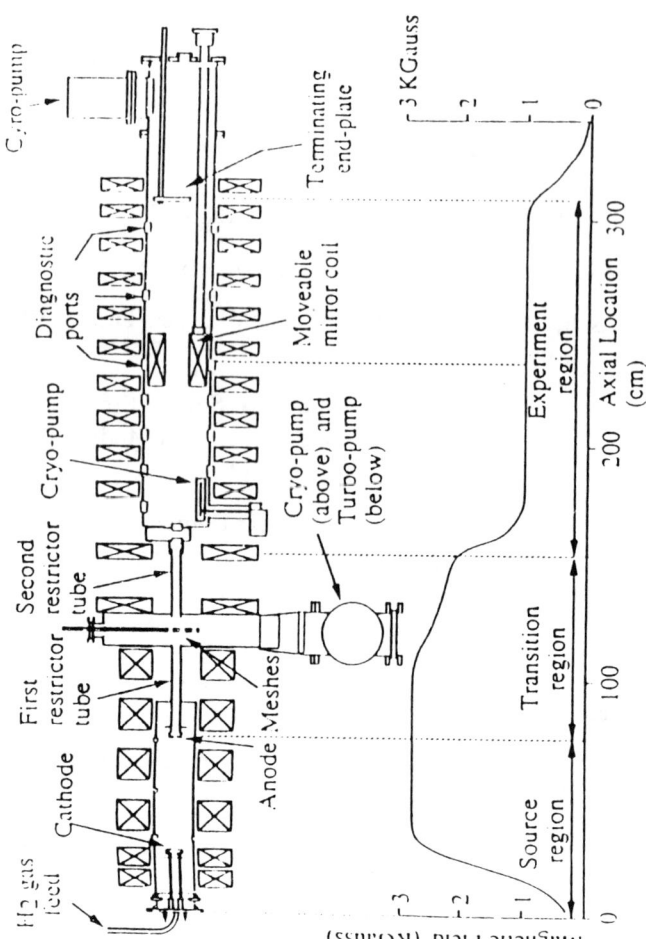

Fig. 1   The Columbia Linear Machine

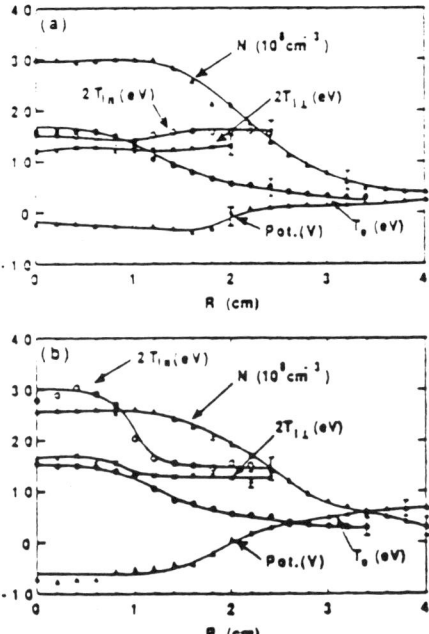

FIG. 2. Radial profiles of plasma parameters. N, $T_{i\parallel}$, $T_{i\perp}$, T_e, and Pot are equilibrium plasma density, parallel ion temperature, transverse ion temperature, electron temperature, and potential, respectively. (a) Profiles without parallel heating of the ions (i.e., zero bias on the accelerating meshes). (b) Profiles with parallel heating of the ions (i.e., −50-V bias on both meshes).

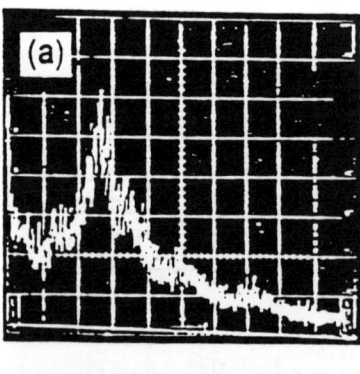
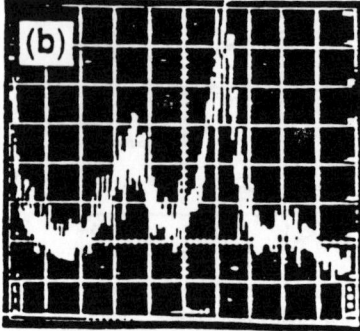

FIG. 3. Density fluctuation spectra. (a) Spectrum without parallel ion heating, corresponding to the profiles of Fig. 2(a). 10 kHz per division. (b) Spectrum with parallel ion heating, corresponding to the profiles of Fig. 2(b). 10 kHz per division.

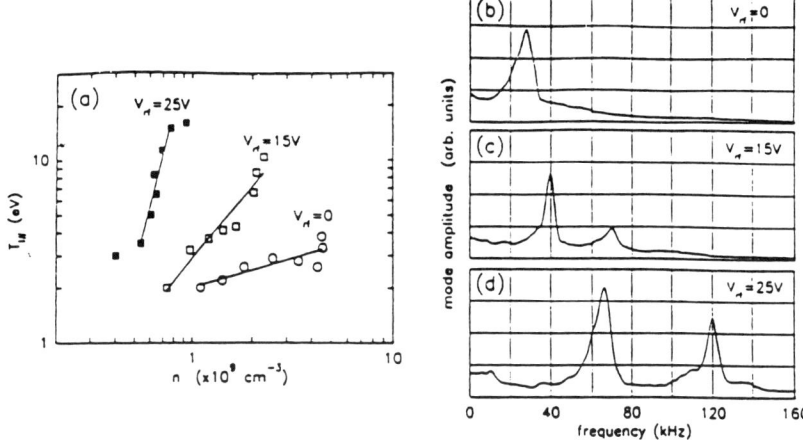

Fig. 4 (a) Parallel ion temperature vs. density showing the increase of $\eta_{i\parallel}$ with r.f. signal level. The respective slopes for the linear regressions are: 0.3, 1.5, and 3.6. (b), (c), (d) Frequency spectra at r = 1.8 cm for V_{rf} = 0, 15, 30 V.

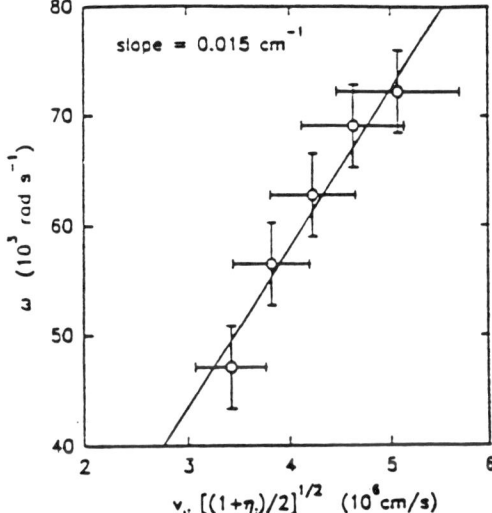

Fig. 5 Mode angular frequency against the product $v_{ti}[(1+\eta_i)/2]^{1/2}$ for different r.f. signal levels at r = 1.8 cm.

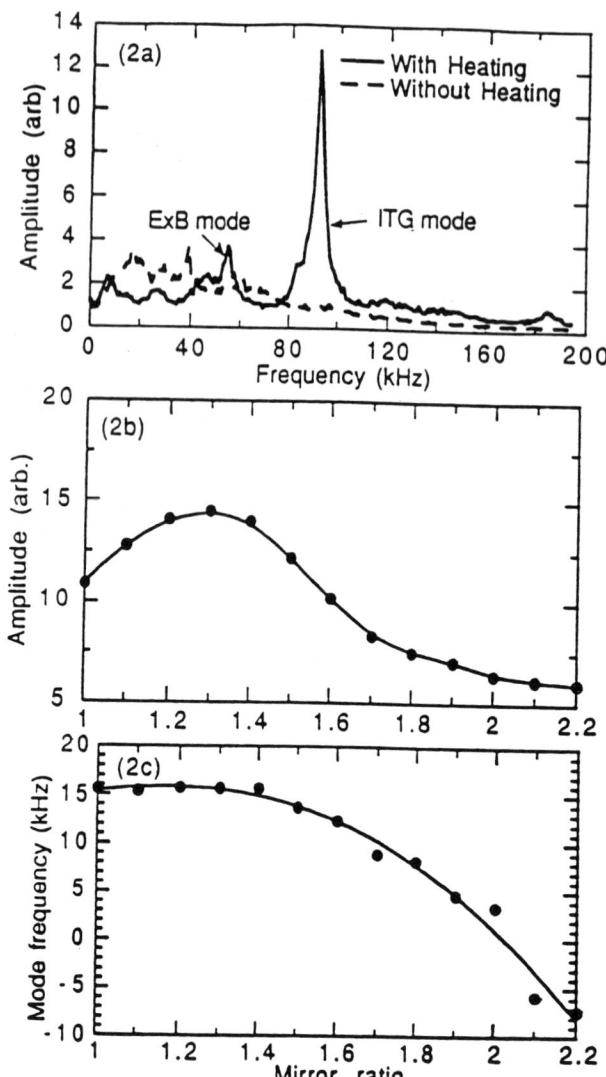

Fig. 6 The transition from the slab to the toroidal ITG mode.
(a) The spectrum of the slab mode (mirror ratio 1)
(b) The amplitude of the mode vs. mirror ratio.
(c) The real frequency of the mode vs. mirror ratio.

A. K. Sen et al. 157

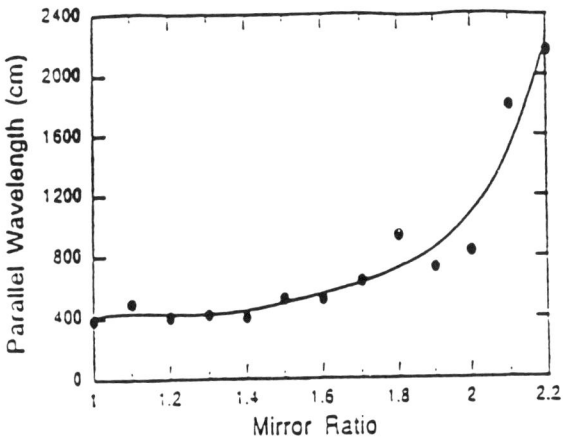

Fig. 7 Parallel wavelength of the ITG mode vs. the mirror ratio at the fixed mirror cell length L_{mir} = 50 cm.

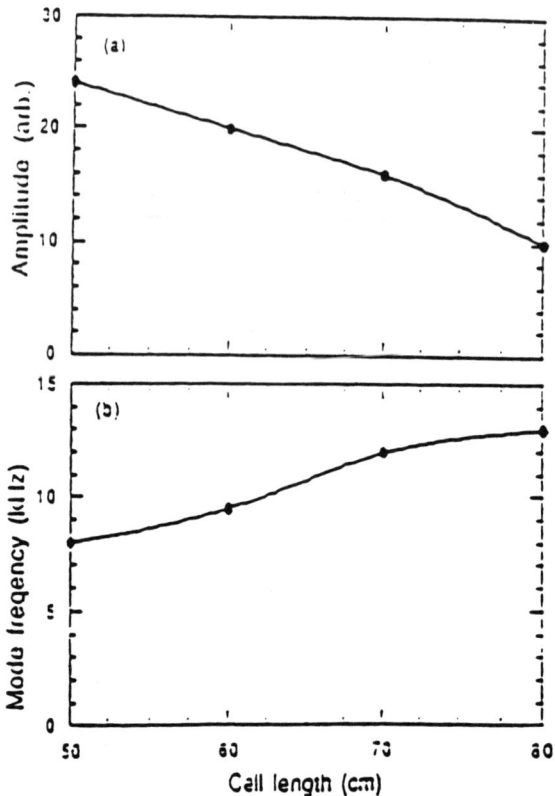

Fig. 8 Parametric dependence of the ITG mode on the mirror cell length at a fixed mirror $R_m = 1.8$. (a) mode amplitude (b) Real frequency.

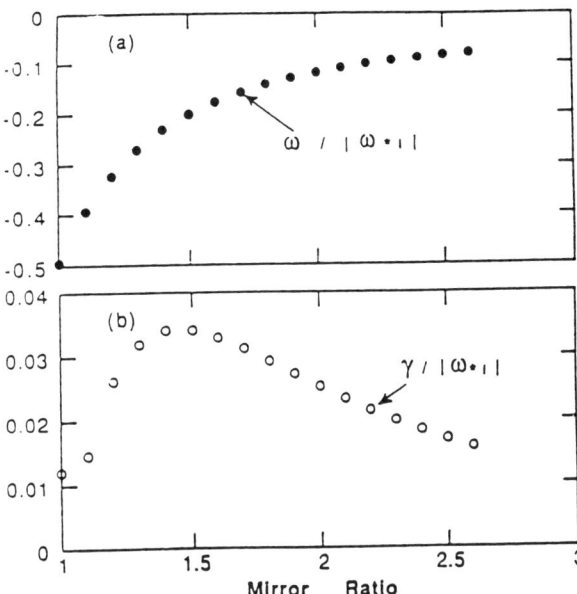

FIG. 9 Growth rate $\gamma/|\omega_{*i}|$ and real frequency $\omega/|\omega_{*i}|$ as a function of mirror ratio.

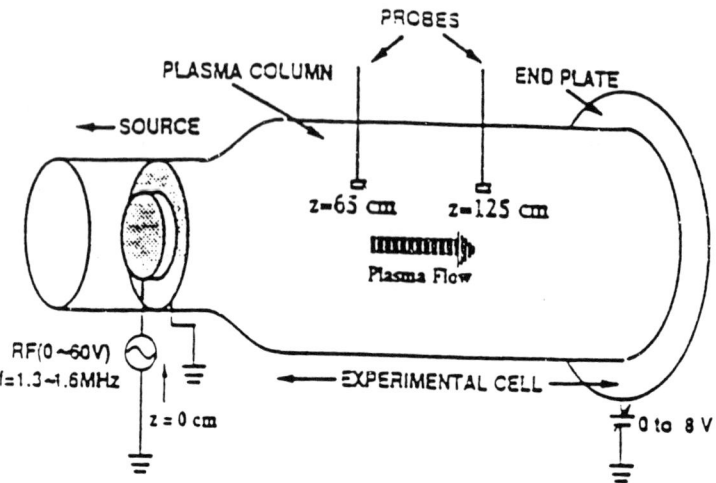

Fig. 10 Schematic of the profile relaxation measurement scheme

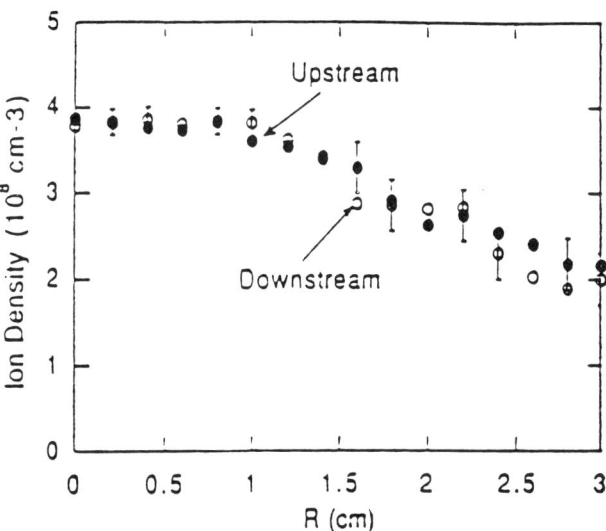

Fig. 11 Upstream and Downstream Ion Density Profiles with heating $V_{rf} = 45$ V

162 ITG Instabilities

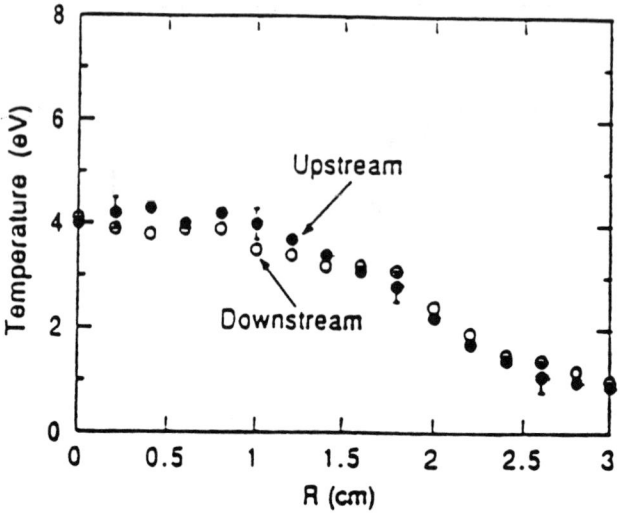

Fig. 12 Upstream and Downstream Ion Temperature Profiles Without Heating

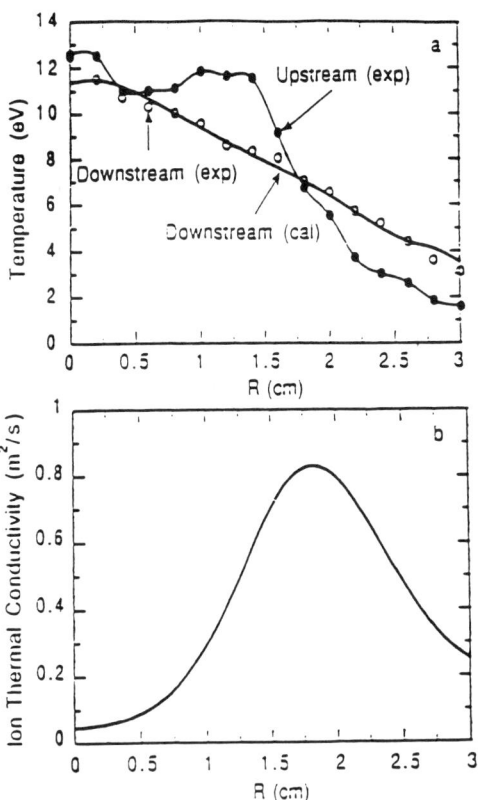

Fig. 13 Measurement of the ion thermal conductivity (slab mode) via profiles relaxation. (a) Upstream and downstream temperature profiles and the best fits from a thermal conduction model. (b) The ion thermal conductivity.

164 ITG Instabilities

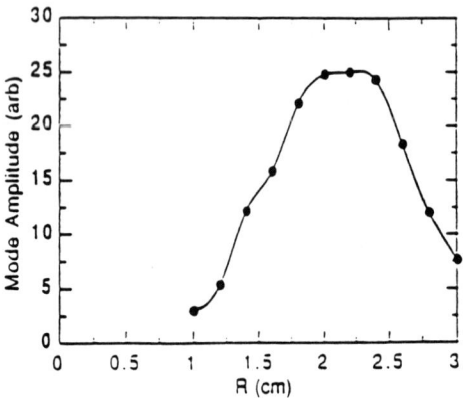

Fig. 14 Measured ITG Mode Radial Profile

3.2 ANOTHER LOOK AT EXPERIMENTAL EVIDENCE FOR ION TEMPERATURE GRADIENT DRIVEN TURBULENCE IN TOKAMAKS

David L. Brower
Department of Electrical Engineering and Institute of Plasma and Fusion Research, University of California, Los Angeles, Los Angeles, California

Abstract. For high-density Ohmic discharges in the TEXT tokamak, a distinct ion mode (i.e. density fluctuations propagating in the ion diamagnetic drift direction) is observed in the microturbulence spectra. The magnitude and spectral characteristics of the mode are identified. Onset of the ion feature occurs at plasma densities where a clear saturation is evident in the global energy confinement time. By injecting pellets, a high-density plasma is created in which the density profile is sharply peaked. Under these conditions the ion feature in the fluctuation spectra is suppressed. Possible connection between this experimentally observed ion mode and the theoretically predicted properties of the ion-temperature-gradient-driven turbulence is explored. Experimental evidence from other devices is also discussed along with a critical evaluation of interpretation issues.

3.2.1 INTRODUCTION

Previous measurements on the Texas Experimental Tokamak (TEXT) have shown that the global energy confinement time (τ_E) saturates with increasing plasma density [1]. This trend, which has been observed in numerous Ohmically-heated tokamaks, can be attributed to enhancement of the ion thermal loss channel over the electron channel at sufficiently high densities. If the ion losses are taken to be governed by neoclassical transport, then transport modeling calculations provide predictions for the density at which τ_E saturates [2]. However, in many discharges refueled by gas puffing (e.g. DOUBLET-III [3], ALCATOR-C [4], and ASDEX [5]), the saturation of τ_E is actually observed to occur at densities considerably lower than the ion neoclassical estimates. As first proposed by Coppi *et al.* [6], this "anomalous" behavior could be caused by the onset of enhanced ion thermal transport due to the excitation of ion-temperature-gradient-driven (η_i mode) instabilities [7]. Using models based on the presence of these modes, transport code calculations have yielded results in reasonable agreement with the density saturation observed in a large number of tokamak experiments [8, 9, 10]. In addition to these high-density Ohmic cases, evidence for anomalous ion thermal transport has also been found in tokamaks heated by neutral beam injection (NBI). Specifically, charge-exchange ion temperature profile measurements on NBI-heated DOUBLET-III discharges indicate that the ion thermal diffusivity is not only much larger but also exhibits a radial dependence dramatically different from that predicted by ion neoclassical theory [11]. Experimental results from ASDEX have shown significant improvement in the global energy confinement time for discharges with pellet injection [12], neutral-beam counterinjection [13], and reduced ratio of neutral-gas fueling to recycling [5]. Strongly peaked density profiles are the common feature of all these operating regimes and transport code calculations indicate the increase in energy confinement time may be due to stabilization of the ion-temperature-gradient-driven instability.

The preceding discussion provides ample motivation to search for direct experimental evidence of ion-temperature-gradient-driven instabilities. Along with anomalous ion thermal transport, theoretical analyses make specific predictions regarding the nature of microturbulence characterizing such modes. If the parameter $\eta_i = L_{n_i}/L_{T_i} = \partial \ln T_i/\partial \ln n_i$ exceeds a threshold value, $\eta_{i|_{\text{crit}}} = 1$ to 2, then drift-type microinstabilities propagating in the *ion* diamagnetic drift direction are predicted to be present with a typical range of $k_\perp \rho_i \leq 1$. This is essentially the same $|k|$-space characteristic of the usual electron drift wave turbulence with $k_\perp$ being the wave vector of the density fluctuation in the plane perpendicular to the toroidal magnetic field and ρ_i being the ion gyroradius.

In this paper, experimental evidence for the existence on ion-temperature-gradient-driven turbulence will be presented. The main result is that an ion mode (i.e. microturbulence propagating in the ion diamagnetic drift direction) has indeed been experimentally observed [14] along with the familiar density fluctuations propagating in the electron diamagnetic drift direction [15, 16, 17]. A unique multichannel far-infrared laser scattering system is employed to measure the entire $S(k_\perp,\omega) \propto \tilde{n}^2(k_\perp,\omega)$ spectra during a single discharge. Implementation of a heterodyne receiver allows the wave propagation direction of simultaneous counter-propagating modes to be resolved. The appearance and growth of the ion mode can be directly correlated with degradation of the global energy confinement time. For high-density pellet-fueled discharges, the density profile is sharply peaked on axis thereby reducing the density scale length (L_n). Under these conditions, the ion feature in the fluctuation spectra is no longer prominent and transport code calculations indicate that η_i is reduced beneath the threshold for instability. Experimental evidence supporting ion-temperature-gradient-driven turbulence from other devices will also be included. In addition to the primary experimental results, issues relating to the interpretation of the turbulence data are addressed. The effects of a radial electric field, edge shear layer, and scattering sample volume size are examined with respect to their potential influence on the turbulence spectra.

3.2.2 EXPERIMENTAL OBSERVATIONS

The density dependence of the global energy confinement time for hydrogen gas-fueled TEXT discharges with $B_T = 2.8\,\text{T}$ and $I_p = 300\,\text{kA}$ is shown in Fig. 1. A clear saturation is evident with increasing density. If an ion-temperature-gradient-driven instability is active, one would expect to observe turbulence propagating in the ion diamagnetic drift direction in the same region of $|k|$-space as the electron drift wave type turbulence. In order to investigate this possibility, we examine the microturbulence poloidal frequency spectra, as measured via collective far-infrared scattering [18], for low ($\bar{n}_e \simeq 2 \times 10^{13}\,\text{cm}^{-3}$), medium ($\bar{n}_e \simeq 4 \times 10^{13}\,\text{cm}^{-3}$), and high ($\bar{n}_e \simeq 8 \times 10^{13}\,\text{cm}^{-3}$) density plasmas, as seen in Fig. 2. The low density regime corresponds to conditions where τ_E is linear with varying density while the medium density discharge represents the transition region to high density where τ_E is saturated. Scattering volumes at each k_θ are located above the midplane along a vertical chord at the major radius $R = 1\,\text{m}$. The heterodyne receiver system permits resolution of the wave propagation direction with negative (positive) frequency corresponding to the electron (ion) diamagnetic drift direction in the laboratory frame of reference. For each wave vector examined, there is a distinct large-amplitude broadband peak ($\Delta\omega/\bar{\omega}$) at negative frequencies which shifts to higher (more negative) frequency as k_θ increases indicating a dispersion for the fluctuations [15, 17]. In contrast, at low plasma densities [Fig. 2(a)], only a low-level contribution to the total scattered power is observed from microturbulence propagating in the ion diamagnetic drift direction which appears to peak at approximately zero frequency for all k_θ. However, at medium plasma densities [Fig. 2(b)], a small-amplitude non-zero peak is observed at positive frequencies in addition to the microturbulence propagating in the electron diamagnetic drift direction.

Finally, at the highest plasma densities [Fig. 2(c)], microturbulence propagating in the ion diamagnetic drift direction is seen to exist with magnitude comparable to that of its negative frequency counterpart and a clear dispersion is observed. The ion feature is observed in the same k_θ space as the electron drift wave type fluctuations (i.e. $1 \leq k_\theta \leq 12\,\text{cm}^{-1}$) with $0.1 \leq \bar{k}_\theta \rho_s \leq 0.5$, where $\bar{k}_\theta = \sum_k k S(k) / \sum_k S(k)$, integrated over all frequencies, and ρ_s is the ion gyroradius times $(T_e/T_i)^{1/2}$. These are the characteristic signatures of ion-temperature-gradient-driven turbulence. Since the fluctuations are measured in the laboratory frame of reference, it should be noted that plasma rotation effects resulting from $\mathbf{E} \times \mathbf{B}$ drifts induced by a negative plasma potential serve to shift the spectra (both the electron and ion features) to more negative frequencies (i.e. electron diamagnetic drift direction) [19]. Raising the plasma density acts to broaden $n_e(r)$ and increase the density scale length in the confinement zone which may drive η_i above the threshold for instability [14]. The appearance of the ion feature in the fluctuation spectra occurs simultaneously with the saturation of τ_E.

By injecting pellets into the plasma, it is possible to obtain high-density discharges with sharply peaked $n_e(r)$ when compared to a gas-fueled equivalent, as shown in Fig. 3. The density limit is extended from $\bar{n}_e \simeq 6 \times 10^{13}\,\text{cm}^{-3}$ to $\bar{n}_e \simeq 8 \times 10^{13}\,\text{cm}^{-3}$ with supplemental pellet fueling when $I_p = 250\,\text{kA}$ and $B_T = 2.8\,\text{T}$. Peaking of the density profile continues up to approximately 50 ms following injection after which a slow decay occurs. Strongly increasing $n_e(r)$ on axis reduces the density scale length L_n and can potentially drive η_i below the

critical level for instability. Such a scenario has often been invoked
to explain the improved energy confinement observed for high-density
pellet-fueled discharges [12, 20]. To be consistent with fluctuation measurements, one would then expect the ion feature to be significantly
reduced for the pellet-fueled case. Figures 4(a) and 4(b) show the fluctuation frequency spectra at times before ($\bar{n}_e \simeq 3 \times 10^{13}\,\text{cm}^{-3}$) and after
($\bar{n}_e \simeq 3 \times 10^{13}\,\text{cm}^{-3}$) pellet injection. For each case, the fluctuations
are observed to propagate predominantly in the electron diamagnetic
drift direction suggesting contributions from the η_i instability are small.
Scattering measurements have been made in the range 15 to 50 ms after
the pellet injection and the results are similar. In contrast, a comparable high-density gas-fueled discharge ($\bar{n}_e \simeq 6 \times 10^{13}\,\text{cm}^{-3}$) contains
appreciable levels of the turbulence propagating in both directions, as
shown in Fig. 4(c). The largest changes in the density scale length
occur away from the plasma edge (see Fig. 3) indicating the ion mode,
which is not observed for the pellet-fueled discharge, would be expected
in this region of the plasma.

In addition to hydrogen plasmas, fluctuation measurements were
made at low and high densities in deuterium and helium plasmas as
shown in Fig. 5. Results in deuterium are similar to that for hydrogen
with a prominent ion mode only observed at the higher densities. In
the case of the helium discharge, a distinct ion mode is not observed in
the fluctuation spectra, even at the highest densities. At present, it is
unclear whether the differences in the fluctuation spectra are due to the
change in the mass-charge ratio or to the change in the plasma profiles.
The measured energy confinement time for the helium plasma is larger
than that for the hydrogen plasma after radiation is subtracted.

The high-density operational limit for the TEXT tokamak is shown
in Fig. 6, where the maximum chord-averaged electron density $\bar{n}_{e,\text{max}}$
is plotted against the plasma current I_p (i.e. Hugill plot) for hydrogen
gas-fueled ohmic discharges with $B_T = 2.8\,\text{T}$. The high-density limit on
TEXT has been extended for two special classes of discharges. These
are helium gas-fueled and hydrogen pellet-fueled plasmas. It is important to note that for both of these discharges, the ion feature in
the fluctuation spectra is suppressed [21]. This correlation once again
supports the notion that the measured fluctuations may be responsible for the changes in transport. In addition, long-time precursors of

order five times the global energy confinement time are observed for disrupting plasmas at the high-density limit on TEXT [22]. These precursors are reflected in increased electron particle and heat transport along with the density fluctuation level. They occur well in advance of any measurable changes in Mirnov activity which typically begins less than one energy confinement time before the discharge termination. For such disruptions, no precursor is observed in the radiated power or its profile. Turbulence changes are manifested in the form of an increased ion feature. Enhanced microturbulence is proposed as the physical mechanism for the confinement degradation and subsequent disruption according to the scenario suggested by Greenwald et al. [23].

3.2.3 INTERPRETATION ISSUES

In the preceding section, the primary experimental evidence for the existence of ion-temperature-gradient-driven turbulence in a tokamak was presented. In this section, specific issues pertaining to the interpretation of these data will be addressed.

Effect of radial electric field

As mentioned earlier, variation in the turbulence mean frequency as a function of wavevector provides a measure of the statistical dispersion. This dispersion, $\bar{\omega}_k$ versus k_θ, for a low and high density discharges is plotted in Fig. 7. The mean frequency for a particular wave vector $\bar{\omega}_k = \sum_\omega \omega S_k(\omega) / \sum_\omega S_k(\omega)$, where the summations are over $-1\,\text{MHz} \leq \omega/2\pi \leq -30\,\text{kHz}$ for fluctuations propagating in the electron diamagnetic drift direction and $30\,\text{kHz} \leq \omega/2\pi \leq 1\,\text{MHz}$ for fluctuations propagating in the ion diamagnetic drift direction. Truncation of the frequency spectra at large negative frequencies, particularly evident for high k_θ at low $\bar{n}_e$ (see Fig. 2), results in an underestimate of $\bar{\omega}_k$. However, this is a small effect causing only minimal changes to the plotted dispersions. For $\bar{n}_e = 2 \times 10^{13}\,\text{cm}^{-3}$, the dispersion of the electron feature is linear with phase velocity $v_{ph,e} \simeq -4.2 \times 10^5\,\text{cm/s}$. Negative here simply denotes the electron diamagnetic drift direction in the laboratory frame of reference. The ion feature is small and has no clear dispersion. At $\bar{n}_e = 8 \times 10^{13}\,\text{cm}^{-3}$, $v_{ph,e}$ is reduced to

$v_{ph,e} \simeq -3.4 \times 10^5$ cm/s and shows evidence of a roll-over at large k_θ while the ion feature phase velocity is $v_{ph,i} \simeq 7 \times 10^4$ cm/s. The phase velocities consist of both mode propagation and bulk plasma rotation with respect to the laboratory frame.

Previous measurements on TEXT have shown that at low densities, where the radial electric field can be determined by a heavy-ion-beam-probe diagnostic [24], E_r accounts for most of the shift in $\overline{\omega}_k$ from zero [25]. The remaining shift is consistent with expectations for electron drift waves. For electron drift waves at large k_θ, i.e. $k_\theta \geq 6\,\mathrm{cm}^{-1}$, essentially all of the shift in $\overline{\omega}_k$ is due to E_r as a consequence of finite ion mass and temperature effects [25]. Applying this result to the high-density data of Fig. 2 would imply that $\omega_{k,\mathrm{ion}} \gg \omega_i^*$, the expected mode frequency for an ion-temperature-gradient-driven turbulence. If the measured electron and ion features come from the same region of the plasma, $|\overline{\omega}_{k,e} - \overline{\omega}_{k,\mathrm{ion}}|$ is independent of E_r. However, since $|\overline{\omega}_{k,e} - \overline{\omega}_{k,\mathrm{ion}}| > \omega_e^* \simeq \omega_i^*$, the measured phase velocity for the ion feature appears to be larger than expectations for an η_i instability. It should be pointed at that E_r is not actually measured for the high density discharges where the ion feature is observed and it is not certain that the spatial distribution of the two modes are the same [27]. Hence, resolution of this connundrum awaits further investigation of these parameters.

On TEXT, it has been observed that a large-amplitude quasi-coherent mode (i.e. narrowband in ω and $k_\perp$) exists in addition to the broadband turbulence on the high field side of the tokamak at large wavenumber, $k_\perp > 4\,\mathrm{cm}^{-1}$, as shown in Fig. 8 [16, 17]. The quasi-coherent mode is always seen at at a lower frequency than the ubiquitous broadband turbulence. If the E_r profile is taken to be poloidally symmetric, which seems a reasonable assumption, the broadband feature is centered at ω_e^* in the plasma frame thereby implying that the quasi-coherent feature may actually be propagating in the ion diamagnetic drift direction. At present, this conclusion is speculative and requires further investigation. The quasi-coherent mode is observed in some form for all plasma conditions on TEXT.

Effect of shear layer

Langmuir probe measurements in the extreme edge and scrape-off regions of TEXT observe a shear layer outside of which the fluctuations propagate in the ion drift direction [19]. This is due to the $\mathbf{E} \times \mathbf{B}$ Doppler shift which changes sign across the shear boundary and is located at the limiter under all conditions. Although the extended interaction length of the scattering volume encompasses this narrow layer (about 2 cm), it is not thought that these fluctuations contribute significantly to the scattering observations of the ion feature for several reasons. First, the phase velocity of the ion feature measured by scattering does not agree with that observed by the probes for fluctuations in the limiter shadow [19]. Secondly, the measurement of the $\tilde{n}$ distribution in the limiter shadow at low and high densities does not exhibit the qualitative change necessary to make the ion feature prominent at high density while being negligible at low density. This is shown in Fig. 9. In addition, since the scrape-off plasma involves a very narrow region compared to the entire scattering volume length and since the density is low there, it is unlikely that fluctuations from this region could dominate the measured spectra. Finally, it is crucial to note that the Langmuir probes see no change in the fluctuation characteristics between high-density gas and pellet fueled discharges [see Fig. 9]. This clearly indicates that the ion feature seen by the scattering and Langmuir probe diagnostics are from different regions of the plasma.

Effect of sample volume size

The extended interaction volume lengths and limited spatial scans available with a heterodyne receiver have prevented accurate measurements of the spatial localization of the ion mode. However, it is known that the ion feature exists in the plasma core as well as the edge through use of two different techniques. First, by perturbing the plasma potential through the use of an applied resonant field, the direction of E_r can be changed over an appreciable region of the plasma edge. Shifts in the density fluctuation frequency spectra can be associated with the changes in E_r allowing one to unambiguously resolve fluctuations in the plasma core and edge for $k_\theta \geq 7\,\mathrm{cm}^{-1}$ [26]. Second, through application of a new two-sample-volume cross-correlation technique [27], it appears that the ion and electron features in the turbulent spectra have different spatial distributions.

Turbulent transport

Earlier, it was pointed out that the presence of an ion feature in the measured fluctuation spectrum is consistent with theoretical predictions for the existence of ion-temperature-gradient-driven turbulence. Moreover, the observed confinement properties for gas-fueled plasmas appear to exhibit significant degradation when the ion feature is present. These arguments are by no means rigorously conclusive in establishing a connection between the measured ion feature and the theory of η_i-mode turbulence. To do so would require actual measurement of the parameter η_i, which in the absence of n_i or Z_{eff} profile measurements, was not feasible for the TEXT device. Unfortunately, it is extremely difficult to accurately determine χ_e and χ_i from power balance analysis of high-density ohmic TEXT discharges due to the collisonal coupling of electron and ions as well as the analysis of the source term. In addition, the fluctuation measurements are only able to characterize the turbulence and do not directly measure the induced transport. Consequently, we are limited to examining correlations between the appearance and amplitude of the ion feature and changes in global transport. Nevertheless, it is quite encouraging that there are no obvious apsects of the experimental results which appear to directly contradict this relationship. It is important to point out that even if ion-temperature-gradient-driven turbulent transport could not explain the measured transport, this instability could still exist in the plasma. Likewise, theoretical predictions for the existence of the instability and the turbulent transport resulting from it are quite independent. In fact, the theoretical treatment and simulations of this instability are still an active area of research that is continually evolving. Hence, it would be premature to conclude from these results whether the measured ion feature verifies or disproves ion-temperature-gradient-driven instability theory.

Review of results from other devices

In addition to the measurements made on TEXT, there have also been experiments on other devices which have provided evidence for ion-temperature-gradient-driven turbulence. For instance, it was noted on ALCATOR-C that the saturation in the global energy confinement correlated with a change in the turbulence group velocity from electron to ion diamagnetic drift direction in the laboratory frame of reference [28]. This change in direction was thought to be due to either an η_i mode or a shear layer effect. Measurements to resolve this issue were never carried out as the machine was shutdown. On JIPP T-IIU, an increase in turbulence propagating in the ion diamagnetic drift direction was also observed at high density. Here, the saturation in τ_E was attributed to ion-temperature-gradient-driven turbulent transport and the increased ion feature was put forward as evidence for this instability [29]. Recent experiments on TORE SUPRA found no evidence for an ion feature in the turbulent spectra at high density when τ_E saturated [30]. However, the saturation in τ_E was attributed to changes in χ_e and not transport in the ion channel. Consequently, one would not expect to observe an ion-temperature-gradient-driven instability in the turbulent spectra. On larger machines such as DIII-D, where a collective scattering system similar to that on TEXT is available, strong spatial variations in the E_r profile broaden the turbulence spectra making identification of a distinct ion mode very difficult. At present, this is an active area of research.

3.2.4 CONCLUSION

In conclusion, a distinct ion mode has been observed in the microturbulence frequency spectra occuring simultaneously with the high-density saturation of the global energy confinement time. The ion feature is seen in the same k_θ space as electron drift wave type fluctuations, possessing the characteristic signatures of ion-temperature-gradient-driven turbulence. There has been no direct identification of the drive mechanism responsible for the ion feature. However, for high-density pellet-fueled discharges, the density profile is sharply peaked on axis thereby reducing the density scale length (L_n). Under these conditions, the ion feature in the fluctuation spectra is no longer prominent, demonstrating that the density profile can be actively used to control the instability.

At high densities, where the ion feature in the fluctuation spectra is strongest, agreement between the microinstability-based transport model and experimentally measured values of the global energy confinement time is realized when anomalous ion effects due to the η_i-mode instability are included [21]. No direct measure of the transport induced by the ion feature has been made. The measured properties of transport and fluctuations appear to be clearly linked to those theoretically predicted for the η_i instability. The scaling of transport and turbulence with mass and ion charge is not presently understood and remains an active area of research.

Even if the observed ion feature does not account for the measured transport, the ion-temperature-gradient-driven instability can still be present in the plasma. Theoretical problems in determining the nature of the instability and the resulting transport are far from resolved. If one remains unconvinced by the evidence put forward for ion-temperature-gradient-driven turbulence in the analysis presented in this work, the question which needs to be resolved is "what instability is responsible for this new mode in the turbulence spectra?"

ACKNOWLEDGMENTS

This research is supported by the U.S. Department of Energy under grants No. DE-AC05-78ET-53043 and DE-FG03-86ER-53225 (Task VIIA).

Bibliography

[1] R.V. Bravenec, K.W. Gentle, P.E. Phillips, *et al.*, Plasma Physics and Controlled Fusion **27**, (1985) 1335.

[2] Waltz, R.E., and Guest, G.E., Phys. Rev. Lett. **42**, (1979) 651.

[3] Ejima, S., Petrie, T.W., Riviere, I.C., Angel, T.R., Armentrout, C.J., *et al.*, Nucl. Fusion **22**, (1982) 1627.

[4] Wolfe, S.M., Greenwald, M., Gandy, R., Granetz, R., Gomez, C., *et al.*, Nucl. Fusion **26**, (1986) 329.

[5] Soldner, F.X., Muller, E.R., Wagner, F., Bosch, H.S., Eberhagen, A., *et al.*, Phys. Rev. Lett. **61**, (1988) 1105.

[6] Coppi, B., Cowley, S., Detragiache, P., Kulsrud, R., Pegoraro, F., Tang, W., in *Proceedings of the Tenth International Conference on Plasma Physics and Controlled Nuclear Fusion Research, London, 1984* (International Atomic Energy Agency, Vienna, 1985), Vol. 2, p. 93.

[7] Rudakov, L.I., and Sagdeev, R.Z., Dokl. Akad. Nauk SSSR **138**, (1961) 581 [Sov. Phys. Dokl. **6**, (1961) 415]; Coppi, B., Rosenbluth, M.N., and Sagdeev, R.Z., Phys. Fluids **10**, (1967) 582.

[8] Romanelli, F., Tang, W.M., and White, R.B., Nucl. Fusion **26**, (1986) 1515.

[9] Dominguez, R.R., and Waltz, R.E., Nucl. Fusion **27**, (1987) 65.

[10] Redi, M.H., Tang, W.M., Efthimion, P.C., Mikkelsen, D.R., and Schmidt, G.L., Nucl. Fusion **27**, (1987) 2001.

[11] Groebner, R.J., Pfieffer, W.W., Blau, F.P., Burrel, K.H., Fairbanks, E.S., and Seraydarian, R.P., Nucl. Fusion **26**, (1986) 543.

[12] Kaufmann, M., Buchl, K., Fussmann, G., Gehre, O., Grassie, K., et al., Nucl. Fusion **28**, (1988) 827.

[13] Gehre, O., Gruber, O., Murmann, H.D., Roberts, D.E., Wagner, F., et al., Phys. Rev. Lett. **60**, (1988) 1502.

[14] Brower, D.L., Peebles, W.A., Kim, S.K., Luhmann, Jr., N.C., Tang, W.M., and Phillips, P.E., Phys. Rev. Lett. **59**, (1987) 48.

[15] Brower, D.L., Peebles, W.A., Luhmann, N.C., and Savage, R.L., Phys. Rev. Lett. **54**, (1985) 689.

[16] Brower, D.L., Peebles, W.A., and Luhmann, Jr., N.C., Phys. Rev. Lett. **55**, 2579 (1985).

[17] Brower. D.L., Peebles, W.A., and Luhmann, Jr., N.C., Nucl. Fusion **27**, (1987) 2055.

[18] Brower, D.L., H.K. Park, W.A. Peebles, and N.C. Luhmann, Jr., "Multichannel Far-Infrared Collective Scattering System for Plasma Wave Studies," chapter in the Academic Press (New York) series on *Topics in Millimeter Wave Technology*, Vol. II, Chapter 3, pp. 83-172, ed. by K.J. Button, 1988.

[19] Ritz, Ch.P., Brower, D.L., Rhodes, T.L., Bengtson, R.D., Levinson, S.J., Luhmann, N.C., Peebles, W.A., and Powers, E.J., Nuclear Fusion **27**, (1987) 1125.

[20] Greenwald, M., Qwinn, D., Milora, S., Parker, J., and Wolfe, S., et al., Phys. Rev. Lett. **53**, (1984) 352.

[21] D.L. Brower, et al., Nuclear Fusion **29**, 1247 (1989).

[22] D.L. Brower, et al., Phys. Rev. Lett. **67**, 200 (1991), Phys. Rev. Lett. **68**, 891 (1992).

[23] M. Greenwald et al., Nucl. Fusion **28**, 2199 (1988).

[24] Schoch., P.M., Carnevali, A., Conner, K.A., Crowley, T.P., Forster, J.C., Hickok, R.L., Lewis, J.F., and Schatz, J.G., Rev. Sci. Instrum. **59**, 1646 (1988).

[25] Yu, C.X., D.L. Brower, S.J. Zhao, *et al.*, Phys. Fluids B **4**, 381 (1992).

[26] Brower, D.L., C.X. Yu, S.J. Zhao, W.A. Peebles, N.C. Luhmann, Jr., X.Z. Yang, P.M. Schoch, and R.L. Hickok, Rev. Sci. Instr. **61**, 3019-3021 (1990).

[27] Brower, D.L., W.A. Peebles, C.L. Rettig, and C.X. Yu, Rev. Sci. Instr. **63**, 4637-4639 (1992).

[28] R.L. Watterson, R.E. Slusher, and C.M. Surko, Phys. Fluids **28**, 2857 (1985).

[29] K. Kawahata *et al.*, in *Proceedings of the Twelfth International Conference on Plasma Physics and Controlled Nuclear Fusion Research, 1988* (Nice, France) (International Atomic Energy Agency, 1989), IAEA-CN-50/A-V-3-1, p. 287.

[30] X. Garbet, *et al.*, in *Controlled Fusion and Plasma Physics* (Proc. 19th Eur. Conf. Innsbruck, 1992), Vol. 16C, Part I, 107.

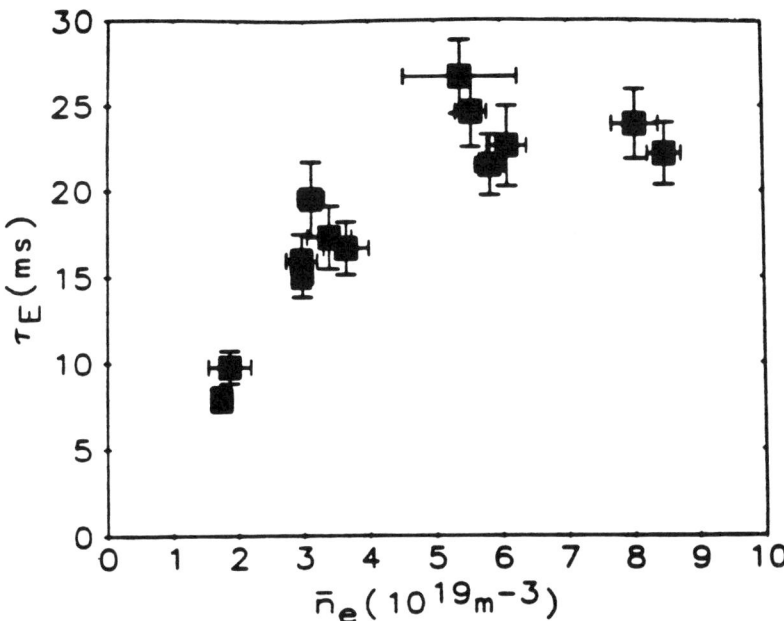

Figure 1. Density dependence of the global energy confinement time for gas fueled TEXT discharges; $I_p = 300\,\text{kA}$, $B_T = 2.8\,\text{T}$.

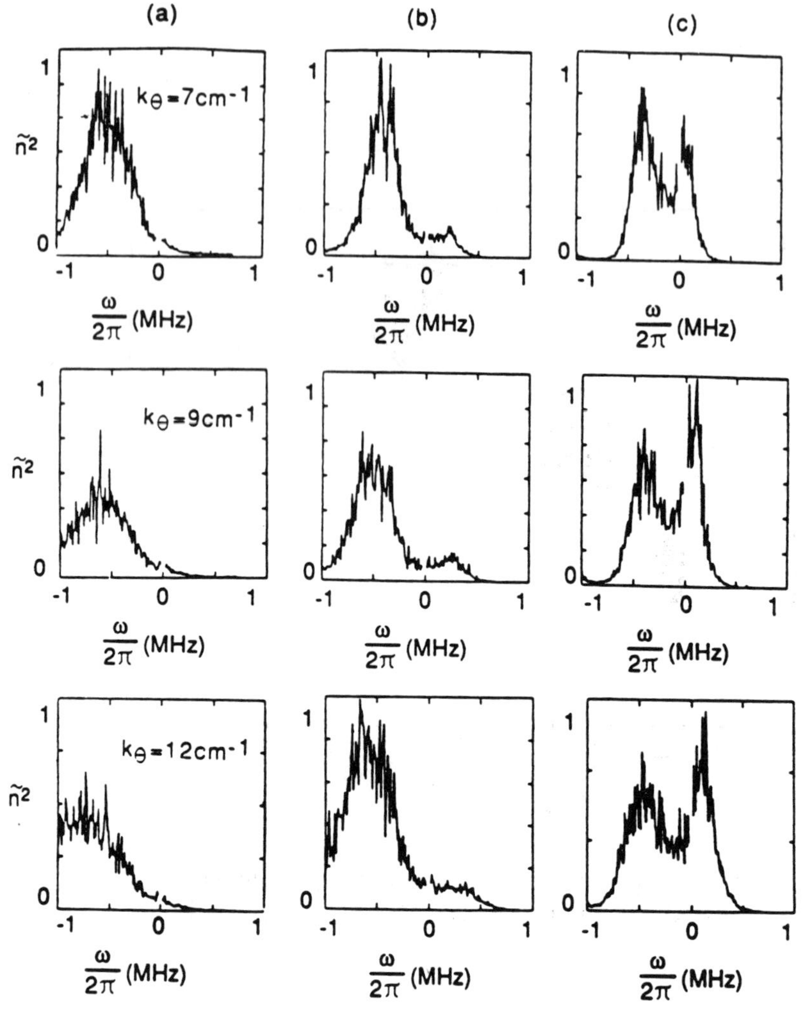

Figure 2. Microturbulence poloidal frequency spectra for $k_\theta = 7, 9$, and $12\,\text{cm}^{-1}$; $I_p = 350\,\text{kA}$, $B_T = 2.8\,\text{T}$, and (a) $\bar{n}_e = 2 \times 10^{13}\,\text{cm}^{-3}$, (b) $\bar{n}_e = 4 \times 10^{13}\,\text{cm}^{-3}$, and (c) $\bar{n}_e = 8 \times 10^{13}\,\text{cm}^{-3}$. Negative (positive) frequency corresponds to the electron (ion) diamagnetic drift direction in the laboratory frame of reference. Vertical axes are in arbitrary units.

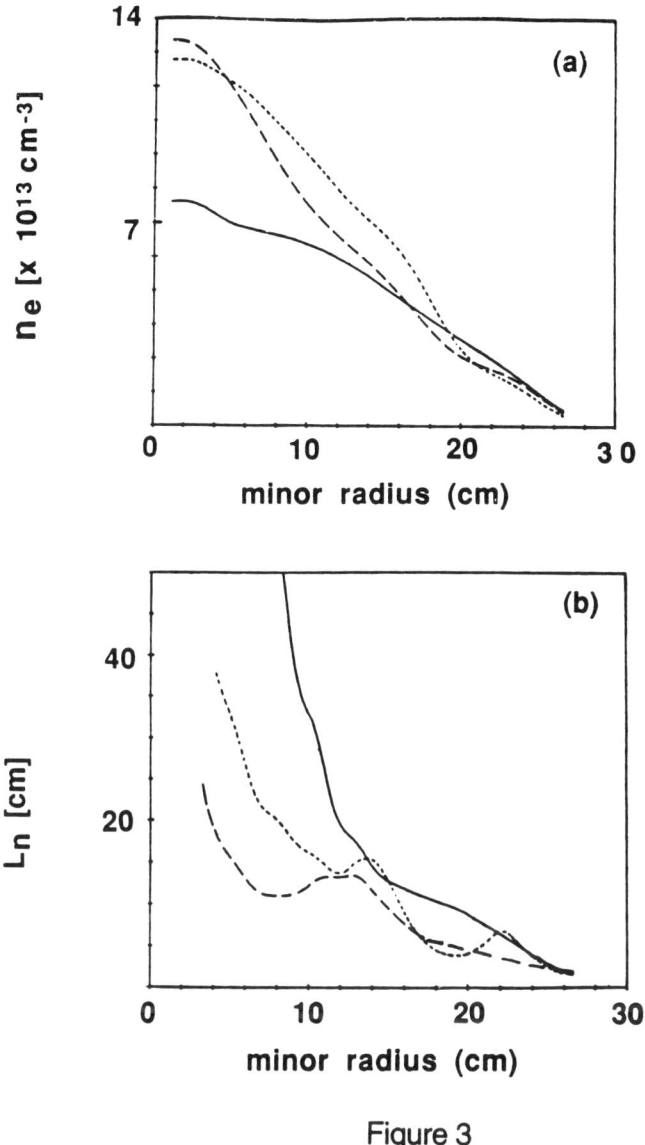

Figure 3

Figure 3. Comparison of (a) density and (b) density scale length profiles for high-density gas-fueled and pellet-fueled discharges ($I_p = 250\,\text{kA}$, $B_T = 2.8\,\text{T}$). The symbols denote (———) gas-fueled, (·······) pellet-fueled (+5 ms), and (— -) pellet-fueled (+50 ms) discharges.

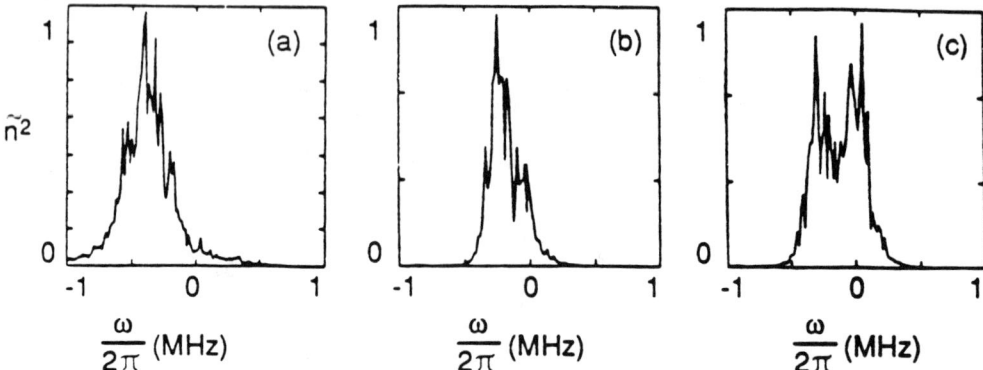

Figure 4. Microturbulence poloidal frequency spectra for $k_\theta = 9\,\mathrm{cm}^{-1}$, $I_p = 250\,\mathrm{kA}$, $B_T = 2.8\,\mathrm{T}$, and (a) pre-pellet; $\bar{n}_e = 3 \times 10^{13}\,\mathrm{cm}^{-3}$, (b) post-pellet; $\bar{n}_e = 7 \times 10^{13}\,\mathrm{cm}^{-3}$, and (c) high-density gas-fueled equivalent discharge; $\bar{n}_e = 6 \times 10^{13}\,\mathrm{cm}^{-3}$. Negative (positive) frequency corresponds to the electron (ion) diamagnetic drift direction in the laboratory frame of reference. Vertical axes are in arbitrary units.

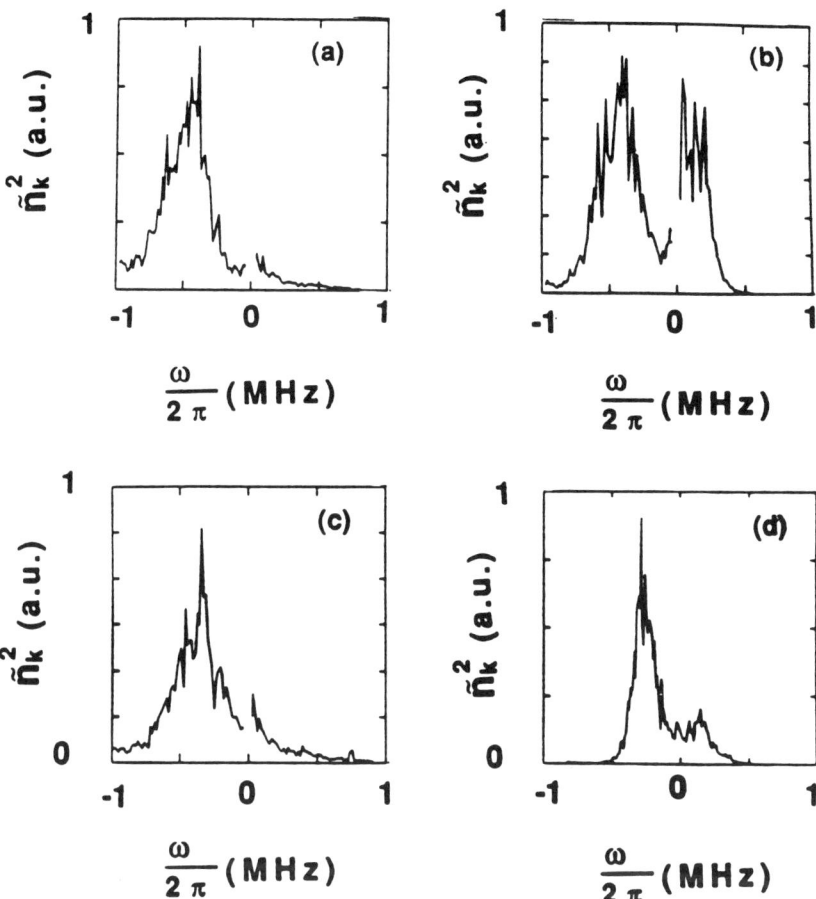

Figure 5. Microturbulence poloidal frequency spectra for $k_\theta = 9\,\text{cm}^{-1}$, (a) Deuterium ($I_p = 400\,\text{kA}$, $B_T = 2.8\,\text{T}$); $\bar{n}_e = 3.5 \times 10^{13}\,\text{cm}^{-3}$, (b) Deuterium ($I_p = 400\,\text{kA}$, $B_T = 2.8\,\text{T}$); $\bar{n}_e = 7 \times 10^{13}\,\text{cm}^{-3}$, (c) Helium ($I_p = 300\,\text{kA}$, $B_T = 2.8\,\text{T}$); $\bar{n}_e = 2 \times 10^{13}\,\text{cm}^{-3}$, and (d) Helium ($I_p = 300\,\text{kA}$, $B_T = 2.8\,\text{T}$); $\bar{n}_e = 9 \times 10^{13}\,\text{cm}^{-3}$. Negative (positive) frequency corresponds to the electron (ion) diamagnetic drift direction in the laboratory frame of reference. Vertical axes are in arbitrary units.

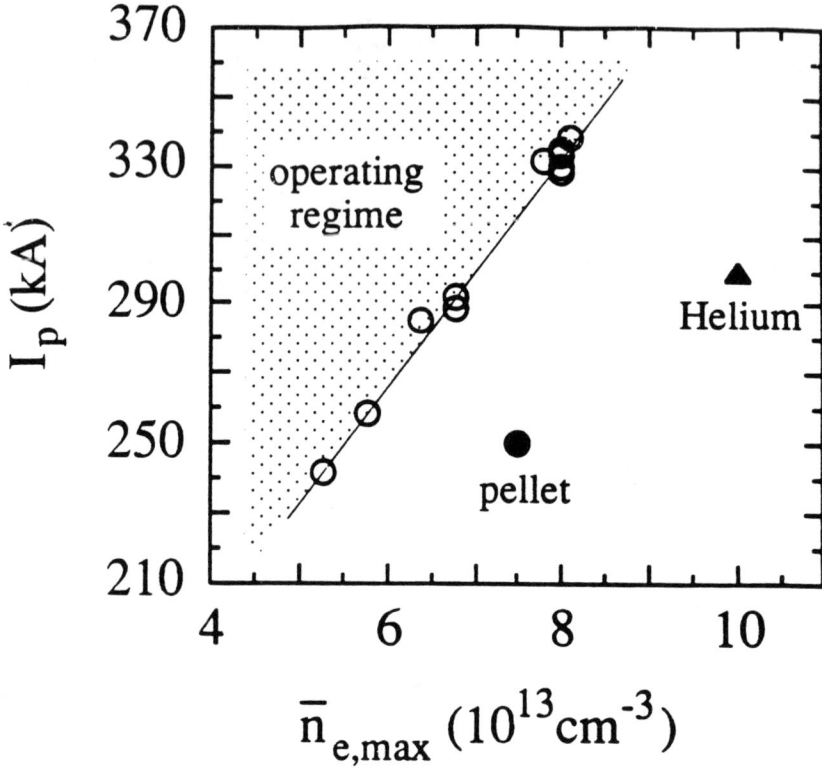

Figure 6. Hugill plot; open circles refer to hydrogen gas fueled plasmas while the solid circle and triangle correspond to hydrogen pellet and helium gas fueled discharges, respectively.

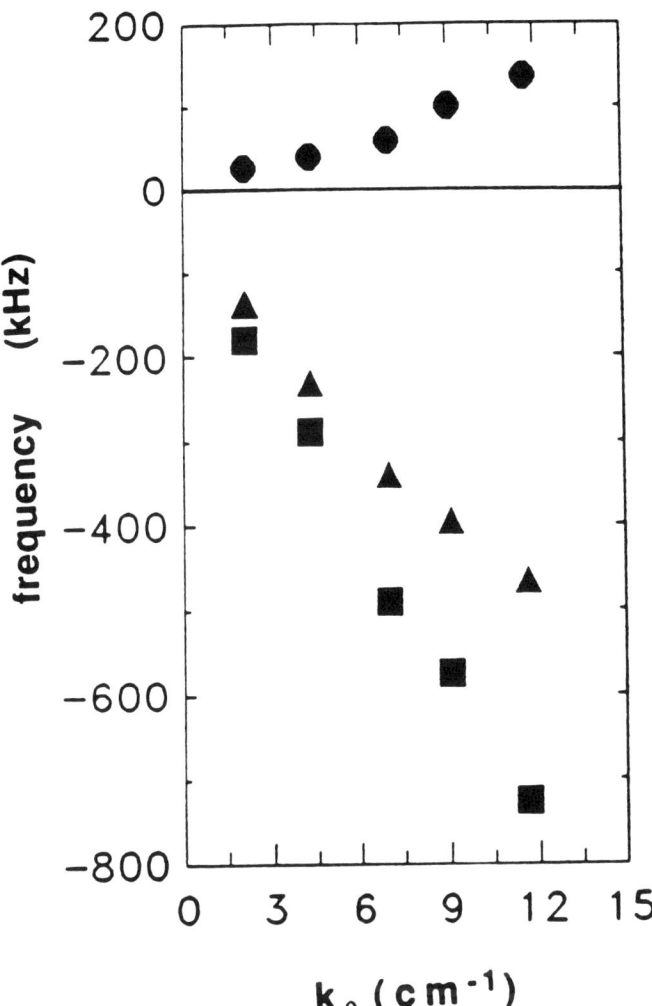

Figure 7. Microturbulence statistical dispersion ($\bar{\omega}_k$ versus k_θ) for electron component (2×10^{13} cm^{-3}; solid squares), electron component (8×10^{13} cm^{-3}; solid triangles) and ion component (8×10^{13} cm^{-3}; solid circles). Negative (positive) frequency corresponds to the electron (ion) diamagnetic drift direction in the laboratory frame of reference.

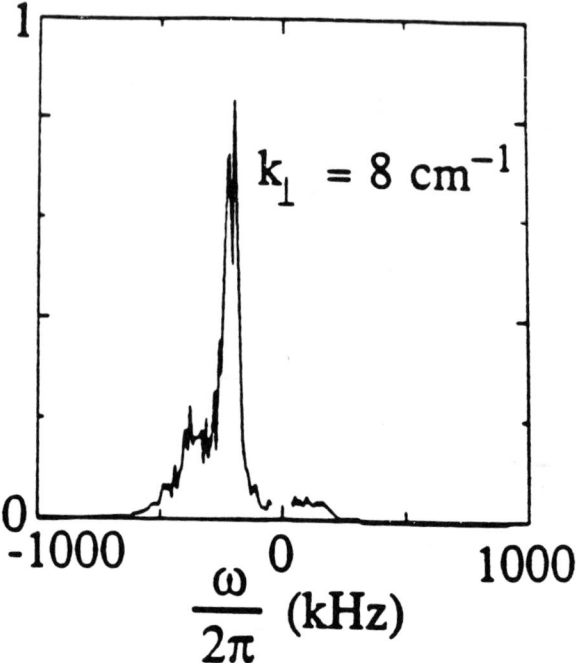

Figure 8. Density fluctuation spectra measured on the high-field side of the torus at $k_\theta = 8\ cm^{-1}$. The large-amplitude quasi-coherent feature ($\bar\omega_{qcm} \simeq -200\,\text{kHz}$) appears in addition to the broadband turbulence ($\bar\omega_{bb} \simeq -350\,\text{kHz}$).

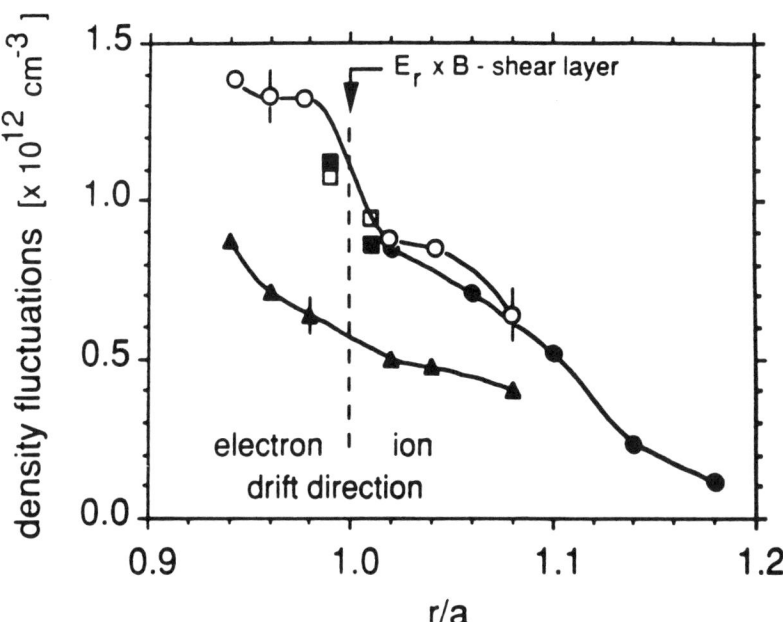

Figure 9. Comparison of gas and pellet fueled discharges in the plasma edge and scrape-off-layer as measured by Langmuir probes. [solid triangle: pre-pellet $\bar{n}_e = 3 \times 10^{13}\,\mathrm{cm}^{-3}$, 200 kA, 2 T, solid circle: gas-fueled $\bar{n}_e = 5 \times 10^{13}\,\mathrm{cm}^{-3}$, 200 kA, 2 T, solid square: gas-fueled $\bar{n}_e = 5 \times 10^{13}\,\mathrm{cm}^{-3}$, 250 kA, 2.8 T, open circle: pellet $\bar{n}_e = 6 \times 10^{13}\,\mathrm{cm}^{-3}$, 200 kA, 2 T, open square: pellet $\bar{n}_e = 6 \times 10^{13}\,\mathrm{cm}^{-3}$, 250 kA, 2.8 T].

3.3 PLASMA TRANSPORT ANALYSIS FOR THE COMPACT HELICAL SYSTEM AND THE LARGE HELICAL DEVICE

K.Y. Watanabe, K. Yamazaki, H. Yamada, and T. Amano
National Institute for Fusion Science, Chikusa-ku, Nagoya, Japan

Abstract. Transport analysis for helical confinement systems are studied based on the electrostatic drift wave turbulence transport and the resistive interchange turbulence transport in addition to the neoclassical transport theory including the effect of radialelectric field and multi-helicity magnetic components. The thermal conductivities predicted by the above theories are compared with CHS (Compact Helical System) experimental data for NBI-heated plasmas. The experimental ion thermal conductivities in the central plasma region agree with the neoclassical value. And the electron thermal transport in the edge plasma region does not contradict the drift wave turbulence transport.

For the plasma parameters in LHD (Large Helical Device), 3-D equilibrium/1-D transport simulations including drift wave or empirical anomalous transport models are carried out, which suggests that the global confinement time of LHD is determined mainly by the electron anomalous transport near the plasma edge region rather than the helical ripple transport in the core region.

3.3.1 INTRODUCTION

Helical confinement configurations have distinct advantages in realizing steady-state operations without current drive and plasma current disruptions. To demonstrate these advantages, the Large Helical Device (LHD) with a superconducting magnetic coil system[1] is under construction. The present confinement properties in these helical systems might be within the so-called "L-mode" range of Tokamak transport, and the confinement improvement is one of the urgent issues for helical confinement systems as well as for tokamak systems. Moreover,

© 1994 American Institute of Physics

helical systems are supposed to suffer from serious helical ripple diffusions in the high temperature plasma regime, different from tokamak systems. So far, the empirical scaling law (so-called "LHD scaling") [2] of the global confinement time was obtained based on several stellarator/heliotron experiments, and the Gyro-reduced Bohm scalings was discussed[3] for the modification of the LHD scaling in the high density regime. The past and present medium-sized experiments[4, 5] suggest that the electron thermal transport outside the half minor radius is anomalous and that near the center is neoclassical for ECH plasma. In NBI-heated plasmas, however, the whole region is anomalous for the electron thermal transport. The spatial dependence of the empirical thermal diffusivity is also attempted to obtain[6, 7]. One of our concerns in this paper is the ion thermal transport. We compare some theoretical models, neoclassical transport, drift wave turbulence and resistive interchange turbulence models, with Compact Helical System (CHS) experimental data.

In next-generation large helical machines with high temperature plasmas, the neoclassical ripple transport might be important even in the core region. In this paper, a simulation model for the equilibrium-transport in the helical confinement systems is developed including the effect of the change in the plasma equilibrium and the magnetic structure on the neoclassical transport. However, without knowing the radial dependence of anomalous transport coefficients, it is impossible to predict whether the effect of this ripple transport is predominant or not, compared with anomalous transport. A empirical model and drift wave turbulence for thermal conductivities are used to forecast the future experiments in LHD (major radius $R_0 \sim 4\,\mathrm{m}$, magnetic field $B \sim 4\,\mathrm{T}$, plasma minor radius $a_p \sim 0.6\mathrm{m}$) in this paper.

In Sec. 2, the neoclassical model of transport coefficients is described. In Sec. 3, the anomalous transport models based on the drift wave turbulences and the resistive interchange turbulence. In Sec. 4, the validity of these anomalous thermal conductivities is checked using CHS experimental data for NBI plasma. A one-dimensional transport simulation modeling coupled with three-dimensional plasma equilibrium and its simulation results for LHD plasmas are given in Sec. 5. In Sec. 6, a summary is given.

3.3.2 NEOCLASSICAL MODELS OF THERMAL CONDUCTIVITIES

The neoclassical transport losses in helical plasma configurations are divided into axisymmetric (SYM) tokamak-like part[10, 9] and asymmetric (ASY) helical-ripple part [10, 11]. Effects of the radial electric field E_ρ ($=-e\partial\Phi/\partial\rho$) are included in the ripple transport simulation[12]. Multiple-helicity effects of the magnetic field configuration are taken into account in the $1/\nu$ regime by introducing the form factor ratio of multi-helicity case to the single helicity case, F_m/F_s. Multiple helicity form factor F_m is calculated by using "GIOTA" code[13]. Asymmetric particle and heat fluxes, Γ_{ASYa} and Q_{ASYa}, of species a (electron ($a = e$) or ion ($a = i$)) as a function of flux-averaged radial variable ρ are given[10] by

$$\Gamma_{ASYa} = -\epsilon_t^2 \epsilon_h^{1/2} v_{da}^2 n_a \int_0^\infty dx\, x^{5/2} e^{-x} \frac{\tilde{\nu}_a(x) A_a(x)}{\omega_a^2(x)}, \qquad (1)$$

$$Q_{ASYa} + \frac{5}{2}\Gamma_{ASYa}T_a = -\epsilon_t^2 \epsilon_h^{1/2} v_{da}^2 n_a T_a \int_0^\infty dx\, x^{7/2} e^{-x} \frac{\tilde{\nu}_a(x) A_a(x)}{\omega_a^2(x)} \quad (2)$$

where

$$A_a(x) = \frac{1}{n_a}\frac{\partial n_a}{\partial \rho} - \frac{Z_a}{T_a}E_\rho + \left(x - \frac{3}{2}\right)\frac{1}{T_a}\frac{\partial T_a}{\partial \rho}, \qquad (3)$$

$$\tilde{\nu}_a(x) = \nu_a^0 x^{-1.5} \epsilon_h^{-1} \left[\left[\left(1 - \frac{1}{2x}\right) \mathrm{erf}\,(x^{1/2}) + \frac{1}{(\pi x)^{1/2}}e^{-x}\right] + \bar{z}_a\right], (4)$$

$$x = \frac{m_a v_{tha}^2}{2T_a}, \qquad (5)$$

$$\omega_a^2(x) = 2.21\frac{\tilde{\nu}_a^2}{F_m/F_s} + 1.5(\epsilon_t/\epsilon_h)^{1/2}(\omega_E + \omega_{Ba})^2$$

$$+(\epsilon_t/\epsilon_h)^{3/2}\left[\frac{\omega_{Ba}}{4} + 0.6|\omega_{Ba}|\tilde{\nu}_a(x)(\epsilon_h/\epsilon_t)^{3/2}\right], \qquad (6)$$

$$\nu_a^0 = \frac{4\pi e^4 n_a \ln \Lambda}{m_a^2 v_{tha}^3}. \qquad (7)$$

Here ϵ_t, ϵ_h, n_a, T_a, v_{da}, v_{tha}, ω_E and ω_{Ba} are toroidal inverse aspect ratio (ρ/R), helical ripple modulation, plasma density, plasma temperature, toroidal drift velocity, thermal velocity, $\mathbf{E} \times \mathbf{B}$ drift and ∇B drift frequencies, respectively. The collision frequency $\tilde{\nu}_a$ is $\tilde{\nu}_e = \nu_{ee} + \nu_{ei}$ with $\bar{z}_e = Z_{\text{eff}}$ for electrons and $\tilde{\nu}_i = \nu_{ii}$ with $\bar{z}_i = 0$ for ions. In the above equations, the ν regime transport was modified according to Ref. [11].

The radial electric profile is determined by the balance between the asymmetric electron and ion loss fluxes,

$$\Gamma_{ASYe}(E_\rho) = \Gamma_{ASYi}(E_\rho) . \tag{8}$$

The validity of these multi-helicity neoclassical transport has been examined by using the DKES code [14].

3.3.3 ANOMALOUS MODELS OF THERMAL CONDUCTIVITIES

Drift wave turbulence (DWT) models

The drift wave turbulence (DWT) models have been successfully applied to simulate tokamak discharges [15, 16]. The electrostatic models related to $\delta \mathbf{E} \times \mathbf{B}/B^2$ turbulent diffusion with long wavelength ($k_\perp \rho_s \sim 1/3$; $k_\perp$ is the perpendicular wave length and ρ_s is the ion Larmor radius with the ion sound velocity $(T_e/M_i)^{1/2}$) consist of electron or ion modes; collisionless or dissipative modes; and cylindrical, toroidal or helical modes. The diffusion coefficient of circulating electron (CE) mode for collisionless (CCE, $\nu_{ei} < \omega_{tet}$) or collisional (XCE, $\omega_{tet} < \nu_{ei}$) regime is given by

$$D_{CE} = \max(D_{CCE}, D_{XCE}) , \tag{9}$$

$$D_{CCE} = \frac{\omega_{*e}}{k_\perp^2} \frac{\omega_{*e}}{\omega_{tet}} , \tag{10}$$

$$D_{XCE} = \frac{\omega_{*e}}{k_\perp^2} \frac{\omega_{*e}}{\omega_{tet}} \frac{\nu_{ei}}{\omega_{tet}} . \tag{11}$$

where $\omega_{*e} = k_\perp T_e / L_n e B$ ($L_n = n/n'$ is the characteristic density length) and $\omega_{tet} = v_{the}(\iota_b/R_0)$ are electron diamagnetic frequency and electron toroidal transit frequency, respectively.

On the other hand, the toroidally trapped electron (TEt) mode diffusion of collisionless (CTEt, $\nu_{\text{efft}} < \omega_{*e}$) or dissipative (DTEt, $\omega_{*e} < \nu_{\text{efft}} < \omega_{tet}$) mode is

$$D_{TEt} = \min(D_{CTEt}, D_{DTEt}) , \tag{12}$$

$$D_{CTEt} = \epsilon_t^{1/2} \frac{\omega_{*e}}{k_\perp^2} , \tag{13}$$

$$D_{DTEt} = \epsilon_t^{1/2} \frac{\omega_{*e}}{k_\perp^2} \frac{\omega_{*e}}{\nu_{\text{efft}}} , \tag{14}$$

where $\nu_{\text{efft}} = \nu_{ei}/\epsilon_t$ is the effective toroidal collision frequency.

In addition to these cylindrical and toroidal electrostatic DWT models, we consider helical ripple contributions to electrostatic DWT. We use the following expression as the helically trapped electron (TEh) mode diffusion of collisionless (CTEh, $\nu_{\text{effh}} < \omega_{*e}$) or dissipative (DTEh, $\omega_{*e} < \nu_{\text{effh}} < \omega_{teh}$) mode:

$$D_{TEh} = \min(D_{CTEh}, D_{DTEh}) , \tag{15}$$

$$D_{CTEh} = \epsilon_h^{1/2} \frac{\omega_{*e}}{k_\perp^2} , \tag{16}$$

$$D_{DTEh} = \epsilon_h^{1/2} \frac{\omega_{*e}}{k_\perp^2} \frac{\omega_{*e}}{\nu_{\text{effh}}} . \tag{17}$$

Here $\nu_{\text{effh}} = \nu_{ei}/\epsilon_h$ is the effective helical collision frequency and $\omega_{teh} = v_{the}(M/R_0)$ is the electron helical transit frequency, where M is pitch number of helical device and $v_{the} = (2T_e/m_e)^{1/2}$.

As for ion mode, the ion temperature gradient (ITG) turbulence is important especially for flat density profiles:

$$D_{\text{ITG}} = \frac{\omega_{*e}}{k_\perp^2} \left(2\frac{T_i}{T_e}\eta_i \frac{L_n}{R}\right)^{1/2} f_{\text{ITGth}} , \tag{18}$$

where

$$f_{\text{ITGth}} = (1 + \exp(-6(\eta_i - \eta_{th})))^{-1} \tag{19}$$

is a threshold function for the onset of ITG modes and ($\eta_i = L_n/L_{Ti} = (n/n')/(T_i/T_i')$).

We adopt the total anomalous transport coefficients due to electrostatic DWT modes as:

$$D_{DWe} = c_{CE}D_{CE} + c_{TEt}D_{TEt} + c_{TEh}D_{TEh}, \tag{20}$$

$$\chi_{DWe} = \frac{5}{2}D_{DWe}(1 + 3c_{ei}\frac{L_n}{R_0}\eta_i f_{ITGth}), \tag{21}$$

$$\chi_{DWi} = \frac{5}{2}(c_{ie}D_{DWe} + c_{ITG}D_{ITG}). \tag{22}$$

The coefficients $c_{CE}, c_{TEt}, c_{TEh}, c_{ei}, c_{ie}$ and c_{ITG} are usually set to unity in this paper.

Resistive interchange (g-mode) turbulence (GMT) models

Since helical system usually has magnetic hill region, the turbulence driven by the resistive interchange mode is considered as a probable candidate for the anomalous transport. We use the following expression as the ion thermal conductivity due to the resistive interchange turbulence[17],

$$\chi_{GMi} = \gamma W^2 \Lambda^2, \tag{23}$$

where

$$\gamma = \frac{1}{S^{1/3}} \left(\frac{\beta}{2} \frac{R_0^2 \kappa_n}{L_p} \frac{\overline{m}}{\iota_b \widehat{S}} \right)^{2/3} \tau_{hp}^{-1}, \tag{24}$$

$$W = \left(\frac{1}{S\iota_b^2 \widehat{S}^2 \overline{m}} \right)^{1/3} \left(\frac{\beta}{2} \frac{R_0^2 \kappa_n}{L_p} \right)^{1/6} \rho, \tag{25}$$

$$\Lambda = \frac{2}{3\pi} \ln \left[\frac{256 S^2}{\beta} \frac{L_p}{R_0^2 \kappa_n} L_p \left(\frac{\iota_b \widehat{S}}{\overline{m}} \right)^4 \right] - \frac{2}{\pi} \ln \Lambda. \tag{26}$$

This coefficient is derived based on the renormalization theory. Here $\beta = p(\rho)/(B^2/2\mu_0)$, $\widehat{S} = |\rho(d\iota_b/d\rho)/\iota_b|$, τ_{hp} is the poloidal Alfvén time, τ_R is the resistive time, $S = \tau_R/\tau_{hp}$ is the magnetic Reynolds number,

$\overline{m}$ is the rms of poloidal mde number and we assume $\overline{m} = 10$ in this paper. $\kappa_n = d\Omega(\rho)/d\rho$ is the normal curvature, and

$$\Omega(\rho) = \frac{1}{2\pi} \int_0^{2\pi} d\theta \frac{M}{2\pi} \int_0^{2\pi/M} d\phi \left(\frac{R^2}{R_0^2} + \frac{|\mathbf{B}^\delta|^2}{B_0^2} \right), \qquad (27)$$

$$\mathbf{B}^\delta = \mathbf{B} - \frac{M}{2\pi} \int_0^{2\pi/M} d\phi \mathbf{B}, \qquad (28)$$

where $R(\rho,\theta,\phi)$, $\mathbf{B}(\rho,\theta,\phi)$ are obtained from VMEC code[18], and θ and ϕ are poloidal and toroidal coordinates, respectively. For the electron thermal conductivity, taking the magnetic fluctuation induced by the electrostatic potential fluctuation into account, we use the following expression[19],

$$\chi_{GMe} = \chi_{GMi} + \frac{\sqrt{\pi}}{2} \frac{\hat{S} v_{the} T R \iota_b}{R_0 \rho^2} \gamma W^4 \Lambda^{4/3}. \qquad (29)$$

3.3.4 COMPARISONS WITH CHS EXPERIMENTAL DATA

The transport coefficients described in the previous section are compared with typical experimental data of NBI-heated discharge in CHS (Compact Helical System, helical period $M = 8$, major radius $R_0 = 1.0$ m, magnetic field $B = 2.0$ T). Figure 1(a) shows the density and temperature profiles experimentally obtained for a typical NBI discharge in CHS ($R_{ax} = 0.99$ m, $P_{NBI}^{abs} = 0.65$ MW). The experimental thermal conductivities in Fig. 1(c), (d) and (e) are obtained by using the PROCTR-MOD code [20, 21]. The radial electric field E_ρ in Fig. 1(b) is calculated from ambipolar condition (8). The thermal conductivities predicted by the neoclassical theory, the DWT models and the GMT models are shown in Fig. 1(c), (d) and (e). In general, the ion root in the whole plasma region is obtained for NBI discharge in CHS. The symmetric part is dominant for electron neoclassical thermal transport, and the asymmetric part is dominant for ion neoclassical thermal transport in NBI discharge in CHS.

Figure 2 shows the dependence of the experimental χ_e normalized by the theoretical χ_e, (a) χ_e^{EX}/χ_{NCe}, (b) χ_e^{EX}/χ_{DWe} and (c) χ_e^{EX}/χ_{GMe}, on the collisionality. Here $\nu_{*e} = \tilde{\nu}_e/\omega_{tet}$. Circles, squares and triangles denote the results at $\rho/a_p = 0.25$, 0.5 and 0.9, respectively. Analyzed experimental data are obtained in NBI discharge in CHS ($B = 0.5 \sim 1.5\,\mathrm{T}$, $P_{\mathrm{NBI}}^{\mathrm{abs}} = 0.21 \sim 0.88\,\mathrm{MW}$, $R_{ax} = 0.92 \sim 1.05\,\mathrm{m}$, $\bar{n}_e = 0.89 \sim 5.91 \times 10^{19}\,\mathrm{m}^{-3}$). As usual, the anomalous transport due to ITG modes is dominant at the central region, and that due to CE modes is dominant at the edge region for the DWT models. From Fig. 2, the DWT models have high probability to explain the experimental electron thermal transport in the edge region. It is difficult to explain the experimental electron thermal transport by the GMT models. The closed triangles in Fig. 2(c) correspond to the inner magnetic axis shift case, $R_{ax} \lesssim 0.98\,\mathrm{m}$. It is expected that the inner shift leads to the magnetic hill configuration and enhances the GMT transport. However, it is not clear for CHS NBI-heated plasma. For $\rho/a_p \lesssim 0.5$, the neoclassical theory, DWT and GMT models cannot explain the experimental electron thermal transport. The values of χ_e predicted by the DWT models are about ten times smaller than experimental data.

Figure 3 shows the dependence of (a) $\chi_i^{\mathrm{EX}}/\chi_{NCi}$, (b) $\chi_i^{\mathrm{EX}}/\chi_{DWi}$ and (c) $\chi_i^{\mathrm{EX}}/\chi_{GMi}$, on the collisionality. Here $\nu_{*i} = \tilde{\nu}_i/(v_{thi}\iota_b/R_0)$. Other parameters are same in Fig. 2. From Fig. 3, we can find that χ_i value predicted by the neoclassical theory and DWT models has the same order of magnitude as χ_i estimated from experiment in the central plasma region, $\rho/a_p \lesssim 0.5$. Considering the dependence of ν_{*i}, the neoclassical theory gives better explanation of the experimental ion thermal transport in the central plasma region. The radial electric field gives the significant effect on the neoclassical transport in helical system. However, the calculation results of E_ρ at $\rho/a_p \lesssim 0.6$ are usually consistent with experimentally estimated E_ρ from the rotation velocity of impurities [22]. It is difficult to explain the experimental ion thermal transport by the DWT and GMT models as well as the neoclassical transport models in the edge region. The dependence of $\chi_i^{\mathrm{EX}}/\chi_{GMi}$ does not depend on the magnetic axis position, the values of χ_i predicted by the GMT models are smaller than experimental data by more than the factor of 2.

3.3.5 TRANSPORT SIMULATION SIMULATIONS FOR THE LARGE HELICAL DEVICE

Simulation model

For the analysis of the LHD transport, a 2.0-dimensional equilibrium-transport code has been developed in which 3D-equilibrium code VMEC [23] and 1D-transport code HTRANS are used. The NBI deposition is calculated by the HFREYA code which is a helical modification of FREYA code[24]) and the slowing-down calculations is done with a Fokker-Planck code FIFPC[25]. The neoclassical transports as described in Secs. 2 are used.

The initial vacuum magnetic surface is calculated by the magnetic field line tracing code HSD[27] with carefully arranged multi-filament currents. In this paper, the fixed boundary version of VMEC code is used. The three-dimensional magnetic field obtained by the finite beta equilibrium of VMEC is used to evaluate the NBI heat deposition and the multiple-helicity neoclassical coefficients. The one-dimensional particle and energy fluid transport equations are of the general form

$$\frac{\partial n_e}{\partial t} = -\frac{1}{V'(\rho)}\frac{\partial}{\partial \rho}[V'(\rho)\Gamma_e] + S_e , \tag{30}$$

$$\frac{\partial}{\partial t}\left(\frac{3}{2}n_e T_e\right) = -\frac{1}{V'(\rho)}\frac{\partial}{\partial \rho}\left[V'(\rho)\left(Q_e + \frac{5}{2}T_e\Gamma_e\right)\right]$$

$$- \Gamma_e E_r - P_{ei} + P_{He} - P_{rad} , \tag{31}$$

$$\frac{\partial}{\partial t}\left(\frac{3}{2}n_i T_i\right) = -\frac{1}{V'(\rho)}\frac{\partial}{\partial \rho}\left[V'(\rho)\left(Q_i + \frac{5}{2}T_i\Gamma_i\right)\right]$$

$$+ \Gamma_i E_r + P_{ei} + P_{Hi} - P_{cx} , \tag{32}$$

where S_e denotes the particle source due to the neutral beam fueling and the feedback controlled gas puffing calculated by the Monte-Carlo code AURORA [26]. The variables P_{He} and P_{Hi} are the input heating power to electrons and ions from the neutral beam calculated by Fokker-Planck code [25] and/or the RF heating. P_{ei}, P_{rad} and P_{cx} are the electron-ion power exchange, the radiation power loss, and the charge exchange power loss, respectively. The particle flux Γ and heat flux Q are defined by using diffusion coefficient D and thermal diffusivity χ.

$$\Gamma_e = \Gamma_{ASYe} - (D_{SYMe} + D_{ANe})\left\langle|\nabla\rho|^2\right\rangle\frac{\partial n_e}{\partial\rho}, \qquad (33)$$

$$Q_e = Q_{ASYe} - (\chi_{SYMe} + \chi_{ANe})n_e\left\langle|\nabla\rho|^2\right\rangle\frac{\partial T_e}{\partial\rho}, \qquad (34)$$

$$Q_i = Q_{ASYi} - (\chi_{SYMi} + \chi_{ANi})n_i\left\langle|\nabla\rho|^2\right\rangle\frac{\partial T_i}{\partial\rho}. \qquad (35)$$

In Figs. 4 and 5, we assume that the empirical anomalous diffusion coefficient and anomalous electron thermal conductivity are [28]

$$D_{ANe}(\rho) = \frac{2}{5}\chi_{ANe}(\rho), \qquad (36)$$

$$\chi_{ANe}(\rho) = \langle\chi_{LHD}(\rho)\rangle g(\rho), \qquad (37)$$

$$\chi_{LHD}(\rho)[m^2/s] = 15.8 B^{-2.0}[T] R_0^{-0.40}[m] T(\rho)^{1.38}[keV]$$
$$n(\rho)^{-0.26}[10^{20}\,\mathrm{m}^{-3}], \qquad (38)$$

$$g(\rho) = \frac{\nu+1}{\mu+\nu+1}(1+\mu\rho^{2\nu}), \qquad (39)$$

where $\langle\ \rangle$ denotes the volume average. The geometrical function g can be chosen to fit the experimental data. Typically, $\mu = 10$ and $\nu = 4$ are used in this paper. $\langle\chi_{LHD}\rangle$ corresponds to $a^2/4\tau_{E,LHD}$. Here $\tau_{E,LHD}$ is the empirical scaling of global confinement time for helical systems (so called "LHD scaling" [2]) given by

$$\tau_{E,LHD}[s] = 0.17 P^{-0.58}[MW]\bar{n}^{0.69}[10^{20}\,\mathrm{m}^{-3}]B^{0.84}[T]R_0^{0.75}[m]a_p^2[m], \qquad (40)$$

where P and $\bar{n}$ are total absorbed heating power and the line averaged plasma density, respectively. And we assume that the anomalous ion thermal conductivity does not exist.

Simulation for LHD plasmas

The LHD magnetic configuration [1] is characterized by the $\ell = 2/M = 10$ heliotron/torsatron with continuous helical coil system. The major radius is 4m (finally determined to 3.9m) and the magnetic field strength is 4 T. The winding law of the helical coil with the major radius R_c and the minor radius a_c is defined by

$$\theta = \frac{m}{\ell}\phi + \alpha_c \sin\left(\frac{m}{\ell}\phi\right) , \qquad \gamma_c = \frac{ma_c}{\ell R_c} ,$$

where γ_c and α_c are coil pitch parameter and pitch modulation parameter, respectively. Three sets of poloidal coils are used to produce various shapes of the plasma cross-section and control the magnetic axis position, and three block layers of helical coils are energized to change γ_c value for the control of the plasma size and he divertor layer.

Typical simulation results of 20MW-NBI+5MW-ECH heated LHD plasmas are shown in Fig. 4 for (a) low density and (b) high density discharges with $\gamma_c = 1.2$, $\alpha_c = 0.1$, $R_c - R_{ax}^V = -0.1$ m and $a_p = 0.55$ m. Here R_{ax}^V is the vacuum magnetic axis. Anomalous particle inward flows are not included in the simulations, and flat or hollow density profiles are obtained. Such hollow density profiles are seen in many existing experiments. In the case of $\langle n_e \rangle = 0.3 \times 10^{20}$ m^{-3}, the electron root is achieved in the plasma central region, $\rho/a_p \lesssim 0.5$, and the high ion temperature is obtained. Due to rather high positive electric potential, ion thermal transport becomes small in the core region. In the case of $\langle n_e \rangle = 1.0 \times 10^{20}$ m^{-3}, the ion root is achieved in the whole plasma region, and χ_i^{NC} is over $5m^2/s$ at $\rho/a_p \sim 0.5$. For electron thermal transport, neoclassical transport cannot be neglected in the central region. On the other hand, the empirical thermal conductivity is dominant at the edge region. This situation is independent of the density. The transport simulation results for various density cases is shown in Fig. 5. The results in Fig. 5 include those in Figs. 4(a) and (b). The electron root is achieved in the central plasma region for $\langle n_e \rangle \lesssim 0.5 \times 10^{20}$ m^{-3}, the ion root is in the whole plasma region for $\langle n_e \rangle \gtrsim 1.0 \times 10^{20}$ m^{-3}. The density profile has the tendency to be peaky for the high density operation.

Figure 6 shows the transport simulation results by using the DWT models for the same device parameter in Figs. 4 and 5. Here we assume that the anomalous diffusion coefficient, electron and ion thermal conductivities are[29]

$$D_{ANe}(\rho) = D_{DWe}, \quad \chi_{ANe}(\rho) = \chi_{DWe}, \quad \chi_{ANi}(\rho) = \chi_{DWi}. \quad (41)$$

In this case, the averaged electron density is factor 1.3 times larger than that in Fig. 4(b). The amplitude of total electron thermal conductivity is similar to that in Fig. 4(b). For ion thermal transport, the ITG mode due to the flat density profile is dominant except the edge region. The asymmetric part of neoclassical transport at $\rho/a_p \sim 0.5$ cannot be neglected as well as the ITG modes.

3.3.6 SUMMARY

Transport analysis for helical confinement systems are studied based on the electrostatic drift wave turbulence transport and the resistive interchange turbulence transport in addition to the neoclassical transport theory including the effect of radial electric field and multi-helicity magnetic components. At first, the thermal conductivities predicted by above theory are compared with CHS (Compact Helical System) experimental data for NBI-heated plasmas. The neoclassical χ_i value fits the experimental data in the core region than the DWT and GMT models. And the electron thermal transport in the edge plasma region agrees with the drift wave turbulence transport. However, other transport models cannot be ruled out. More detailed comparison between experiments and theories, for example using experimental data of density fluctuation, are required.

In order to predict LHD (Large Helical Device) plasmas, a new 3-D (dimensional) equilibrium/1-D transport simulation model for helical confinement systems has been developed. For the transport processes, we have considered the neoclassical transport theory including the effect of radial electric field and multi-helicity magnetic components with the anomalous semi-empirical transport. These simulation results show that the neoclassical ripple transport is large for ion thermal transport at the high density regime with the ion root. The global confinement time of LHD is determined mainly by the electron anomalous transport near the plasma edge region rather than the helical ripple transport in the core region. However, for the helical device the reduction of this neoclassical asymmetric ion transport loss is important because this transport reduction is effective for improving ion confinement and raising the central ion temperature. The electron root transport with neoclassical χ_i value allows us to get high temperature low density plasmas in LHD, and the effects of ion anomalous transport will be clarified in the future experiment.

ACKNOWLEDGMENTS

The authors would like to thank Drs. S.P. Hirshman and C.L. Hedric for providing us with the VMEC and GIOTA code. We are grateful to CHS and LHD group for discussion and to Drs. O. Motojima, M. Fujiwara, and A. Iiyoshi for continuous encouragements.

Bibliography

[1] IIYOSHI, A., FUJIWARA, M., MOTOJIMA, O., OHYABU,N., YAMAZAKI, K., Fusion Tech. **17** (1990) 169.

[2] SUDO, S., TAKEIRI, Y., ZUSHI, H., SANO, F., ITOH, K., KONDO, K., IIYOSHI, A., Nucl. Fusion **30** (1990) 11.

[3] MURAKAMI, M., ACETO, S.C., ANABITARTE, E., ANDERSON, D.T., et al., in Plasma Physics and Controlled Nuclear Fusion Research 1990 (Proc. 13th Int. Conf. Washington, 1990) IAEA-CN-53/C-1-3.

[4] ZUSHI, H., et al., Nucl. Fusion **24** (1984) 305.

[5] IGUCHI, H., HOSOKAWA, M., HOWE, H.C., IDA, K., IDEI, H., KANEKO, O., et al., in Controlled Fusion and Plasma Heating (Proc. 16th Europ. Conf. Amsterdam, 1990), Vol. 2, European Physical Society (1990) 451.

[6] SANO, F., TAKEIRI, Y., HANATANI, K., ZUSHI, H., SATO, M., SUDO, S., et al., Nucl. Fusion **30** (1990) 81.

[7] RINGLER, H., GASPARINO, U., KUHER, G., MAASSBERG, H., RENNER, H., et al., Plasma Phys. Controlled Fusion **32** (1990) 933.

[8] HINTON, F.L., HAZELTINE, R.D., Rev. Mod. Phys. **48** (1976) 239.

[9] CHANG, C.S., HINTON, F.L., Phys. Fluids **25** (1982) 1493.

[10] SHAING, K.C., CALLEN, J.D., Phys. Fluids **26** (1983) 3315.

[11] CRUME, E.C., SHAIN, K.C., HIRSHMAN, S.P., van RIJ, W.I., Phys. Fluids **31** (1988) 11.

[12] HASTING, D.E., HOULBERG, W.A., SHAING, K.C., Nucl. Fusion **25** (1985) 445.

[13] HEDRIC, C.L., private communication on the GIOTA code (Oak Ridge National Laboratory, 1988).

[14] HIRSHMAN, S.P., SHAING, K.C., van RIJ, W.I., BEASLEY, C.O., CRUME, E.C., Phys. Fluids **29** (1986) 2951.

[15] PERKINS, F.W, in Heating in Toroidal Plasmas (Proc. 4th Int. Symp. Rome, 1984), Vol. 1, Int. School of Plasma Physics, Varena (1984) 97.

[16] DOMINGUEZ, R.R., WALTZ, R.E., Nucl. Fusion **27** (1987) 65.

[17] CARRERAS, B.A., GARCIA, L., DIAMOND, P.H., Phys. Fluids **30** (1987) 1388.

[18] NAKAMURA, Y., ICHIGUCHI, K., WAKATANI, M., et al., J. Phys. Soc. Jpn. **58** (1989) 3157.

[19] CARRERAS, B.A., DIAMOND, P.H., Phys. Fluids B **1** (1989) 1011.

[20] HOWE, H.C., HORTON, L.D., CRUME, E.C., et al., in Controlled Fusion and Plasma Physics (Proc. 16th Eur. Conf. Venice, 1989), Vol. 13B, Part II, European Society (1889) 683.

[21] YAMADA, H., IDA, K., IGUCHI, H., et al., in Controlled Fusion and Plasma Physics (Proc. 18th Eur. Conf. Berlin, 1991), Vol. 15C, Part II, European Society (1991) 137.

[22] IDA, K., YAMADA, H., IGUCHI, H., et al., Phys. Fluids B **4** (1992) 1360.

[23] HIRSHMAN, S.P., van RIJ, W.I., MERKEL, P., Comput. Phys. Commun. **43** (1986) 143.

[24] LISTER, G.G., POST, D.E., GOLDSTON, R., in Third Symp. Plasma Heating in Toroidal Devices, (Proc. Inter. School of Plasma Physics, Varenna, Italy, 1976), p. 303, (ed by E. Sindoni) Editrice Compositori, Bologna (1976).

[25] FOWLER, R.H., SMITH, J., ROME, J.A., Comput. Phys. Commun. **13** (1978) 323.

[26] HUGHES, M.H., POST, D.E., J. Comput. Phys. **28** (1978) 43.

[27] YAMAZAKI, K., MOTOJIMA, O., ASAO, M., Fusion Tech. bf21 (1992) 147.

[28] YAMAZAKI, K., AMANO, T., Nucl. Fusion **32** (1992) 633.

[29] YAMAZAKI, K., OHYABU, N., OKAMOTO, M., *et al.*, in Plasma Physics and Controlled Nuclear Fusion Research 1990 (Proc. 13th Int. Conf. Washington, 1990) IAEA-CN-53/C-4-11.

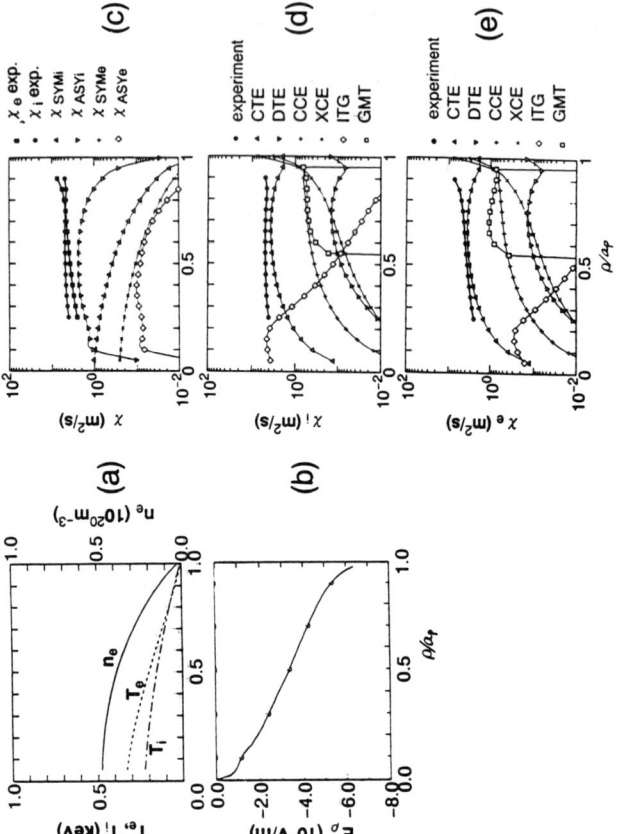

Figure 1. Comparisons of modeled thermal conductivities with CHS experimental data for a typical NBI discharge ($R_{ax} = 0.99m$, $P_{NBI}^{abs} = 0.65MW$). E_ρ in Fig. 1(b) is calculated from ambipolar condition (8).

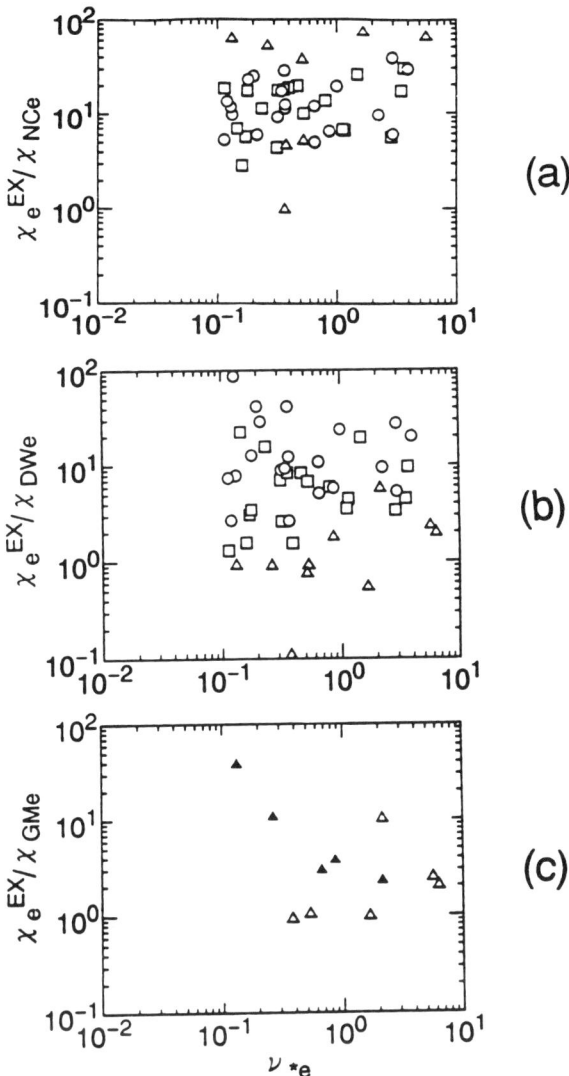

Figure 2. Dependence of the experimental χ_e normalized by the theoretical χ_e, (a) χ_e^{EX}/χ_{NCe}, (b) χ_e^{EX}/χ_{DWe} and (c) χ_e^{EX}/χ_{GMe}, on the collisionality in NBI discharges in CHS ($B = 0.5 \sim 1.5\,\text{T}$ $P_{NBI}^{abs} = 0.21 \sim 0.88\,MW$, $R_{ax} = 0.92 \sim 1.05\,\text{m}$, $\bar{n}_e = 0.89 \sim 5.91 \times 10^{-19}\,m^{-3}$). Here $\nu_{*e} = \tilde{\nu}_e/\omega_{tet}$.

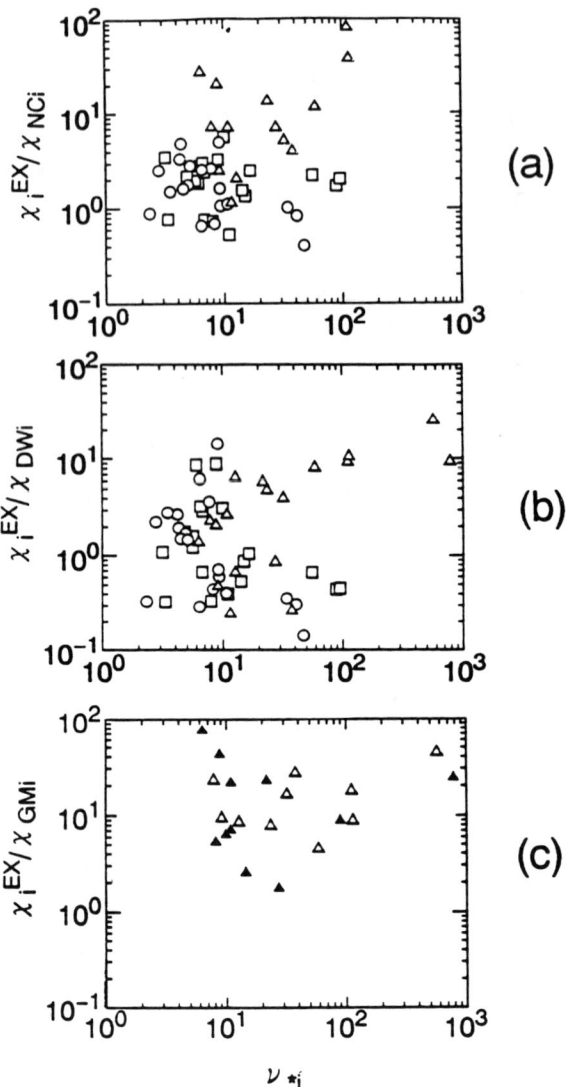

Figure 3. Dependence of the experimental χ_i normalized by the theoretical χ_i, (a) χ_i^{EX}/χ_{NCi}, (b) χ_i^{EX}/χ_{DWi} and (c) χ_i^{EX}/χ_{GMi}, on the collisionality in NBI discharges in CHS. Here $\nu_{*i} = \tilde{\nu}_i/(v_{thi}\iota_b/R_0)$.

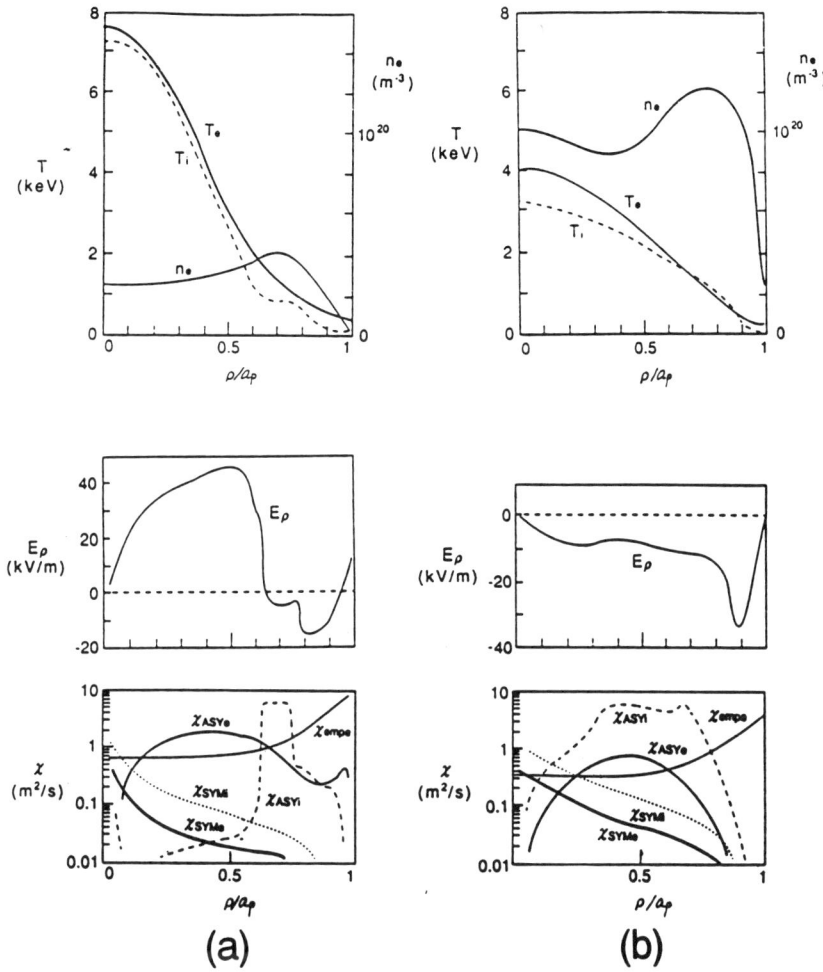

Figure 4. Transport simulation results of LHD plasmas in 20 MW NBI heating case using empirical anomalous transport model for electrons. (a) Low density case with positive electric field (Electron Root), (average electron density: $3.0 \times 10^{19} \mathrm{m}^{-3}$). (b) High density case with negative electric field (Ion Root), (average electron density: $1.0 \times 10^{20} \mathrm{m}^{-3}$). Here $R_c = 4.0\,\mathrm{m}$, $B = 4\,\mathrm{T}$, $\alpha_c = 0.1$ and $\gamma_c = 1.2$. χ_{empe} corresponds to χ_{ANe}.

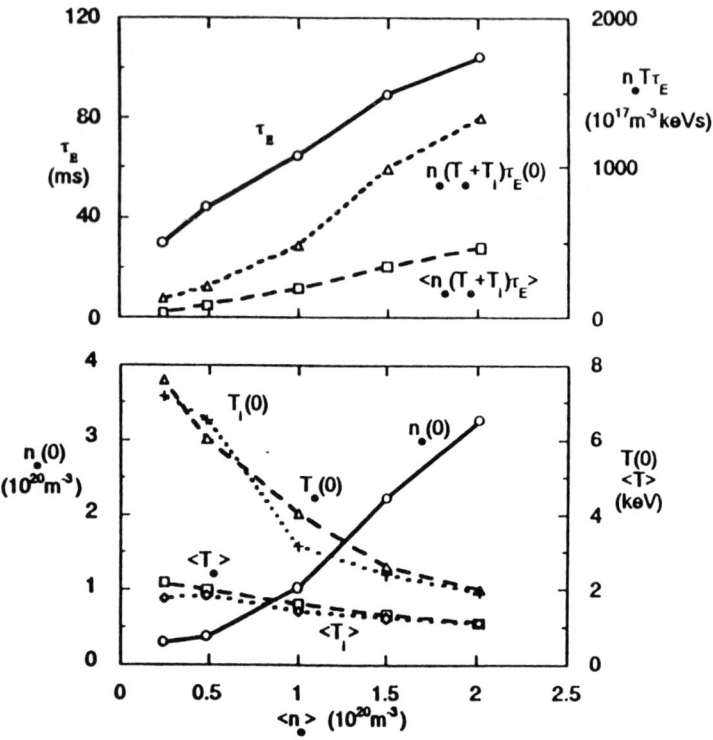

Figure 5. Transport simulation results for various density regime of 20MW-NBI heated LHD plasmas. Results of Fig. 4(a) and (b) are included in Fig. 5 and the magnetic configuration.

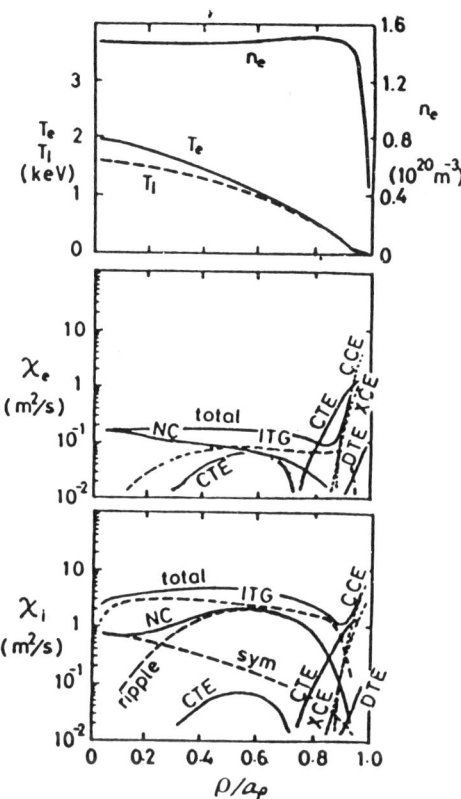

Figure 6. Transport simulation results of 20MW-NBI heated LHD plasmas using DWT models for anomalous transport with $\langle n_e \rangle = 1.30 \times 10^{20} \mathrm{m}^{-3}$.

Chapter 4
PARTICLE SIMULATIONS

4.1 KINETIC SIMULATION OF MICROINSTABILITIES IN TOKAMAK PLASMAS

W.W. Lee, S.E. Parker, R.A. Santoro, and J.C. Cummings
Princeton Plasma Physics Laboratory, Princeton, New Jersey

Abstract. A comprehensive program for the development and use of particle simulation techniques for solving the gyrokinetic Vlasov-Maxwell equations on massively parallel computers has been carried out at Princeton Plasma Physics Laboratory. This is a key element of our ongoing theoretical efforts to systematically investigate physics issues vital to understanding tokamak plasmas. In this work, our focus is on the various aspects of spatial-gradient-driven microinstabilities. Although their presence is consistent with a number of significant confinement trends, results from high temperature tokamaks such as TFTR have highlighted the need for better insight into the linear and nonlinear properties of such instabilities in long-mean-free-path plasmas. In addressing this general issue, we report important new results including (i) the first fully toroidal 3D gyrokinetic simulation of ITG modes and (ii) the comparisons of the fluctuation spectra in the simulation with those from the recent BES measurements of the TFTR core plasma for the supershot discharges. We also discuss (iii) the effects of ITG modes on the inward pinch of impurities in 3D slab geometry and (iv) the finite-β stabilization of microinstabilities in 1D simulation. Finally, (v) the relationship between energy flux and entropy production in the simulation is explored.

With the advent of simulation techniques [1, 2, 3] for the gyrokinetic Vlasov-Maxwell equations and the progress made in simulation algorithms utilizing massively parallel architecture, we have developed an efficient 3D gyrokinetic particle code in toroidal ($x = r\cos\theta$, $y = r\sin\theta$, $\zeta = -R_0\varphi$) geometry implemented on the Connection Machine (CM200) for the systematic kinetic investigation of ITG instabilities. In order to gain a better understanding of the linear and nonlinear behavior of these modes, we have carried out three types of simulations: linearized [2], partially linearized [2] and fully nonlinear [3]. In the linearized simulation the most unstable modes with ballooning-type mode structures of moderate (m, n) with $k_r \ll k_\theta$ and $k_\theta \rho_s \leq 0.5$ have been observed to dominate the time evolution of the instability as shown in Fig. 1. This is basically a benchmarking procedure in that it involves comparisons with results from realistic linear toroidal eigenmode calculations [4]. Good agreement is found with respect to both the magnitude of the eigenvalues (within 25%) and the characteristic features of the eigenfunctions, which exhibit strong ballooning in the poloidal direction and radially extend over many rational surfaces. This is in contrast to the long-accepted picture predicted by 1D radial eigenmode calculations and supported by earlier idealized (very weak magnetic shear) 2D calculations that the eigenmodes are strongly localized between rational surfaces. In fact, even for very long wavelengths modes, the eigenfunctions generally exhibit a strong ballooning character with the associated radial structure relatively insensitive to ion Landau damping at the rational surfaces. In the partially linearized simulation, the $\mathbf{E} \times \mathbf{B}$ advection is the only nonlinearity kept and is the mechanism responsible for the nonlinear saturation of the instability. The results in the nonlinear stage of development indicate that

(i) the fluctuations become more isotropic as shown in Fig. 1, but the ballooning structure persists;

(ii) the energy cascades to both longer and shorter wavelength modes with clear evidence of enhanced growths for the small $k_\theta \rho_s$ modes;

(iii) the steady-state fluctuation spectrum in the ballooning region peaks at $k_\theta \rho_i \approx 0.14$ and $k_r \rho_i \approx 0$, as shown in Fig. 2, which resembles the recent fluctuation measurements [5] of the TFTR supershot discharge using beam emission spectroscopy (BES); and

(iv) the energy flux from the toroidal simulation is greatly enhanced over its slab counterpart and the resulting thermal diffusivities, χ_i, scale like gyro-Bohm.

Perkins et al. [6] has recently argued that the existence of gyro-Bohm scaling, as observed in (iv), requires that the fluctuation spectra for $k_\perp \rho_s$ be independent of the minor radius. This is consistent with the comparison made in (iii), where $(a/\rho_s)_{\text{TFTR}} \geq 5 - 10(a/\rho_s)$ simulation. Fully nonlinear simulations are also carried out with results comparable to those from the partially linearized simulation. The objective here is to look for insights into the possible origins of ITG transport in steady state. Results indicate that stochastic particle orbits associated with the $\mathbf{E} \times \mathbf{B}$ advection are the essential ingredient, which we will discuss later. Furthermore, the present studies are of particular relevance to the development of nonlinear models of microturbulence with a significantly revised picture for the radially-extended long-wavelength ITG instabilities and a demonstration of the capability to provide a realistic picture of the driven versus damped (inertial) regions of the spectrum.

Impurity transport properties in the presence of microturbulence have also been studied using gyrokinetic particle simulations. In particular, results from 3D sheared slab simulations have provided insight into the physics responsible for the inward pinch of the impurities in the presence of ITG modes. It is found that the parallel acceleration in the poloidal direction and the $\mathbf{E} \times \mathbf{B}$ advection in the radial direction give rise to the observed inward pinch of the impurity ions in both the linear and nonlinear stages of development. Specifically, the parallel electric field acts to set up a phase difference between the impurity density fluctuations and the background potential fluctuations and thereby induce inward transport through the $\mathbf{E} \times \mathbf{B}$ motion. This is true even when both the bulk ions and impurity ions have a flat density profile. In general, the pinch is sensitive to the temperature gradient of the impurities as well as the $\eta_i (\equiv L_n/L_{Ti})$ values for the bulk ions. For example, for the warm impurities that have been equilibrated with the background ions through collisions, it becomes necessary to include a impurity temperature gradient for the pinch to occur. Moreover, the direction of the pinch for the impurity ions depends critically on the η_i values of the bulk ions. Because particle diffusion is intrinsically ambipolar, the inward flux for the impurity ions is compensated by the outward flux of the primary ions for an instability in which the electron response is nearly adiabatic. Since the concentration for the impurities is usually very low, their average inward velocity is, therefore, much higher than that of the primary ions. Figure 3 gives the simulations results with cold and heavier impurities, where the Z/M scaling is clearly shown. Comparisons with results from ongoing toroidal simulations of this type as well as with those from experimental studies of impurity transport [7] are encouraging.

In order to make further progress in simulating realistic tokamak plasmas, it is necessary to include finite-β physics as well as nonadiabatic trapped-electron dynamics in our toroidal code. Presently, electromagnetic (finite-β) codes in 1D and 2D slab geometry have been developed to study the influence of magnetic perturbations on the stability and transport of microinstabilities in general and on ITG modes in particular. Two types of formulations for studying the finite-β effects have been used. The first uses the usual $v_\parallel$ as the phase space variable, while the other is based on the $p_\parallel \equiv v_\parallel + (q/mc)A_\parallel$ formulation. The basic difference is that the $v_\parallel$-formulation requires the solution of a generalized Ohm's law, in addition to the usual Poisson's equation and Ampére's law. The results, as shown in Fig. 4, have confirmed the theoretical prediction of finite-β stabilization of microinstabilities. The stabilization can be viewed as coming from the reduction of the parallel electric field, $E_\parallel \equiv E^L + E^T = -\partial \Phi/\partial x_\parallel - (1/c)\partial A_\parallel/\partial t$, due to the presence of the induction (transverse) electric field, E^T. Another essential ingredient is the finite-Larmor-radius (FLR) effects, which provide the necessary coupling between the microinstabilities and shear-Alfvén waves. Since drift instabilites usually have higher frequencies than those of the ITG modes, they can interact easier with the higher frequency shear-Alfvén modes and become stabilized at lower β values. The degree of stabilization is also proportional to the magnitude of $k_\perp \rho_i$. Furthermore, in the case of shear slab, it is found that higher radial harmonics of the ITG eigenmodes cannot be fully stabilized by the finite-β effects, which, consequently, results in the formation of microtearing modes near the rational surfaces in the simulation [2]. Thus, anomalous transport due to magnetic reconnection may be as important as the aforementioned $\mathbf{E} \times \mathbf{B}$ advection. To this end, we are now in the process of developing efficient numerical algorithms to account for the nonadiabatic electrons response and finite-β effects in toroidal geometry. One important aspect is the implementation of the trapped-electron dynamics, where a bounce-averaged response and a multi-stepping scheme are under active consideration.

Finally, the origin of the observed steady-state energy transport in the simulation due to ITG turbulence has been investigated. Let $F_i(x, v_\parallel, t) = F_{Mi} + f_i$, the entropy balance equation for the governing gyrokinetic equations in slab geometry can be written as [9]

$$\left(\frac{\partial}{\partial t}\right)\left\langle \int \left(\frac{f_i^2}{\tau F_{Mi}}\right) dv_\parallel + |\nabla_\perp \Phi|^2 + \Phi^2 \right\rangle + \left\langle \left(\frac{\partial \Phi}{\partial x_\parallel}\right) \int \left(\frac{f_i^2}{F_{Mi}}\right) v_\parallel dv_\parallel \right\rangle$$

$$= \frac{\kappa_{Ti} \langle Q_{ix}\rangle}{\tau},$$

where Φ is the electrostatic potential, $\tau \equiv T_e/T_i$, $Q_{ix} \equiv p_i v_{Ex}$ is the ion energy flux, p_i is the perturbed ion pressure and v_{Ex} is the $\mathbf{E} \times \mathbf{B}$ velocity in the direction of ion temperature gradient, $\kappa_{Ti} \equiv -d\ln T_i/dx$, and $\langle \ \ \rangle$ denotes spatial average. The first term associated with f_i^2 on the left-hand side of the equation describes the time rate of change of entropy, whereas the second f_i^2 term is related to the collisionless dissipation due to nonlinear velocity space effects. The term on the right-hand side has been identified by Horton [10] as the entropy production term. In the simulation, in the steady-state stage of the evolution long after the saturation of the ITG modes, we have the indication that the balance of this equation is between the time rate of change of entropy and the entropy production term associated with the inhomogeneity. Contribution from all other terms are negligible, including the nonlinear velocity space effects. (One more interesting aspect of this equation is that the all-important $\mathbf{E} \times \mathbf{B}$ nonlinearity also does not contribute at all in this balance.) The implication here is that the energy flux observed in the simulation is the result of stochastic particle motion in response to the turbulent electrostatic potential field. Consequently, stochasticity produces diffusion and diffusion, in turn, increases entropy. This is the most likely origin of the energy flux observed in all of our gyrokinetic collisionless simulations. However, the corresponding picture for the fluid description of the ITG modes is quite different. The entropy balance equation in this case becomes [9]

$$\left(\frac{\partial}{\partial t}\right)\left\langle \frac{p_i^2}{\tau + u_i^2} + |\nabla_\perp \Phi|^2 + \Phi^2 \right\rangle /2 = \frac{\kappa_{Ti} Q_{ix}}{\tau}$$

where u_i is the perturbed ion velocity. In the steady state, the left-hand side of the equation vanishes and the energy flux should, therefore, also disappear. This is not really surprising, since we do not expect that single particle stochastic motion can be captured in the fluid description with truncated velocity moments. Thus, any observable energy flux in the fluid simulation must come from the externally introduced dissipation. This is in agreement with the conventional wisdom that fluid description is only valid in the collisional limit. In this limit, the two approaches should converge, when collisions are also introduced into the gyrokinetic particle simulation.

In this section, we have outlined the recent gyrokinetic particle simulation effort at PPPL and, hopefully, we have demonstrated the usefulness of such an approach in understanding anomalous transport due to ion temperature gradient drift modes in tokamaks. Our future plan includes the investigations of trapped electron instabilities, finite-β effects on microinstabilities, and energetic particle effects on MHD modes.

ACKNOWLEDGMENT

This work is supported by the U.S. Department of Energy Contract No. DE-AC02-76CHO3073.

Bibliography

[1] W.W. Lee, J. Comp. Phys. **72**, 243 (1987).

[2] A.M. Dimits and W.W. Lee (to appear in J. Comput. Phys.).

[3] S.E. Parker and W.W. Lee, Phys. Fluids B **5**, 77 (1993).

[4] W.M. Tang and G. Rewoldt (to appear in Phys. Fluids).

[5] R.J. Fonck, R.D. Durst, and S.F. Paul (to appear in Phys. Rev. Lett.).

[6] F.W. Perkins, *et al.*, Phys. Fluids B **5**, 477 (1993).

[7] E.J. Synakowski, *et al.*, Phys. Rev. Lett. **65**, 2255 (1990).

[8] J.V.W. Reynders, Ph.D. Thesis (1992).

[9] W.W. Lee and W.M. Tang, Phys. Fluids **31**, 612 (1989).

[10] W. Horton, Plasma Phys. **22**, 345 (1980).

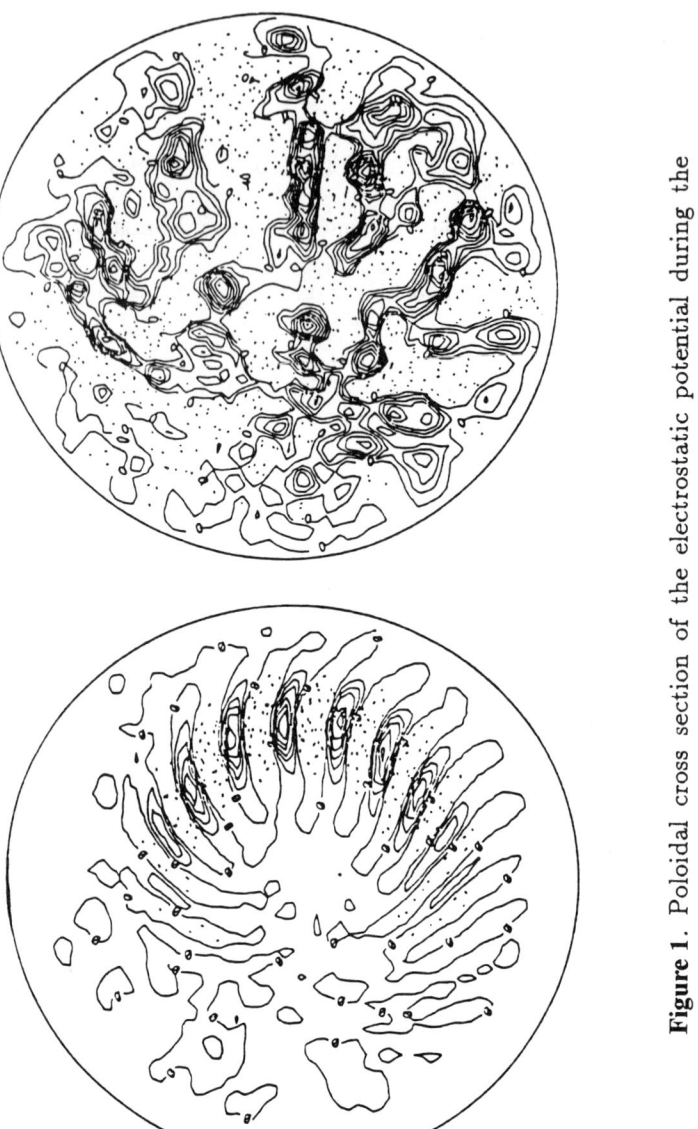

Figure 1. Poloidal cross section of the electrostatic potential during the linear phase $\Omega_i t = 500$ (left) and the saturated turbulent state $\Omega_i t = 12,500$ (right) of the ITG instability for the $128 \times 128 \times 64$ grid with 1,048,576 particles ($L_T = 50\rho_i$, $\Delta x = \Delta y = 1.5\rho_i$, $a = 96\rho_i$) in toroidal geometry.

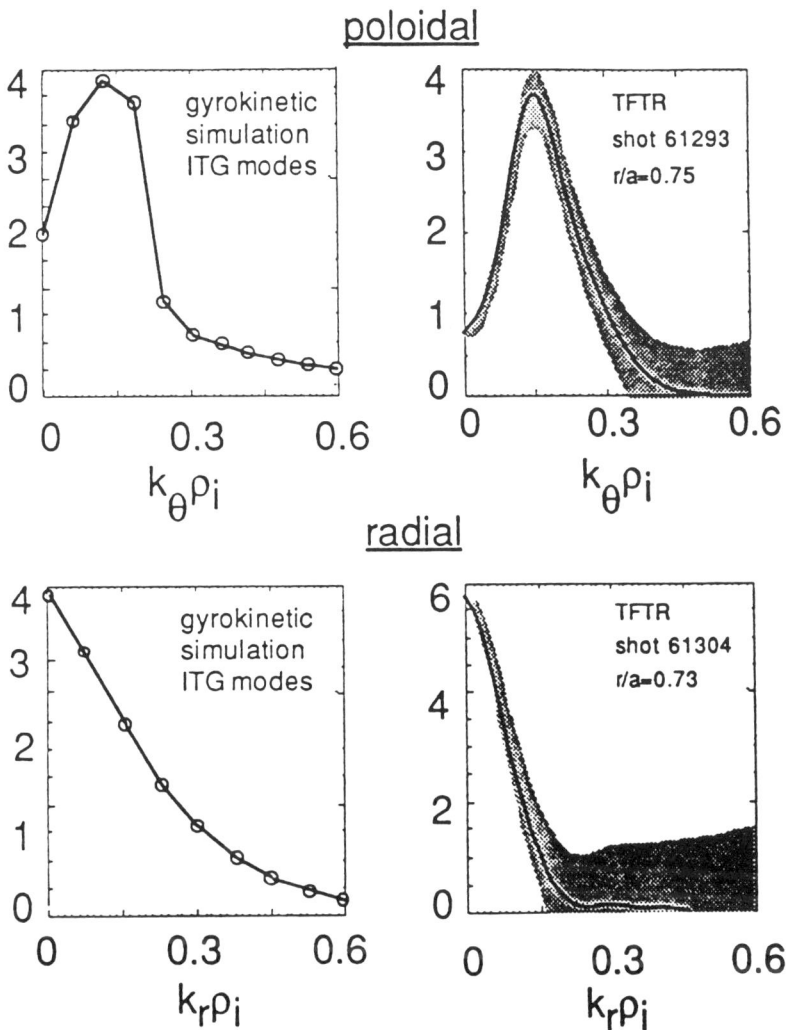

Figure 2. Fluctuation density spectra from the BES measurements of the TFTR supershot discharges and from the ITG simulation shown in Fig. 1.

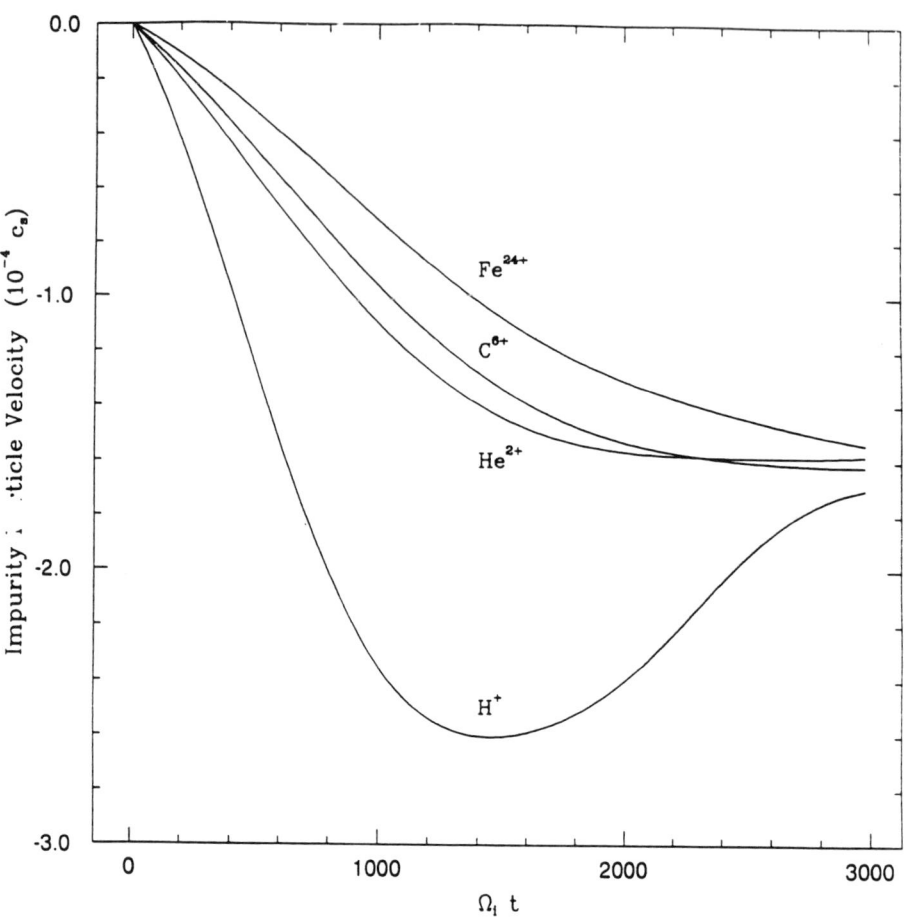

Figure 3. Inward pinch velocity for various cold impurities due to ion temperature gradient drift instabilities in 3D sheared slab simulation.

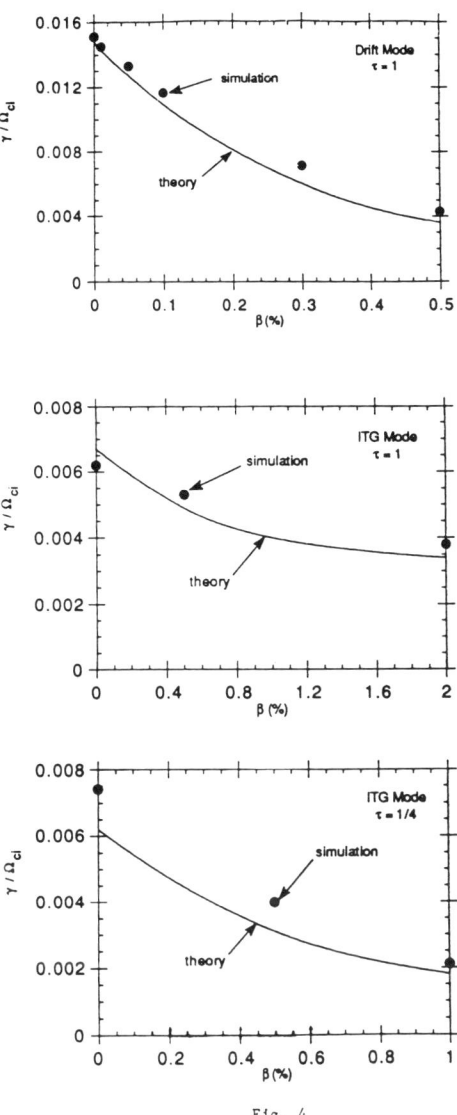

Figure 4. Finite-β stabilization of drift instabilities and ITG modes from 1D simulation.

4.2 3D GYROKINETIC PARTICLE SIMULATIONS OF ION TEMPERATURE GRADIENT-DRIVEN TURBULENCE AND TRANSPORT

R.D. Sydora
Department of Physics, University of California at Los Angeles, Los Angeles, California 90024-1547

Abstract. The ion temperature gradient-driven mode, in three dimensional sheared-slab geometry, is investigated using fully nonlinear and low noise, weighted-particle gyrokinetic particle-in-cell models. Comparisons between the two models are made and in certain regimes the agreement is good. The adiabatic, Boltzman response model is used for the electron density but alternative treatments are discussed in relation to the development of equilibrium poloidal shear flows. Single and multiple rational surface results from nonlinear simulations are presented for parameters relevant to present experimental regimes. Quasilinear versus turbulent saturation dynamics is contrasted and the ion thermal diffusivity is enhanced by nonlinear interaction between modes with different rational surfaces. Kinetic mixing length diffusivities give good estimates of the radial ion thermal flux measured in the simulations.

4.2.1 INTRODUCTION

Experiments have now firmly established that the ion thermal transport in high temperature tokamak plasmas is anomalous [1, 2, 3, 4]. This transport remains anomalous even in improved confinement regimes such as H-mode [5, 6, 7] and Supershot. [8] Based on experimental studies of the ion thermal confinement [9] and the fluctuation spectrum [10] it is widely believed that the presence of ion temperature gradient (ITG) drift instabilities is determining part of the heat confinement in tokamaks. Theoretical and numerical simulation studies of the ion temperature gradient drift mode, which becomes unstable when a threshold value of the steepness of the density and temperature gradient is reached, have focused on two main branches of the instability. One is the local (or slab) unstable branch and the stability properties depend mainly on the local magnetic shear as well as local gradient steepness. [11] The toroidal branch of instability is determined by these parameters as well but also the toroidal curvature. [12] Since the curvature varies from favorable to unfavorable in a tokamak, the global mode structure varies in such a manner as to give it a ballooning structure. Both branches of instability could co-exist in a tokamak plasma.

In this work we focus on the slab model of ITG-driven turbulence and consider approaches to simulating the linear and nonlinear behavior of these instabilities. Since the high temperature operating regime of present day tokamaks leads to more collisionless plasma behavior, intrinsic kinetic effects such as wave-particle resonances and finite gyroradius effects of the charged particles must be taken into account. Therefore, this paper utilizes the particle simulation approach where large collections of particles are advanced in time according to gyrophase-averaged drift equations of motion and a Poisson-type equation, which contains the low frequency dielectric response, is solved at each step to give the self-consistent electric field. We neglect the effects of plasma currents and self-consistent magnetic fields in this work. The drift equations of motion are used for the dynamics of the particles perpendicular to the equilibrium magnetic field and the exact motion is followed along the magnetic field. This is the main content of the fully nonlinear gyrokinetic model. In order to assess the importance of the nonlinear parallel dynamics, it is possible to construct a simulation model which linearizes about this parallel motion and hence verify the role played by individual nonlinearities. One of the goals of this paper is to compare these fully nonlinear and "partially" linearized models for an ITG turbulence model.

Two other aspects of linear and nonlinear ITG turbulence formation are addressed in this work. The first involves the interpretation of quasilinear versus turbulent saturation of ITG instabilities in three dimensional, multiple rational surface simulations. This has implications as to whether true steady state radial energy fluxes can be accurately simulated and what bounds we are able to get on the ion thermal confinement. The second issue considered here is the local generation of sheared $\mathbf{E} \times \mathbf{B}$ flows and whether this will have an effect of the saturation dynamics of ITG turbulence. These flows are observed in the simulations and their growth and development is partly determined by the choice made for the electron dynamics. The Boltzman response is used for the adiabatic electron dynamics, however, the choice for the electron number density is not a unique one. Two different models are presented in this paper: the high electron mobility model where the electron density is taken to be the average ion density and the low electron mobility where the electron density is given by its initial value.

Since the original suggestion that ITG drift modes could play a role in anomalous ion thermal transport [13, 14] there has been a growing body of theoretical, numerical simulation and experimental work in this area. Rather than attempt to summarize all of these developments, certain studies are mentioned which have some bearing on the subject of this paper. The kinetic integral equation describing the stability of radial ITG drift wave eigenmodes in a sheared slab geometry was first developed by Coppi *et al.* [11] and the most reliable and accurate numerical solutions were made by Waltz *et al.*[15], using the differential formulation, and by Linsker [16] who solved the integral eigenmode equation. The real frequency, growth rate, and threshold for instability as well as scaling with parameters were well known from the linear theory. The slab model of ITG turbulence was first studied with simple fluid models, numerically [17], and later analytically [18]. Early particle simulations [19] verified the presence of the localized eigenmodes in the sheared slab and further studies [20] emphasized the importance of higher radial mode numbers. The nonlinear initial-value fluid simulation model was used over a wider parameter regime and scaling runs in 2 and 3-dimensions yielded important parametric dependencies [21]. A three dimensional gyrokinetic particle simulation study [22] showed these same parametric variations but the magnitude of the thermal fluxes were much lower. Subsequent work with low noise gyrokinetic models designed to operate near marginal stability also confirmed that the magnitude of the thermal fluxes were lower due to kinetic effects localizing the electrostatic perturbations, hence reducing the effective step size involved in the radial thermal diffusion.[23] Gyro-Landau fluid models with proper closure to match the kinetic plasma response have also confirmed this trend [24]. The toroidal branch of instability has also been analytically investigated previously [12, 25] and nonlinear toroidal simulations are currently being carried out [26] Since the toroidal branch contains lower thresholds for instability as well as larger linear growth rates and radial mode widths it is possible that these ITG drift modes could dominate the ion thermal transport.

This paper is organized in the following way. Section 2 describes the equations of motion used in the particle simulations as well as the renormalized electrostatic potential equation. Section 3 outlines the parameter regime and simulation configuration. Section 4 contains the results from two and three dimensional simulations and Sec. 5 gives a discussion of the results. Finally, a summary is made in Sec. 6.

4.2.2 GYROKINETIC-VLASOV AND POISSON EQUATIONS

The derivation for the equations of motion for the particles following the characteristics of the nonlinear electrostatic gyrokinetic-Vlasov equation is given in Ref. [27]. The equations used in this study are given here for completeness.

Particle Equations of Motion

The equations of motion for the ith gyrokinetic particle in gyrocenter coordinates, $(\mathbf{R}, \mu, v_\parallel)$, are given by:

$$\frac{d\mathbf{R}_i}{dt} = v_{\parallel i}\hat{\mathbf{b}} - \left(\frac{q}{m\Omega}\right)\left(\frac{\partial \Psi}{\partial \mathbf{R}} \times \hat{\mathbf{b}}\right) \tag{1}$$

$$\frac{dv_{\parallel i}}{dt} = -\left(\frac{q}{m}\right)\frac{\partial \Psi}{\partial \mathbf{R}} \cdot \hat{\mathbf{b}} \tag{2}$$

$$\mu = \frac{mv_\perp^2}{2B} \tag{3}$$

where $\mathbf{R} = \mathbf{x} + \boldsymbol{\rho} = \mathbf{x} + \mathbf{v}_\perp \times \hat{\mathbf{b}}/\Omega$,

$$\hat{\mathbf{b}} = \mathbf{B}/B, \quad \Omega = qB/mc,$$

$$\Psi \equiv \overline{\Phi}(\mathbf{R}) - \left(\frac{q}{2T}\right)\left(\frac{v_t}{\Omega}\right)^2 \left|\frac{\partial \Phi}{\partial R_\perp}\right|^2,$$

$$\overline{\Phi}(R) = \Sigma_k \Phi(k) J_o(k_\perp v_\perp/\Omega) e^{i\mathbf{k}\cdot\mathbf{R}}$$

$$= \left\langle \int \Phi(x)\delta(x - R - \rho)dx \right\rangle_\varphi$$

and $\langle\ \rangle_\varphi \equiv \oint d\varphi/2\pi$. The gyroradius is denoted by ρ and φ is the gyrophase angle. These equations of motion are advanced in time using a second order predictor-corrector differencing scheme. The total distribution function, which is gyrophase-independent, can be constructed from $F(\mathbf{R}, \mu, v_\parallel, t) = \Sigma_i \delta(\mathbf{R}-\mathbf{R}_i)\delta(v_\parallel - v_{\parallel i})\delta(\mu - \mu_i)$ and this can be used to construct the density of guiding centers. The usual low frequency gyrokinetic ordering has been assumed [28] These equations have been written to second order in the gyrokinetic expansion parameter. The second order terms generally had little effect on the nonlinear evolution of the ITG drift instabilities considered in this paper. A Maxwellian distribution of perpendicular velocities is assumed.

In strongly magnetized plasmas, characteristic of tokamaks, the parallel and perpendicular dynamics naturally separate and hence the linear and nonlinear properties of the low frequency microinstabilities reflect this separation. Depending on the type of instability, parallel and/or perpendicular nonlinearities may become important. This property is useful in constructing particle simulation models, again working with the characteristics of the intrinsic particle orbits, which may have a component of it's evolution linear such as linear ballistic particle motion along the magnetic field or perhaps fully linearized equations of motion could be advanced in time. Generally, however, a nonlinearity such as the $\mathbf{E} \times \mathbf{B}$ is needed for the saturation dynamics of the instability. Several partially-linearized schemes have been proposed and have been useful for linear and saturation studies of systems with weak pressure gradients since they possess low noise properties [29].

Linearizing the parallel motion of the ions the modified characteristic equations become:

$$\frac{d\mathbf{R}_i^{(o)}}{dt} = v_{\parallel i}^{(o)} \hat{\mathbf{b}} - \left(\frac{q}{m\Omega}\right)\left(\frac{\partial \Psi}{\partial \mathbf{R}^{(o)}} \times \hat{\mathbf{b}}\right) \qquad (4)$$

$$\frac{dv_{\|i}^{(o)}}{dt} = 0 \tag{5}$$

$$\frac{dW_i}{dt} = -\left(\frac{q}{m}\right)\left(v_{\|i}^{(o)}/v_{th}^2\right)\frac{\partial \Psi}{\partial \mathbf{R}^{(o)}} \cdot \hat{\mathbf{b}} \tag{6}$$

$$\Delta \mathbf{R}_i = \mathbf{R}_i^{(o)}(t) - \mathbf{R}_i^{(o)}(t=0) \tag{7}$$

where $F(\mathbf{R}, \mu, v_\|, t) = f^{(o)} + f^{(1)} = \Sigma_i (1 + W_i + \kappa \Delta \mathbf{R}_j) \delta(\mathbf{R} - \mathbf{R}_i^{(o)}) \delta(v_\| - v_{\|i}^{(o)}) \delta(\mu - \mu_i)$ and $\kappa = \kappa_n - (3/2 - v^2/2v_{th}^2)\kappa_T$ where $\kappa_n \equiv -d \ln n_o/dx$ and $\kappa_T \equiv -d \ln T_o/dx$ are the zeroth-order spatial inhomogeneities which gives the well-known stability parameter $\eta_i = \kappa_T/\kappa_n$. If a Maxwellian distribution function is assumed for the initial distribution it will remain one for all time and therefore this will be a solution of the zero-order Gyrokinetic-Vlasov equation. This scheme has also been implemented in two and three dimensions. The results are compared in Sec. 4, with the fully nonlinear model given by Eqs. (1)–(3).

Field Equation

The self-consistent determination of the plasma evolution requires the electrostatic potential be determined from the macroscopic charge density accumulated from the charged particles. The Poisson relation for the low frequency gyrokinetic system in the particle coordinates is:

$$\nabla^2 \Phi - \frac{\tau}{\lambda_D^2}(\Phi - \tilde{\Phi}) + \left(\frac{\rho_s}{\lambda_D}\right)^2 \nabla_\perp \cdot \{(n_i - n_o)\nabla_\perp \Phi/n_o\} = -4\pi|e|(\bar{n}_i - n_e) \tag{8}$$

with the variables defined as: $\tilde{\Phi} = \Sigma_k \Phi(k) I_o(b) e^{-b} e^{i\mathbf{k}\cdot\mathbf{x}}$, $b = k_\perp^2 \rho_i^2$, $\tau = T_e/T_i$, $\lambda_D^2 = T_e/4\pi n_o e^2$, $\rho_s = \sqrt{\tau}\rho_i$, and

$$\bar{n}(x) = \Sigma_k \int F(k) J_o(k_\perp v_\perp/\Omega) e^{i\mathbf{k}\cdot\mathbf{x}} d\mu \, dv_\|$$

$$= \left\langle \int F(\mathbf{R}, \mu, v_\|) \times \delta(\mathbf{R} - x + \rho) d\mathbf{R} d\mu \, dv_\| \right\rangle_\varphi$$

where F is the discrete particle distribution. The equation emphasizes the dominant effect of the ion polarization shielding response over the Debye shielding effect. The form of the equation can be interpreted as being a normal Poisson equation but with an effective dielectric response for low frequency plasmas determining the physical cutoff scale of the potential response. This equation is solved iteratively in k-space and the convolutions, such as the third term on the left-hand side of Eq. (8) are done in real space. In the gyrophase averages, four points are adequate to resolve the finite Larmor radius effects [29] for $k_\perp \rho_i \leq 2$. The boundary conditions on the electrostatic potential are periodic in the y- and z-directions and the potential is set equal to zero at the boundaries in the x-direction. However, a periodic boundary condition in the x-direction was also implemented in the three dimensional sheared slab case and will be discussed later.

The electron density response is take to be adiabatic and hence the Boltzman relation can be used to relate the density and electrostatic potential. This appears to be an adequate model for ITG drift modes since the electrostatic perturbations are quasi-neutral type. The phase velocity of the ITG modes is much smaller than the electron thermal speed. Two models for the electron density response have been implemented and investigated since the specification of the average number density in the Boltzman relation is not unique [19]. However, physically realizable choices can be made when considering the wavelengths included in the simulation plasma. The first is the high electron mobility model where the electron number density is set equal to the average ion density. In this case the electrons move with the ions and the model can be used when the $k_\parallel = 0$ wavenumber is included in the simulation calculation. A second choice is to set the electron number density equal to the initial electron number density and this choice can quite naturally lead to the development of "ambipolar" modes which are associated with electron and ion charge separation on ion Larmor radii scales. We note that these modes can be large if the perpendicular energy of the ions is large and could play a role in the saturation and transport levels of ITG-driven fluctuations.

4.2.3 SIMULATION CONFIGURATION AND PARAMETERS

In the local or slab model, used extensively in microturbulence simulations and analytic models, the radial position corresponds to the x-direction, y is the poloidal direction and z the toroidal direction. The ambient magnetic field is given by $\mathbf{B} = B_o(\mathbf{z} + \mathbf{y}(x - x_o)/L_s)$, where x_o is a reference mode rational surface and L_s is the magnetic shear scale length. The parallel wavenumber for this magnetic field is given by $k_\parallel(x) = k_z + k_y(x - x_o)/L_s$ and the mode rational surface, for a given (k_y, k_z), is determined by the condition $k_\parallel = 0$. Defining $k_y = 2\pi m/L_y$ and $k_z = 2\pi n/L_z$ the radial position of the mode rational surfaces is given by $x_{mn} = x_o \pm (n/m)(L_s L_y/L_z)$ in the three dimensional slab model. The ITG drift wave localizes around the mode rational surface because of the small parallel wavenumber there, and the wave eigenmode also has finite radial extent because of ion Landau damping and the linear radial mode width is determined by the place where $k_\parallel(x) v_{ti} \approx \omega$ and ω is the ITG drift mode frequency. This localization width is roughly 5-10 ion gyroradii in radial width, for typical shear scale lengths in tokamaks and since the mode rational surface spacing is much less than this, three dimensional effects play an important role in the radial transport of ion thermal energy. A crude estimate of the radial mode rational surface spacing is given by the formula $\Delta x_{mn} = (L_y L_s/L_z)n/m(m+1)$ where overlap is easily satisfied for high m (or shorter poloidal wavelength) modes.

The linear stability properties of the ITG drift wave eigenmodes have been previously studied using an integral eigenvalue code.[16] We use such a code to determine the stability of the various wavenumbers used in our particle simulations. The real frequency, growth rate and radial mode structure of the ITG drift modes are compared with the linear modes which grow initially in the simulation calculations. The simulations give us information about the saturation dynamics and subsequent thermal fluxes. The linear analysis is important for helping us delineate the regions of stability and instability for various (k_y, k_z) modes because one of the principal mechanisms of saturation involves coupling of unstable modes to damped ones. Therefore, one must have an accurate picture of where the most unstable modes occur and where the stable wave spectrum begins.

The numerical eigenmode analysis gives the most reliable estimates of the growth rate, real frequency, and radial mode structure. A typical result from this analysis is shown in Fig. 1. We illustrate here the growth rate and real frequency versus $k_y \rho_i$ and the corresponding radial variation of the electrostatic potential versus x/ρ_i for the fastest growing normal modes. The parameters used for this linear analysis correspond to $L_s/L_n = 5$, $\eta_i = 4$, $T_e/T_i = 1$, $m_e/m_i = 0.000544$. From Fig. 1a we note that the dominant mode is the lowest even parity eigenmode and the lowest odd parity mode is only weakly unstable. As the shear becomes weaker higher order radial eigenmodes are more unstable. Fig. 1b shows that the eigenmodes are localized to within a few ion gyroradii of the mode rational surface located at $x = 0$.

4.2.4 SIMULATION RESULTS

The results of two and three dimensional simulations using the fully nonlinear gyrokinetic model and the partially linearized model are presented in this section. A comparison is made between the two methods for the single rational surface case. Simulation results for the three dimensional, multiple rational surface configuration are also given and the saturation dynamics is contrasted with the two dimensional case.

2D-Single Rational Surface Model

We first consider the following test case in order to facilitate a comparison between the nonlinear and partially linearized gyrokinetic model. The ion temperature profile is initialized to be exponential because this gives a constant temperature scale length which simplifies the analysis. The initial ion density is taken to be uniform and the ratio of magnetic shear to temperature gradient scale length is given by $L_s/L_T = 40$ where $L_T \equiv 1/\kappa_T$. Other parameters used were: $T_e/T_i = 1$, $m_e/m_i = 0.000544$, $L_x \times L_y = 64\rho_s \times 32\rho_s$, $n_o = 1500$ particles/cell, and time step $\Omega_{ci}\Delta t = 8.0$. The simulations were run from $\Omega_{ci}t = 0 - 40000$ which corresponded to a time $t = 1075 L_T/c_s$. The most unstable radial eigenmode, corresponding to wavenumber

$k_y \rho_i = 0.2$, was the lowest odd parity ($l = 1$) mode with complex frequency $\omega/\omega_{*T} = -0.3 + 0.025i$.

The time evolution of the thermal profiles for this single mode rational surface case are given in Fig. 2 using the nonlinear gyrokinetic model. The arrow indicates the position of the mode rational surface. In this case there are no heat sources present and the ion temperature gradient mixes over the fastest growing radial eigenmode and ultimately saturates by quasilinear modification of the background thermal gradient. Subtracting the initial temperature profile from the profile attained at saturation gives the perturbed thermal profile versus radius and this is illustrated in Fig. 2b. The density profile is shown in Fig. 2c.

Using the same set of parameters as the previous nonlinear simulation, the partially linearized model was used to compare the evolution of the thermal profiles and the electrostatic energy. The partially linearized model only followed the perturbed distribution and only 16 particles/cell were required. Fig. 3a shows a comparison of the electrostatic energy time history in the two models. Good agreement is found for the saturated amplitudes and a more careful comparison of the time evolution of individual Fourier modes showed excellent agreement in the two cases. The radial ion temperature perturbation, averaged over the y-direction, is illustrated in Fig. 3b and it also shows good agreement between the two simulation models in the saturated state. In these two dimensional simulations with single rational surface, in the absence of heat sources, the $m = 0$ (or $k_y = 0$) temperature perturbation grows locally and cancels out the background temperature gradient. The behavior of this mode in three dimensional simulations will be addressed later.

Previously we just considered a few unstable k_y modes and examined saturation under conditions of background profile modification. We next examine the single rational surface case but where many k_y modes are present so that there are growing modes and a sufficient number of damped modes at short wavelength to allow for wave energy saturation by coupling to these damped modes. Let us first examine, however, the evolution of a single unstable ITG drift wave eigenmode in the presence of both density and temperature gradients. The parameters chosen were the same as the linear eigenmode analysis discussed in the previous section and summarized in Fig. 1. In this case

$L_s/L_n = 5$ and $\eta_i = 4$. The evolution of a single mode from the partially linearized simulation model is show in Fig. 4 and corresponds to wavelength $k_y \rho_s = 0.5$. The radial eigenmode structure is most clearly 'visualized' by plotting the v_x versus x particle 'phase space', where v_x corresponds to an $\mathbf{E} \times \mathbf{B}$ drift component for each particle. This is given in Fig. 4a and the particles serve as markers for the radial electric potential structure. The corresponding electric potential contours are shown in Fig. 4b and the results of the linear mode properties agree very well with the kinetic eigenmode analysis based on the integral equation solution. During the evolution of this single mode, saturation occurs by local flattening of the temperature gradient over the eigenmode width. We have attempted to minimize this effect on saturation by using a localized heat source. This was done by using Monte Carlo methods where particles were re-thermalized, consistent with the background thermal profile, on a characteristic time scale which was short compared to the background relaxation time scale. Using this method we were able to prevent the saturation of the single or few unstable modes and the simulation eventually terminated because of the lack of a sink for the wave energy.

We increased the wavenumber regime to include growing as well as damped ITG drift modes and the wavenumber range $0.01 \leq k_y \rho_i \leq 1.5$ was used. The system size in this case was $L_x \times L_y = 32\,\rho_s \times 512\rho_s$. This corresponded to the inclusion of over 120 modes in the y-direction. Without the localized heat source saturation resulted from the competition between coupling to short wavelengths and localized relaxation of the background ion temperature gradient. The potential contours for this case are shown in Fig. 5 and the development of smaller and smaller scales was observed as the modes saturated. When the heat source was applied to maintain the background gradient this saturation process was made very clear and a quasi-steady state could be maintained.

The radial heat flux in this two dimensional case remains localized to a small region around the rational surface. Even though a saturated spectrum is present with a mean gradient the heat can only be convected a small distance radially. The radial step size of the heat diffusion coefficient is approximately the linear radial eigenmode width. As will be seen in the next section the coupling of radial eigenmodes can lead to a turbulent radial diffusion process which can enhance the anomalous radial ion heat flux.

One final result on the evolution of ITG drift modes for a single rational surface concerns the spontaneous development of a localized radial electric field. This is given in Fig. 6. The $m = 0$ or $k_y = 0$ component of the electrostatic potential versus radius is shown at two different time steps. The localized radial electric field accompanies the saturation of the unstable modes and is of sufficient magnitude to affect the saturation level. It is more pronounced in the low electron mobility model where the electron density response is set equal to its initial value. This electric field remains present in the saturation phase of the fluctuations and appears to be generated by a mode coupling process.

3D-Multiple Rational Surface Model

The addition of three dimensional electrostatic perturbations is accomplished by adding finite k_z wavenumbers to the previous simulations. The number of toroidal modes was taken to be 8 and there were 120-k_y modes which gave a total of 960 modes in the 3D simulations. The length of the simulation region in x was $64\rho_s$. The number of grid points in the z-direction was equal to 32 and a stretched grid was used so that the spacing of the mode rational surfaces could be adjusted and the rational surfaces filled the entire region of the simulation domain. Two sets of boundary conditions were imposed on the electrostatic potential at the left and right boundaries. The first involved matching the left and right x-boundary mode rational surfaces giving a periodic radial boundary condition. In this case the heat flows continuously and circulates through the system. The second involved fixing the electrostatic potential at both boundaries and adding a heat source over the radial length of the simulation domain. We present results from the latter case only in this paper. The parameters were chosen to be the same as the linear eigenmode analysis presented in Fig. 1.

The time evolution of the total electrostatic energy is shown in Fig. 7a. and saturation is achieved at approximately $t = 500 L_n/c_s$. The time evolution of the radial thermal flux is given in Fig. 7b and shows an oscillatory behavior in the post-saturation phase. The radial heat flux is determined by the correlation of the fluctuating radial velocity with the pressure fluctuation and is written as

$$Q_i = \langle \hat{v}_x \hat{p}_i \rangle \ . \tag{9}$$

In Fig. 8a the electrostatic ITG drift eigenmodes are shown to develop on the closely spaced, mode rational surfaces and the v_x versus x phase space clearly illustrates the radial mode variation and coupling. The resultant radial temperature variation averaged over the y- and z-directions, corresponding to the ($k_y = 0, k_z = 0$) mode of the temperature, is shown in Fig. 8b. It is evident that local flattening and steepening of the temperature on many overlapping rational surfaces leads to a fine-grained turbulent oscillations but an overall mean gradient persists. This is the signature of turbulent saturation. If no heat source is present this saturation process still occurs but over later times the smallest radial wavenumbers dominate and the mean temperature gradient profile relaxes.

This can be observed in the following case where the rational surfaces are located more in the interior of the simulation domain and no heat sources are present. The time evolution of the total electrostatic energy is shown in Fig. 9a and the corresponding mode structure using the v_x versus x particle phase space, is shown in Fig. 9b. The rational surfaces are concentrated only within about $10\rho_s$ from the reference rational surface located in the middle of the simulation domain. Fig. 10a gives the temperature perturbation profile at the time of saturation and turbulent saturation is achieved. However, at a later time one can observe in Fig. 10b that the the smallest radial wavenumber, which roughly corresponds to the total radial width that the rational surfaces occupy, dominates and the mean temperature gradient has relaxed.

Finally, in Fig. 11 the three dimensional electrostatic potential surfaces of constant potential are presented to illustrate the breakup of the linear modes into shorter wavelength turbulence. In the linear phase, shown in Fig. 11a, the coherent linear eigenmode structures dominate but at later time, after mode coupling has occurred, the potential develops smaller scale structure and coherent vortices break up.

4.2.5 DISCUSSION OF RESULTS

The results of the two and three dimensional study of the linear and nonlinear evolution of the ITG drift modes indicates that the saturation dynamics and resultant thermal fluxes tend to be quite different. In the single rational surface simulations without heat sources leads to simple local profile relaxation and hence there is no net steady state radial thermal flux. When the profile is not allowed to relax then poloidal mode coupling leads to transfer of wave energy to shorter wavelengths where damped modes are present. This indicates that the $\mathbf{E} \times \mathbf{B}$ nonlinearity is dominant since it is the main effect which can cause this transfer of energy.

In the three dimensional case a net thermal energy flux is observed due to interaction among various unstable ITG drift eigenmodes centered about different mode rational surfaces. Turbulent saturation is observed prior to relaxation of the mean ion thermal gradient and in the transition to turbulence smaller scale potential structures develop. We also note here that a detailed examination of the spectrum in (k_y, k_z) wavenumber space reveals that some fraction of the energy in the most unstable wavelengths also transfers to longer wavelengths in addition to the transfer of energy to the shorter, more stable modes. The additional degrees of freedom inherent in the three dimensional system make this re-arrangement of the wave energy possible. The final relaxed state or quasi-steady state spectrum is observed to be a balance between wave growth at the intermediate poloidal wavenumbers, $k_\theta \rho_s (1 + \eta_i)/\tau \approx 1$, and damped modes at the shorter wavelengths.

The heat transport coefficient, χ_i, is defined by

$$\chi_i = -Q_i/(NdT_i/dx) \tag{10}$$

where Q_i is given in Eq. (9) and it can also be expressed in the following manner,

$$\chi_i = -L_T \, \mathrm{Im}(\Sigma_k (ck_\theta \hat{\phi}) \delta T_i / B_o T_i) \ . \tag{11}$$

The time evolution of the radial ion heat flux is shown in Fig. 7b and in the saturation phase of the instability the time-averaged value is determined to be $\langle Q_i/c_s \rangle = 0.37 \times 10^{-4}$. This gives a normalized ion heat conductivity $\chi_i/(\rho_s^2 c_s/L_n) \simeq 0.19$, where the simulation values

$\rho_s/L_n = 0.00625$ and $\eta_i = 4$ were used. Let us compare this value to the linear estimate of χ_i using the kinetic mixing length results. Defining the average radial mode width as

$$\Delta_x^2 = \langle \phi^2 \rangle / \langle (d\phi/dx)^2 \rangle \qquad (11)$$

where $\langle\ \rangle$ is a volume average, the mixing length estimate of $\gamma \Delta_x^2$ gives $\chi_i \equiv \gamma L_n/(c_s \langle k_\perp^2 \rho_i^2 \rangle) = 0.26$. The lowest order, even parity radial eigenmode mainly determined the linear mode width and the growth rate. This kinetic mixing length estimate compared favorably with the nonlinear three dimensional simulation result. We note here that these thermal fluxes are about an order of magnitude below the experimentally determined ion thermal conductivity.

The presence of a finite radial electric field which gives rise to sheared poloidal flows can also regulate the stability of the system in the saturated state. The mean flows observed in the simulations accompany the saturation phase. If these flows can be damped by viscosity or other mechanisms this can affect the fluctuations and hence the anomalous transport rates may fluctuate greatly. The role of this radial electric field in the nonlinear evolution of ITG drift modes is currently being explored.

4.2.6 SUMMARY

The results obtained in this paper can be summarized in the following manner. First, we have attempted to benchmark the fully nonlinear gyrokinetic particle simulation model with more approximate models of the plasma dynamics in order to gain confidence that we are not leaving out crucial elements in the nonlinear behavior of ITG drift instabilities. A select test case is presented but studies in other parameter regimes and in three dimensions have been made and the limitations of the partially linearized simulation models are now more clear. Approximate models with low noise properties have the unique advantage that fewer particles may be used to simulate the growth and saturation of unstable modes and also the behavior of instabilities near marginal stability can be more easily studied.

Second, the saturation dynamics of ITG drift modes in single and multiple rational surface, sheared slab geometry has been investigated and we have found that the nonlinear interaction time scale can be much shorter, and in fact much less, than the relaxation time scale of mean thermal gradients. The turbulent saturation or transition to turbulence of the ITG drift modes is accompanied by the generation of small scale fluctuations which are heavily damped and the generation of localized radial electric fields giving rise to mean poloidal flows. The saturated state of the electrostatic potential fluctuations reveals that the dominant energy is in the wavenumbers $\langle k_y \rho_s \rangle^{rms} \simeq 0.5$. The thermal flux in the saturated state is in rough quantitative agreement with the kinetic mixing length estimates.

Finally, the ion thermal flux in the kinetic regime we consider here is about one order of magnitude lower than is observed in present day, high temperature tokamaks which motivates one to seriously consider toroidal effects on these instabilities. The larger radial correlation lengths and lower threshold values for the toroidal branch of the ITG instability could make this a more dominant process in governing the ion thermal confinement properties in tokamaks. Also, the presence of other instabilities and finite electron-ion collisions may be important as has been found in a previous study using the two dimensional shearless slab model [31].

ACKNOWLEDGMENTS

The author would like to thank Prof. John M. Dawson for many helpful discussions as well as encouragement and Dr. W.W. Lee for computational advice regarding the partially linearized simulation model and many aspects of ITG instabilities. The author would also like to thank B. Dorland for helpful discussions on the linear kinetic integral eigenvalue solutions.

Bibliography

[1] S. Ejima, T.W. Petrie, A.C. Riviere, T.R. Angel, C.J. Armentrout, D.R. Baker, F.P. Blau, G. Bramson, N.H. Brooks, R.W. Callis, R.P. Chase, J.C. DeBoo, J.S. DeGrassie, E.S. Fairbanks, R.K. Fisher, R.J. Groebner, C.L. Hsieh, J. Hugill, G.L. Jahns, J.M. Lohr, J.L. Luxon, M.A. Mahdavi, F.B. Marcus, N. Ohyabu, P.I. Petersen, W.W. Pfeiffer, R.P. Seraydarian, A.M. Sleeper, R.T. Snider, R.D. Stambaugh, T. Tamano, T.S. Taylor, R.E. Waltz, J.C. Wesley, and S.S. Wojtowic, Nucl. Fusion **22**, 1627 (1982).

[2] B. Blackwell and ALCATOR Group, in *Plasma Physics and Controlled Nuclear Fusion Research 1982*, Proceedings of the 9th International Conference, Baltimore (IAEA, Vienna, 1983), Vol. 2, p. 27.

[3] M. Greenwald and ALCATOR Group, Phys. Rev. Lett. **53**, 352 (1984).

[4] R.J. Groebner, W.W. Pfeiffer, F.P. Blau, and K.H. Burrell, Nucl. Fusion **26**, 543 (1986).

[5] K.H. Burrell, S. Ejima, D.P. Schissel, N.H. Brooks, R.W. Callis, T. Carlstrom, A.P. Colleraine, J.C. DeBoo, H. Fukumoto, R.J. Groebner, D.N. Hill, R.-M. Hong, N. Hosgane, G.L. Jackson, G.L. Jahns, G. Janeschitz, A.G. Kellman, J. Kim, L. Lao, P. Lee, J. Lohr, J.L. Luxon, M. Ali Mahdavi, C. P. Moeller, N. Ohyabu, T.H. Osborne, D. Overskei, P.I. Petersen, T.W. Petrie, J.C. Phillips, R. Prater, J.T. Scoville, R.P. Seravdarian, M. Shimada, B.W. Sleaford, R.T. Snider, R.D. Stambaugh, R.D. Stav, H. St. John, R.E. Stockdale,

E.J. Strait, T.S. Taylor, J.F. Tooker, and S. Yamaguchi, Phys. Rev. Lett. **59**, 1432 (1987).

[6] C. Gowers, D. Barlett, A. Boileau, S. Corti, A. Edwards, N. Gottardi, K. Hirsch, M. Keilhacker, E. Lazzaro, P. Morgan, P. Nielsen, J. O'Rourke, H. Salzmann, P. Smeulders, A. Tanga, and M. von Hellermann, in *Proceedings of the 15th European Conference on Controlled Fusion and Plasma Heating* (European Physical Society, Budapest, 1988), Vol. 12B, Pt 1, p. 2239.

[7] D.P. Schissel, R.E. Stockdale, H. St. John, and W.M. Tang, Phys. Fluids **31**, 3738 (1988).

[8] S.D. Scott, P.H. Diamond, R.J. Fonck, R.J. Goldston, R.B. Howell, K.P. Jaehnig, G. Schilling, E.J. Synakowski, M.C. Zarnstorff, C.E. Bush, E. Fredrichson, K.W. Hill, A.C. Janos, D.K. Mansfield, D.K. Owens, H. Park, G. Pautasso, A.T. Ramsey, J. Schivell, G.D. Tait, W.M. Tang, and G. Taylor, Phys. Rev. Lett. **64**, 531 (1990).

[9] M.C. Zarnstorff, C.W. Barnes, P.C. Efthimion, G.W. Hammett, W. Horton, R.A. Hulse, D.K. Mansfield, E.S. Marmar, K. McGuire, G. Rewoldt, B.C. Stratton, E.J. Synakowski, W. Tang, J. Terry, X.Q. Xu, M.G. Bell, M. Bitter, N.L. Bretz, R. Budny, C.E. Bush, G.J. Fonck, E.D. Fredrichson, H.P. Furth, R.J. Goldston, B. Grek, R.J. Hawryluk, K.W. Hill, H. Hsuan, D.W. Johnson, D.C. McCune, D.M. Meade, D. Mueller, D.K. Owens, H.K. Park, A.T. Ramsey, M.N. Rosenbluth, J. Schivell, G.L. Schmidt, S.D. Scott, G. Taylor, and R.M. Wieland, in *Plasma Physics and Controlled Nuclear Fusion Research, 1990*, Proceedings of the 12th International Conference, Washington (IAEA, Vienna, 1991), Vol. I, p. 109.

[10] D.L. Brower, W.A. Peebles, S.K. Kim, N.C. Luhmann,Jr., W.M. Tang, and P.E. Phillips, Phys. Rev. Lett. **59**,49 (1987).

[11] B. Coppi, M.N. Rosenbluth, and R.Z. Sagdeev, Phys. Fluids **10**, 582 (1967).

[12] 2F. Romanelli, Phys. Fluids B **1**, 1018 (1989).

[13] B.B. Kadomtsev and O.P. Pogutse in *Reviews of Plasma Physics* (Consultants Bureau, New York, 1970), Vol. 5, p. 249.

[14] L.I. Rudakov and R.Z. Sagdeev, Dokl. Akad. Nauk SSSR **138**, 581 (1961) [Sov. Phys. Dokl. **6**, 415 (1965)].

[15] R.E. Waltz, W. Pfeiffer, and R.R. Dominguez, Nucl. Fusion **20**, 43 (1980).

[16] R. Linsker, Phys. Fluids **24**, 1485 (1981).

[17] W. Horton, R.D. Estes, and D. Biskamp, Plasma Phys. **22**, 663 (1980).

[18] G.S. Lee and P.H. Diamond, Phys. Fluids **29**, 3291 (1987).

[19] W.W. Lee, W.M. Tang, and H. Okuda, Phys. Fluids **23**, 2007 (1980).

[20] P.W. Terry, J.N. Leboeuf, P.H. Diamond, D.R. Thayer, J.E. Sedlak, and G.S. Lee, Phys. Fluids **31**, 2920 (1988).

[21] S. Hamaguchi and W. Horton, Phys. Fluids B **2**, 1833 (1990).

[22] R.D. Sydora, T.S. Hahm, W.W. Lee, and J.M. Dawson, Phys. Rev. Lett. **64**, 2015 (1990).

[23] M. Kotschenreuther, H.L. Berk, R. Denton, S. Hamaguchi, W. Horton, C.-B. Kim, M. Lebrun, P. Lyster, S. Mahajan, W.H. Miner, P.J. Morrison, D. Ross, T. Tajima, J.B. Taylor, P.M. Valanju, H.V. Wong, S.Y. Xiao, and Y.-Z. Zhang, in *Plasma Physics and Controlled Nuclear Fusion Research, 1990*, Vol. 2, p. 361, IAEA, Vienna, 1991.

[24] B. Dorland and G.W. Hammett, Princeton Plasma Physics Laboratory Report, PPPL-2874, 1992, Phys. Fluids B (in press).

[25] S.C. Guo, L. Chen, S.T. Tsai, and P.N. Guzdar, Plasma Phys. Controlled Fusion **31**, 423 (1989).

[26] M. Beer, G.W. Hammett, W. Dorland, and S.C. Cowley, Bull. Amer. Phys. Soc. **37**, 1478 (1992).

[27] D.E. Dubin, J.A. Krommes, C. Oberman, and W.W. Lee, Phys. Fluids **26**, 3524 (1983).

[28] E. Frieman and L. Chen, Phys. Fluids **25**, 502 (1982).

[29] A. Dimits and W.W. Lee, Princeton Plasma Physics Report, PPPL-2718, 1990, (to be published in J. Comp. Phys.).

[30] W.W. Lee, J. Comp. Phys. **72**, 243 (1987).

[31] W.W. Lee and W.M. Tang, Phys. Fluids **31**, 612 (1988).

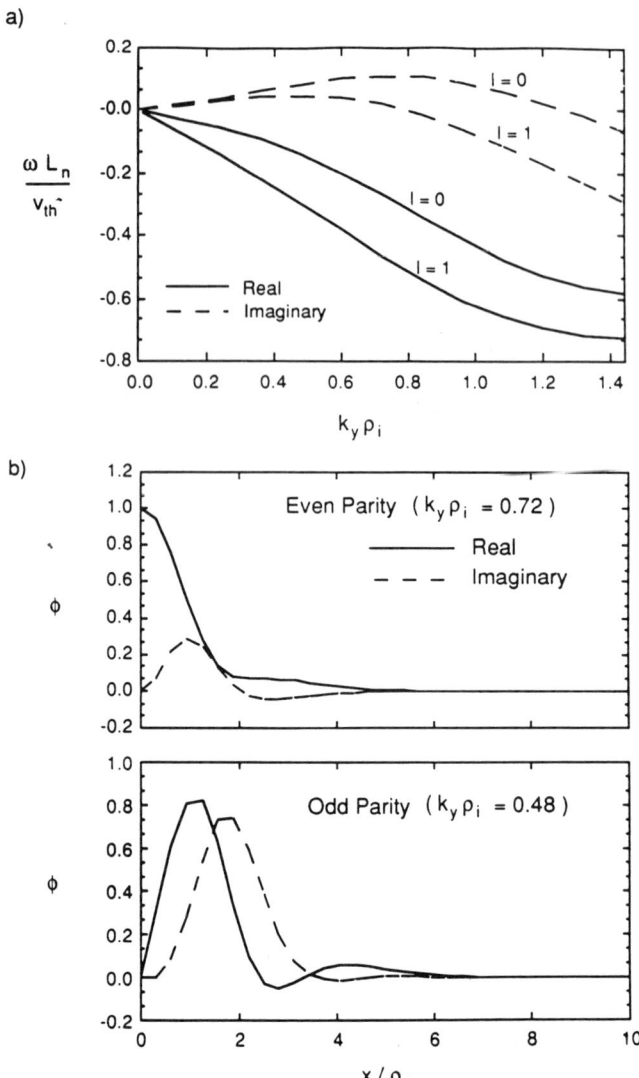

Figure 1. Real frequency and growth rate and radial eigenmode variation for even and odd parity modes with $\eta_i = 4$, $L_s/L_n = 5$ and $T_e/T_i = 1$. a) Real and imaginary frequency for most unstable even ($l = 0$) and odd ($l = 1$) eigenmode as a function of wavenumber. b) Radial eigenmode variation for most unstable wavelengths. Rational surface is located at $x = 0$.

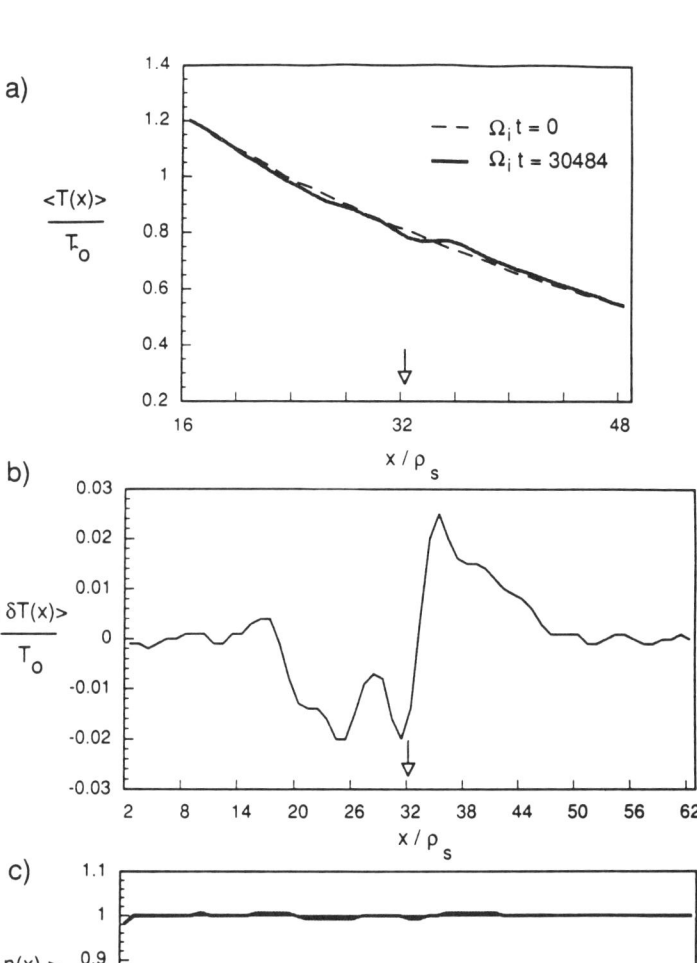

Figure 2. Temperature and density profiles, averaged over y at different time steps from fully nonlinear gyrokinetic model with $L_s/L_T = 40$. a) Initial ion temperature profile and at saturation. The solid arrow denotes the position of the mode rational surface. b) Perturbed ion temperature profile at saturation of the instability. c) Initial density profile and at saturation.

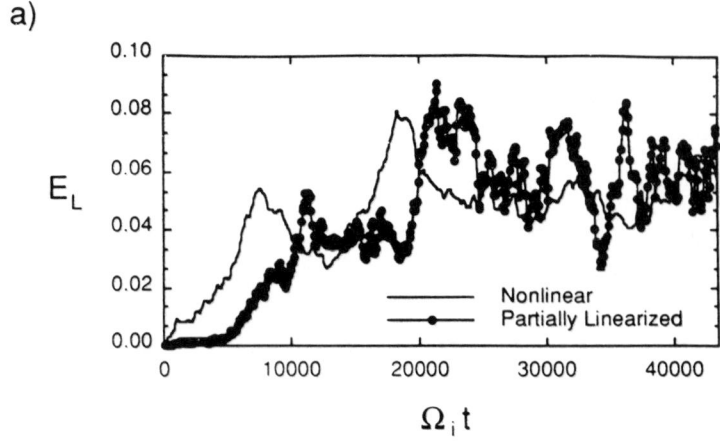

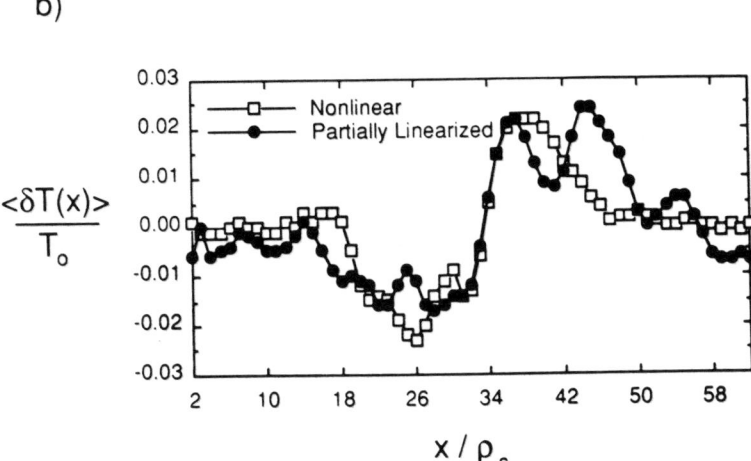

Figure 3. Comparison between fully nonlinear and partially linearized simulations illustrating the total electrostatic energy time evolution and y-averaged temperature perturbation from both models. a) Total electrostatic energy versus time. b) Temperature perturbation versus x at various times.

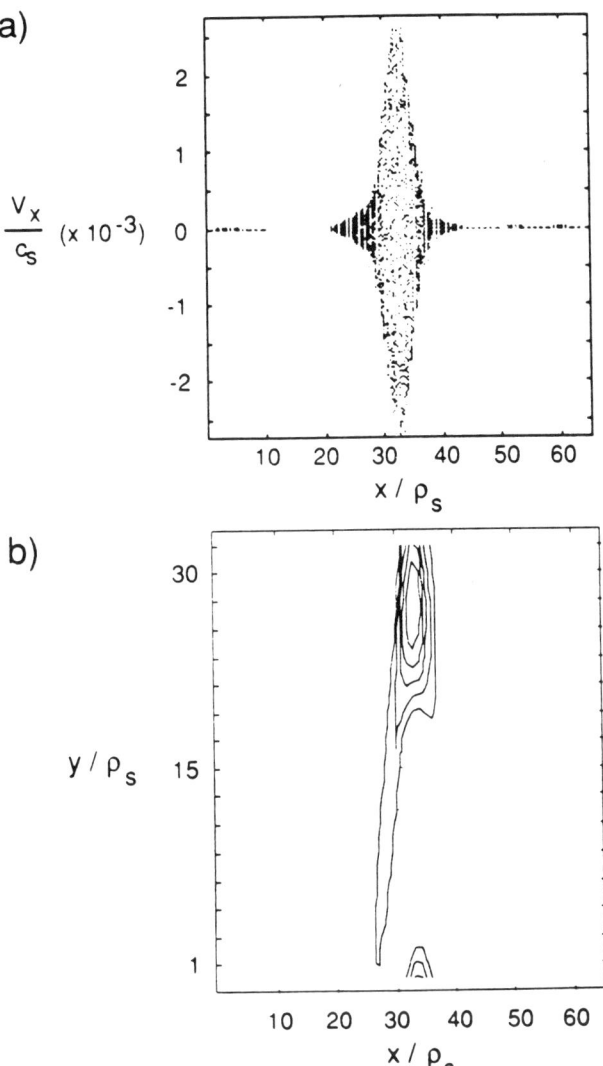

Figure 4. $x - v_x$ phase space illustrating the radial electrostatic potential mode structure and potential contours taken at $\Omega_{ci} t = 52,253$ for $L_s/L_n = 5$, $L_x \times L_y = 64\rho_s \times 32\rho_s$, $\eta_i = 4$ and $T_e/T_i = 1$. a) Ion $x - v_x$ phase space and rational surface is at $x/\rho_s = 32$. b) Electrostatic potential contours.

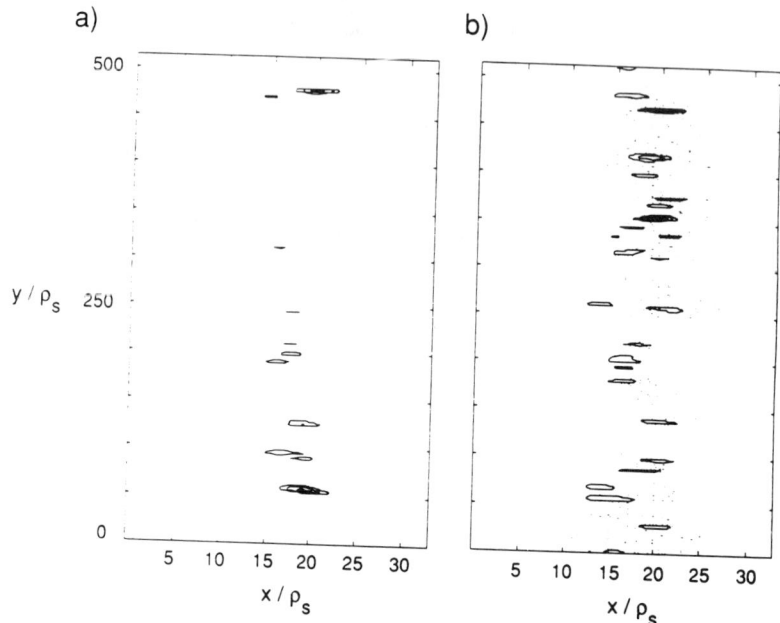

Figure 5. Electrostatic potential contour surfaces for $L_x \times L_y = 32\rho_s \times 512\rho_s$, $L_s/L_n = 5$ and $\eta_i = 4$ at time level a) $\Omega_{ci}t = 69,670$ and b) $\Omega_{ci}t = 87,088$.

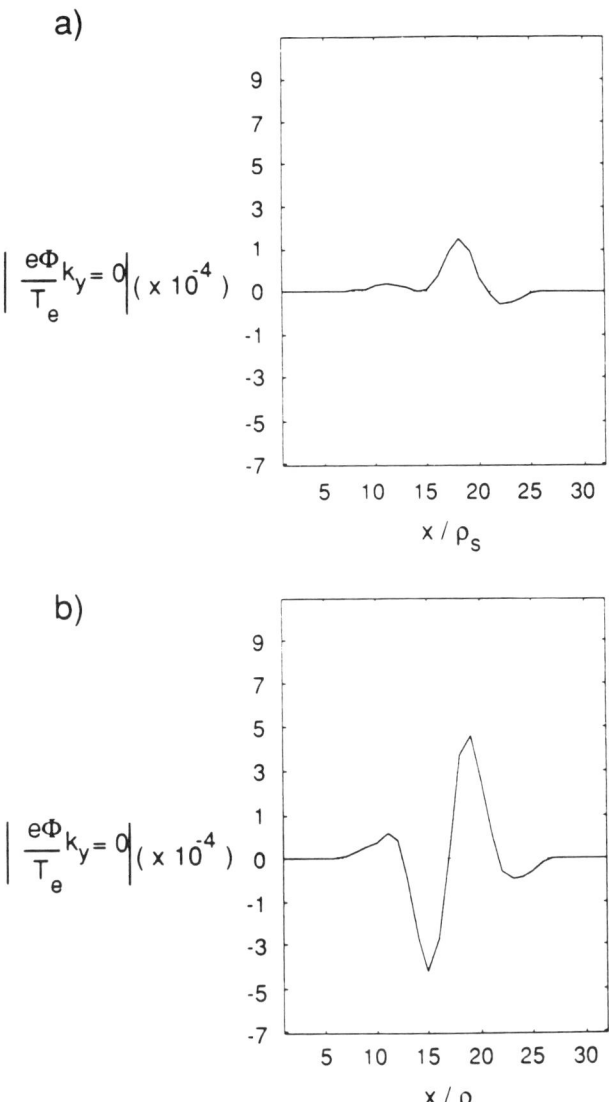

Figure 6. Electrostatic potential versus x for $k_y = 0$ mode at time levels a) $\Omega_{ci}t = 52,253$ and b) $\Omega_{ci}t = 60,962$.

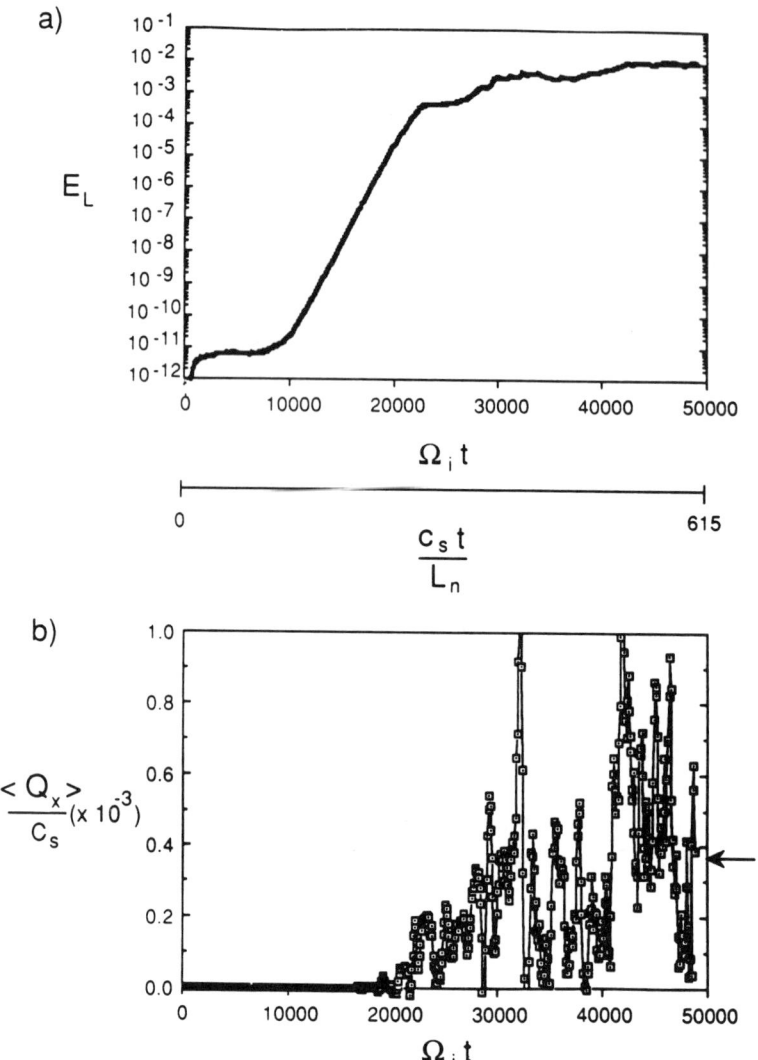

Figure 7. Time evolution of the total electrostatic energy and the radial thermal flux for the three dimensional case with $\eta_i = 4$ and $L_s/L_n = 5$. a) Total electrostatic energy versus time. b) Radial ion thermal flux versus time.

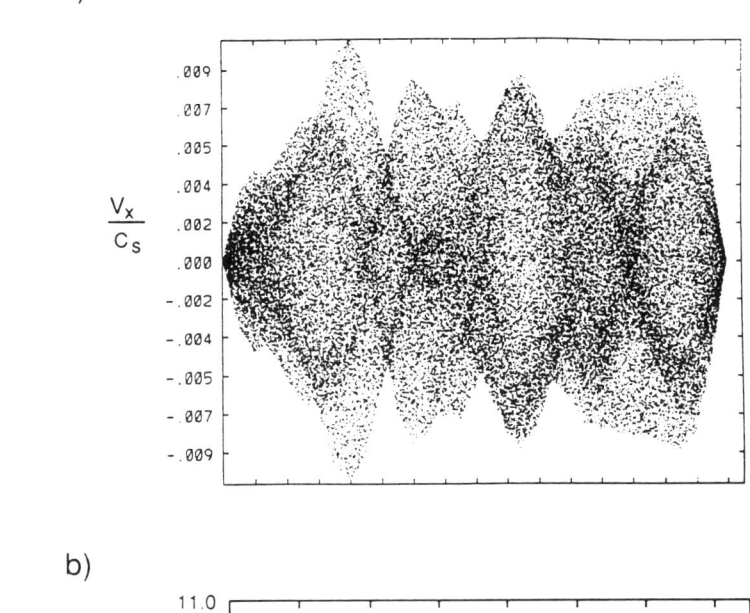

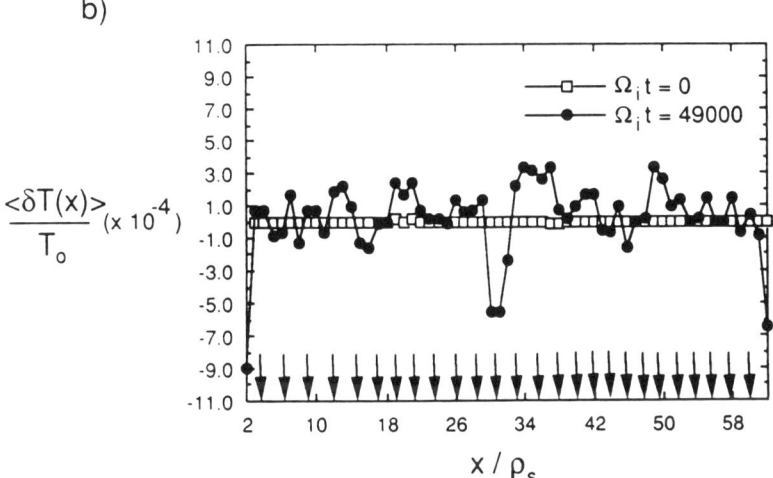

Figure 8. Ion $x - v_x$ phase space taken at saturation and average perturbed temperature profile versus x taken initially and at saturation. a) $x - v_x$ phase space and rational surface locations indicated by the arrows. b) Perturbed temperature profile averaged over y- and z-directions as a function of x.

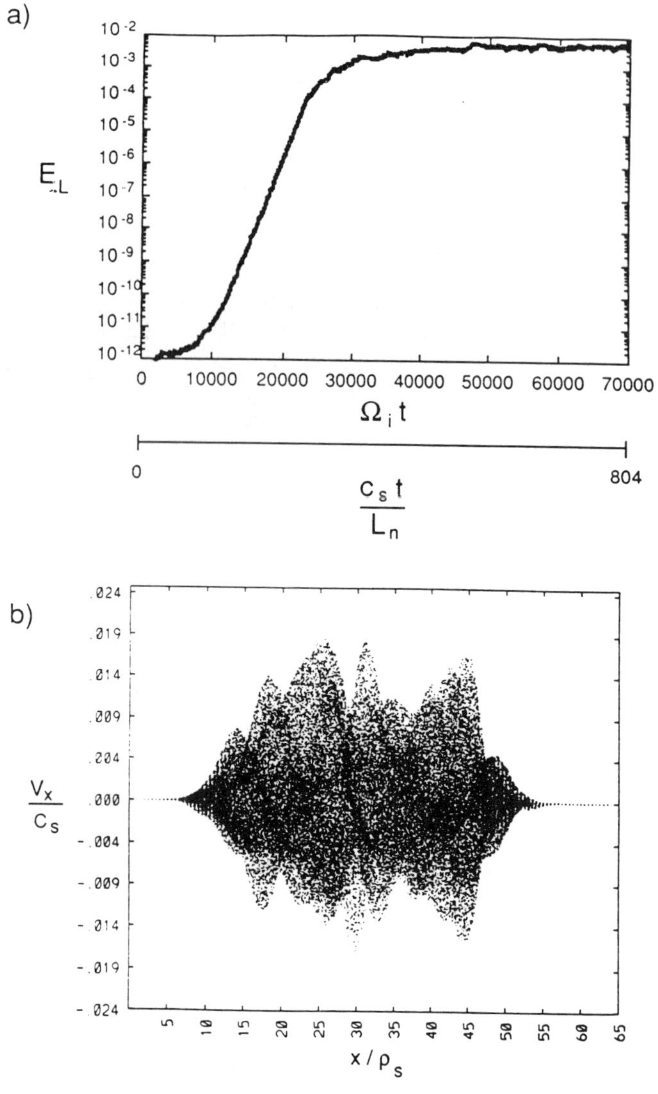

Figure 9. Time evolution of the total electrostatic energy and the ion $x - v_x$ phase space for the three dimensional case with $\eta_i = 4$ and $L_s/L_n = 5$. The rational surfaces are more centrally located, away from the boundaries. a) Total electrostatic energy versus time. b) Ion $x - v_x$ phase space taken at the time of saturation.

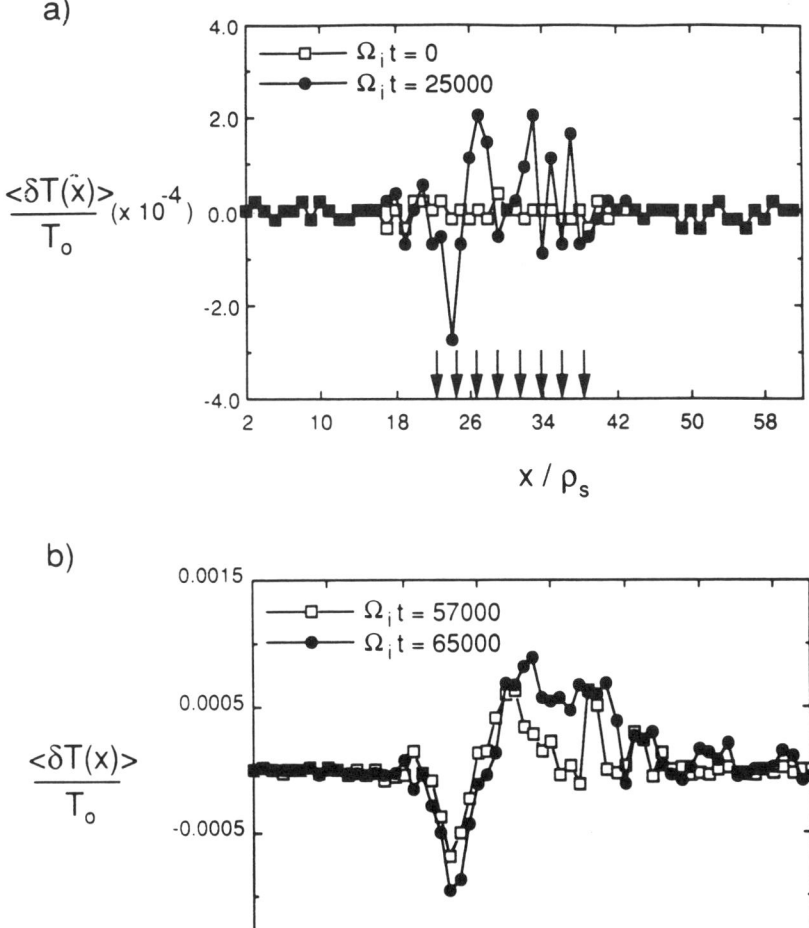

Figure 10. Average perturbed temperature profile versus x taken at saturation and in the late, post-saturation phase. a) Average temperature perturbation versus x at saturation and rational surface locations indicated by the arrows. b) Perturbed temperature profile taken at the end of the simulation.

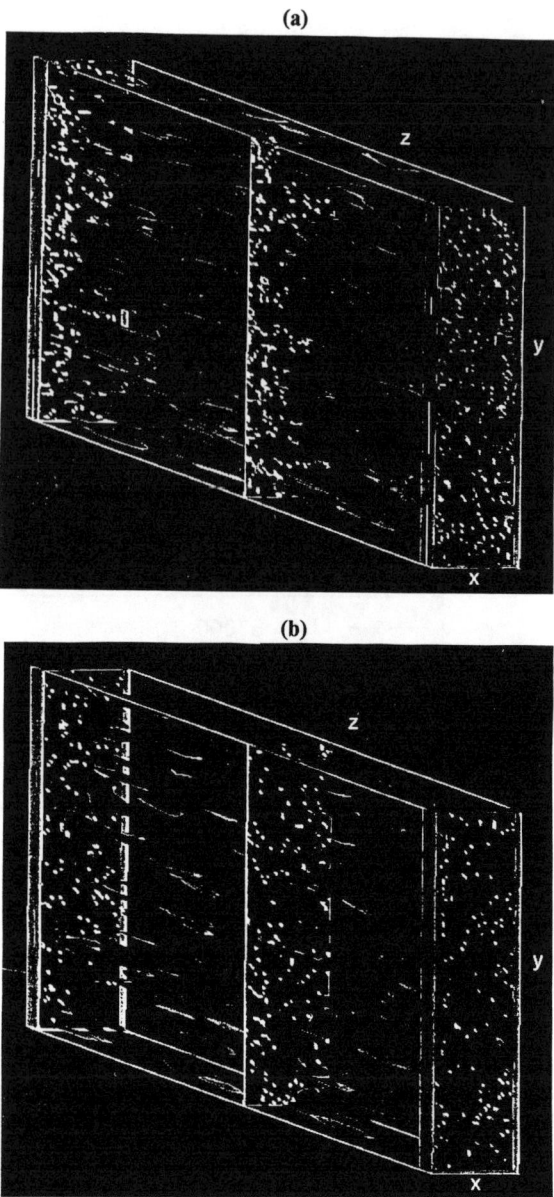

Figure 11. Three dimensional electrostatic potential contour surfaces at two different times. a) Contour surfaces during linear phase of instability. b) Same contour surfaces taken at saturation during the transition to turbulence.

4.3 TRANSPORT IN THE SELF-ORGANIZED RELAXED STATE OF ION TEMPERATURE GRADIENT INSTABILITY

T. Tajima, Y. Kishimoto,[1] M.J. LeBrun, M.G. Gray, J-Y. Kim, W. Horton, V. Wong, and M. Kotschenreuther
Institute for Fusion Studies, The University of Texas, Austin, Texas 78712

Abstract. We investigate the anomalous heat conduction in a tokamak plasma analytically and computationally. Our toroidal particle simulation shows a new emerging physical picture that the toroidal plasma exhibits marked properties distinct from a cylindrical plasma: (i) the development of radially extended potential streamers localized to the outside of the torus, (ii) more robust ion temperature gradient instability, (iii) radially constant eigenfrequency, (iv) global temperature relaxation, and (v) radially increasing heat conductivity χ_i. These results are analyzed by linear and quasilinear kinetic theory. A relaxation theory based on the reductive perturbation theory in the quasilinear equation is developed. The theory constrains the thermal flux so that χ_i increases radially. The Bohm-like scaling is found in connection with the radially extended mode structure.

4.3.1 INTRODUCTION

The η_i instability [1] as well as the ion temperature gradient (ITG) instability [2] are possible candidates to explain the anomalous ion thermal transport observed in many large tokamak plasma [3, 4, 5, 6]. In the previous reports [7, 8] we have found that the properties of the η_i and ITG instabilities in a slab (or cylindrical) plasma and those in a toroidal (tokamak) plasma are quite distinct. Simulation runs of the toroidal plasma have shown that the main differences of toroidal modes as compared with cylindrical cases are: (i) the eigenmodes are

---
[1]Naka Fusion Research Establishment, JAERI, Naka, Japan

elongated in the radial direction and nearly global, while cylindrical ones are localized around the rational surfaces; (ii) the toroidal modes have higher growth rates than the cylindrical counterparts; (iii) the eigenfrequencies are nearly constant over an extended radial region; (iv) a global temperature relaxation and enhanced thermal conduction are observed; (v) in particular, the heat conductivity χ_i is an increasing function of the minor radius as opposed to a decreasing one in a cylindrical counterpart. These signatures are quite striking, as they are all in the right direction to match more closely with experimental tendencies than the cylindrical characteristics. Thus the importance of the toroidal geometric effects on the physics of these modes and associated transport should be emphasized. These differences are primarily due to the toroidicity-induced coupling of rational surfaces over many poloidal mode numbers.

In the present article we shall characterize the temporal evolution of toroidal plasma and the η_i and ITG modes and the resultant properties of the modes and transport. We study this problem through toroidal particle simulation and linear and nonlinear analysis.

4.3.2 TOROIDAL PARTICLE SIMULATION

Employing the Toroidal Particle Code [9], we investigate the η_i and ITG instability in a toroidal plasma, simulating a tokamak plasma. In the present investigation we use the fully dynamic ions with adiabatic electrons with the following typical parameters: the aspect ratio $R_0/a = 4$, the average ratio of the ion plasma frequency to ion cyclotron frequency $\langle \omega_{pi}/\omega_{ci} \rangle = 1$, the average ratio of the ion Larmor radius to minor radius $\langle \rho_i/a \rangle \approx 0.02$, the magnetic shear parameter $\hat{s} = rq'/q \sim 1.0$, the Debye length $\lambda_{Di} \approx \rho_i$, and the toroidal mode number(s) $n = 9$; 7 and 9; 7, 9, and 16; 7, 8, and 9, while the poloidal mode numbers vary from $m = 4$ to 39. The profiles of the safety factor $q(r)$, $\hat{s}(r)$, $\rho_i(r)$, and the density scale length parameter $\epsilon_n(r)$ are identical or similar to those shown in Fig. 5 of Ref. 7. For the η_i mode runs, we choose the initial η_i parameter $\eta_i(r) \equiv \partial \ell n\, T_i(r)/\partial \ell n\, n_i(r) =$ constant, varying from 0 to 4 with $n_i(r)$ being a Gaussian. For the ITG ($\eta_i \to \infty$) mode runs, we choose the parameter $\epsilon_T \equiv L_T/R$ (where L_T is the ion temperature gradient length and R the major radius) typically ranging around 0.07.

In this case the initial temperature profile may be a Gaussian or $\tan^{-1}$ profile.

Typical temperature relaxation in time is shown in Fig. 1(a) for the ITG modes. A radial portion of an exponential shoulder ($T_i(r) \propto e^{-r/L_T}$) shows up in a rapid time scale (in a time comparable to the wave growth). The extent of that portion (with the exponential profile) gradually expands until almost all radial portions that encompass the rational surfaces. For the η_i modes the radial profile relaxes from $\tan^{-1}$ towards that of the Gaussian, for which the initial $\eta_i(r; t = 0)$ changes (overall decreases) to become a global constant $\eta_i(t_0)$. This is shown in Fig. 2(a). During these relaxations (either for ITG or for η_i modes) the respective parameter ϵ_T or η_i that characterizes the instability tends to become a global constant. Once these self-similar (or universal) profiles are achieved, the system evolves much more slowly with the profile functional form kept intact (this may be called the profile consistency [10]) but the value of ϵ_T^{-1} or η_i gradually decreasing toward the marginal stability value. This trend is shown in Fig. 1(c) for ϵ_T and in Fig. 2(b) for η_i respectively. Corresponding to the globally (nearly) constant ϵ_T (or η_i), the (nearly) globally constant eigenfrequency appears as shown in Fig. 1(b). Our observation of the global eigenvalue and eigenfunction is consistent with 2D linear theory results [11]. The step-like feature in Fig. 1(b) is due to another eigenvalue. Following the change of global constant ϵ_T, the eigenfrequency also changes accordingly, as shown in Fig. 1(d).

The ion heat conductivity χ_i for the value $\eta_i = 4$ at the end of runs is plotted as a function of the minor radius in Fig. 3 (with single n runs as well as multiple n's). In all cases the χ_i is an increasing function of r (nearly exponential). As noted in Ref. 7 this is a salient feature of toroidal plasma in contrast to that of cylindrical case in which χ_i decreases as a function of r. The latter would be the case if the heat diffusivity is determined by a local diffusion step Δr that is tied to the local temperature (and thus the local Larmor radius). On the other hand, the appearance of the radially increasing χ_i indicates some mechanism that is global in nature. This indication coincides with the implications of the toroidal mode characteristics we describe in the above. Furthermore, in many large tokamak high temperature discharges produced by auxiliary heating such radially increasing χ_i

behaviors are universally reported [3, 4, 5, 6].

In Fig. 4 we show typical temporal evolution of the amplitude of electric potential ($m = 9, n = 9$) and three snapshots of potential contours in r and θ, corresponding to the three stages marked by I, II, and III. Stage I is the linear state, II around the saturation (early nonlinear), and III in the steady state (fully developed nonlinear). The radially extended nearly global mode structure is evident throughout those three stages, including the latest steady nonlinear stage. The potential vortices are strongly twisted and choppier away from $\theta = 0$ (torus outside), while near $\theta = 0$ the prominent ballooning structure is evident.

4.3.3 LINEAR AND QUASILINEAR THEORY

As we described above, the temperature gradient decreases toward a marginal stable value. In the TPC simulation even in the stage when the mode amplitude is saturated, the temperature profile appears to be slightly on the unstable side so that there occurs constant excitation of waves whose energy is expended on the transport maintaining a level of heat flux. In order to theoretically describe the toroidal ITG modes, we calculate the perturbed circulating nonadiabatic ion distribution $\hat{f}_i^{na} = \hat{g}\exp(i\mathcal{L})$ in the ballooning space at $\theta = 0$ as

$$\hat{g} = \frac{e\hat{\phi}}{T_i} J_0(u) \frac{\Omega - \Omega_{*i}^T \left(\tilde{v}_\parallel^2 + \tilde{v}_\perp^2 + \eta_i^{-1} - 3/2\right)}{\Omega - (\tilde{v}_\parallel^2 + \tilde{v}_\perp^2/2) - \Omega_t \tilde{v}_\parallel} f_M , \qquad (1)$$

and $\mathcal{L} = \mathbf{v} \times \mathbf{b} \cdot \mathbf{k}_\perp/\omega_{ci}$ ($\mathbf{b} \equiv \mathbf{B}/B$) for Im $\omega > 0$. In Eq. (1), f_M represents the local Maxwellian, and $\Omega = \omega/\omega_d$, $\Omega_{*i}^T = \omega_{*i}^T/\omega_d$, $\Omega_t = k_\parallel v_i/\omega_d$, $u = k_\theta \rho_i \tilde{v}_\perp$, $\tilde{v}_\parallel = v_\parallel/v_i$, and $\tilde{v}_\perp = v_\perp/v_i$ ($v_i \equiv (2T_i/M_i)^{1/2}$) where $\omega_{*i}^T = -(cT_i/eB)(k_\theta/L_T)$, and $\omega_d = 2\epsilon_T \omega_{*i}^T$ is magnetic drift frequency. When the mode approaches the marginal stability, the denominator in Eq. (1) also approaches zero, giving rise to kinetic resonant effects. This underlines the importance of kinetic effects in the present problem. Using Eq. (1), the linear theoretical dispersion relation was derived and the real and imaginary parts of eigenfrequency were computed. The result [12] shows that as the ϵ_T increases and approaches the marginal stability ϵ_{Tc}, that is, when the growth rate γ decreases toward zero, the real frequency of the eigenvalue first rapidly decreases toward a value roughly ω_d (the magnetic drift frequency) and then tends to be a constant (of order unity) times ω_d. As far as $k_\parallel v_i/\omega_d \ll 1$, the real frequency ω_r approaches a constant $\omega_r \sim \omega_d$, where the growth rate γ is around a fraction of ω_d, where the fraction is estimated about $\frac{1}{4}$.

The evolution of the background distribution function f_0 may be described by a quasilinear theory: $\partial f_0/\partial t = \mathcal{L}[E(\hat{f}), \hat{f}]$, where $\hat{f}$ is the perturbed distribution, E the electric field generated by $\hat{f}$, and $\mathcal{L}$ is the usual Vlasov operator. When we make a second moment of velocity of this quasilinear equation, we obtain an equation describing the background temperature evolution:

$$\frac{3}{2} n \frac{\partial T}{\partial t} + \nabla \cdot \mathbf{Q} = 0, \qquad (2)$$

where the source term is neglected (consistent with the simulation situation), T refers to the background temperature and Q is the second moment of velocity of the term $\mathcal{L}[E(\hat{f}), \hat{f}]$ and it takes the form

$$rQ = \text{Re} \int d\mathbf{v} \, \frac{1}{2} m v^2 \, \frac{c\mathbf{E}^* \times \mathbf{B} \cdot \hat{\mathbf{e}}_r}{B^2} \cdot g \exp(i\mathcal{L}),$$

$$= \frac{n_0 \, r \, k_\theta c e}{B} |\phi(r)|^2 G(\Omega, \eta_i, L_T, r), \qquad (3)$$

where Q has the meaning of heat flux due to the fluctuating electric fields E and G is the normalized heat flux (of the order unity). From Eq. (1) G is given as

$$G = -\frac{i}{\sqrt{\pi}} \int_0^{+\infty} d\tilde{v}_\perp^2 \int_{-\infty}^{+\infty} d\tilde{v}_\parallel \, J_0^2(u)(\tilde{v}_\parallel^2 + \tilde{v}_\perp^2) \frac{\Omega - \Omega_{*i}^T(\tilde{v}_\parallel^2 + \tilde{v}_\perp^2 + \eta_i^{-1} - 3/2)}{\Omega - (\tilde{v}_\parallel^2 + \tilde{v}_\perp^2/2) - \Omega_t \tilde{v}_\parallel}$$

$$\cdot \exp(-\tilde{v}_\parallel^2 - \tilde{v}_\perp^2) \qquad (4)$$

for $\text{Im}\,\omega > 0$. For the stable region the velocity integrals must be carried out with analytical continuation from $\text{Im}\,\omega > 0$ to $\text{Im}\,\omega < 0$ across the marginal stability. Then, the resonant residual part of G is calculated as

$$G = 4\sqrt{\pi}\,\alpha \left(\Omega + \frac{\Omega_t^2}{4}\right)^{1/2} \int_{-1}^{+1} dy \int_C dt \, J_0^2(u)(\tilde{v}_\parallel^2 + 2t) \exp(-\tilde{v}_\parallel^2 - 2t)$$

$$\cdot \left[\Omega - \Omega_{*i}^T \left(\bar{v}_\|^2 + 2at + \eta_i^{-1} - 3/2\right)\right] \cdot \delta\left[t - \left(\Omega + \frac{\Omega_t^2}{4}\right)(1 - y^2)\right] \quad (4')$$

where $\bar{v}_\| = (\Omega + \Omega_t^2/4)^{1/2} y - \Omega_t/2$, and $\alpha = 1/2$ for $\mathrm{Im}\,\omega = 0$ and $\alpha = 1$ for $\mathrm{Im}\,\omega < 0$. Here the integration contour C is chosen so that the complex variable t goes around the pole $t = (\Omega + \Omega_t^2/4)(1 - y^2)$.

It should be noted that the heat flux G in Eq. (4) does not vanish even when the mode becomes marginally stable $\mathrm{Im}\,\omega \to 0$ and stays positive in the neighborhood of $\mathrm{Im}\,\omega = 0$. Numerical integrated value of Eq. (4) is given in Ref. 12. This is in sharp contrast to a fluid theoretical calculation of quasilinear heat flux [13], where $Q = 0$ at marginal stability.

4.3.4 RELAXATION THEORY FOR TOROIDAL PLASMA

We discussed the relaxation process of the temperature of the toroidal plasma and modes evolution in Sec. 2. In the early stage of the simulation before saturation (roughly corresponding to Stage I in Fig. 4) the temperature profile rapidly (in a fluid time scale) relaxes toward an exponential profile (ITG case). In the later stage the temperature profile relaxes slowly, maintaining the functional form, exponential (ITG case), and its parameter ϵ_T only gradually increasing toward the marginal stability, though ϵ_T is still less than ϵ_{Tc}. In this stage the physics is kinetic, as it is near marginal stability. Since there exist these two distinct time scales in relaxation, we introduce a multiple time expansion in our quasilinear equation (2). A systematic expansion may be carried out by the well-known reductive perturbation method [14], introducing a smallness parameter ε (which will be determined below). We expand the time scales, temperature, and heat flux as follows:

$$\frac{\partial}{\partial t} = \frac{\partial}{\partial t_0} + \varepsilon^2 \frac{\partial}{\partial t_1} + \varepsilon^4 \frac{\partial}{\partial t_2} + \cdots,$$

$$T = T_0 + \varepsilon T_1 + \cdots,$$

$$Q = \varepsilon Q_0 + \varepsilon^2 Q_1 + \cdots, \quad (5)$$

where $Q_0 = -\chi_i^0 \nabla T_0$, $Q_1 = -\chi_i^0 \nabla T_1 - \chi_i^1 \nabla T_0$, and T_0 is the global temperature profile, which, as we showed, exhibits a self-similar profile (exponential for ITG, Gaussian for η_i) and is only slowly varying, i.e. $\partial T_0/\partial t_0 = 0$. The present reductive perturbation theory is crucially based on the fundamental properties of relaxation of the tokamak-like toroidal plasma we described. The important point is that the instability is radially global and vigorous in toroidal plasmas so that the plasma parameters have to relax rapidly until they approach sufficiently near the stable profile. (In a cylindrical plasma, therefore, there is no such relaxation theory). However, as long as there is enough heat reservoir (as in our simulation case) or there is energy input, the sustained finite fluctuations will cause a certain amount of heat flux and thus dissipation, which as to be compensated by the weak but still unstable wave activities. The amount of sustained fluctuations thus is a function of the energy (or power) input (or the availability of the heat reservoir).

From this expansion theory we obtain a series of transport equations (as commented on in Eq. (2) we have neglected source terms for clarity of theory. However, it is readily possible to generalize this to include source terms):

$$\mathcal{O}(\varepsilon): \qquad \frac{3}{2} n_0 \frac{\partial T_1}{\partial t_0} + \nabla \cdot Q_0 = 0 , \qquad (6)$$

$$\mathcal{O}(\varepsilon^2): \qquad \frac{3}{2} n_0 \frac{\partial T_0}{\partial t_1} + \nabla \cdot Q_1 = 0 . \qquad (7)$$

In Eq. (6) if there is a finite $\nabla \cdot Q_0(r)$, it will force T_1 to rapidly relax such that $T_1 \to 0$. Thus on average we obtain $\partial T_1/\partial t_0 = 0$ and, therefore, $\nabla \cdot Q_0 = 0$ and

$$rQ_0(r) = r\chi_i^{(0)}(r) \frac{\partial T_0}{\partial r} = \text{const} . \qquad (8)$$

Thus for ITG in which the relaxed T_0 profile is $\sim \exp(-r/L_T)$ and for η_i the relaxed profile is $T_0 \sim \exp(-r^2/2L_T^2)$, for which the $\mathcal{O}(\varepsilon)$ constraint (8) requires that the thermal conductivity vary as

$$\chi_i^0(\text{ITG}) \sim \frac{\text{const}}{r} \exp(r/L_T) , \qquad (9)$$

$$\chi_i^0(\eta_i) \sim \frac{\text{const}}{r^2} \exp\left(\frac{r^2}{2L_T^2}\right) . \tag{9'}$$

These formulae apply away from the center and edge where the source term effects can never be negligible. These radially increasing χ_i profiles are in agreement with our simulation results, as well as the experimental results.

In the next order in ε, $\mathcal{O}(\varepsilon^2)$, we can obtain the slow evolution of the background temperature $T_0(t_1)$. In the case when $k_\theta \rho_i \ll 1$ (i.e. $J_0^2(u) \sim 1$), and $k_\parallel \sim 0$, we can write $G(\Omega, L_T, \eta_i, r) = G(\Omega, L_T)$, where Ω and L_T become a global constant due to toroidal plasma properties. Thus

$$rQ(r) = \frac{n_0 \, ce}{B} n \, q(r) |\phi(r)|^2 G(\Omega, L_T) , \tag{10}$$

where n is a given toroidal mode number and use is made of $k_\theta^n = nq/r$. Since $rQ(r) = \text{const}$, except for a weak r-dependence of q, Eq. (10) is nearly r-independent, from which we conclude that $|\phi_0(r)|^2 = |\phi_0|^2 + \varepsilon |\phi_1(r)|^2 + \cdots$, $Q_1(r)$ becomes

$$Q_1 = \frac{n_0 \, ce}{B} n \, q(r) |\phi_1(r)|^2 G . \tag{11}$$

Note that Eq. (6) will not determine the overall strength of Q, although the radial profile is determined. In order to fix the strength of Q or χ_i, we have to employ Eq. (7) as well.

4.3.5 HEAT CONDUCTIVITY

We investigate the heat conductivity in the presence of ITG (or η_i) driven plasma fluctuations. As we discussed in the previous section, the initial temperature profile evolves rapidly into a self-similar functional form. After this is established, the temperature profile relaxes so that the global parameter (ϵ_T or η_i) slowly change, but the temperature functional form is maintained. After this the profile smoothly tends to the steady-state profile. In this section we try to describe the heat conductivity in each stage. Typically in our initial value simulation the potential fluctuation level shoots up as the linear instability grows until saturation. Then the amplitude tops and later decays or stagnates. The heat conductivity $\chi_i(t)$ evolves in time in a similar fashion; it increases until the saturation time, after which time it decreases and eventually stagnates at a roughly steady value. Using Eq. (10), the ion heat conductivity can be written as

$$\chi_i(r) = \frac{1}{2} \frac{(k_\theta \, \rho_i(r)) v_i(r) \, L_T}{T_i(r)^2} \left\langle (e\phi(r))^2 \right\rangle \text{Re } G(\Omega, L_T) \,, \quad (12)$$

where $\left\langle |\phi(r)|^2 \right\rangle$ to the first approximation may be regarded as constant $\langle |\phi_0|^2 \rangle$. It is evident that the equation (12) has a monotonically increasing radial dependence, as given in Eq. (9). In the following we evaluate the heat conductivity in two different regimes, one in the fluid regime and the other in the kinetic one.

Hydrodynamic Regime

In the present TPC simulation in stages I and II the dynamics is hydrodynamic, while in stage III it is kinetic as discussed in Sec. 4. Hydrodynamic or kinetic is defined by $|\omega + i\gamma| \gtrless \max(k_\parallel v_i, \omega_d)$. In this subsection we investigate the heat conductivity in the hydrodynamic regime. However, it is noted that if we had an external power source strong enough, even in the steady saturation stage III we might have the hydrodynamic regime. In this hydrodynamic regime we assume that the saturation mechanism is the wave breaking (i.e. wave turnover effect), since the effect of multiples of the modes (n's) (i.e. the nonlinear mode coupling among different toroidal eigenmodes) is not strong in the present investigations in which the external power source is absent and the heat reservoir is provided only by the central plasma.

The displacement of the plasma fluid by the excited wave k_θ in the radial direction is roughly

$$\xi = \frac{c \, k_\theta \, \phi}{\gamma_h B} \,, \quad (13)$$

where γ_h is the growth rates in the hydrodynamic regime. When the radial displacement ξ becomes approximately the half wavelength in the radial direction, $\xi \sim \pi/k_r$, the wavebreaking takes place and thus the wave growth saturates. This condition yields the approximate saturated potential amplitude as

$$\phi_{\text{sat}} = \frac{\pi \gamma_h B}{c k_r k_\theta}. \tag{14}$$

In the hydrodynamic regime the growth rate is estimated to be

$$\gamma_h^2 = \omega_*^T \omega_d. \tag{15}$$

Thus in the hydrodynamic regime we obtain χ_i from Eq. (12) and approximate expressions (14), (15) as

$$\chi_i^h = 4\pi^2 \frac{k_\theta \rho_i v_i R \epsilon^2}{(k_r a)^2} G_h, \tag{16}$$

where $\epsilon = a/R$. The hydrodynamic normalized flux G_h is evaluated as

$$G_h = -\frac{2}{\sqrt{\pi}} i \int_0^\infty dv_\perp \, v_\perp \, J_0^2(u) e^{-v_\perp^2} \int_{-\infty}^\infty dv_\parallel \left[1 + \frac{2}{(2\epsilon_T)^{1/2}} \left(v_\parallel^2 + v_\perp^2 + \frac{1}{\eta_i} - \frac{3}{2} \right) \right]$$
$$\times \left[1 - i\sqrt{2\epsilon_T} \left(v_\parallel^2 + \frac{v_\perp^2}{2} \right) \right] (v_\parallel^2 + v_\perp^2) e^{-v_\parallel^2} \cong \frac{C_1}{\epsilon_T^{1/2}}, \tag{17}$$

where expansion in terms of $1/2\epsilon_T \ll 1$ was carried out and C_1 is a constant of order unity. With Eq. (17), the χ_i expression (16) becomes

$$\chi_i^h = \frac{4\pi^2 C_1}{\epsilon_T^{1/2}} \frac{k_\theta \rho_i v_i R \epsilon^2}{(k_r a)^2}. \tag{18}$$

Choi and Horton [15] and Romanelli [16] among others have shown that the global linear radial mode width is a geometrical mean of ρ_i and a so that approximately we have

$$(k_r a)^2 = \frac{a}{\rho_i}. \tag{19}$$

If the mode width in the nonlinear stage is not too different from the linear stage [this is not inconsistent with our simulation], then Eq. (19) may be used as an estimate in Eq. (18). This yields

$$\chi_i^h = C_2 \left(\frac{cT}{vB} \right) \left(a L_T^{-1/2} R^{-1/2} \right), \tag{20}$$

where C_2 is a constant and $\langle k_\theta \, \rho_i \rangle \sim 0.2$ is taken. In our simulation the radial size of the streamer (the potential vortex structure) is about the same as in the linear stage (Fig. 4). In the simulation in particular the ratio $\rho_i/a \sim 1/40$ is approximately eight times larger than the typical experimental value of $\sim 1/300$. This makes the radial size scaling in our simulation about three times greater in terms of the fraction of the minor radius. In the future smaller ratios ρ_i/a will be used at in simulation to definitely resolve the scaling. If we take scalings for the radial streamer size as $k_r^{(s)} \sim 1/\rho_i$ and $k_r^{(\ell)} \sim 1/a$, the corresponding χ_i scalings would be

$$\chi_i^{h(s)} \sim C_2' \left(\frac{cT}{eB} \right) \left(\rho_i \, L_T^{-1/2} \, R^{-1/2} \right) = \sqrt{2} \, C_2' \left(\frac{M^{1/2} \, c^2 \, T^{3/2}}{e^2 \, B^2} \right) \left(L_T^{-1/2} \, R^{-1/2} \right)$$

[corresponding to the gyroBohm scaling] and

$$\chi_i^{h(\ell)} \sim C_2''(v_i \, a) \left(a \, L_T^{-1/2} \, R^{-1/2} \right) ,$$

[corresponding to the hydrodynamic transport scaling] respectively. The ion energy confinement time from Eq. (20) is approximately obtained from $\tau_i^h \sim L_T^2 / \chi_i^h$ as

$$\tau_i^h = C_3 \frac{eB}{cT} \, a^{-1} \, L_T^{5/2} \, R^{1/2} , \qquad (21)$$

where C_3 is a constant.

Kinetic Regime

We find in stage III in our simulation that the system reaches close to marginal stability (though slightly above the marginal). This implies that the dynamics is kinetic in this stage. From a closer scrutiny of stage III, we see that the mode amplitude of various waves increases or decreases incessantly as shown in Fig. 4(a). Sometimes a sudden crash of amplitude appears, which usually coincides with reconnection of one vortex occupying radial interval $r \in [r_1, r_2]$ with another at $r \in [r_2, r_3]$. Some resurgence and decay of the mode amplitude are related to a local temperature bump or hole. Detailed description of these processes will be reported elsewhere in the future. The overall behavior of the temperature profile and wave activities is as follows: if there appears a local temperature bump steeper than the global gradient characterized by ϵ_T, the wave activity in the neighborhood is elevated so that the locally enhanced heat conduction will smooth the bump. On the other hand, if there appears a local hole less steep than the global gradient, the wave activity in the vicinity is depressed so that the locally reduced heat conduction is overwhelmed by the overall global rate of heat conduction, which leads to smoothing of the hole. Thus the overall growth rate γ is sustained around a constant γ_c (critical growth rate), which is above the marginal value of zero in order to sustain the overall global energy dissipation (heat transport). The smallness parameter ε is determined as $\sqrt{\gamma_c/\omega_r}$. If γ was less than γ_c, all wave activity would have diminished and no global heat conduction would be maintained. The heat reservoir provides heat and then lets the temperature profile steepen, which lends to an increase in γ to approach γ_c, where the critical parameter γ_c is much smaller than γ_h or the real frequency ω_r. We call this behavior of the tokamak plasma the self-organized criticality. This is not unlike the situation discussed on the avalanche of a sand pile by Bak et al. [17] However, their theory does not have distinction between the marginal stability and critical threshold as ours does, nor do they provide a self-consistent analytic expression for transport. In order to explain the profile consistency, others have resorted to the plasma's self-organizing tendency [10, 17].

We now wish to derive the ion heat conductivity in the self-organized critical state by evaluating the residual part (4), though there is non-residual contribution. In the kinetic regime, we can approximate $\Omega = 1$ and $\Omega_t = 0$ to estimate the normalized flux G as

$$G_k(\Omega = 1) = \int_{-1}^{1} dx \exp(-2+x^2)(2-x^2)\left[1 - \Omega_{*i}^T\left(2 - x^2 - \frac{3}{2}\right)\right] \approx C_4 + C_5 \Omega_{*i}^T, \quad (22)$$

where $\Omega_* = \frac{1}{2\epsilon_T}$ and C_4, C_5 are constants of order unity. The excited waves can scatter particles in this regime to balance the wave growth [19]

$$\gamma - k_r^2 D = 0 , \quad (23)$$

where the diffusion coefficient due to wave-particle interaction is estimated as

$$D = \frac{\delta \xi^2}{\delta t} = \frac{c^2 k_\theta^2 |\phi|^2}{\delta t \, \omega^2 \, B^2}, \tag{24}$$

and

$$\delta t \sim \frac{\pi a}{2 \omega_d \, \hat{s} \, \Delta r} \propto \frac{1}{\omega_d} \tag{25}$$

where Δr is typical radial extension of global eigenmode and δt is estimated from the typical time scale that the global eigenmode is affected by the strong sheared phase rotation in the poloidal direction. Equations (23), (24), and (25) give

$$|\phi|^2 = \frac{B^2 \gamma \omega}{(k_\theta \, k_r)^2 c^2}. \tag{26}$$

As remarked in Sec. 2, when the system approaches from the hydrodynamic regime toward the kinetic one, the growth rate decreases accompanied by a rapid decrease of ω_r first followed by a near plateau value $\omega_r \sim \omega_d$. If we call the entry to the plateau regime (kinetic) begins when ω_r becomes insensitive to the change of ϵ_T and thus γ, the entrance is evaluated at $\gamma = f \omega_d$, and $f \cong 1/4$. Equation (26) with $\gamma = f \omega_d$ and Eq. (22) [and taking the second term in (22)] give rise to the ion heat conductivity estimate in the kinetic regime as

$$\chi_i^k = C_6 \left(\frac{cT}{eB}\right) (a \, L_T \, R^{-2}), \tag{27}$$

where once again we took $(k_r \, a)^2 = a/\rho_i$ [Eq. (19)] and C_6 is a constant. If, as before, $k_r^{(s)} \sim 1/\rho_i$ and $k_r^{(\ell)} \sim 1/a$ scalings are taken instead, we obtain the corresponding χ_i scalings as

$$\chi_i^{k(s)} \sim C_6' \left(\frac{cT}{eB}\right) (\rho_i \, L_T \, R^{-2})$$

and

$$\chi_i^{k(\ell)} \sim C_6''(v_i \, a)(a \, L_T \, R^{-2})$$

respectively. Note that in the kinetic regime the choice of $(k_r \, a)^2 \cong a/\rho_i$ should be even better justified, as the system is close to the linear stage. The ion energy confinement time derived from Eq. (27) is

$$\tau_i^k = C_7 \frac{eB}{cT} a^{-1} L_T R^2, \tag{28}$$

where C_7 is a constant. The energy confinement time in Eq. (28) is the so-called Bohm scaling proportional to the toroidal field B.

4.3.6 DISCUSSIONS AND CONCLUSIONS

A characteristic feature of these simulations is an initial nonlinear phase during which linear wave growth saturates and the temperature profile relaxes comparatively rapidly to a self-similar profile not far from marginal stability. In the later nonlinear phase, the self-similar profile relaxes more slowly towards marginal stability. The radial size of the potential vortex structure is approximately the same as in the linear stage and we estimate radial correlation lengths to be $\sim (\rho_i a)^{1/2}$. The thermal diffusivity χ_i scales as (Eq. (16)):

$$\chi_i \sim \frac{\langle k_\theta \rho_i \rangle v_i a^2}{(k_r^2 a^2) R}$$

and if we take $\langle k_\theta \rho_i \rangle \sim \mathcal{O}(1)$ and $k_r(\rho_i a)^{1/2} \sim \mathcal{O}(1)$, $\chi_i \propto \frac{a}{R} v_i \rho_i$, a Bohm-like scaling.

As stated earlier, our simulations and discussion are limited to no (or weak) power input in the radial region of consideration. Thus the scalings Eqs. (21) and (28) do not have an explicit dependence on the input power P. [The input power P might appear through other parameters, as has been done in Ref. 21]. Also implicit in our calculation is the plasma current I, which appear, for example, in $\hat{s}$ but was in our discussion tucked away in such approximations as $\hat{s} \sim 1$. Furthermore, the average ratio of ion Larmor radius to minor radius is $\left\langle \frac{\rho_i}{a} \right\rangle \sim 0.02$, much larger than typical experimental values. Thus a direct one-to-one comparison with experimental data cannot be made. It is, however, interesting to see in light of Eq. (28) that the experimental confinement time goes like $\tau_E \propto B^{0.9} R^{2.3} \epsilon^{-0.2}$ in Ref. 21 for example.

Other simulation codes [22] have shown nonlinear states with vortex structures significantly smaller than the radial scale lengths of the linear eigenmode. In these simulations, $\frac{\rho_i}{a} \sim \frac{1}{400}$, the initial temperature and density profiles were not close to marginal stability and were "maintained on the average" by imposing periodic boundary conditions. Global change of $\omega_*(r)$, for example, is not allowed in those models [22], as the shell thickness is much smaller than a. Linear growth does not saturate by profile relaxation but by the onset of a "nonlinear" instability resulting in the "break-up" of the linear eigenmode and the formation of vortex structures with characteristic scale lengths of many ion Larmor radii. When this occurs, the potential amplitudes at saturation of linear growth are not determined by Eq. (14), and Eq. (16) for χ_i is no longer applicable. A limited number of runs with these other simulation codes suggest that the χ_i scaling tends to be gyroBohm.

It is evident that we need many more simulation runs (particularly those of scaling physics) and more detailed theoretical analysis. Such investigations will tell us how much is model dependent and which scalings of χ_i applies. It is the purpose of the present paper to foster the stimulus to such investigations and to give some insight to processes that might be important to consider.

An earlier paper [23] deals with a view that plasma profiles would remain close to marginal stability.

ACKNOWLEDGMENTS

This work was supported by the U.S. Department of Energy contract #DE-FG05-80ET-53088.

Bibliography

[1] W. Horton, D-I. Choi, and W.M. Tang, Phys. Fluids **24**, 1077 (1981).

[2] B. Coppi, M.N. Rosenbluth, and R.Z. Sagdeev, Phys. Fluids **10**, 582 (1967).

[3] S.D. Scott, *et al.*, Phys. Fluids B **2**, 1300 (1990).

[4] T. Kurki-Suonio, R.J. Groebner, and K.H. Burrell, Nucl. Fusion **32**, 138 (1992).

[5] B. Balet, *et al.*, Nucl. Fusion **30**, 2034 (1990).

[6] JT-60 TEAM, in *Plasma Physics and Controlled Nuclear Fusion Research, 1990*, Vol. 1, p. 53 (IAEA, Vienna, 1991).

[7] M.J. LeBrun, T. Tajima, M. Gray, G. Furnish, and W. Horton, Phys. Fluids B **5**, 752 (1993).

[8] M. Kotschenreuther, *et al.*, in *Plasma Physics and Controlled Thermonuclear Research* (IAEA, Vienna, 1993), to be published.

[9] M.J. LeBrun and T. Tajima, submitted to J. Comput. Phys.

[10] B. Coppi, Comm. Plas. Phys. Contr. Fus. **5**, 261 (1980).

[11] W.M. Tang and G. Rewoldt, to be published in Phys. Fluids B.

[12] Y. Kishimoto, T. Tajima, M.J. LeBrun, M. Gray, J-Y. Kim, and W. Horton, IFSR#589, submitted to Phys. Rev. Lett. (1993).

[13] F. Romanelli, Phys. FluidsB **1**, 1018 (1989).

[14] H. Washimi and T. Taniuti, Phys. Rev. Lett. **17**, 996 (1966).

[15] D-I. Choi and W. Horton, Phys. Fluids **23**, 356 (1980).

[16] F. Romanelli, in this Proceedings.

[17] P. Bak, C. Tang, and K. Wiesenfeld, Phys. Rev. A. **38**, 364 (1988).

[18] B.B. Kadomtsev, Sov. J. Plasma Phys. **13**, 443 (1987).

[19] T.H. Dupree, Phys. Fluids **11**, 2680 (1968).

[20] J. Christiansen, *et al.*, Nucl. Fusion **32**, 316 (1992).

[21] F.W. Perkins, *et al.*, Phys. Fluids B **5**, 477 (1993).

[22] M. Kotschenreuther and V. Wong (private communication, 1993).

[23] W. Manheimer and J.P. Boris, Comments Plasma Phys. Contr. Fusion **3**, 15 (1977).

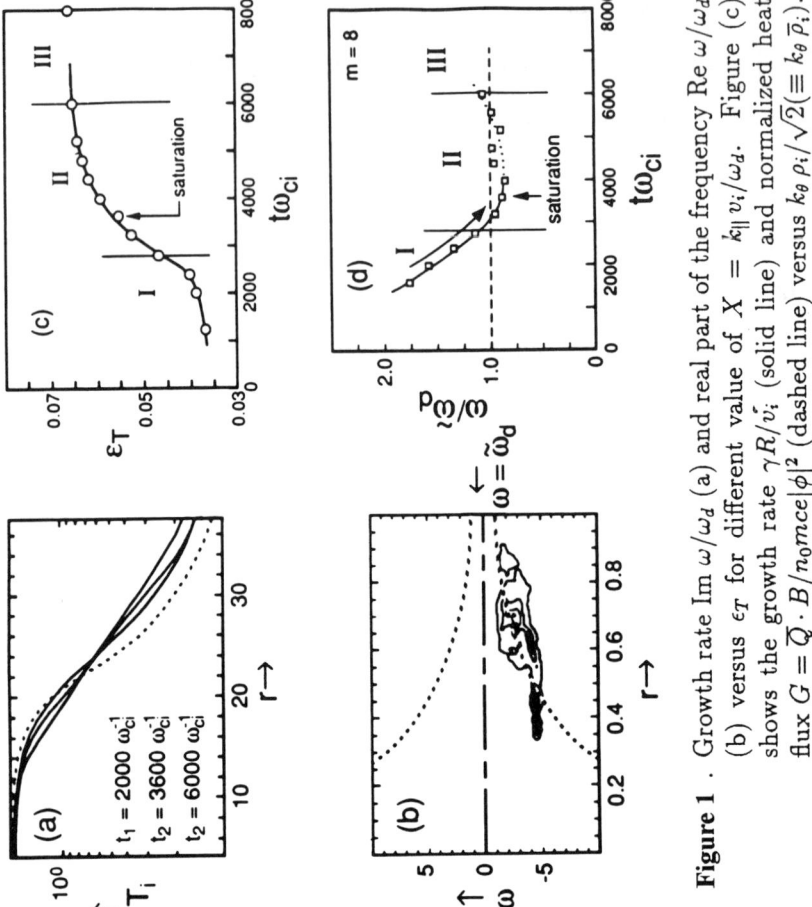

Figure 1. Growth rate Im ω/ω_d (a) and real part of the frequency Re ω/ω_d (b) versus ϵ_T for different value of $X = k_\parallel v_i/\tilde{v}_i$ (solid line) and normalized heat flux $G = \bar{Q} \cdot B/n_0 mce|\phi|^2$ (dashed line) versus $k_\theta \rho_i/\sqrt{2} (\equiv k_\theta \bar{\rho}_i)$.

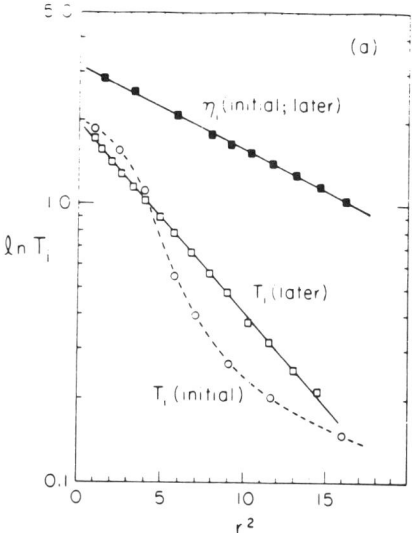

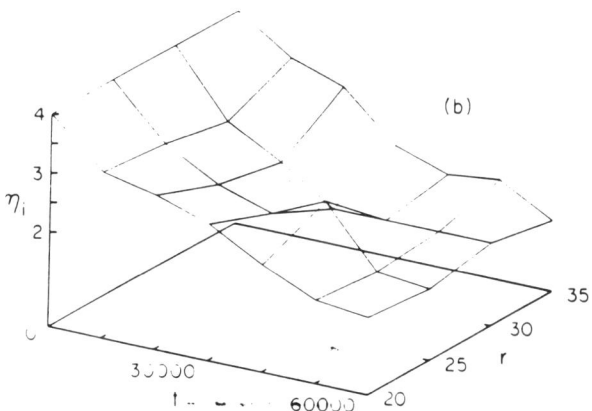

Figure 2. Temporal change of the plasma temperature profile as observed in our TPC simulation. (a) The initial arctan profile relaxes toward the gaussian with which η_i becomes a global constant. (b) The evolution of η_i as a function of time and radial position (initially gaussian temperature profile).

274 Ion Temperature Gradient Instability

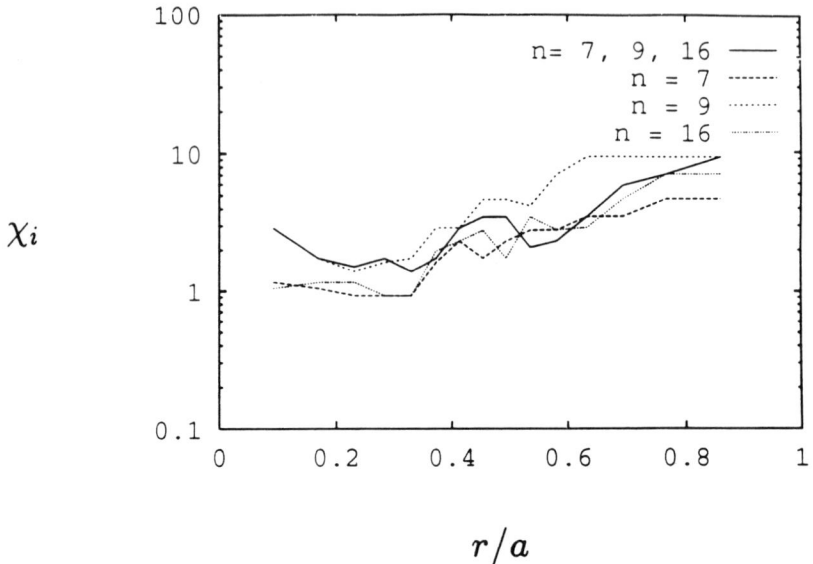

Figure 3. The measured heat conductivity (in the unit of $\rho_i^2 c_s/L_n$) vs. the main radius of the toroidal plasma from the TPC simulation. Both single helicity runs as well as multihelicity runs show similar qualitative tendencies.

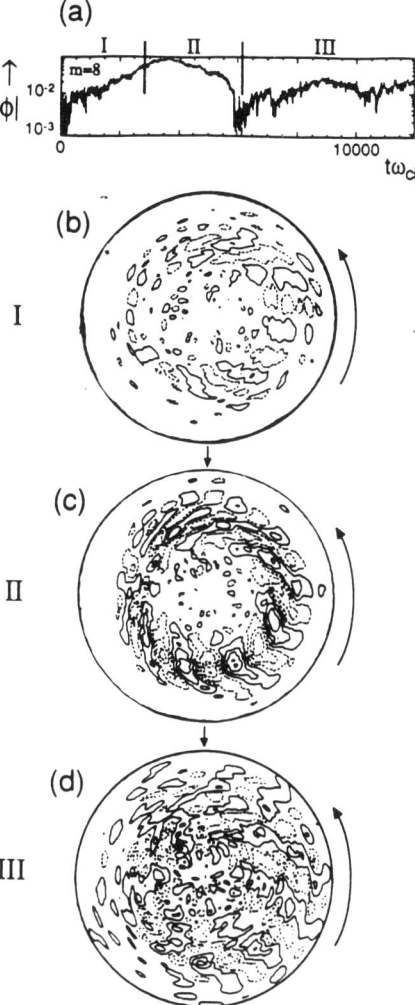

Figure 4. Temporal evolution of the potential energy and potential structures. (a) The potential energy vs. time for three stages I (linear stage), II (saturation), and III (steady). (b) Potential structure in the poloidal cross-section at stage I, (c) at stage II, (d) at stage III. The toroidal outward direction is to the right.

4.4 THE MARGINAL STABILITY OF EXPERIMENTAL PROFILES

M. Kotschenreuther
Institute for Fusion Studies, The University of Texas at Austin,
Austin, Texas

Abstract. The marginal stability of the ion temperature profiles for both L-mode and supershots is analyzed using a fully implicit δf gyrokinetic code including most relevant kinetic and collisional effects. The stability results tend to indicate that the L-mode discharges are in agreement with the marginal stability hypothesis in the core while the pellet shot is rather well above the critical temperature gradient. the linear stability results are highly inconsistent with the experimentally inferred transport, which implies that effects beyond lowest order ballooning theory must be important.

The subject of the marginal stability of experimental profiles to ion temperature gradient driven (ITGD) instabilities has a long history [1, 2, 3, 4]. It had been widely thought that the plasma was close to marginal stability until recent experiments by Zarnstorff et al. [2, 3] which show that the profiles can be perturbed to be seemingly far from marginal stability without increasing the thermal diffusivity χ. However, it is necessary to make an accurate calculation of instability growth rates and thresholds before definitive conclusions can be drawn. A comprehensive study [5] on the correlation between the ITGD turbulence and the experiments mentioned above is carried out, recently, with some η_i-mode theories. It is pointed out [5] that those theories seem to be insufficient to explain some experimental observations, and a more complete study of the ITGD mode is anticipated. For example, these analyese neglected important effects such as impurities and beams. Here we present results from a very complete code for several TFTR experimental data sets to examine marginal stability of the experimental profiles. Many of these effects are included in the codes of Rewoldt and Tang [6] and Dong, Horton, and Kim [7] but not all of them at the same level of completeness as developed here.

The simulation code is a fully gyrokinetic initial value code which solves the gyrokinetic equation in ballooning coordinates for five species: the bulk ions, electrons, two impurities (carbon and Iron/Nickel), and a high energy beam species. The fully kinetic treatment of all species naturally includes all effects from trapped particles, toroidal curvature, and finite Larmor radius-averaging of the fields. The code uses a Lorentz pitch angle scattering operator for all species, thus reproducing the diffusive nature of the Coulomb collisions, unlike the Krook relaxation collision models. Also, the code is electromagnetic (it includes the parallel vector potential $A_{\|}$). The code uses a grid in velocity space and in the ballooning angle, and uses a fully implicit technique to advance both the distribution functions and the electromagnetic fields ($\phi, A_{\|}$). Thus, the code has no stability time step restrictions from parallel electron motion or Alfvén waves, which greatly increases the allowable time step for these primarily electrostatic instabilities with drift wave frequencies $\omega \sim \omega_*$, where ω_* is the drift wave frequency. The code routinely runs at tens to hundreds of times the CFL stability criterion for electrons and shear Alfvén modes. Furthermore, the perturbed electron distribution functions accurately reproduce the adiabatic response (for passing particles) and the bounce averaged response (for trapped particles) for such large time steps. For experimental parameters, the simulation code yields an eigenfunction (with all species taken into account) in about 2-3 minutes of Cray CPU time. This is considerably faster than other comparably complete simulation codes.

The code has been run for several TFTR data sets. Here, we will present runs from two data sets: 1) the set of L-mode shots in the ρ_* scan used to test the Bohm-GyroBohm scaling hypothesis and 2) the pellet shot experiment by Zarnstorff et al. [2, 3] to test the marginal stability hypothesis. Here $\rho_* = \rho_s/a$ with $\rho_s = (2m_i T_e)^{1/2}/eB$ and a is the minor radius

The stability of the profiles in the ρ_* scan may be summarized as follows. For virtually all shots, the temperature gradient within the half radius ($r/a < 0.5$) is almost always within 20% of the value needed for marginal stability to ITGD modes. No other modes with $\omega \sim \omega_*$ other than ITGD modes are unstable. Beyond $r/a = 0.5$, the temperature profile becomes progressively further away from marginal stability as r increases. These results are shown for a particular shot in Fig. 1, which gives the ratio of the experimental temperature gradient scale length L_{T_i} to the value computed for marginal stability for the most unstable $k_\theta \rho_i$.

The degree to which the temperature gradient is above the critical gradient is shown in Fig. 2 versus $k_\theta \rho_i$ for $r/a = 0.5$, for the shot in Fig. 1. Note that the most unstable mode has $k_\theta \rho_i \sim 0.2$. Also note that modes with $k_\theta \rho_i \ll 0.1$ have frequencies less than the deuterium bounce frequency, and are trapped ion modes. For all shots, the trapped ion modes are stable, and become strongly stable as $k_\theta \rho_i$ is decreased. Trapped ion modes are often discussed as a possible explanation for the Bohm transport scaling observed in these shots, but their linear stability suggests they are not the cause.

These results for the growth rate spectrum are in good agreement with beam emission spectroscopy (BES) measurements of the density fluctuation spectrum. The BES measurements show a peak in the spectrum at $k_\theta \rho_i \sim 0.2$, and the BES measurements show very little activity for small $k_\theta \rho_i$.

It is instructive to discuss the reasons why the longer wavelength modes are not highly unstable for these shots. Three stabilizing effects have been identified which are important in many TFTR shots, including those discussed here. These effects are impurities, beams and collisions. We now examine these one at a time.

First, it is useful to write down the dispersion relation schematically. Though electromagnetic effects have been included in code results, the modes under discussion are primarily electrostatic, and the magnetic fluctuation corrections are small. Therefore the dispersion relation is primarily the quasinuetrality equation $\sum_s Z_s e \, \delta n_s = 0$. Splitting the density response into the adiabatic and non-adiabatic parts, we have:

$$\sum_s \frac{n_s Z_s^2}{T_s} \phi = \sum_s \frac{n_s Z_s^2}{T_s} \int f_s \, J_0(k_\perp v_\perp/\Omega_s) \, d\mathbf{v} \qquad (1)$$

where f_s is the solution of the drift kinetic equation

$$\frac{\partial f}{\partial t} + i\omega_D f + \frac{v_\parallel}{qR} \frac{\partial f}{\partial \theta} - C(f) = \left\{ i(\omega_* - \omega_D) J_0 \phi + \frac{1}{qR} \frac{\partial}{\partial \theta} (J_0 \phi) \right\} F_M \qquad (2)$$

where $v_\parallel = \pm(2\epsilon/m)^{1/2}(1 - \lambda B(\theta))^{1/2}$ and $B = B_0/(1 - r\cos\theta/R)$. In Eq. (1) the species index s runs over all elements including the beam ions and the impurities. Here ϵ and λ are the energy per unit mass and the pitch angle variable $\lambda = \mu/\epsilon$ with $\mu = \frac{1}{2}v_\perp^2/B$ = constant.

Note that impurities enter the dispersion relation Eq. (1) as $n_{0s}Z_s$, just as in Z_{eff} when T_s are comparable. Since experiments typically have $Z_{\text{eff}} > 2$, the impurities can potentially be as important as the bulk thermal ions in the dispersion relation, even though they only constitute a small percentage of the plasma. Also, note that for impurities $\omega_{*z} \sim \omega_{*d}/Z \ll \omega_{*d}$ where ω_{*d} is the deuterium drift frequency. For Maxwellian impurities, $\omega \gg \omega_{*z}$ implies that the impurities can only contribute damping. Furthermore, for hot ion mode shots and supershots on TFTR, impurities can remove up to roughly 50% of the thermal ions which are driving the instability. This dilution stabilization effect is important for such shots.

Also note that the beam contribution to the dispersion relation is proportional to $e\phi/T_{\text{beam}} \ll e\phi/T_{\text{ion}}$. Therefore, the beam contribution is much less than the thermal ion contribution. Because of this, the beams make little contribution to the dynamics of these essentially electrostatic modes. However, the beams affect the equilibrium by diluting the bulk ions. Thus the beams give another example of dilution stabilization. These effects are shown in Figs. 2 and 3, which give the marginal stability points of ITGD modes for experimental parameters. Figure 3 gives the marginal results for an L-mode, both with and without impurities and beams. Without these effects, the shot is roughly 50% above the marginally stable point (which is well above the error bars for the experimental temperature gradient), whereas with these effects the shot is within roughly 10% of marginal stability. Figure 4 gives results for a hot ion mode shot on TFTR which has a large fraction of impurities and beams which give a strong stabilizing effect. As can be seen, without these effects the shot appears to be at roughly 2.5 times the critical temperature gradient, whereas with these effects the shot is close to being marginally stable.

Collisions can affect the dispersion relation by modifying the trapped electron response and the impurity response. In has been found previously by Rewoldt and Tang [6] that the trapped electron effects can add an important destabilization to ITGD modes, especially near marginal stability. Collisions generally reduce the trapped electron effects, because when the effective collision frequency for the trapped electron region, $\nu_{\text{eff}} = \nu_e/\epsilon$, is of the order of the mode frequency ω, collisions diffuse trapped electrons out of the trapped region into the passing region, where the electrons primarily behave adiabatically. This reduction of the non-adiabatic electron response can be stabilizing to the ITGD modes. We have found that some Krook relaxation collision operators can substantially underestimate collisional effects on trapped particles, particularly at large ν_{eff}/ω. This underestimate is true for the ingenious and elaborate Krook models which interpolate between the asymptotic analytic results, such as used by Rewoldt et al. [8] An example of collisional trapped electron effects on ITGD modes is shown in Fig. 4, which has parameters close the experimental parameters of Fig. 2 (but for simplicity we have taken $Z_{\text{eff}} = 1$ and no electron temperature gradient $\eta_e = 0$). The results show the growth rate γ/ω_* versus the normalized collision frequency $\gamma_{\text{eff}}/\omega$. We expect from the previous discussion that when $\nu_{\text{eff}}/\omega \gg 1$, that γ/ω_* should approach the value for an adiabatic case, and that the transition should occur at $\nu_{\text{eff}}/\omega \sim 1$. This is indeed the case for the Lorentz collision operator. Figure 4 also shows the same results for the Krook model of Rewoldt et al. The Krook model trends the same way, but it greatly underestimates the amount of collisions needed to reduce the trapped electron effects. (An alternative interpolation formula might improve that model's accuracy.)

In addition, the impurity-impurity collisional effects can be high, since $\nu_{z-z} \sim Z^2 \nu_{i-i}(Z_{\text{eff}}-1)$ which is often only a few times less than the electron collision frequency ν_e. Since impurities only damp the ITGD mode driven by the bulk ions, increased impurity collisional damping is stabilizing. This effect is primarily important for low frequency modes such as trapped ion modes.

A pellet injection shot experiment was performed by Zarnstorff et al.[2] to test the marginal stability hypothesis. This shot has also been examined with the simulation code. In this shot, a deuterium pellet was injected to flatten the density profile, which increased η_i by a large factor, and the plasma temperature response to the pellet substantially increased the ion temperature gradient. Thus the pellet should presumably make the profile much more unstable, however the inferred X_i did not increase.

This shot had been previously examined with the simulation code [4] and in the Horton *et al.*'s kinetic ballooning stability analysis [5]. Reference [4] used the assumption that the impurity Z_{eff} was flat before the pellet, and that the pellet did not affect the impurity density. However, recent experimental measurements which radially resolve Z_{eff} are now available, and these new measurements strongly change the results. In particular, the measurements show that there is a strong dip in the impurity density (both before and after the pellet) precisely where the temperature profile is the steepest. Thus the impurity stabilization is weak there.

The results for stability from this shot are shown in Fig. 5. For a considerable fraction of the minor radius, $(0.2 < r/a < 0.5)$ the shot is typically almost 3 times above the marginal stability threshold. Though one may expect some error in the experimental measurements, the fact that the ITGD mode is grossly unstable over a substantial range of radius leads to the conclusion that the mode must indeed be well above the stability threshold.

Despite the fact that the modes are strongly above marginality, the experimentally inferred X_i actually decreased. Even more surprising, the X_i in this shot is almost identical to the experimental X_i in the ρ_* scan L-mode shots analyzed above, even though most of the other plasma parameters are similar. Consider the comparison table of these two TFTR shots at $r/a = 0.25$, ($a = 80\,\text{cm}$ in both cases):

PARAMETER	PELLET SHOT AFTER PELLET	L-MODE SHOT IN ρ_* SCAN
$L_{T_\text{crit}}/L_{T_\text{expt}}$	2.89	1.1
Experimental χ_i	$1.17 m^2/\text{sec}$	$1.29 m^2/\text{sec}$
Gyro-Bohm normalized experimental χ_i	0.27	0.49
T_i	5.25 keV	3.76 keV
Safety factor q	1.32	0.82
T_e/T_i	0.89	0.71
Z_eff	1.5	2.3
shear parameter s	0.81	0.71
L_{n_e}	3.04 m	1.39 m
L_{T_i}	0.16 m	0.39 m

Note that the equilibrium parameters of these shots are not greatly different except for the scale lengths L_{n_e} and L_{T_i}, which are quite different and are the primary reason why the pellet shot is so much further from marginality. For short wavelength modes of this type, the gyro-Bohm normalized conductivity $\chi_i/(v_{t_i}\rho_i^2/a)$ is the theoretically appropriate dimensionless conductivity to compare for these two shots. Notice that the gyro-Bohm normalized experimental conductivity is actually less for the shot which is *grossly above criticality* when compared with the shot which is close to marginality. The only other dimensionless parameters which are slightly different are q and Z_eff. Both of these parameters differ in a way which would be expected to make the pellet shot have greater transport, which makes this result all the more surprising. Z_{eff} is somewhat less for the pellet shot, which should make it more unstable. Also, q is somewhat higher; to the extent that I_p scaling holds, one would expect that this would tend to make χ_i larger for the pellet shot. Thus it is extremely difficult to understand how the L-mode shot 10% above criticality could have worse transport than the grossly unstable pellet shot.

There are several possible interpretations of this result:

1) The experimental data is in error. Though this is being investigated further, on first examination the discrepancy seems too large to be accounted for by the error bars.

2) The code is in error. To check this, the code has been compared with other kinetic codes by other authors. The code has compared well (within 10% to 15%) with results for marginal stability and frequency for several runs with each of the following three codes: an eigenfunction code by Dong [7], *et al.*, a δf particle code by Xu and Rosenbluth [9] and a grid code by Wong [10]. The codes by Wong [7] and Xu [6] do not have impurities, trapped electrons and beams. Some comparisons have been made with Dong, *et al.*[7] including these effect, but neglecting trapped ion effects. Comparisons are under way with the code by Rewoldt and Tang [6].

In addition, a δf particle code in ballooning coordinates has been constructed which includes all the same physics as the grid code, to make detailed comparisons using different algorithms. Agreement between these codes over a range of parameters is a strong check of accuracy, since it is virtually impossible that an error in one code would produce the same answer as the error in a different code with a completely different algorithm. To date, the codes have been checked by comparing the perturbed density response of each species to a given ϕ_{k_θ} with a frequency ω and shape similar to the known eigenfunctions. There is extremely good agreement for the collisionless case, and preliminary results indicate good agreement for the including collisions. Tests for the self-consistent ϕ will be done in the future.

3) If the experimental data and the codes are correct, it indicates that the experimental χ_i is essentially independent of whether the plasma is very close to linear marginal stability or very far past marginal stability (as computed from lowest order ballooning mode theory.) This

implies that the widespread theoretical framework used to investigate transport from ITGD instabilities is missing a very important factor. For example, linear mixing length estimates of X_i indicate that the pellet shot should have much greater transport.

In conclusion, the L-mode shots appear in agreement with the marginal stability hypothesis in the core. However, the pellet shot is *much* more unstable, with much steeper temperature gradients, and yet has better confinement. Pending further checks, this indicates that lowest order linear ballooning theory is missing an extremely important factor needed to explain transport.

ACKNOWLEDGMENTS

The author thanks W. Horton, J.-Y. Kim, and J.Q. Dong and M. Zarnstorff for useful communications during the course of this work. The work was supported by the Department of Energy contract #DE-FG05-80ET-53088.

Bibliography

[1] S.D. Scott, P.H. Diamond, R.J. Fonck, R.J. Goldston, R.B. Howell, K.P. Jaehnig, G. Schilling, E.J. Synakowski, M.C. Zarnstorff, C.E. Bush, E. Fredrickson, K.W. Hill, A.C. Janos, D.K. Mansfield, D.K. Owens, H. Park, G. Pautasso, A.T. Ramsey, J. Schivell, G.D. Tait, W.M. Tang, and G. Taylor, Phys. Rev. Lett. **64**, 531 (1990).

[2] M.C. Zarnstorff, C.W. Barnes, P.C. Efthimion, G.W. Hammett, W. Horton, R.A. Hulse, D.K. Mansfield, E.S. Marmar, K. McGuire, G. Rewoldt, B.C. Stratton, E.J. Synakowski, W. Tang, J. Terry, X.Q. Xu, M.G. Bell, M. Bitter, N.L. Bretz, R. Budny, C.E. Bush, G.J. Fonck, E.D. Fredrickson, H.P. Furth, R.J. Goldston, B. Grek, R.J. Hawryluk, K.W. Hill, H. Hsuan, D.W. Johnson, D.C. McCune, D.M. Meade, D. Mueller, D.K. Owens, H.K. Park, A.T. Ramsey, M.N. Rosenbluth, J. Schivell, G.L. Schmidt, S.D. Scott, G. Taylor, and R.M. Wieland, in *Plasma Physics and Controlled Nuclear Fusion Research*, 1990, Proceedings of the 12th International Conference, Washington (IAEA, Vienna, 1991), Vol. I, p. 109.

[3] M.C. Zarnstorff, N.L. Bretz, P.C. Efthimian, G. Hammett, W. Horton, R. Hulse, D. Mansfield, E. Marmar, K. McGuire, G. Rewoldt, B. Stratton, E. Synakowski, W. Tang, J. Terry, X. Xu, M. Bell, M. Bitter, N. Bretz, R. Budny, C. Bush, R. Fonck, E. Fredrickson, H. Furth, R. Goldston, B. Grek, R. Hawryluk, K. Hill, H. Hsuan, R.A. Hulse, D. Johnson, M. McCune, D. Meade, D. Mueller, D. Owens, H. Park, A. Ramsey, M. Rosenbluth, J. Schivell, G. Schmidt, S. Scott, G. Taylor, and R. Wieland, in *Proceedings of the 17th European Physical Society Conference on Con-*

trolled Fusion and Plasma Heating, Amsterdam (European Physical Society, Budapest, 1990), Vol. I, p. 39.

[4] M. Kotschenreuther, H.L. Berk, R. Denton, S. Hamaguchi, W. Horton, C.-B. Kim, M. Lebrun, P. Lyster, S. Mahajan, W.H. Miner, P.J. Morrison, D.W. Ross, R.D. Sydora, T. Tajima, J.B. Taylor, P.M. Valanju, H.V. Wong, S.Y. Xiao, Y.Z. Zhang, in *Plasma Physics and Controlled Nuclear Fusion Research*, 1990, Proceedings of the 13th International Conference, Washington (IAEA, Vienna, 1991) Vol. II, p. 361.

[5] W. Horton, D. Lindberg, J-Y. Kim, J.Q. Dong, G.W. Hammett, S.D. Scott, M.C. Zarnstorff, and S. Hamaguchi, Phys. Fluids **4**, 953 (1992).

[6] G. Rewoldt and W.M. Tang, Phys. Fluids B **2**, 318 (1990) and Chapter X, Sec. Y of this volume.

[7] J.Q. Dong, W. Horton, and J-Y. Kim, Phys. Fluids B **4**, 1867 (1992).

[8] X.Q. Xu and M.N. Rosenbluth, Phys. Fluids B **3**, 627 (1991).

[9] V. Wong, private communication (1992).

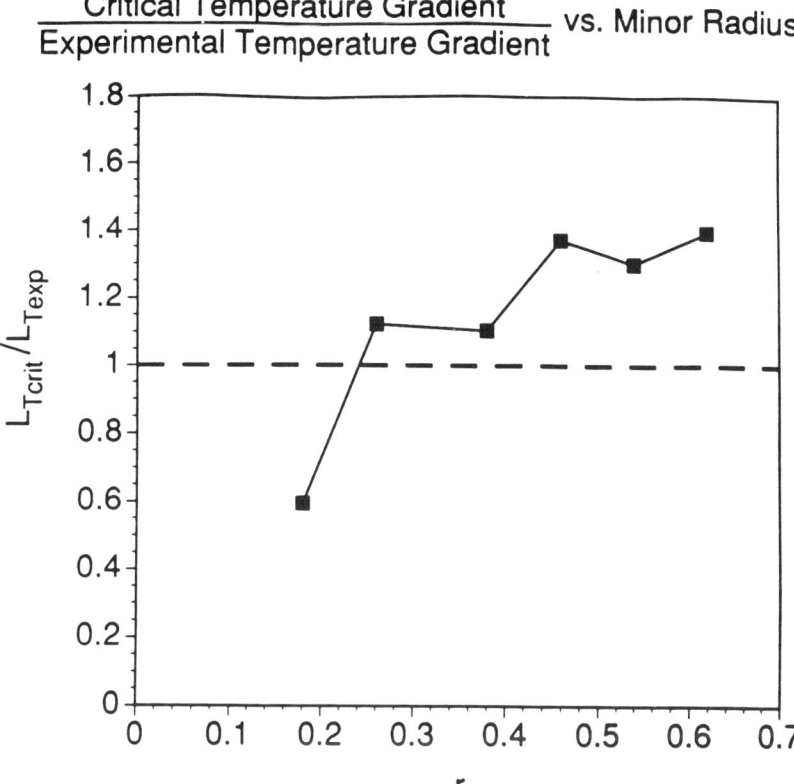

Figure 1. The ratio of the experimental ion temperature gradient to the computed critical gradient, versus minor radius, for an L-mode shot.

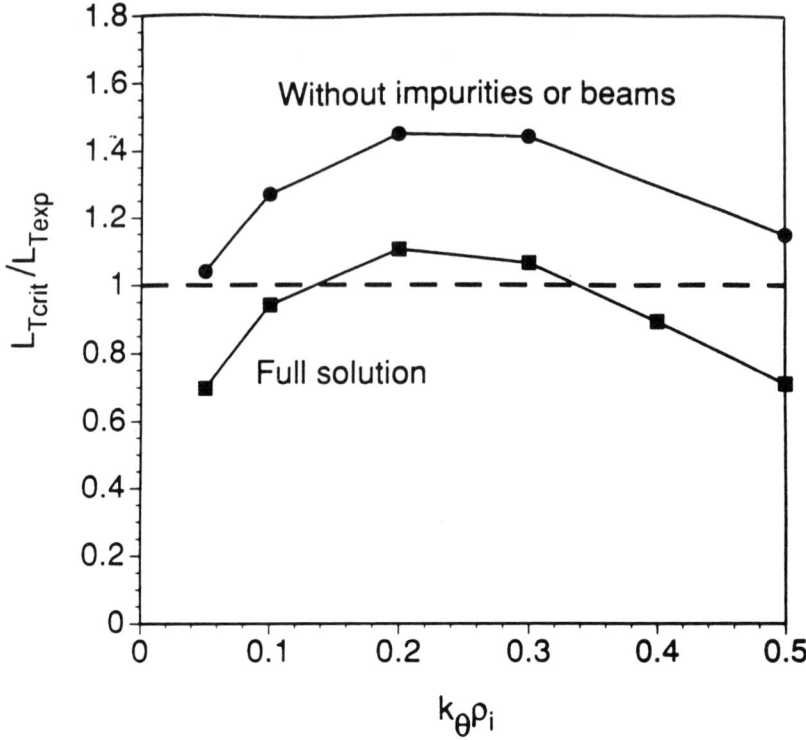

Figure 2. The ratio of the experimental ion temperature gradient to the critical gradient for the shot in Fig. 1 at $r/a = 0.5$, versus $k_\theta \rho_i$. Trapped ion modes with $\omega_{\text{bounce}} \geq \omega$ begin at $k_\theta \rho_i \simeq 0.09$. Also shown are results neglecting impurities and beams.

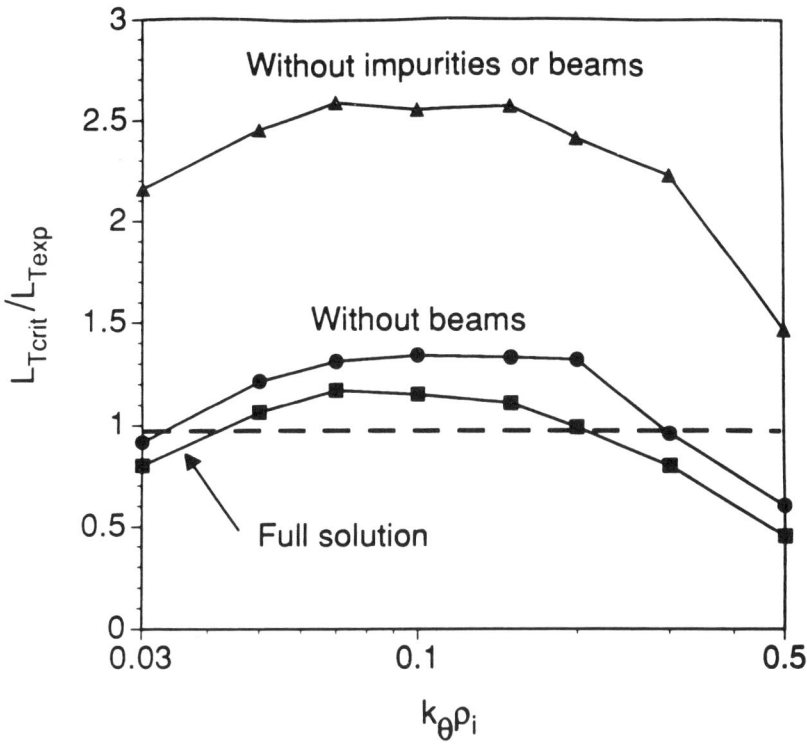

Figure 3. The ratio of the experimental ion temperature gradient to the critical gradient for a hot ion mode shot with a large beam fraction and impurity fraction. The results without impurities and beams are also shown.

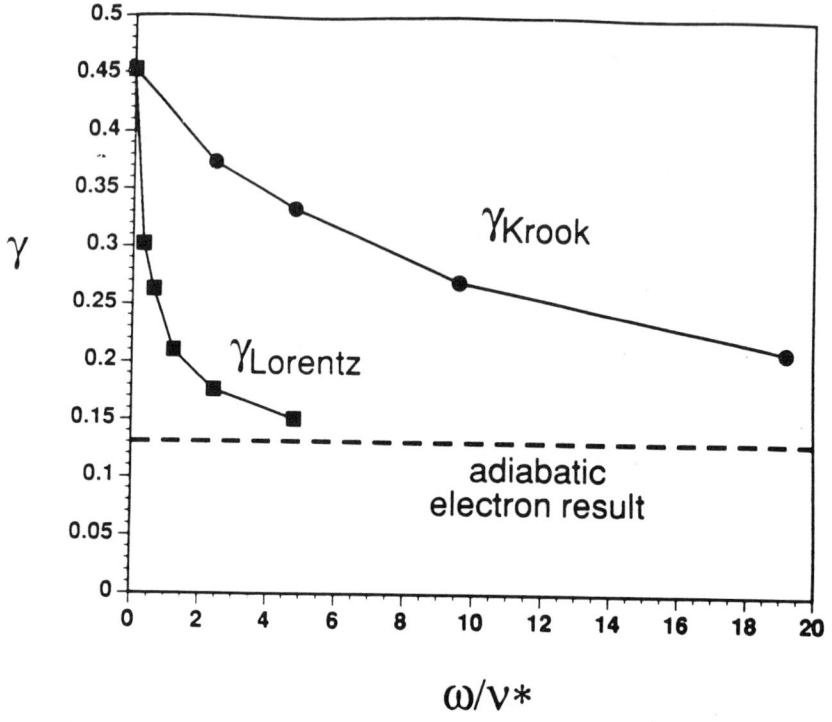

Figure 4. Growth rate of an ITG mode versus the effective collision frequency for trapped electrons, for the Lorentz model and an elaborate Krook model.

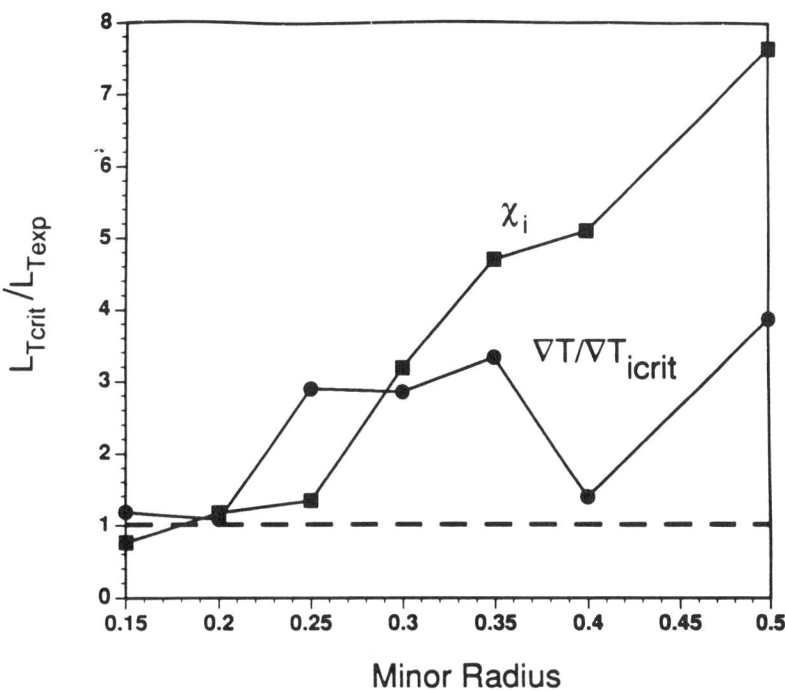

Figure 5. The ratio of the experimental ion temperature gradient to the critical gradient versus minor radius, after pellet injection in an experiment by Zarnstorff to test the marginal stability hypothesis. Also shown in the experimentally inferred χ_i.

Chapter 5
FLUID AND GYROFLUID SIMULATIONS

5.1 REACTIVE DRIFT WAVE MODEL FOR TOKAMAK TRANSPORT

H. Nordman and J. Weiland
Institute for Electromagnetic Field Theory and EURATOM/NFR Association Chalmers University of Technology, S-41296 Göteborg, Sweden

Abstract. Transport due to toroidal η_i and collisionless trapped electron modes is studied using an advanced reactive fluid model. Transport code simulations of DIII-D and JET discharges are analyzed. The recent experimental results with off-axis electron cyclotron heating in the DIII-D tokamak [Phys. Rev. Lett. 68, 52 (1992)] are recovered with inward flow of electron energy and peaked electron temperature profile. For the JET equilibria the model gives heat diffusivities of the correct order of magnitude which grow with radius up to $r/a = 0.8$.

5.1.1 INTRODUCTION

The evidence for turbulence as the main cause of anomalous transport in magnetic confinement systems is steadily increasing. There is, however, still no consensus concerning the detailed nature of the driving mechanisms of the turbulence. It now seems that the gyrokinetic simulations may soon be able to resolve these questions. One particular feature, which has now been verified by the gyrokinetic simulations [1, 2], is the dominance of toroidal effects. Toroidal effects are clearly very fundamental since they are associated with the basic property of the magnetic field of confining the plasma only in two dimensions. By bending the system to a torus we seem to eliminate the third dimension but instead we introduce the curvature effects. The most fundamental property of a magnetic confinement system, its ability to confine plasma pressure, is clearly limited by the toroidal effects through the magnetohydrodynamic (MHD) ballooning mode which limits the total pressure gradient. On the next level enter the reactive drift modes, which in their toroidal version are very similar to the MHD ballooning mode. These modes resolve the degeneracy in ∇p and specify that the density and temperature scale lengths will tend to equilibrate on a time scale shorter than the confinement time. These modes are accordingly very strong candidates for explaining the anomalous transport in tokamaks. Due to the dominance of toroidal effects it is essential to have a physics model that includes the toroidal effects with sufficient accuracy. Thus while the inclusion of only the toroidicity drive for η_i modes essentially gives a slab mode with an enhanced growth rate, a more complete model which consistently includes also the isothermal limit where $\omega_D \gg \omega$ introduces a new regime of resonant modes ($\omega \sim \omega_D$) where also stabilizing effects of toroidicity are present [3, 4]. In this new regime the diffusion coefficients tend to grow with radius, at least up to 0.8 of the small radius and in the flat density limit ($\epsilon_n \gg 1$) the density scale length (L_n) scales out.

Also the effect of toroidicity on the radial variation of χ_i has now been verified by the gyrokinetic simulations [1] which in the cylinder geometry produce a χ_i that decreases with radius in agreement with simpler drift wave models but in toroidal geometry gives a χ_i which grows with radius and is of the same order of magnitude as observed in the experiments. Both the tendency for the dominance of toroidal effects and the radial growth of χ_i are emphasized in the reactive limit when the kinetic wave particle resonances are averaged out by nonlinear effects in velocity space [5]. The influence of parallel ion motion, including the slab driving effect, is in fact usually negligible when ion Landau damping is absent. In this limit the description becomes two dimensional. The radial profile of χ_i is improved by the stabilizing effects of toroidicity which are strongest close to the axis. Nonlinearly these stabilizing effects appear as pinch fluxes which reduce the effective A typical example of a nondiffusive phenomenon is the offset linear scalings of fluxes with gradients obtained in many experiments [6, 7]. This is often interpreted as due to a convection and can also be described as an off-diagonal term in a diffusion matrix. Such a term usually corresponds to a flux along the gradient of the considered quantity, i.e. opposite to the normal diffusion. This is referred to as a pinch effect.

Another phenomenon which also often has been considered to be nondiffusive is the insensitivity of the temperature profiles to the location of the heat sources. This is called profile consistency [8] and is closely related to the marginal stability hypothesis [9] according to which the profiles of density and temperature have to stay close to those corresponding to marginal stability of a strong mode. The importance of heat pinch effects for profile consistency in connection with pellet fuelling was discussed in detail in Ref. [10].

Theoretically, particle pinch effects at the edge were obtained for drift wave systems several years ago [11, 12]. Heat pinch effects have also recently been obtained [13, 14, 15, 16, 17] as well as particle pinches in the bulk region [13, 14].

A new experimental result during 1992 which challenges the estimated level of the transport in most physics based models of transport is the observation that the k-spectrum peaks at considerably lower mode numbers than seen before [18]. This is particularly pronounced for k_r where the peak can not be distinguished from the fundamental mode. We note here that the scale-length used in our transport estimate corresponds to the average space variation and should rather be compared to the correlation length which is isotropic and of comparable magnitude.

In this work we present transport code simulations of reactive drift wave systems with special emphasis on pinch flows and the radial variation of the diffusion coefficients. In particular, the off-axis heating experiment in DIII-D is simulated self-consistently and some recent JET equilibria are investigated with the interpretative transport code.

5.1.2 FLUID MODEL OF TOROIDAL DRIFT WAVE TRANSPORT

An improved two-dimensional toroidal fluid model is used to derive the density perturbation for ions, trapped electrons and free electrons as (for details see Refs. [13] and [14])

$$\frac{\delta n_i}{n} = \omega_{*e} \frac{\omega(1 - \epsilon_n) - (7/3 - \eta_i - 5\epsilon_n/3)\omega_{Di} + k^2 \rho_s^2 f(\omega)}{N_i} \frac{e\phi}{T_e} \quad (1)$$

$$\frac{\delta n_{et}}{n_{et}} = \omega_{*e} \frac{\omega(1 - \epsilon_n) - (7/3 - \eta_e - 5\epsilon_n/3)\omega_{De}}{N_e} \frac{e\phi}{T_e} \quad (2)$$

$$\frac{\delta n_{ef}}{n_{ef}} = \frac{e\phi}{T_e} \quad (3)$$

where

$$f(\omega) = (\omega - \omega_{*ip})\left(\frac{\omega}{\omega_{*e}} + 5\epsilon_n/3\tau\right),$$

and

$$N_j = \omega^2 - \frac{10}{3}\omega\omega_{Dj} + \frac{5}{3}\omega_{Dj}^2.$$

ω_{*j} is the diamagnetic drift frequency of species j, $\omega_{*ip} = \omega_{*i}(1 + \eta_i)$, $f_t = \sqrt{2\epsilon/(1+\epsilon)}$ where $\epsilon = r/R$, $\tau = T_e/T_i$, $\epsilon_n = \omega_D/\omega_* = 2L_n/L_B$, $\eta_j = L_n/L_{Tj}$. The collisionless model includes first order Finite-Larmor-radii effects (FLR), polarization drift effects, compressibility due to field curvature and perpendicular diamagnetic heat flow in the energy balance. We note that Eqs. (1)–(3) agree exactly with kinetic theory in both the adiabatic and isothermal limits. Recently, Waltz et al. [19] rederived the ion density perturbation (1) from a more general gyrofluid theory in the relevant limit.

The quasineutrality condition

$$\frac{\delta n_i}{n} = f_t \frac{\delta n_{et}}{n_{et}} + (1 - f_t)\frac{\delta n_{ef}}{n_{ef}} \quad (4)$$

leads to a quartic dispersion relation describing the toroidal η_i-mode and an essentially symmetric collisionless trapped electron mode (CTE-mode). These modes are driven by fluid resonances when $\omega \sim \omega_D$. The resonances are similar to he fluid resonance in a two-stream instability. We note the very close similarity between the trapped ion mode and the toroidal η_i-mode in the two dimensional limit as used here. This limit becomes relevant for the toroidal η_i-mode when wave particle resonances are averaged out by nonlinear effects in velocity space [5]. In realistic cases it seems that both free and trapped ions may contribute to the same type of flute mode response.

5.1.3 TRANSPORT CODE SIMULATIONS

The transport code self-consistently solves the time-dependent radial diffusion equations

$$\frac{\partial}{\partial t}\begin{pmatrix}T_i\\T_e\\n\end{pmatrix} = \frac{1}{r}\frac{\partial}{\partial r}r\begin{pmatrix}\chi_{ii} & \chi_{ie} & \chi_{in}\\ 0 & \chi_{ee} & \chi_{en}\\ 0 & D_{ne} & D_{nn}\end{pmatrix}\frac{\partial}{\partial r}\begin{pmatrix}T_i\\T_e\\n\end{pmatrix} + \begin{pmatrix}S_i\\S_e\\S_n\end{pmatrix} + R\begin{pmatrix}T_e - T_i\\T_i - T_e\\0\end{pmatrix} \tag{5}$$

where $R(r,t) = 2\nu_{ei}m_e/m_i$ is the resistive temperature equilibration term and $S_j(r,t)$ represents other sources and sinks.

The effective quasilinear diffusion coefficients, originally derived in Ref. [13] are defined as $\chi_j = -\langle \delta T_j v_{Er}\rangle/dT_j/dr$ and $D=-\langle \delta n v_{Er}\rangle/dn/dr$ and become

$$\chi_i = \frac{1}{\eta_i}[\eta_i - \frac{2}{3} - (1-f_t)\frac{10}{9\tau}\epsilon_n - \frac{2}{3}f_t\Delta_i]\frac{\gamma^3/k_x^2}{(\omega_r - 5\omega_{Di}/3)^2 + \gamma^2} \tag{6}$$

$$\chi_e = f_t\frac{1}{\eta_e}\left[\eta_e - \frac{2}{3} - \frac{2}{3}\Delta_e\right]\frac{\gamma^3/k_x^2}{(\omega_r - 5\omega_{De}/3)^2 + \gamma^2} \tag{7}$$

$$D = f_t\Delta_n\frac{\gamma^3/k_x^2}{\omega_{*e}^2} \tag{8}$$

where $k_x = k_y$ and $k^2\rho_s^2 = 0.1$ are used and where the contributions from the trapped electron mode and the η_i-mode are added together. The complicated expressions for Δ_j appear due to the trapped electron response and are given in Ref. [13] and [14]. In Eqs. (6)–(8), a modified mixing length estimate of the saturation level is used, i.e. $e\phi/T_e = \gamma/\omega_{*e} \cdot 1/k_x L_n$ (obtained by balancing the linear growth with the main convective nonlinearity). This estimate of the saturation amplitude is in better agreement with experiments (giving a fluctuation level which increase with radius) than the usual mixing length level. In addition, it leads to quasilinear expressions which are in good agreement with nonlinear mode coupling simulations [14] except very close to the marginal stability limit, where the transport is underestimated. We emphasize that the diffusivities may become negative. This corresponds to a net pinch effect which tends to equilibrate the length scales L_{Ti} The system (5) has been solved in the region $0.1 < r/a < 0.9$. At the inner boundary the profiles were kept flat and at the outer boundary the amplitudes were fixed. The neoclassical contribution to the transport was neglected.

5.1.4 Simulation of the DIII-D Off-Axis ECH Experiment

The most recent experimental evidence and the first measurement of energy pinch effects was obtained for off-axis electron cyclotron heating (ECH) in DIII-D [20]. In this experiment a peaked electron temperature profile was obtained although the heat source was located just outside half of the minor radius. Power balance calculations showed an inward electron energy flux for radii smaller than that of the heating location. The DIII-D parameters used in the simulations are $a = 0.62$ m, $R = 1.7$ m, $B = 2.2$ T and elongation $\kappa = 1.9$. We also use sources in accordance with the experiment, i.e. a localized ECH source and weak Ohmic heating for electrons and no ion heat sources. The only particle fuelling is located at the edge (gas puffing). The total energy flux, which is always outward in our model, is written

$$Q = q_e + q_i + (T_e + T_i)\Gamma_n > 0 \qquad (9)$$

where $q = n \langle \delta T \, v_{Er} \rangle$ and $\Gamma_n = \langle \delta n v_{Er} \rangle$.

Since the particle source is located at the edge it follows that $\Gamma_n = 0$ in the interior in stationary state. This means that the off-diagonal pinch fluxes exactly balance the diagonal outward fluxes. Due to the strong pinch fluxes in our model this occurs for comparatively steep density gradients. The main case of interest in the present study is that of off-axis electron heating. Inside the electron heat source the only source is then the small Ohmic heating. Neglecting this we then have $Q = 0$ or

$$q_e = -q_i \,. \qquad (10)$$

The positive ion heat flux is driven by the resistive electron-ion temperature equilibration term in Eq. (5) and an electron heat pinch is required to maintain the electron temperature in the centre. Since $\Gamma_n = 0$ this situation corresponds to an electron energy pinch. The larger this pinch is, the larger electron temperature can be maintained in the center. The condition $\Gamma_n = 0$ and $q_e = -q_i$ inside the source have to be fulfilled in equilibrium both in the experiment and in the code. This is a natural consequence of the location of the sources and stationarity.

In Figs. 1a,b the stationary temperature (obtained from self consistent simulation and experiment) and density profiles are given for a 1.2 MW localized ECH source centered at 0.55 of the minor radius. In Fig. 2 the corresponding energy fluxes from the simulations are shown.

At the resonance location, r_s, a rather sharp transition to an electron heat pinch occurs. For $r > r_s$ the electron heat flux is larger than the ion heat flux. This is in qualitative agreement with the experiment. We also note that the neoclassical flux (shown in Ref. [20]) is negligible in the present case. As was pointed out in Ref. [13] the transport model used here requires a rather small η_e to give an electron heat pinch. This is accomplished by a local flattening of the electron temperature within and just inside the ECH source region as seen in the transport code results.

In Table I the peakedness of the electron temperature profile and the electron confinement time τ_E are given for various choices of heating location r_s. We note the weak dependence of the peakedness with r_s corresponding to strong profile consistency. This is made possible by the pinch effect. We have verified that the heat pinch appears for $r \leq r_s$ in all cases presented in Table I.

The very sharp transition of the electron heat flux is due to the requirement that $q_e = -q_i$ inside the source and $q_e > q_i$ outside the heat source. The latter condition follows from the fact that all the heating is on the electrons and the equilibration term becomes inefficient when T_i approaches T_e. (The overall flux is given by the sources in equilibrium). In the drift wave model, q_e is given by

$$q_e = -n\chi_e \frac{dT_e}{dr} = -f_t \left\{ n\frac{dT_e}{dr} - \frac{2}{3}T_e(1+\Delta_e)\frac{dn}{dr} \right\} \frac{\gamma^3/k_x^2}{(\omega_r - 5\omega_{De}/3)^2 + \gamma^2} \tag{11}$$

where Δ_e is a very complicated expression due to the trapped electron response and f_t is the fraction of trapped electrons. A similar expression holds for the particle flux. We note the presence of heat pinch terms proportional to dn/dr in the expression for q_e. In realistic cases, however, the eigenfrequency is a (strongly nonlinear) function of density and temperature gradients, making a separation of fluxes into diffusive and nondiffusive parts intractable.

Simulation of JET experiments

In a recent JET report [21], theoretically predicted transport coefficients due to ∇T_i instabilities were compared with those inferred from experimental measurements. The results emphasized the theoretical problem of reproducing the increase of χ_i with radius as observed in the experiments. In this section the interpretative (time independent) version of the transport code is used to compare the radial profiles of the heat diffusivities with the two typical JET equilibria presented in Ref. [21]. In these equilibria a separation between the electron and ion heat fluxes were possible. The fluid model used here has earlier shown its ability to give radial profiles of the diffusion coefficients growing with radius up to $r/a = 0.8$, also for JET equilibria [14].

The results for the L-mode discharge (19699) are shown in Fig. 3a. The results are within the experimental errorbars for $0.5 \leq r/a \leq 0.8$. For $r/a \leq 0.5$ and $r/a \geq 0.8$ the experimental diffusivities are larger than the theoretical. The increase of the theoretical diffusion coefficients with radius up to $r/a = 0.8$ is both due to the pinch terms (which increase towards the center) and the choice of saturation amplitude.

The corresponding result for the H-mode discharge (24737) is shown in Fig. 3b. Also here, the agreement is good in the main part of the confinement region.

5.1.5 CONCLUSIONS

In this work we have described the most recent results obtained by a transport model based on reactive drift modes [4, 13, 14]. New evidence for the relevance of this model has recently been obtained both from a new derivation by gyrofluid theory [19] and gyrokinetic simulations [1] showing that the radial growth of χ_i is due to toroidal effects. We have shown the compatibility of drift wave transport with the electron energy pinch obtained in the off-axis ECH experiments on DIII-D. This is the firt demonstration of an energy pinch in drift wave transport models.

In order to reproduce the present electron energy pinch in DIII-D a transport model must be able to give both heat and particle pinches. Although there is no net particle pinch in this case, the pinch flux is able to cancel the usual outward particle flux so that what remains is only the inward electron heat flow and outward ion heat flow. The fluid model used here has these properties. As is clear from Ref. [16] fully kinetic models that linearize in velocity space give far too weak pinches. Not even the two temperature models used in Ref. [17] is sufficient to obtain an electron energy pinch. This is mainly because of too weak particle pinch effects.

Our pinches occur mainly due to the inclusion of the diamagnetic heat flow in the energy equation which introduces pinch terms proportional to L_n/L_B at the same time as it makes the model agree with kinetic theory also for $\omega_D \gg \omega$. Kinetic models, of course, are valid for arbitrary ω/ω_D but if these are used, it is necessary to consider the full nonlinear relaxation problem also in velocity space. In the dissipative regime ($\nu_{ei} \geq \omega_*$) the velocity space dynamics is linearized and the pinch effects are considerably weakened [22]. In a collisionless system as the present ($\nu_{ei} \simeq 0.2\omega_*$), however, the resonances in velocity space tend to be averaged out by the nonlinearities [5].

The fluid model used here seems to be the only present drift wave model that gives strong enough pinch effects to be able to explain experimental results of the type considered here. It also gives realistic radial profiles of diffusion coefficients and ratios of χ_e/χ_i. The strong sensitivity of the diffusion coefficients to the temperature and density profiles makes the local diffusion coefficients compatible with the strong profile consistency.

Bibliography

[1] Y. Kishimoto, T. Tajima, M.J. LeBrun, M.G. Gray, J.-Y. Kim, and W. Horton, IFSR#589; T. Tajima, Y. Kishimoto, M.J. LeBrun, M.G. Gray, J-Y. Kim, and W. Horton, "Transport in the Self-Organized Relaxed State of Ion Temperature Gradient Instability, Sec. 3.2 of this book.

[2] S.E. Parker, R.A. Santoro, and W.W. Lee, Bull. Amer. Phys. Soc. **37**, 1555 (1992).

[3] P. Andersson and J. Weiland, Phys. Fluids **31**, 359 (1988).

[4] A. Jarmen, P. Andersson, and J. Weiland, Nucl. Fusion **27**, 941 (1987).

[5] J. Weiland, Phys. Fluids B **4**, 1388 (1992).

[6] The JET Team, Presented by M. Keilhacker, Plasma Phys. and Controlled Fusion **33**, 1453 (1991).

[7] J.D. Callen, J.P. Christiansen, J.G. Cordey, P.R. Thomas, and K. Thomsen, Nucl. Fusion **27**, 1857 (1987).

[8] B. Coppi, Comm. Plasma Phys. and Contr. Fusion **5**, 261 (1980).

[9] W.M. Manheimer and T.M. Antonsen, Phys. Fluids **22**, 957 (1979).

[10] B. Coppi, Phys. Lett A **128**, 193 (1988).

[11] B. Coppi and C. Spight, Phys. Rev. Lett. **41**, 551 (1978).

[12] T.M. Antonsen, B. Coppi, and R. Englade, Nucl. Fusion **19**, 641 (1979).

[13] J. Weiland, A. Jarmen, and H. Nordman, Nucl. Fusion **29**, 1810 (1989).

[14] H. Nordman, J. Weiland, and A. Jarmen, Nucl. Fusion **30**, 983 (1990).

[15] R.E. Waltz, Phys. Fluids **29**, 3684 (1986).

[16] P.W. Terry, Phys. Fluids B **1**, 1932 (1989).

[17] O.T. Kingsbury and R.E. Waltz, Phys. Fluids B **3**, 3539 (1991).

[18] R.J. Fonck, G. Cosby, R. Dust, and S.F. Paul, Bull. Amer. Phys. Soc. **37**, 1483 (1992).

[19] R.E. Waltz, R.R. Dominguez, and G.W. Hammett, Phys. Fluids B **4**, 3138 (1992).

[20] T.C. Luce, C.C. Petty, and J.C.M. de Haas, Phys. Rev. Lett. **68**, 52 (1992).

[21] J.W. Connor, G.P. Maddison, H.R. Wilson, G. Corrigan, T.E. Stringer, and F. Tibone, JET-P(92) **63**, (1992).

[22] R.E. Waltz and R.R. Dominguez, Phys. Fluids B **1**, 1935 (1989).

r_s/a	$\tau_E(10^{-1}s)$	$T_e(r/a=0.1)$ (keV)
0.25	1.14	2.6
0.35	1.12	2.3
0.45	1.12	2.15
0.55	1.10	2.0

Table I. Peakedness of the electron temperature profile and the electron confinement time τ_E as obtained from numerical simulations for various choices of heating location r_s.

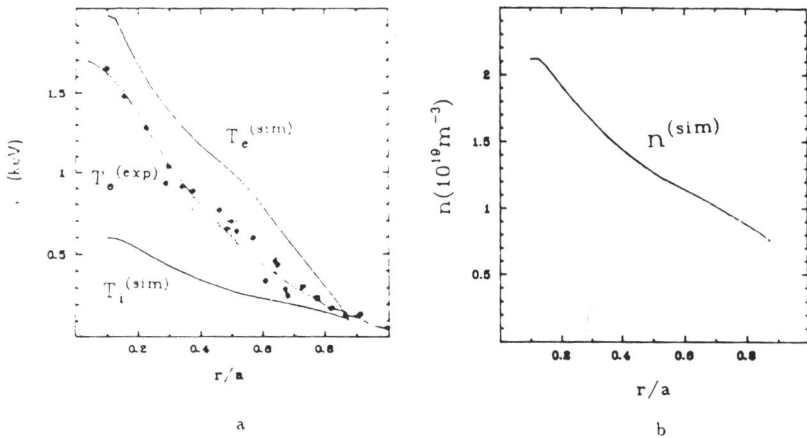

Figure 1. Stationary density and temperature profiles as a function of radius as obtained from transport code simulations of off-axis electron cyclotron heating in DIII-D geometry. In (a) the ion and electron temperature profiles are given. The 1.2 MW ECH surce is located at $r/a = 0.55$ with a halfwidth of 0.1. Also shown is the experimental electron temperature profile (taken from Ref. [20]). In (b) the density profile is shown. The particle source is centered at $r/a = 0.65$ (gas puffing).

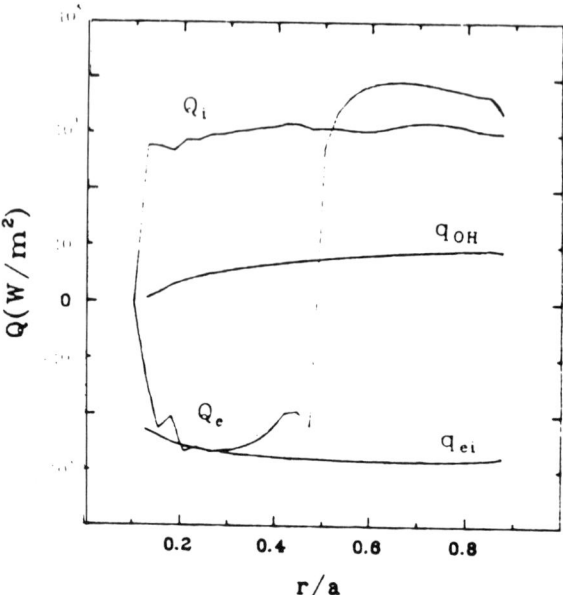

Figure 2. Stationary electron and ion energy fluxes as a function of radius as obtained from transport code simulations for the same discharge as in Fig. 1. The fluxes associated with ohmic heating (q_{OH}) and equilibration (q_{ei}) are also shown.

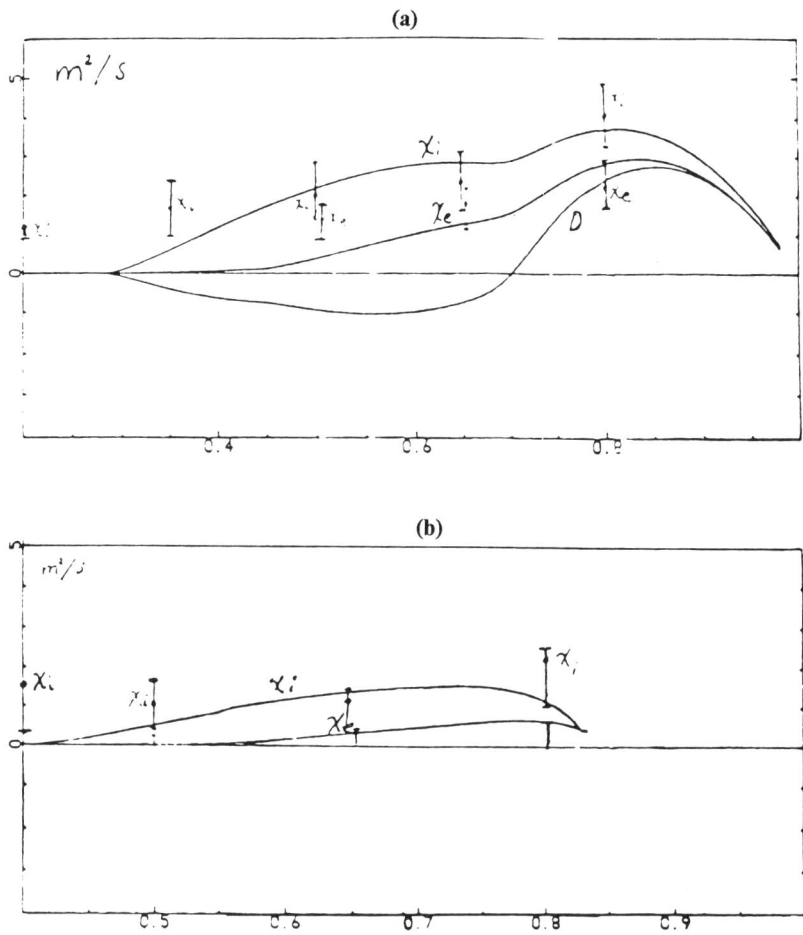

Figure 3. Comparison of the theoretical and experimental heat diffusion coefficients for JET L-mode shot (a) 19699 and (b) 24737 (experimental data taken from Ref. [21]).

5.2 TOROIDAL TURBULENCE SIMULATIONS WITH GYRO-LANDAU FLUID MODELS IN A NONLINEAR BALLOONING MODE REPRESENTATION

R.E. Waltz
General Atomics, P.O. Box 85608, San Diego, California 92138-5608,
G.D. Kerbel
NERSC at Lawrence Livermore National Laboratory, Livermore, California

Abstract. The method of Hammett and Perkins [Phys. Rev. Lett. **64**, 3019 (1990)] to model Landau damping has been recently applied to the moments of the gyro-kinetic equation with curvature drift by Waltz, Dominguez, and Hammett [Phys. Fluids B **4**, 3138 (1992)]. The higher moments are truncated in terms of the lower moments (density, parallel velocity, and parallel and perpendicular pressure) by modeling the deviation from a perturbed Maxwellian to fit the kinetic response function at all values of the kinetic parameters: $k_\parallel v_{th}/\omega$, $b = (k_\perp \rho)^2/2$, and ω_D/ω. Here the resulting gyro-Landau fluid equations are applied to the simulation of ion temperature gradient (ITG) mode turbulence in toroidal geometry using a novel 3D nonlinear ballooning mode representation. The representation is a Fourier transform of the Cowley et al. [Phys. Fluids B **3**, 2767 (1991)] field line following twisted eddy basis $(k_{x'}, k_{y'}, z')$ with periodicity in toroidal and poloidal angles. Particular emphasis is given to the role of nonlinearly generated $n = 0$ ($k_{y'} = 0$, $k_{x'} \neq 0$) "radial modes" in stabilizing the transport from the finite-n ITG ballooning modes.

5.2.1 INTRODUCTION

Turbulence among ion temperature gradient (ITG) modes is thought to be a useful paradigm for understanding some of the most important features of transport in tokamaks. While such modes have long been regarded as essentially fluid-like in nature, kinetic effect such as Landau resonances in the parallel and curvature drift motion as well as finite Larmor radius effects have some important quantitative effects. Fluid models have found wide use in simulating plasma turbulence because of their simplicity and tractability. Such models advance a few moments of a kinetic equation rather than the full velocity distribution function. Until the work of Hammett and Perkins [1], it was not widely appreciated that kinetic effects such as Landau damping could be represented by truncating the fluid hierarchy with addition of a collisionless heat conductivity $[\chi = (2/\sqrt{\pi})v_{th}/|k_{\|}|]$. Taking moments of the nonlinear electrostatic gyrokinetic equation this earlier $\mathbf{E} \times \mathbf{B}$ and parallel motion result has been generalized to include large gyro motion [2, 3] and in particular Waltz, Dominguez, and Hammett [3] have added the curvature drift and resonance effects required for toroidal geometry. This yields nonlinear fluid equations for the gyro-density, parallel motion, and pressures. The procedure for truncating the higher fluid moments involves a highly constrained fit to the linear kinetic response function at both large and small values of the kinetic parameters for parallel, gyro, and curvature motion: $k_{\|}v_{th}/\omega$, $b = (k_{\perp}\rho)^2/2$, and ω_D/ω. Taking adiabatic electrons, such gyro-Landau fluid (GLF) models for the ions show excellent agreement with kinetic calculations of ITG local (constant $k_{\|}$ and ω_D) linear stability [3].

This paper focuses on the application of GLF models for ITG turbulence in toroidal geometry using a novel 3D nonlinear ballooning mode representation. The most important parametric scaling features of ITG turbulence in a "slab" limit without curvature are probably well described by the simple fluid model simulations of Hamaguchi and Horton [4]. However apart from some early and very limited work with simple fluid models by Horton et al. [5] and also by Waltz [6] there has not been enough work simulating ITG turbulence in toroidal geometry to understand its parametric dependence. As useful as preliminary slab studies have been, the lower threshold, larger linear growth rates and radial mode widths imply that ITG turbulent transport in actual toroidal geometry should be considerably larger and perhaps have different parametric dependencies. The previous work in toroidal geometry has been done using the conventional multiple helicity (m, n, r) representation [6]. While this is a natural representation for slab or interchange-like modes, the ballooning mode representation would seem to offers a clear advantage when the turbulence is strongly ballooning.

In outline Sec. 2 precisely defines the GLF model equations for toroidal geometry. Sec. 3 describes the nonlinear ballooning mode representation and Sec. 4 treats the physics of "radial modes" or $n = 0$ ballooning modes. Sec. 5 briefly describes the effects of equilibrium scale sheared radial electric field in the context of ballooning modes. Sec. 6 provides the main body of the work with many numerical illustrations. Our conclusions are summarized in Sec. 7.

5.2.2 GLF EQUATIONS FOR TOROIDAL GEOMETRY

A derivation of the GLF model equations for toroidal geometry is given in Ref. [3]. For completeness we summarize it here to define the model. The model follows from taking moments of nonlinear gyro-kinetic equations for perturbed density N_k, parallel motion U_k, and pressures $P_{\|k}$, $P_{\perp k}$ of the gyro-centers. The higher order perturbed moments are closed as linear combinations of the lower moments by first assuming a perturbed Maxwellian and then adding "Maxwellian deviations" fitted to the kinetic response function. The resulting equations of ion motion are

$$dN_k/dt = -i\omega_*[(\hat{1}-\hat{\eta})\phi_{1k} + \hat{\eta}\phi_{2k}] + i\omega_D \phi_{12k} - i\hat{k}_\| U_k$$
$$+ i\omega_D \tau^{-1}(P_{\|k} + P_{\perp k})/2 , \tag{1}$$

$$dU_k/dt = -i\hat{k}_\|(\tau^{-1}P_{\|k} + \phi_{1k}) + i\omega_D \tau^{-1}\left[(\Gamma_\| + \Gamma_\perp)/2 \cdot U_k - i\sigma_t \mu U_k\right] , \tag{2}$$

$$dP_{\|k}/dt = -i\omega_*(\hat{1}\phi_{1k} + \hat{\eta}\phi_{2k}) + iX_\| \omega_D \phi_{12k} - i\hat{k}_\|\left[\Gamma_\| U_k - i\sigma_s \chi_1^\| T_{\perp k}\right]]$$
$$+ i\omega_D \tau^{-1}\left[\mathbf{X}_\| \mathbf{P}_{\|k} + (3/2) \mathbf{T}_{\|k} + (1/2)\mathbf{T}_{\perp k} - i\sigma_t(\nu_\|^\| \mathbf{T}_{\|k} + \nu_\perp^\| \mathbf{T}_{\perp k})\right] \tag{3}$$

$$dP_{\perp k}/dt = -i\omega_*[(\hat{1}-\hat{\eta})\phi_{2k} + 2\hat{\eta}\phi_{3k}] + iX_\perp \omega_D \phi_{23k} - i\hat{k}_\|\left[\Gamma_\perp \mathbf{U}_k - i\sigma_s \chi_1^\perp \mathbf{T}_{\perp k}\right]$$
$$+ i\omega_D \tau^{-1}\left[\mathbf{X}_\perp \mathbf{P}_{\perp k} + \mathbf{T}_{\perp k} + (1/2)\mathbf{T}_{\|k} - i\sigma_t(\nu_\|^\perp \mathbf{T}_{\|k} + \nu_\perp^\perp \mathbf{T}_{\perp k})\right] \tag{4}$$

where the terms with bold brackets $[\ ]$ represent higher moments with the Maxwellian deviations. The latter have coefficients $\chi_1^{\|} = 2\chi_1^{\perp} = 2/\sqrt{\pi}$ for parallel motion, and $\mu = (0.80 - 0.57\,i\sigma_t)$ [7], $\nu_{\|}^{\|} = \nu_{\perp}^{\perp} = (1 - i\sigma_t)$, $\nu_{\perp}^{\|} = \nu_{\|}^{\perp} = 0$ for curvature motion (with $\sigma_s = k_{\|}/|k_{\|}|$, $\sigma_t = \omega_D/|\omega_D|$). The gyro-averaged electric potentials are given in terms of Bessel functions

$$\phi_{1k} = \Gamma_0 \phi_k \tag{5}$$

$$\phi_{2k} = [\Gamma_0 - b(\Gamma_0 - \Gamma_1)]\phi_k \tag{6}$$

$$\phi_{3k} = (1/2)[(2 - 4b + b^2)\Gamma_0 + (5b - 2b^2)\Gamma_1 + b^2\Gamma_2]\,\phi_k \tag{7}$$

with $\phi_{12k} = (\phi_{1k} + \phi_{2k})/2$ and $\phi_{23k} = (\phi_{2k} + \phi_{3k})/2$ and $\phi_k = (e/T_e)\Phi_k$. The gyro-density is given by a "Poisson relation"

$$N_k = n_k/n_0 + \tau(\phi_k - \phi_{1k})\,, \tag{8}$$

where n_k is the physical ion density perturbation and the second term is the polarization drift. The perturbed temperature is $T_k = P_k - N_k$. To model ITG or η_i-mode turbulence we take $n_k/n_0 = R_k\tau^{-1}\phi_k$ with electron response function $R_k = 1$ to model adiabatic electrons ($\tau^{-1} = T_i/T_e$).

The adiabatic compression indices are $\Gamma_{\|} = 3$, $\Gamma_{\perp} = 1$, $\chi_{\|} = 2$, $\chi_{\perp} = 3/2$. It is particularly important to retain the detailed compression model since at low $k_{\perp}$, the ITG threshold is sensitive to these indices: $\eta_{i\text{crit}} = \Gamma - 1$. For one dimensional parallel motion only (i.e. restricting to slab geometry) $\Gamma_{\|} = 3$ suggests $\eta_{i\text{crit}} = 2$ whereas for three dimensional curvature drift motion important for toroidal geometry we have $\Gamma_{\|} + (2/3)\Gamma_{\perp} = 5/3$ [and $(1/3)\chi_{\|} + (2/3)\chi_{\perp} = 5/3$] so $\eta_i\text{crit} = 2/3$ which is much lower. Keeping all these different compression indices requires a two pressure model.

The equations of motion require further definition. Here and throughout we use a system of units with $c_s = (T_e/M)^{1/2}$ the unit of velocity, $\rho_s = c_s/\Omega$ the unit of micro-length, a (the minor radius) the unit of macro-length, and a/c_s the unit of time. The cross field wave numbers in and across the flux surface $k = (k_y, k_x)$ are normed to ρ_s. The parallel wavenumber $k_{\|}$ is normed to a^{-1} and the major radius R to a.

The density gradient is given by $\hat{1} = a/L_n$, the temperature gradient by $\hat{\eta} = a/L_T$. The normalized electron diamagnetic frequency is $\omega_* \hat{1}$ with $\omega_* = k_y$. The normed curvature drift frequency $\omega_D = \omega_* \mathbf{g}$ where $\mathbf{g} = (2/R)(\cos\theta + k_x/k_y \sin\theta)$ represents the local curvature with θ the poloidal angle.

We have presented a "simple gyro-average" model valid linearly to all orders in b in shearless geometry. In a sheared magnetic field the parallel wave number has a first order in b gyro-correction given by

$$\hat{k}_\| = k_\| - i\, k_s(k_y k_x/k_\perp^2) b \Gamma_0'(b)/\Gamma_0(b) \,, \tag{9}$$

(where $k_s = \hat{s}/Rq$ and $\hat{s} = d\ln q/d\ln r$) to approximate the Linsker effect [8]. Simple nonlinear $\mathbf{E} \times \mathbf{B}$ convective time derivatives are given by

$$df_k/dt = \partial f_k/\partial t - \sum_{k_1}(k_{x1}\, k_{y2} - k_{x2} k_{y1})\, \phi_{k1}\, f_{k2} \,, \tag{10}$$

with $\mathbf{k}_2 = \mathbf{k} - \mathbf{k}_1$ and all perturbed fields normed to ρ_s/a. This approximate nonlinear coupling is valid at low $k_\perp$ but probably over estimates the coupling with gyro-effects by 20%–30% at $k_\perp = 2$. A more accurate "full gyro-average" model valid through first order in b is described in Ref. [2]. It handles the Linsker effect implicitly and contains additional nonlinear FLR effects including a nonlinear hyper-viscosity due to the Landau resonance in the gyro-motion.

In physical units, the plasma flux is given by

$$\Gamma_{px} = n_0 \sum_k (n_k/n_0)^*(-i\, k_y\, c\, \Phi_k/B) \,, \tag{11}$$

and the energy flux

$$\Gamma_{Ex} = (3/2)\, p_0 \sum_k (p_k^a/p_0)^* (-i k_y\, c\Phi_k/B) \,, \tag{12}$$

where $p_k^a/p_0 = (1/3)\, p_{\|k}/p_0 + (2/3)\, p_{\perp k}/p_0$ with the physical pressures given by $p_{\|k}/p_0 = P_{\|k} - \tau(\phi_k - \phi_{1k})$ and $p_{\perp k}/p_0 = P_{\perp k} - \tau(\phi_k - \phi_{2k})$. Note energy continuity equation must have an exchange term:

$$\partial E/\partial t + \partial \Gamma_{Ex}/\partial x = p_0 \sum_k \partial(n_k/n_0)^*/\partial t \cdot (e/T)\, \Phi_k \,. \tag{13}$$

For adiabatic electrons used here there is no plasma flow or exchange term and we need concern ourselves only with the ion heat flow which we summarize by quoting $\chi = \Gamma_{Ex}/(p_0/L_T)$ in gyroBohm units $(c_s/a)\rho_s^2$.

As shown in Ref. [3], the GLF model very accurately describes the local (constant $k_\parallel$ and **g**) linear growth rates (as well as frequencies) for the ITG mode over both large and small kinetic parameters. The exception is a small spurious destabilizing effect on the electron branch in the toroidal limit (large $\omega_D/k_\parallel v_{th}$) for small but negative ω/ω_D.

5.2.3 3-D NONLINEAR BALLOONING MODE REPRESENTATION

Previous nonlinear ballooning mode representations (BMR) have resulted in very awkward and impractical 2-D formulations of the nonlinearity which not always properly accounted for the third dimension ballooning mode angle $\hat{\theta}$. By using the "twisted eddy basis" of Cowley et al. [9], we have formulated a simple and practical nonlinear BMR. Starting from the lab fixed basis (x, y, z) we transform to the twisted eddy basis (x', y', z'):

$$x = x'$$

$$y = y' + \hat{s}\, x'\, (z'/Rq)$$

$$z = z' \tag{14}$$

with

$$\partial/\partial x = \partial/\partial x' - \hat{s}(z'/Rq)\partial/\partial y'$$

$$\partial/\partial y = \partial/\partial y'$$

$$\partial/\partial z = \partial/\partial z' - (\hat{s}/Rq)x'\partial/\partial y' \ . \tag{15}$$

Thus z' stays along the sheared field as x' changes $[\partial/\partial z' = \partial/\partial z + (\hat{s}/Rq)x\partial/\partial y]$ and (x', y') directions remain perpendicular to field. Fourier transforming in the perpendicular directions from (x', y', z') to $(k_{x'}, k_{y'}, z')$

we have

$$k_x = k_{x'} - \hat{s}(z'/Rq)k_{y'}$$
$$k_y = k_{y'}$$
$$z = z' . \tag{16}$$

Writing the curvature drift frequency as

$$\omega_D = (2/R)\{k_{y'}\cos(z'/Rq) + [\hat{s}(z'/Rq)k_{y'} - k_{x'}]\sin(z'/Rq)\}$$
$$= (2/R)(nq/r)[\cos(\theta) + \hat{s}(\theta - \hat{\theta})\sin(\theta)] \tag{17}$$

it is easy to identify the $(k_{x'}, k_{y'}, z')$ basis with the usual BMR:

$$k_{y'} = k_\theta = (nq/r) \tag{18}$$

is the good quantum number toroidal label n with q and r the safety factor and radius of the reference flux surface;

$$z'/Rq = \theta \tag{19}$$

is the extended poloidal angle along the field line; and

$$k_{x'}/(k_{y'}\hat{s}) = \hat{\theta} \tag{20}$$

is the ballooning angle or phase between poloidal harmonics starting at a reference rational surface $m = nq$. The gradient parallel to the field line is

$$ik_\| = \partial/\partial z' = (1/Rq)\partial/\partial\theta . \tag{21}$$

A crucial point emphasized in Ref. [9] is that the nonlinear convolution has a simple invariant form in the twisted eddy basis:

$$df_{k'}(z')/dt = \partial f_{k'}(z')/\partial t - \sum_{k'1}(k_{x'1}k_{y'2} - k_{x'2}k_{y'1})\phi_{k'1}(z')f_{k'}(z') \tag{22}$$

with $\mathbf{k}'_2 = \mathbf{k}' - \mathbf{k}'_1$.

Fourier amplitudes in the transformed twisted eddy basis $(k_{x'}, k_{y'}, z')$ as they stand are not periodic in the physical poloidal angle. To impose periodicity, it convenient to distinguish "primary" modes and "image" modes. Linear mode labels are given by $(n, \hat{\theta})$ or $(k_{y'}, k_{x'})$. The linear modes will be centered about $\theta = \hat{\theta}$, with $\hat{\theta}$ near 0 ballooning in the bad curvature region and $\hat{\theta}$ near π in the good curvature region. Primary modes have $-\pi < \hat{\theta} \leq \pi$. Each primary mode has an infinite series of image modes labeled by $-\infty < p < \infty$ with the same n (or $k_{y'}$) but with $\hat{\theta} \to \hat{\theta} + p\,2\,\pi$ [or $k_{x'} = k_{x'} + p\hat{s}k_{y'}\,(2\,\pi)$]. See Fig. f-3-1. As linear modes, the image modes have the same linear growth rate as the primary except they are centered about $\theta \to \theta + p\,2\,\pi$. On an infinite grid, the statistical or time average of intensity of the image modes must be identical θ displaced copies of the primaries. Cowley $et\ al.$ [9] advocate forgoing amplitude periodicity. This amounts to advancing the image modes independent of the primaries. They argue that sufficiently short correlation length along the field line makes periodicity of amplitudes irrelevant. This indeed my be the case.

However, practically we must numerically truncate the grid in θ and $\hat{\theta}$ (as well as n). Imposing instantaneous cyclic symmetry of the images with the primaries as in the original BMR [10] not only preserves periodicity of amplitudes, it naturally truncated the $\hat{\theta}$ grid by explicitly advancing only the primaries on a limited θ grid. Thus we take the view that a $\hat{\theta}$ mode should not be included unless the θ grid is extended to cover it as a linear normal mode. Taking images as cyclic copies allows us to do this. (Beer $et\ al.$ [7] have also developed a nonlinear BMR code, although they have taken a somewhat different approach to periodicity constraints.) The BMR periodicity condition is

$$\phi_n^{\hat{\theta}+2\pi p}(\theta) = \exp(-ip\,2\pi n q)\phi_n^{\hat{\theta}}(\theta - 2\,\pi p) \tag{23}$$

or

$$\phi_{(k_{x'}+2\pi p k_{y'}),k_{y'}}(z') = \exp(-ip\,2\pi nq)\phi_{k_{x'},k_{y'}}(z' - 2\,\pi p R q)\,. \tag{24}$$

From $y = r(\theta - \phi/q)$ it is easy to reconstruct the real space physical perturbations periodic in θ and ϕ as

$$\phi(x,\,\theta,\,\phi) = \sum_p \sum_n \int_{-\pi}^{\pi}(d\hat{\theta}/2\,\pi)\phi_n^{\hat{\theta}}(\theta - 2\pi p)$$

$$\times \exp\left[-in\phi + inq(\theta - p2\pi) + inq\hat{s}(\theta - p2\pi - \hat{\theta})x/r\right] \tag{25}$$

The plasma energy flow at poloidal angle θ for example is given by the time average $\langle \ \rangle$,

$$\Gamma_{Ex}(\theta) = (3/2)p_0 \sum_p \sum_n \int_{-\pi}^{\pi} (d\hat{\theta}/2\pi)$$

$$\left\langle (p^a/p_0)_n^{*\hat{\theta}}(\theta - 2\pi p)[-i\,(nq/r)c\Phi_n^{\hat{\theta}}(\theta - 2\pi p)/B] \right\rangle . \quad (26)$$

If the turbulence is sufficiently ballooning, the primaries can be advanced on $-2\pi = \theta_{\min} \leq \theta < \theta_{\max} = 2\pi$ with nonlinear coupling only to nearest neighbor images $p = +1$ and $p = -1$ as θ displaced cyclic copies. In effect the nonlinear coupling grid ($\mathbf{k}'_1$ and $\mathbf{k}'_2$) extends only from $-3\pi < \hat{\theta} \leq 3\pi$ with $\mathbf{k}$ restricted to $-\pi < \hat{\theta} \leq \pi$. No other coupling is possible since only these images overlap in θ with the primaries. The *a' posteriori* validity condition is that the excitation of primary amplitude are sufficiently small near the chosen boundaries at $\theta = \pm 2\pi$. For very strong ballooning it may be sufficient to ignore coupling to the images entirely. Of course if the turbulence becomes interchange like extending to larger θ's well beyond $\pm 2\pi$ and requiring $p = \pm 2$ and so on, the method becomes impractical.

It is possible to show that this is an energy conserving procedure. Defining the system primary "energy" from $\theta_{\min} \leq \theta < \theta_{\max}$ and $\mathbf{k}$ within $-\pi < \hat{\theta} \leq \pi$ as $E = (1/2) \int d\theta \sum_k \sum_i f_k^{i*} f_k^i$ with $i = 1, 4$ denoted the four time advanced moment fields, then energy enters and leaves the system only linearly with the nonlinear terms only passing energy from one mode to another. Energy is passed nonlinearly from one θ location to another only by interaction of the primaries with the images (i.e. retaining only primaries with no interaction with images nonlinearly preserves energy at each θ).

We note in passing that the Fourier representation in k_x and k_y is particularly convenient for exact evaluation of the Bessel function forms and the toroidal dissipative terms with $|\omega_D|$. In fact advancement of the parallel gradient terms, and in particular the dissipative terms with $|k_\parallel|$, is most easily done by a "time-split" algorithm Fourier transform from θ to $k_\parallel$ space.

5.2.4 RADIAL MODES

From Fig. 1 it is readily apparent that this procedure spans all "ballooning modes" with finite n, but does not include what could be called "radial modes" with n (and $m = nq$)=0. These have $k_{y'} = 0$ but $k_x \equiv -k_{x'} \neq 0$. Their contribution to the amplitude Eq. (25) is $\sum_{k_x} \phi_0^{k_x}(\theta)\exp(+ik_x x)$ with $\phi_0^{k_x}(\theta + 2\pi) = \phi_0^{k_x}(\theta)$ i.e. periodicity is automatic. The $(k_{y'}, k_{x'}) = (0, 0)$ mode is the equilibrium which advances on a transport time scale with gradients maintained by sources. In effect the radial modes are "near equilibria" having no fast variations over the flux surface other than a possible slow in-out variation ($m \approx \pm 1$) from a small but finite $k_\parallel \approx 1/Rq$. Electron density responds only to deviations from a flux surface average potential $\langle\phi\rangle$: $n_k/n_0 = \phi - \langle\phi\rangle$. For components of the motion with $k_\parallel \equiv 0$ and $\phi \equiv \langle\phi\rangle$, it is reasonable to take the effective electron response $R_k = 0$ for these radial modes [7, 11] and N_k becomes the ion vorticity. However for components of the motion with finite $k_\parallel \approx 1/Rq$, $R_k = 1$ for motions slower than the transit frequency. Apart from possible ion Landau damping which we shall discuss momentarily, the radial modes as presently formulated would have a null dispersion relation $\omega \approx 0$. They are however driven and damped by their nonlinear coupling to the ballooning modes. In turn the radial modes act back on the ballooning modes. The finite radial mode pressure perturbations cause local steepening and flattening of the driving pressure gradient. This effect probably cancels out to some degree. More importantly, the radial mode potential perturbations cause locally sheared $\mathbf{E} \times \mathbf{B}$ rotations in the flux surfaces. Diamond and Kim [12] have called this nonlinear Reynold's stress effect "eddy sheared flow generation." Sheared $\mathbf{E} \times \mathbf{B}$ rotation of sufficient strength *and either sign* has a stabilizing effect on the ballooning mode turbulence. (We shall discuss equilibrium scale sheared $\mathbf{E} \times \mathbf{B}$ rotation below). In general the strength condition [13] is roughly $cE'_r/B \approx \gamma_k(k_x/k_y)$ where γ_k and k_x, k_y are typical linear mode growth rates and wave numbers for the finite n ballooning modes. The strength condition for radial modes with wave number k_{x0} in normalized units is $k_{x0}^2|\phi_0| \approx \gamma_k(k_x/k_y)$. This could be easily satisfied if the radial modes acquire amplitudes comparable to the ballooning modes at similar cross field wave numbers ($k_{x0} \approx k_y$). The nonlinear driving and·damping of radial modes is clear. [The nonlinear coupling is simply given by Eq. (22).] However the actual amplitude of the radial modes depends also on their linear damping mechanism.

The linear damping of radial modes is not well understood and requires special discussion. For these near equilibria modes, it seem plausible and perhaps crucial to add neoclassical magnetic pumping terms to the ion dynamics. Following Kim et al. [14] we add $-\mu_2(B/B_\theta)U_{\theta k}$ to the right hand side of the parallel momentum equation Eq. (2) and $+\mu_1(B_\phi/B_\theta)(ik_x)(B/B_\theta)U_{\theta k}$ to the right hand side of the continuity equation Eq. (1). In the banana regime $\mu_1 = \mu_2 = \mu_{mp} = \nu_{ii}\sqrt{\epsilon}$ with μ_2 often referred to as direct magnetic pumping and whereas μ_1 gives rise to neoclassical polarization drift induced by magnetic pumping. The poloidal velocity is given in terms of the parallel velocity, and $\mathbf{E} \times \mathbf{B}$ and diamagnetic velocity in the flux surface as

$$(B/B_\theta)U_{\theta k} = U_{\|k} + (B_\phi/B_\theta)(ik_x)[\phi_k + (1/3)P_{\|k} + (2/3)P_{\perp k}] . \quad (27)$$

Assuming for the moment $k_\| = 0$ and ignoring any residual geodesic curvature effects, the pressure perturbations are decoupled from Eqs. (1) and (2). The quadratic dispersion relation has two solution: $\omega = [-\mu_2 - \mu_1(B_\phi/B_\theta)^2 k_x^2/(R_k + k_x^2)]$ and $\omega = 0$. The first represents poloidal flow damped by magnetic pumping. For $k_\| = 0$ and $R_k = 0$, the μ_1 term is dominant since $(B_\phi/B_\theta)^2 \approx 10^2$. Thus even small magnetic pumping rates of order 0.01 or even 0.001 could be very powerful. For motions with small but finite $k_\|$, $R_k = 1$ and since $k_x^2 \ll 1$, the μ_2 term is dominant and magnetic pumping may not be very strong. However it is the second solution which represents undamped sheared toroidal flows that we must control. Otherwise very large eddy generated $\mathbf{E} \times \mathbf{B}$ shear flows could build up. For $k_\| \equiv 0$, it is not clear what damps shear flow in the direction of symmetry, i.e. the toroidal direction. If there is some coupling to finite $k_\|$ motion, it is possible to have some Landau damping. This could provide a component of damping orthogonal to the poloidal damping magnetic pumping. As we illustrate below this appears to be the case.

5.2.5 EQUILIBRIUM SCALE SHEARED E × B FLOWS

The BMR as presented here is a Fourier transform representation which intrinsically assumes that the turbulence is homogeneous from one flux surface to another or that the correlation length is shorter than any equilibrium length scale. Profile effects or gradual inhomogeneities are generally ρ_s/a small and ignorable. The exception is when the plasma is spun to have large equilibrium potentials $e\Phi \gg T$ or when sharp gradients in the radial electric field build up near the edge. In this case the equilibrium sheared flow rate cE'_x/B can become comparable to the mode growth rates $\gamma_k(k_x/k_y)$ [13]. Homogeneity is broken. However it is relatively straight forward to include these linear inhomogeneities in the BMR formalism. The time derivative in Eq. (10) merely acquires a linear mode coupling term from the $\mathbf{E} \times \mathbf{B}$ shear:

$$\partial f_{k_{y'} k_{x'}}(\theta)/\partial t \to \partial f_{k_{y'} k_{x'}}(\theta)/\partial t + \gamma_E k_{y'} \partial f_{k_{y'} k_{x'}}(\theta)/\partial k_{x'} \qquad (28)$$

where $\gamma_E = (cE'_x/B)/(c_s/a)$ is the shear rate in normalized units. [We have simply replaced the $x = x'$ variation in the $\mathbf{E} \times \mathbf{B}$ flow with $-i\partial/\partial k_{x'}$]. In this radially nonuniform system $k_{x'}$ or $\hat{\theta}$ is no longer a good quantum number. Without shear there is a system of linearly independent modes with those having $\hat{\theta}$ near 0 growing much faster than modes with $\hat{\theta}$ near π which in fact are typically damped. The shear couples these modes together. The stronger the shear, the more tightly coupled the growing $k_{x'} = 0(\hat{\theta} = 0)$ component is coupled to the damped components at larger $k_{x'}(\hat{\theta})$. The overall fastest mode becomes more weakly growing and finally damped at sufficiently large shear. The $k_{x'}$ spectrum of this "sheared" mode becomes broader. In x space, the ballooning mode of infinite extent or "envelope" without electric field shear (or other profile effects), acquires an radial envelope of increasingly smaller length as the electric shear increases. This means that near $\theta = 0$ the ballooning mode eddies whose large radial mixing lengths are truncated by the turbulent nonlinear interaction now have an additional linear length truncation from the shear in addition to smaller linear formation (growth) rates. Since both growth and length scales are effected, $\mathbf{E} \times \mathbf{B}$ shear is expected to have a strong effect on the turbulent diffusion.

5.2.6 NUMERICAL ILLUSTRATIONS

The numerical illustrations presented here are organized around a toroidal standard case with physical parameters $R/a = 3$, $a/L_n = 1$, $a/L_T = 3$, $\tau^{-1} = T_i/T_e = 1$, $q = 2$, and $\hat{s} = 1$. The parameters are meant to be typical of $r/a = 0.5$. The numerical parameters are $(nq/r)\rho_{s\max} = k_\theta \rho_{s\max} = 1$ or in normalized units $k_{y'\max} = 1$ with $\Delta k_{y'} = 0.1$ and $\Delta k_{x'} = (\hat{s}\pi/4)\Delta k_{y'} = 0.078$, and $-2\pi < \theta < 2\pi$ with $\Delta\theta = 4\pi/64$. This spans 450 complex primary ballooning modes with $-\pi \le \hat{\theta} \le \pi$ (upper primary wedge in Fig. f-3-1). An small artificial viscous damping rate $\mu_x k_x^2$ with $\mu_x = 0.05$ is put on the fields. A typical time step is $0.05\,(c_s/a)$ and good statistical averages appear to require run times of order $1000\,(a/c_s)$.

This mode spacing is relatively course but hopefully adequate. ρ_s/a which determines the maximum toroidal n number simulated does not enter the calculation. However to put this in perspective, if $\rho_s/r = 0.0025$ then the maximum physical (m, n) is $(400, 200)$. If $\Delta k_{y'}/k_{y'\max} = 1/10$ this implies that Δn is 20. This corresponds to simulating 1/20 of the torus or a field line following band in the flux surface of toroidal width $\Delta\phi = 2\pi/20$, perpendicular width $\Delta y = \Delta\phi r/q$ in the flux surface and thickness across the flux surface $\Delta x = \Delta y(4/\hat{s}\pi) = 0.2\,r$ or $80\,\rho_s$. In effect the cross field correlation lengths are assumed to be shorter than the band width and thickness. To test a two-fold scaling with gyroradius, comparisons with simulations at $\Delta k_{y'}/k_{y\max'} = 1/10 \to 1/20$ would be needed.

While we now have some considerable experience with this type of nonlinear BMR, the results presented here should be considered exploratory and detailed comparison with experiment maybe premature. Several problems remain. To explore these, we shall first consider simulations retaining only the primary ballooning modes with no images and no coupling to radial modes. We then discuss simulations with images and finally we explore the coupling to radial modes which appear to have potentially very important but poorly understood effects on the level of turbulence.

Simulations with primary ballooning modes

Figure 2 illustrates the linear growth rate spectrum and normal ballooning mode structure at various $\hat{\theta}$ for the standard case. The maximum growth rate is 0.083 at $k_{y'} = 0.3$ and $\hat{\theta} = 0$. It is readily apparent that modes with $\hat{\theta}$ beyond $\pi/2$ are damped and that as expected, modes are localized about $\theta = \hat{\theta}$. The lower $k_{y'}$ modes have an increasingly wider extent along the field line. Modes at $\hat{\theta} = 0$ with $k_{y'} < 0.1$ (not included in the simulations) appear to go stable at $k_{y'} < 0.07$ with damping rates of -0.02 at $k_{y'} = 0.04$. Such modes require a θ grid extending to $\pm 4\pi$ or more. We have found no evidence of low $k_{y'}$ unstable interchange like modes extending over many (10's) 2π. On the other hand to access the importance of these very low $k_{y'}$ stable modes requires simulations at much smaller $\Delta k_{y'}/k_{ymax'}$. The linear growth rates and mode widths from GLF models compare [15] reasonably well with the kinetic theory calculations of Dong, Horton, Kim [16] although not as well as may have been expected from the excellent agreement in local kinetic theory [3]. Peak growth rates and critical gradients can be off by 20%–30%.

Figure 3 illustrates the nonlinear simulation of the standard case showing the time history of the flux surface average diffusion in Fig. 3(a). The spectrum of pressure perturbations and contributions to diffusion at $\theta = 0$ are shown in Fig. 3(b) with peaks at $k_{y'} = 0.3$. The diffusion and turbulence level as a function or poloidal angle θ are shown in Fig. 3(c) and 3(d) respectively. The dashed lines add the $\pm 2\pi$ displaced copies of the diffusion and turbulence with the physical values restricted to $-\pi < \theta \leq \pi$. The diffusion is highly asymmetrical with an out/in ratio of 8/1 and the turbulence level is less asymmetrical at 5/3. We have made some numerical fidelity checks. For example extending the θ grid from $\pm 2\pi$ to $\pm 3\pi$, doubling the θ grid, or halving the time steps have little effect. Changing the boundary conditions from zero at $\theta = \pm 2\pi$ to periodic on $-2\pi < \theta < 2\pi$ has little effect. Some tests with decreasing $\Delta k_{x'}$ have been done; a case with $\Delta k_{x'} = 0.052$ in place of 0.078 (i.e. 670 complex mode in place of 450) is somewhat less noisy as expected and but the average transport and turbulence levels are about only about 10% higher. No studies with smaller $\Delta k_{y'}$ have been done.

Figure 4 shows instantaneous random intensity plots for the pressure perturbations along the extended poloidal angle at several $k_{y'}$ values. Each line corresponds to a different $k_{x'}$. It is clear from this that there are relatively few large unstable outward ballooning modes pumping the tails of many stable inward ballooning modes. The rather poor statistics shown in Fig. 2(a) is a reflection of the presence of a few large amplitude modes.

Figure 5(a) shows a gray shade contour plot of the instantaneous potential over a constant toroidal angle $\phi = 0$ annulus. Although no nonlinear interaction with the images was retained in the simulation, the images are added using Eq. (25) to reconstruct the physical potential in real space appropriately periodic in toroidal and poloidal angle. In order to make the eddies visible we have displayed the result using a band size $\Delta \phi = 2\pi/8$ or $\Delta x = 0.5\,r$ at $r = 0.5\,a$. (In the context of the earlier discussion we have made the gyroradius 20/8 times larger but the strip is still $80\,\rho_s$ across.) The annulus is composed of eightfold intersections with the band. Figure 5(b) shows a corresponding plot of the dominant $k_{y'} = 0.3$ linear ballooning mode. It is readily apparent that many features of the dominant linear mode survives in the turbulent state. There are radially long (but not infinite) eddies in the bad curvature $\theta = 0$ region. They appear to span a sizeable fraction of the $80\,\rho_s$ strip. The eddies are much less intense toward the good curvature region and "twist" from perpendicular to alignment with the flux surfaces.

The level of toroidal ITG transport from the primary mode acting alone is large but probably not excessive compared to experimental levels. (See Ref. [15] for some preliminary comparison to a TFTR standard L–mode discharge.) It is very clear however that the transport levels from toroidal simulations can be well in excess of "slab" simulations. For example turning off the curvature terms in the standard case reduced the transport from $17\,(c_s\,\rho_s^2/a)$ to about 0.5. Although our studies of toroidal GLF models using a conventional $(m,\,n,\,r)$ code were unsuccessful, extensive work has been done with slab GLF models. Figure 6 shows a nonlinear BMR code slab simulation in excellent comparison to $(m,\,n,\,r)$ code results. (The parameters given in the caption differ somewhat from the standard case. In particular η_i is 4 not 3. Also Pade' approximates were used to evaluate the Bessel functions here.)

In the slab case each $\hat{\theta}$ mode at a given $k_{y'}$ has the same growth rate and of course there is no ballooning. A variety of various θ and $\hat{\theta}$ grids give the same result. The (m, n, r) cases covered a large variety of grid spacing and cut-offs and most significantly spanned threefold variations in ρ_s/a with $\Delta n = 1$ spacing and up to tenfold variation with $\Delta n = 5$ and 10. *No measurable deviation from gyroBohm scaling was found in (m, n, r) slab cases.*

The critical a/L_T for the parameters of the standard case is 1.8. We have as yet seen no evidence of subcritical turbulence. The transport is generally described by $\chi \propto (c_s/R)\rho_s^2 \, F\,(R/L_T, R/L_n, q, \hat{s})$ assuming gyroBohm scaling. In the weak density gradient limit (roughly $R/L_n < 3$), and near threshold, F is well modeled by $F \propto (R/L_T - R/L_{Tcrit})$. A persistent research goal is to characterize the dependence on the remaining parameter particularly q and $\hat{s}$. Linear growth and mixing length model may offer some guide. For this we simply take the linear growth rate γ and mixing length given by the linear radial mode width $\Delta_x = 1/(k_{y'}\hat{s}\,\theta_{\rm rms})$ of the fastest growing mode: $D_{\rm ML} = \gamma \Delta_x^2$. $\theta_{\rm rms}$ is the characteristic θ extent of a linear mode as shown in Fig. 2. Figure 7 shows variations of $D_{\rm ML}$ about the standard case at $a/L_T = 3$ (as well as 4). The top panels verify that $R/L_n = 3$ is already in the weak density gradient regime. The bottom two panels show that the dependence on q and $\hat{s}$ is not well characterized by power law dependence. Scaling away from the standard case ($q = 2 \to 1.5$, $\hat{s} = 1 \to 0.5$) one variable at a time gives $D_{\rm ML} \propto (q^{1.8}/\hat{s}^6)$ where as simulations show $\chi \propto (q^{2.9}/\hat{s}^{5.6})$. However variations with $2\hat{s} = q$ show $D_{\rm ML}$ constant (see dashed lines) whereas simulation ($1 < q < 3$) show $\chi \propto q^{-0.8}$. By comparison the "slab" limit simulations of Ref. [4] can be characterized by $(q/\hat{s})^{1/2}$ at weak shear and $(q/\hat{s})^2$ at strong shear. Note that in the slab limit q and $\hat{s}$ are not independent but can only enter in the combination $L_s = Rq/\hat{s}$.

While the scaling of these simulations without primaries are in very rough parametric agreement with the mixing length model, the most striking disagreement is in the level of transport. For the standard case $D_{\rm ML} \approx 0.8\,[(c_s/a)\,\rho_s^2]$, whereas the simulation showed $\chi \approx 17\,[(c_s/a)\rho_s^2]$ with peak values at $\theta = 0$ twice as large. Redefinition of mode widths by $\sqrt{2} \to \sqrt{4}$ could explain a factor of $2 \to 4$ disagreement at most.

The predicted mixing length is $\Delta_x \approx 3\rho_s$ or $\theta_{rms} \approx 1$. The largest radial eddy lengths in Fig. 5(a) are much larger by at least tenfold. The length of the eddies is more characteristic of a nonlinear correlation length or coupling length. We can only speculate that a better recipe for the mixing length is $\Delta_{\hat{x}} = 1/(k_{y'}\hat{s}\Delta\hat{\theta})$ where $\Delta\hat{\theta}$ is the largest ballooning angle with significant nonlinear amplitude relative to the peak mode at $k_{y'} = 0.3$ and $\hat{\theta} = 0$. This angle could be much smaller than θ_{rms} and yet it seems reasonable that it would scale proportional to θ_{rms}. From Fig. 4, the smallest $\Delta\hat{\theta}$ is $\pi/12 \approx 1/(3.8)$. This would then account for an enhancement of $(3.8)^2 \approx 14$. Another view is to note that apart from some small nonlinear coupling by θ overlap with the images, retaining only the $\hat{\theta} = 0$ ballooning modes has no nonlinear coupling or saturation and hence infinite transport. Thus it is the $\hat{\theta}$ width of the nonlinear spectrum not the linear extended angle width θ_{rms} which controls the transport level.

Finally we briefly explore the effects of equilibrium length sheared radial electric fields. Expressing the shear rate as $\gamma_E = (cE'_x/B)/(c_s/a)$, $\gamma_E \approx 1.0$ was required to stabilize the fastest growing ballooning mode. This is O(10) larger than the simplest linear damping formula $\gamma_E k_y \Delta_x \gg \gamma$ would suggest. On the other hand while $\gamma_E = 1.0$ indeed completely stopped turbulent transport, $\gamma_E = 0.1$ decreased the transport level by fourfold. This seems to confirm the idea that decrease in the eddy radial length by the linear shear coupling maybe more important than the actual linear stabilization. We note in passing that the destabilizing effects of parallel shear flow $\gamma_p = V'_{\parallel}/(c_s/a)$ can be accounted by a term $-i\gamma_p k_{y'}\phi_{1k}$ to the righthand side of Eq. (2). γ_p values exceeding 1 and approaching 2 are required for significant change in the stability toroidal stability.

Simulations with image modes

For the strongly ballooning ITG mode turbulence presented here, it would seem a' priori likely that the nonlinear interaction of the primaries with the images should not be important. From Fig. 2 (and also Fig. 4), it is apparent that very little of the transport flow and only some of the pressure perturbation extends beyond $\theta = \pm\pi$. The mode half-widths are nearly all within $\pm\pi$. Without significant overlap, the nonlinear coupling should be small. It is expensive to couple the images. The nonlinear Courant condition on time steps is $\Delta t < 1/(v_{Ex} k_{x\max})$. Including the nearest neighbor coupling ($p = \pm 1$) increases $k_{x\max}$ three-fold and the added computing per time step is roughly twofold, hence runs are six times more expensive. A limited number of runs have been made. Figure 8(a) illustrates the instantaneous ($t = 300\, c_s/a$) primary modes on the right hand side panels and how they are imbedded in the image modes on the left hand side panels. Unfortunately, the image cases we have run do not have long time saturation. They initially track primary cases with similar diffusion levels as we would have expected, but then eventually run to much higher levels as illustrated in Fig. 8(b). The fidelity of energy conservation is excellent and gives no indication of numerical instability. The instantaneous fluctuation plots are similar in all respects to those in Fig. 8(a). There is some indication from Fig. 8(a) right-hand panels, that the lower $k_{y'}$ modes have an extended interchange like tail and are not well confined to $-2\pi < \theta \leq 2\pi$. This may be the source of the problem.

Simulations with radial modes

Radial modes can have a pronounced effect on controlling the level of turbulence and transport. They are nonlinearly pumped by the ballooning modes and feed back to stabilize them. This effect has been seen in a number of recent simulations [7, 11]. (For example see the slab GLF simulations by Dorland *et al.* at this workshop.) As we have noted the key question is what is the linear damping mechanism for such modes. Figure 9(a) illustrates the standard case of primary ballooning modes (without images) but with radial modes. The simplest assumption is to take the electrons with no response ($R_k = 0$) and allow no linear driving or damping of the ion motion, i.e. the vorticity, parallel ion motion, and pressure perturbations are only non linearly $\mathbf{E} \times \mathbf{B}$ pumped. Trace a shows that the radial modes are pumped to levels which almost completely eliminates the transport leaving only small and rather intermittent levels. Trace b shows the addition of the ion physics as given by Eqs. (1) through (4). This involves $k_\parallel$ terms as well as residual geodesic curvature terms [$\propto (2/R) k_{x'} \sin(\theta)$]. This provides some Landau damping but undamped $\mathbf{E} \times \mathbf{B}$ poloidal sheared poloidal rotations apparently remain. Tracec, show the addition of magnetic pumping with ($\mu_{mp} = 0.1$) as described in Sec. 4. The diffusion level at $\chi \approx 6\left[(c_s/a)\rho_s^2\right]$ is considerably smaller than 17, the value with only the primary ballooning modes (see Fig. 2). The dashed line, traced adds the images modes with radial modes periodic on $\theta \to \theta + 2\pi$. Here the simulation with images appears to tract the simulation without images but the long time saturation is not entirely clear. Since the radial modes have components with $k_\parallel$ not necessarily zero, the dotted curve, tracee shows a (no image) case with $R_k = 1$. Considerably higher transport levels result making it clear that only detailed electron models with a $k_\parallel$ dependence will suffice. Figure 9(b) shows the pressure and transport spectrum at $\theta = 0$ [corresponding to tracec in Fig. 9(a)]. While the radial modes themselves at $k_{y'} = 0$ give no transport, their pressure (or density) fluctuations should be measurable as the longwave limit in beam emission spectroscopy (BES) measurements [17]. Indeed the BES k_y spectra show a peak in k_y but do not fall to 0 as $k_y \to 0$. The damping rate spectrum for these zero frequency radial modes is shown in Fig. 9(c) for $R_k = 0$. Even the lowest rate of $-0.034\,(c_s/a)$ at $k_{x'} = 0.078$ is significant given that the peak ballooning mode growth is 0.083. The damping rate is virtually independent of the magnetic pumping rate from $\mu_{mp} = 0.1$ to 0.001 again allowing the interpretation that the magnetic pumping damps only the poloidal rotations and Landau damping controls the damping rate for the remaining toroidal rotation. Thus the turbulence level may not depend on the collisional scaling of μ_{mp}. For example simultaneous with $\mu_{mp} = 0.01$ give the same result as with 0.1. (The damping spectrum for $R_k = 1$ is virtually the same at high magnetic pumping rates but is somewhat less at very low values.) However note that the damping is progressively smaller at low $k_{x'}$. Thus it may not be surprising that the spectrum of radial modes falls to the longest available wave number as shown in Fig. 9(d). The equilibrium ($k_{x'} = 0$) is not evolved but no scale separation is apparent between the equilibrium and the radial modes. It is tempting to speculate that this maybe a source of Bohm scaling. However note that the k_x spectrum of the finite-n ballooning modes (not shown) is also peaked at $k_x = 0$. Indeed the BES shows spectra peaked at $k_x = 0$ [17]. More importantly, the effective strength for $\mathbf{E} \times \mathbf{B}$ shear effects on the ballooning modes should be proportional to $cE'_x/B \propto k^2_{x'}|\phi_k|$. A spectrum of this is shown as a dashed line in Fig. 9(d). It suggests that the lowest $k_{x'}$ modes may not be the most effective or of more importance than the $k_{x'} \approx 1$ modes. Only comparison with simulations at much finer grid spacing will resolve the scaling with gyroradius.

5.2.7 SUMMARY

The nonlinear ballooning mode representation would seem to offer a clear advantage over previous multiple helicity (m, n, r) representations if the turbulence is sufficiently ballooning with very limited interchange-like extensions along the field beyond $\theta = \pm\pi$. In this case it should be possible to retain only the primary modes with ballooning angles $-\pi < \hat{\theta} \le \pi$ and very limited interaction with the image modes having $\hat{\theta}$ beyond $\pm\pi$. Adding a nearest neighbor image $(p = \pm 1)$ should have at most a small effect. Even though ITG mode turbulence is strongly ballooning and well localized in θ for the moderate to high $k_{y'}$ modes, this may not be the case for the lowest $k_{y'}$ modes and we have not yet been able to conclusively demonstrate that the interaction with the images is unimportant. Even though simulations with images initially track simulations with primaries alone, we have not yet shown long time saturation with images included.

With this caution in mind, we can make several tentative conclusions about BMR simulations retaining only primaries. The simulations show a rather simple threshold behavior: assuming a gyroBohm scaling $\chi \propto (c_s/R)\rho_s^2 F(R/L_t, R/L_n, q, \hat{s})$ we find $F \propto (R/L_t - R/L_{T\text{crit}})$ in the weak density gradient regime $R/L_n < 3$. There is no evidence of subcritical turbulence. Variations about the standard case with $a/L_T = 3$, $a/L_n = 1$, $a/R = 1/3$, $T_i/T_e = 1$, $q = 2$, and $\hat{s} = 1$ show roughly $F \propto q^{(2\to 3)}/\hat{s}^{(5\to 6)}$ although it is doubtful that the scaling can be described in simple power law form. These parametric dependencies are very roughly consistent with those of linear growth and mixing length scaling. However the size of the transport and the observed eddies suggests that the actual turbulent mixing length is much larger than that associated with a linear mode width. The mixing length may be better given by the nonlinear spectral width in the ballooning angle $\Delta\hat{\theta}$ [$\Delta_{\hat{x}} = 1/(k_{y'}\hat{s}\Delta\hat{\theta})$] rather than the linear width in extended poloidal angle θ_{rms} of the most unstable mode [$\Delta_x = 1/(k_{y'}\hat{s}\,\theta_{rms})$]. The smaller $\Delta\hat{\theta}$ may only scale as the larger θ_{rms}.

Perhaps our most important result is that the sheared flows of the nonlinearly driven near-equilibrium radial modes ($n = 0$ ballooning modes) were found to have a very strong stabilizing effect on the finite-n ballooning modes, confirming the computational results of Dorland et al. [11] and of Beer et al. [7]. The general mechanism for turbulence-generation of poloidal flow has been discussed by Diamond and Kim [12] (who emphasized its possible role on the L–H transition at the edge) and has been seen in numerical simulations of resistive pressure-driven edge turbulence [18, 19], while our results (along with Dorland et al. and Beer et al.) show this can be important process for ITG turbulence which may be more relevant in the plasma interior. Turbulence-driven rotational shear stabilization has been missed in previous studies of ITG turbulence either because an incorrect form of the adiabatic electron response was used (dropping the $\langle\phi\rangle$ term in $n_k/n_0 = \phi - \langle\phi\rangle$, thus allowing the electrons to artificially short-out the axisymmetric electrostatic potential), or because the boundary conditions were chosen in such a way as to freeze the radial modes along with the equilibrium gradients. The nonlinear ballooning mode representation used here (and also in Beer et al.) is particularly convenient in allowing the radial modes to be calculated self-consistently, independent of the quasilinear relaxation of the equilibrium itself.

The nonlinear Reynolds stress provided by the ballooning modes can drive the sheared flows of the radial modes to levels sufficient to virtually stop the transport unless there is a competing linear damping mechanism. A unique contribution of our work is to show that neoclassical magnetic pumping may provide a sufficiently strong damping mechanism. We have found that classical magnetic pumping completely damps the poloidal $\mathbf{E} \times \mathbf{B}$ flow in the radial modes and the remaining toroidal component is apparently controlled by coupling to the Landau damping by its finite $k_\parallel$ component. The remaining radial mode levels are still sufficient to make the transport from the ballooning modes 2–3 times less. More work is needed to accurately model the linear damping processes for the radial modes. In particular a more physical model of the electron response is required which interpolates between no response ($R_k = 0$) at $k_\parallel$ exactly 0 or the $n = 0$, $m = 0$ component and an adiabatic response ($R_k = 1$) at $n = 0$ but $k_\parallel$ finite.

ACKNOWLEDGMENTS

It is a pleasure to acknowledge useful discussions with G.W. Hammett and his co-workers M. Beer, and W. Dorland on many aspects of GLF models and nonlinear BMR formulations. One of us (R.E.W.) wishes to acknowledge useful discussions with Y.B. Kim on applications of magnetic pumping. We wish to thank the Los Alamos Advanced Computing Laboratory for access to the CONNECTION MACHINE as part of the Numerical Tokamak Project.

This work is sponsored by the U.S. Department of Energy under Contract Nos. DE-FG03-92ER54150 and W-7405-ENG-48.

Bibliography

[1] G.W. Hammett and F.W. Perkins, Phys. Rev. Lett. **64**, 3019 (1990).

[2] G.W. Hammett, W. Dorland, and F.W. Perkins, Phys. Fluids B **4**, 2052 (1992); see also W. Dorland and G.W. Hammett, "Gyrofluid Turbulence Models With Kinetic Effects," Princeton Plasma Physics Laboratory report PPPL-2874 (1992) submitted to Phys. Fluids].

[3] R.E. Waltz, R.R. Dominguez, and G.W. Hammett, Phys. Fluids B **4**, 3138 (1992).

[4] S. Hamaguchi and W. Horton, Phys. Fluids B **2**, 3040 (1990).

[5] W. Horton, R.D. Estes, and D. Bischamp, Plasma Phys. **22**, 663 (1980); see also W. Horton, D. Choi, and W. Tang, Phys. Fluids **24**, 1085 (1981).

[6] R.E. Waltz, Phys. Fluids **31**, 1962 (1988) [see also Phys. Fluids B **2**, 2118 (1990)].

[7] M. Beer, G.W. Hammett, W. Dorland, and S.C. Cowley, BAPS **37**, 1478 (1992) 5P 23.

[8] R. Linsker, Phys. Fluids **24**, 1485 (1981).

[9] S.C. Cowley, R.M. Kulsrud, R. Sudan, Phys. Fluids B **3**, 2767 (1991).

[10] J.W. Connor, R.J. Hastie, J.B. Taylor, Proc. R. Soc. London, Ser. A **365**, 1 (1979).

[11] W. Dorland, G.W. Hammett, T.S. Hahm, and M.A. Beer, BAPS **37**, 1478 (1992) 5P 22.

[12] P.H. Diamond and Y.B. Kim, Phys. Fluids B **3**, 1626 (1991).

[13] H. Biglari, P.H. Diamond, and P. Terry, Phys. Fluids B **2**, 1 (1990).

[14] Y.B. Kim, P.H. Diamond, and H. Biglari, Phys. Fluids B **3**, 384 (1991).

[15] G. Staebler, R.E. Waltz, *et al.*, Proceedings of the Fourteenth Int. Conf. on Plasma Physics and Controlled Nuclear Fusion Research, Wurzburg, Germany, October 1992 IAEA CN-56/D-4-17.

[16] J.Q. Dong, W. Horton, J.Y. Kim, Phys. Fluids B **4**, 1867 (1992).

[17] R.J. Fonck, N. Bretz, G. Cosby, R. Durst, E. Mazzucato, R. Nazikian, S. Paul, S. Scott, W. Tang, M. Zarnstorff, Nineteenth European Conference on Controlled Fusion and Plasma Physics, Innsbruck, Austria June 29-July 3, 1992, to be published in Controlled Fusion and Plasm Physics.

[18] A. Hasegawa and M. Wakatani, Phys. Rev. Lett. **59**, (1987) 1581.

[19] B. Carreras, V. Lynch, and L. Garcia, Phys. Fluids B **3**, 1438 (1991).

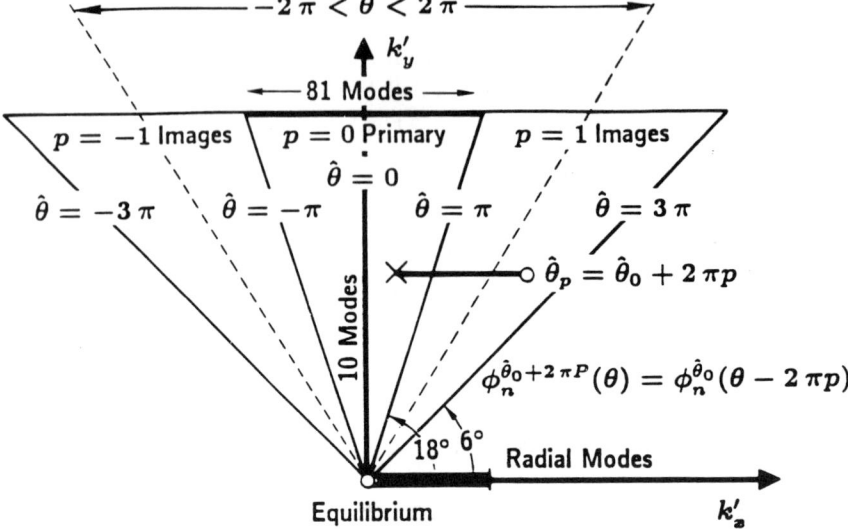

Figure 1 Ballooning mode and radial mode grid in $(k_{x'}, k_{y'})$. $18°$ and $6°$ indicate actual k_y/k_x slope at $\hat{s} = 1$.

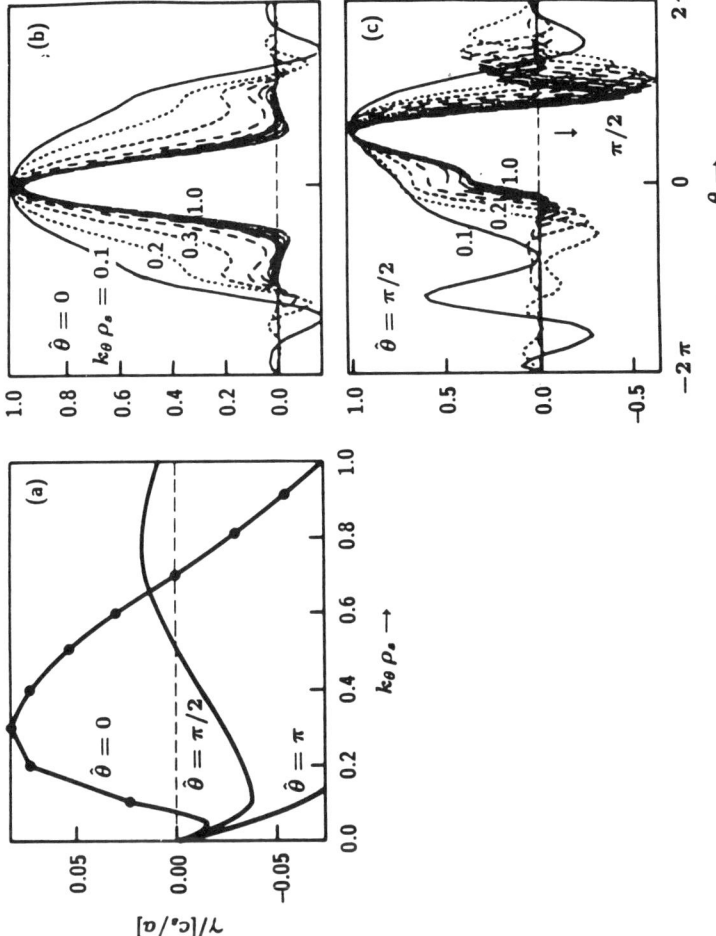

Figure 2 Linear ballooning mode growth rate spectrum (a); extended angle θ mode structure for $\hat{\theta} = 0$ (b) and $\hat{\theta} = p/2$ (c) for standard case.

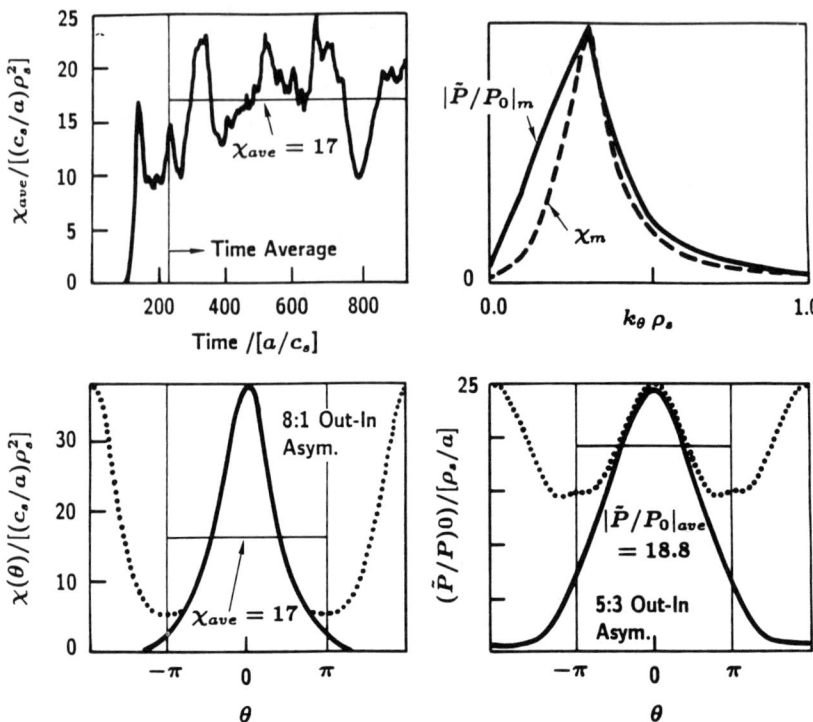

Figure 3 Standard case nonlinear simulation. (a) Flux surface average diffusivity vs. time; (b) Pressure perturbation and diffusivity spectrum at $\theta = 0$ vs. poloidal wave number; (c) Time average diffusivity vs. extended poloidal angle θ; (d) Time average pressure perturbation vs. extended poloidal angle θ.

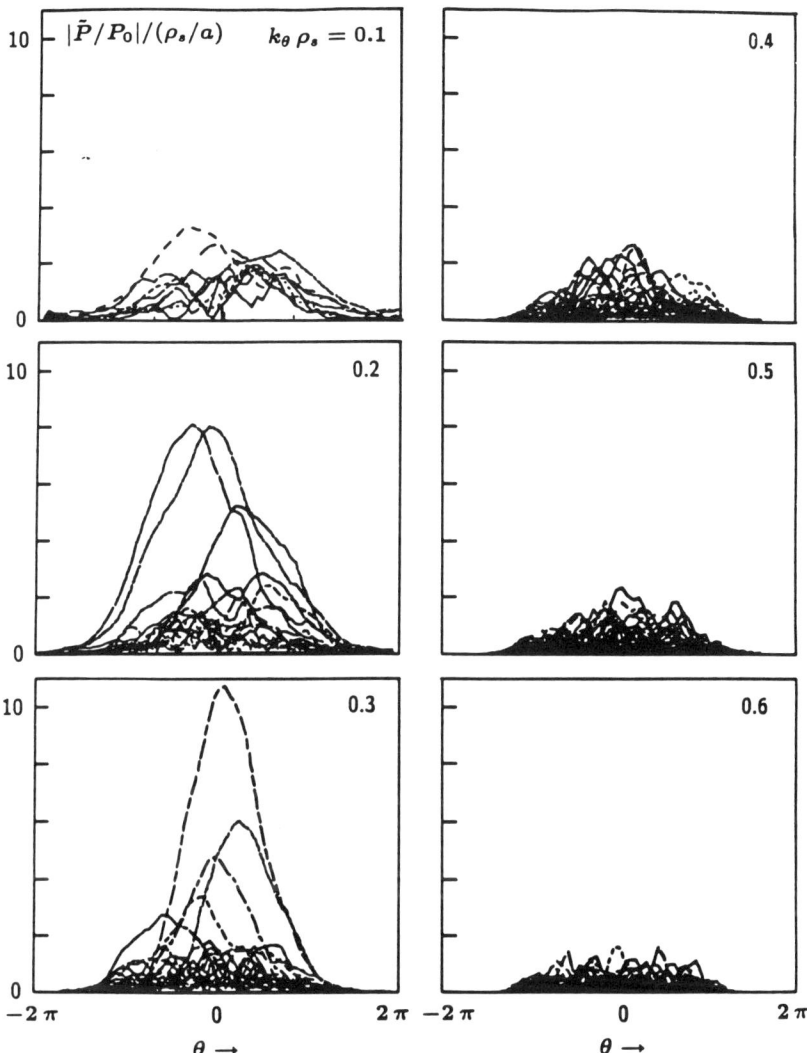

Figure 4 Instantaneous pressure fluctuations for each $k_{x'}$ for several $k_{y'}$ vs. extended poloidal angle in standard case.

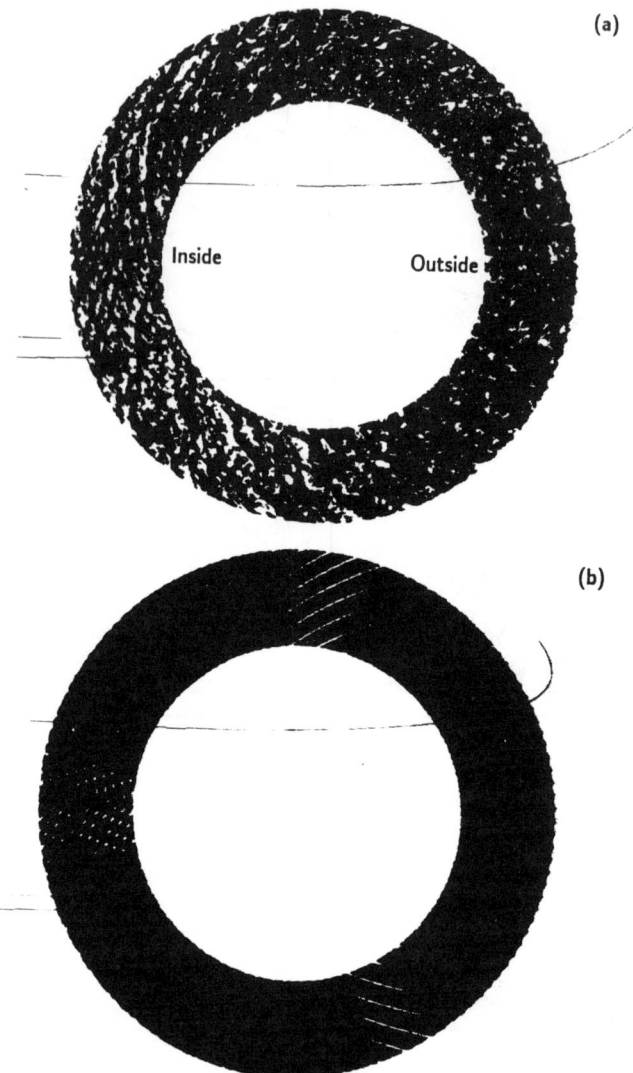

Figure 5 Instantaneous potential contour plots on a constant toroidal angle slice annulus formed by eight intersections with simulation band for standard case. (a) turbulent state; (b) $k_{\theta \rho_s} = 0.3$, $\hat{\theta} = 0$ linear ballooning mode.

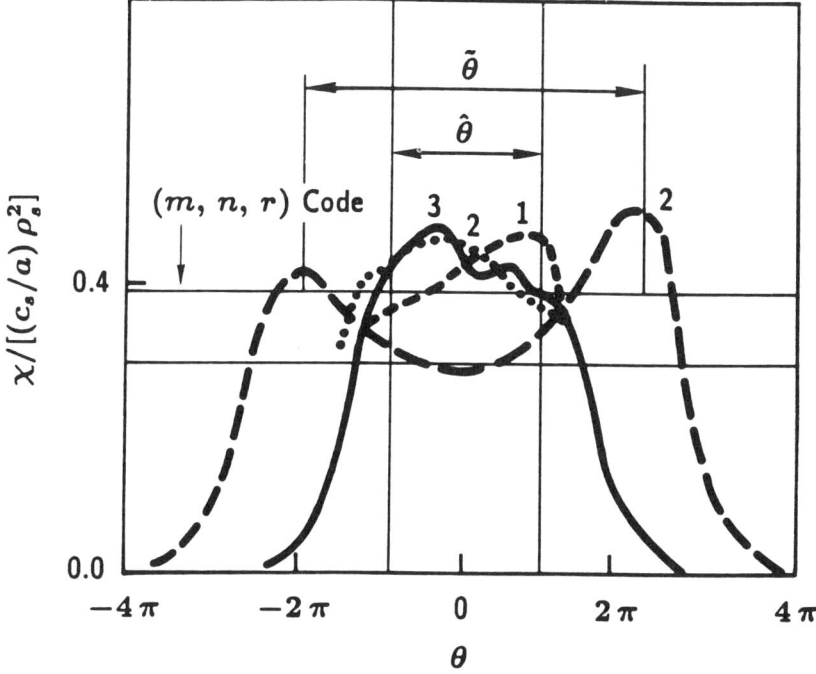

Figure 6 Time average diffusivity vs. extended poloidal angle for slab limit simulation with BMR code with various grids (1_120, 2_230, 3_340 modes) compared to multiple helicity (m, n, r) code simulations over various grids (band). $[a/L_n = 1,\ a/L_T = 4,\ a/R = 1/3,\ q = 2.1,\ \hat{s} = 1$, i.e. $L_n/L_s = 0.16,\ k_{\theta\rho_{s\_max}} = 1.5,\ \mu_x = 0.05]$.

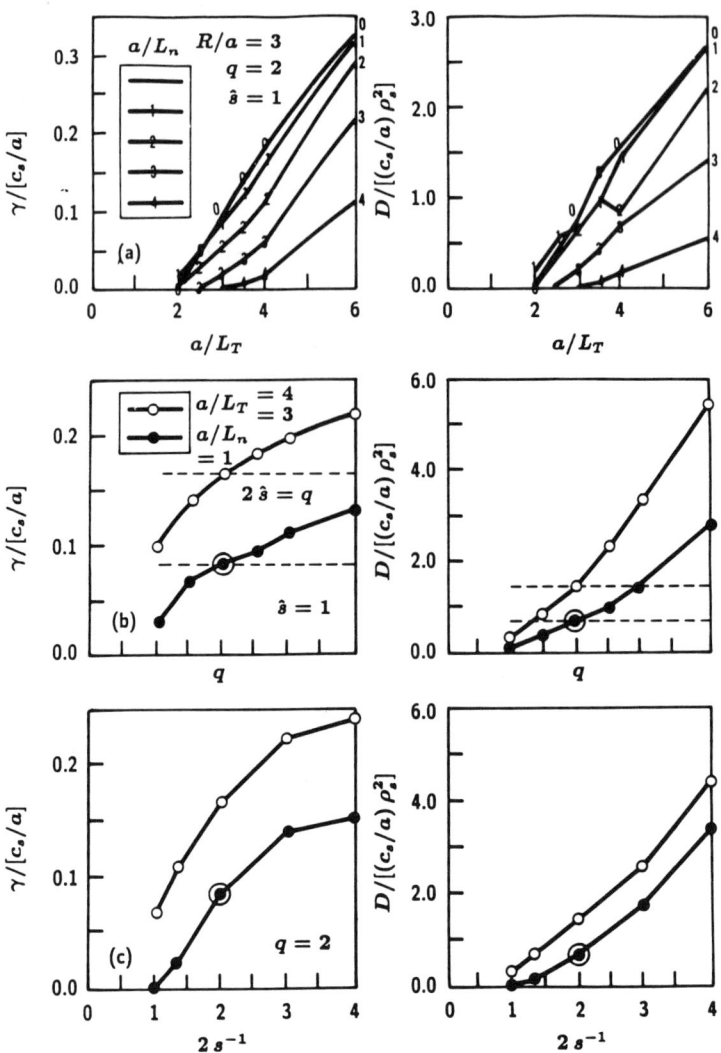

Figure 7 Parametric dependence of growth rate and mixing model diffusivity. Maximum growth rate and mixing model diffusivity vs. temperature gradient (a); vs. safety factor q at $\hat{s} = 1$ (b); vs. inverse shear parameter $(\hat{s})^{-1}$ at $q = 2$ (c). Variations with $2\hat{s} = 1$ in dashed lines.

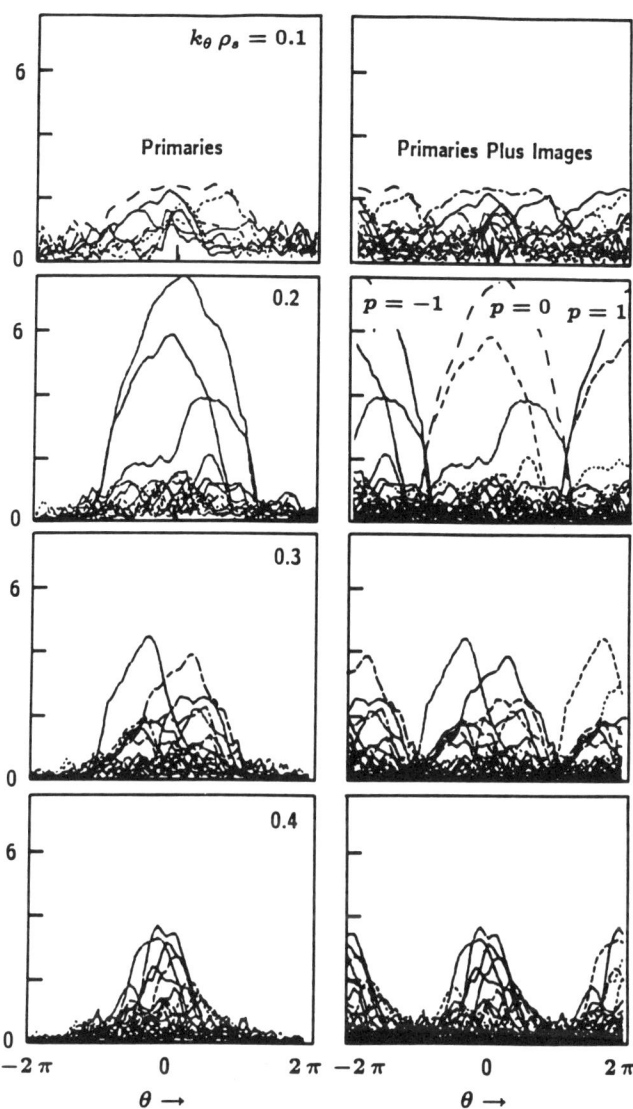

Figure 8 (a) Instantaneous pressure fluctuations for each $k_{x'}$ for several $k_{y'}$ vs. extended poloidal angle in standard case with images. Primaries in left panels and primaries embedded in images in right panels.

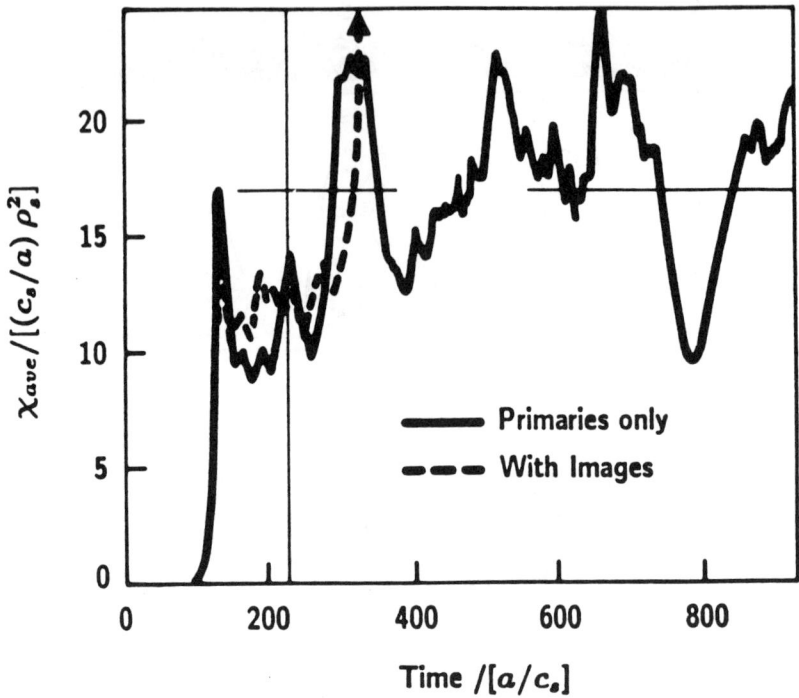

Figure 8 (b) Flux surface average diffusion vs. time for standard case (solid line) and with images added (dashed line).

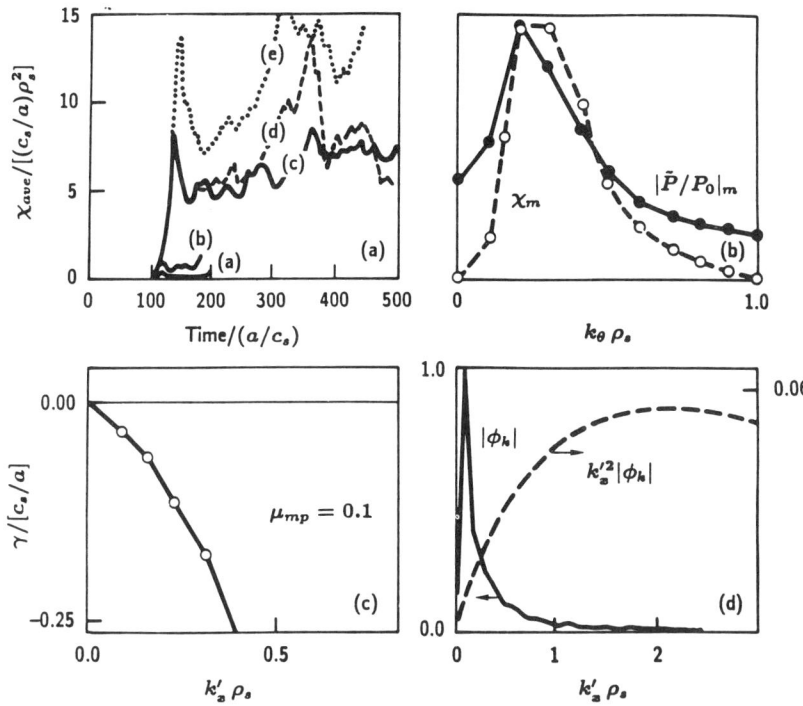

Figure 9 Effect of radial modes on ballooning mode transport. (a) Flux surface average diffusion vs. time with different radial mode linear dynamics: trace_a no linear ion dynamics, trace_b linear ion dynamics added, trace_c magnetic pumping added, trace_d images added, trace_e $R_k = 1$ in place of $R_k = 0$ for radial modes and no images. (b) Pressure and diffusion spectrum vs. poloidal wave number corresponding to trace_c. (c) Linear damping rate spectrum for radial modes with magnetic pumping $\mu_{mp} = 0.1$ and cyclic boundary conditions on $-\pi < \theta < \pi$. (d) θ average spectrum of radial mode amplitudes corresponding to trace_c.

5.3 NONLINEAR GYROFLUID MODEL OF ITG TURBULENCE

W. Dorland, G.W. Hammett, T.S. Hahm and M.A. Beer
P.O. Box 451, Princeton, NJ, 08543, Princeton University Plasma Physics Laboratory

A b s t r a c t. *Early results from nonlinear simulations and analysis based on a recently derived nonlinear gyrofluid model [W. Dorland and G.W. Hammett, Phys. Fluids B, 812 (1993)] of electrostatic ion-temperature-gradient driven turbulence are presented. Comparisons with gyrokinetic particle simulations reveal a few important simulation requirements (such as enforcing radial periodicity), and indicate that the gyrofluid description is probably adequate to describe three-dimensional, low-frequency drift-type turbulence. Results from a detailed weak-turbulence analysis of drift wave turbulence are presented which support this conclusion. The importance of keeping the proper adiabatic electron response is also discussed. In particular, perpendicular velocity shear is greatly enhanced when the magnetic shear is weak if the nonphysical radial transport of electrons is disallowed.*

5.3.1 INTRODUCTION

Recent tokamak experiments [1, 2, 3] indicate that ion-temperature-gradient driven turbulence is a candidate to explain the observed heat transport. To facilitate numerical and analytical studies in more realistic geometries, we have recently developed a nonlinear "gyrofluid" model [4] of low-frequency ($\omega \ll \Omega_{ci}$) electrostatic plasma turbulence. This model improves over previous fluid models by including "kinetic" effects such as Landau damping and its inverse ($\omega \sim k_\parallel v_{ti}$), finite Larmor radius (FLR) averaging ($k_\perp \rho_i \sim 1$) and nonlinear FLR phase-mixing ($\propto k_\perp^2 |\Phi_{\mathbf{k}} \mathbf{k} \times \mathbf{k}'|$). Previously, we presented linear benchmarks of this model, comparing with Linsker's electrostatic, fully gyrokinetic integral eigenvalue code [5]. Here, we present nonlinear results from our model, and discuss a few numerical, or simulation-related issues that should be considered carefully regardless of the basic dynamic model (particle or fluid) being used. One interesting physical result that we have found is the robust nonlinear generation of sheared velocity flows by ITG turbulence [6]. This effect is most pronounced if the magnetic shear is weak, or in toroidal geometry [7, 8].

The most important numerical consideration we mention here concerns the relation between the background profile and the simulation box-size. In sheared-slab simulations, the dimensions of the simulation domain determine the spacing of the rational surfaces. If the rational surfaces are thinly spread or clustered in one small region of the simulation domain, the results obtained from the simulation can be misleading, since a nonlinearly saturated state may be obtained *via* local flattening of the driving gradients. Horton *et al.* [9] and Hamaguchi and Horton [10] established clearly that two-dimensional sheared-slab simulations predict much lower transport levels than three-dimensional simulations. Here, we point out that three-dimensional simulations can be quasi-two-dimensional if the rational surfaces do not fill the simulation domain, since the flattening of the background gradients driving the turbulence around a given rational surface results in steepened gradients where there are no other rational surfaces.

We also discuss how to include periodic radial boundary conditions in the presence of magnetic shear, using a method related the Cowley's field-line coordinate approach [23] which turns out to be equivalent to a method used by Kotschenreuther [11]. This prevents steep gradients from forming at the edges of the simulation domain, and preserves the average temperature gradient throughout the box.

In Sec. 2 we outline our "3+1" gyrofluid model. In Sec. 3 we discuss the simulation-related issues mentioned above. Section 4 contains results from a typical nonlinear simulation which shows the role of the self-generated perpendicular velocity shear. Section 5 briefly presents results from a detailed weak turbulence analysis of drift waves with a short discussion. Finally, we conclude with a discussion of the relevance of these results to magnetic confinement fusion experiments in Sec. 6.

5.3.2 GYROFLUID EQUATIONS

Our model equations are obtained by taking moments of the gyrokinetic equation [12, 13] directly. The resulting set of fluid equations is closed with the Landau damping model of Ref. [14] generalized to include FLR effects in Ref. [4]. Here we present the simplest sheared-slab "3+1" model (so-called because we evolve three parallel moments, $n, u_\parallel$, and $T_\parallel$ and one perpendicular moment, $T_\perp$) which is accurate enough for the present purposes, and is the least complicated model we have derived which nevertheless contains the physics relevant to the present paper.

The usual gyrokinetic ordering is $\omega/\Omega \sim \rho/L \sim k_\parallel \rho \sim F_1/F_0 \sim e\Phi/T_e \sim \epsilon \ll 1$ and $k_\perp \rho \sim 1$, where ω is a typical frequency of the fluctuation spectrum, Ω is the ion cyclotron frequency, ρ is the ion Larmor radius, L is a typical scale length of the system, $k_\parallel$ and $k_\perp$ are typical parallel and perpendicular wavenumbers of the fluctuation spectrum, and Φ is the electrostatic potential. F_1 is the gyrophase-independent part of the perturbed guiding-center distribution function, and $F_0(\mathbf{R}, v_\parallel, v_\perp, t)$ is the equilibrium distribution of guiding centers. Note that the total F is not necessarily Maxwellian; the linear perturbation is in fact quite non-Maxwellian. Using this ordering, we also expand the fluid variables into background and fluctuating components ($n = n_0 + n_1(t), \ldots$) and then rescale the fluctuating parts to reflect the ordering most naturally. That is, the non-dimensional perturbed variables ($\tilde{n}, \tilde{u}_\parallel, \ldots$) are given in terms of the dimensional perturbed variables ($n_1, u_{\parallel 1}, \ldots$) by

$$\left(\frac{n_1}{n_0}, \frac{u_{\parallel 1}}{v_t}, \frac{T_{\parallel 1}}{T_i}, \frac{T_{\perp 1}}{T_i}, \frac{q_{\parallel 1}}{v_t p_{\parallel 0}}, \frac{q_{\perp 1}}{v_t p_{\perp 0}}, \frac{e\Phi_1}{T_i}\right) = \frac{\rho_i}{L_n}(\tilde{n}, \tilde{u}_\parallel, \tilde{T}_\parallel, \tilde{T}_\perp, \tilde{q}_\parallel, \tilde{q}_\perp, \tilde{\Phi}). \tag{1}$$

Furthermore, we normalize (x, y, z, t) according to

$$\tilde{x} = \frac{x - x_0}{\rho_i}, \quad \tilde{y} = \frac{y}{\rho_i}, \quad \tilde{z} = \frac{z}{L_n}, \quad \text{and} \quad \tilde{t} = \frac{tv_t}{L_n}. \tag{2}$$

We have used the definitions $L_n^{-1} = -\partial \ln n_0/\partial x$, $v_t = \sqrt{T_{0i}/m_i}$ and $\rho_i = v_t/\Omega_i$. Also, $x_0 = L_x/2$, where L_x is the box size in the x direction. For convenience, we do not write the tildes over the non-dimensional variables except where confusion might otherwise be generated. Thus, throughout most of this paper, $(n, u_\parallel, \ldots)$ are the same as the non-dimensional variables $(\tilde{n}, \tilde{u}_\parallel, \ldots)$ in Eq. (1).

We include FLR effects with a model which is rigorously second-order accurate in $k_\perp \rho$ and well-behaved for $k_\perp \rho$ arbitrarily large. For convenience, we define $\Psi \equiv \langle J_0 \rangle \Phi$ and $\mathbf{v}_\Psi \equiv \langle J_0 \rangle \mathbf{v}_E$, where $\langle \cdots \rangle$ is an average over velocity space. We also introduce two modified Laplacian operators $\widehat{\nabla}_\perp^2$ and $\widehat{\widehat{\nabla}}_\perp^2$:

$$\frac{1}{2}\widehat{\nabla}_\perp^2 \langle J_0 \rangle \equiv b\frac{\partial \langle J_0 \rangle}{\partial b} \quad \widehat{\widehat{\nabla}}_\perp^2 \langle J_0 \rangle \equiv b\frac{\partial}{\partial b}\left(\langle J_0 \rangle + b\frac{\partial \langle J_0 \rangle}{\partial b}\right). \quad (3)$$

The FLR parameter $b \equiv k_\perp^2 \rho_i^2$. As $b \to 0$, the operators $\widehat{\nabla}_\perp^2, \widehat{\widehat{\nabla}}_\perp^2 \to \nabla_\perp^2$.

Good approximations for $\langle J_0 \rangle$ are the subject of much discussion in Ref. [4], where we found discernible differences in the linear growth rates and eigenmodes obtained with the various models. Nonlinearly, we have found that any of the models presented as alternatives to the usual Taylor series approximation ($\langle J_0 \rangle = 1 - b/2$) produce results which are virtually indistinguishable. This is probably because the results are not sensitive to the details of the high-$k_\perp \rho$ part of the fluctuation spectrum, as long as the qualitative features of the FLR orbit-averaging are retained. The Taylor series model fails to meet this single criterion. Results reported here use either the $\langle J_0 \rangle = (1 + b/2)^{-1}$, $\langle J_0^2 \rangle = (1 + b)^{-1}$ or the $\langle J_0 \rangle = \Gamma_0^{1/2}$, $\langle J_0^2 \rangle = \Gamma_0$ FLR models.

An effect introduced in Ref. [4] as "nonlinear FLR phase-mixing" is included here with the "$\mathcal{N}$" operators. Physically, particles with different Larmor radii $\mathbf{E} \times \mathbf{B}$ drift with different velocities in the turbulent electrostatic potential, and consequently phase-mix perturbations perpendicular to the magnetic field at a rate approximately $\propto k_\perp^2 |\Phi_\mathbf{k} \mathbf{k} \times \mathbf{k}'|$. We define a one-pole FLR phase-mixing operator $\mathcal{N}_1$ operating on some moment M by

$$\mathcal{N}_1 M = \nu_1 \left|\left[\frac{1}{2}\widehat{\nabla}_\perp^2 \mathbf{v}_\Psi\right] \cdot \nabla\right| M - \lambda_1 \left[\frac{1}{2}\widehat{\nabla}_\perp^2 \mathbf{v}_\Psi\right] \cdot \nabla M. \quad (4)$$

where $(\nu_1, \lambda_1) = (0.4, 0.6)$ and a two-pole FLR phase-mixing operator by

$$\mathcal{N}_{21} = -\frac{\nu_2}{4}\left|\left[\frac{1}{4}\widehat{\nabla}_\perp^4 \mathbf{v}_\Psi\right] \cdot \nabla\right| - \frac{\lambda_2}{4}\left[\frac{1}{4}\widehat{\nabla}_\perp^4 \mathbf{v}_\Psi\right] \cdot \nabla$$

$$\mathcal{N}_{22} = \nu_2 \left|\left[\frac{1}{2}\widehat{\nabla}_\perp^2 \mathbf{v}_\Psi\right] \cdot \nabla\right| - \lambda_2 \left[\frac{1}{2}\widehat{\nabla}_\perp^2 \mathbf{v}_\Psi\right] \cdot \nabla.$$

with $(\nu_2, \lambda_2) = (1.6, 1.3)$.

We apply our model to the usual sheared-slab geometry. With the normalizations of Eqs. (2), this leads to

$$\nabla_\perp \equiv \hat{\mathbf{x}}\frac{\partial}{\partial x} + \hat{\mathbf{y}}\frac{\partial}{\partial y} \quad \hat{\mathbf{b}}\cdot\nabla \equiv \frac{\partial}{\partial z} + sx\frac{\partial}{\partial y}$$

$$s \equiv L_n/L_s \,. \tag{5}$$

Finally, we also define:

$$\frac{d}{dt} = \frac{\partial}{\partial t} + \mathbf{v}_\Psi \cdot \nabla \,, \quad \mathbf{v}_\Psi = \hat{\mathbf{b}} \times \nabla\Psi \,.$$

The equations we solve are:

$$\frac{dn}{dt} + \left[\frac{1}{2}\widehat{\nabla}_\perp^2 \mathbf{v}_\Psi\right]\cdot\nabla T_\perp + \mathcal{N}_{21}T_\perp + \hat{\mathbf{b}}\cdot\nabla u_\| + \left[1 + \eta_{i\perp}\frac{1}{2}\widehat{\nabla}_\perp^2\right]\frac{\partial\Psi}{\partial y} = 0 \tag{6}$$

$$\frac{du_\|}{dt} + \mathcal{N}_1 u_\| + \hat{\mathbf{b}}\cdot\nabla(T_\| + n + \Psi) = 0 \tag{7}$$

$$\frac{dT_\|}{dt} + \mathcal{N}_1 T_\| + 2\hat{\mathbf{b}}\cdot\nabla u_\| + \eta_{i\|}\frac{\partial\Psi}{\partial y} + 2\sqrt{\frac{2}{\pi}}|k_\|| T_\| = 0 \tag{8}$$

$$\frac{dT_\perp}{dt} + \left[\frac{1}{2}\widehat{\nabla}_\perp^2 \mathbf{v}_\Psi\right]\cdot\nabla n + \left[\widehat{\nabla}_\perp^2 \mathbf{v}_\Psi\right]\cdot\nabla T_\perp + \mathcal{N}_{22}T_\perp$$

$$+ \left[\frac{1}{2}\widehat{\nabla}_\perp^2 + \eta_i(1+\widehat{\nabla}_\perp^2)\right]\frac{\partial\Psi}{\partial y} + \sqrt{\frac{2}{\pi}}|k_\||(T_\perp + \frac{1}{2}\widehat{\nabla}_\perp^2 \Psi) = 0 \tag{9}$$

Poisson's equation (assuming quasineutrality and adiabatic electrons) is:

$$\tau(\Psi - \langle\langle\Psi\rangle\rangle) = \langle J_0\rangle^2\left(n + \frac{1}{2}\widehat{\nabla}_\perp^2 T_\perp\right) + (\langle J_0^2\rangle - 1)\Psi \,, \tag{10}$$

where we denote the flux-surface average by

$$\langle\langle\cdots\rangle\rangle \equiv \frac{1}{L_y L_z}\int dy\, dz \cdots \,.$$

This term must be included to prevent non-physical electron transport across flux surfaces. It results in the strong generation of perpendicular velocity shear, as discussed in more detail below and in Ref. [6]. The quasineutrality constraint $n_e = n_i$ is applied to the *particle* densities, not the *guiding center* densities. Eq. (10) expresses the transformation from the ion guiding center density n and temperature $T_\perp$ to the local particle density, including a term proportional to Ψ which gives the "polarization" contribution. Other parameters are $\tau = T_{i0}/T_{e0}$, $\eta_i = L_n/L_T$, and $L_T^{-1} = -\partial \ln T_0/\partial x$.

Equations (6–9) contain two types of damping: Landau damping and nonlinear phase-mixing. This dissipation is sufficient to reach and maintain a nonlinearly saturated steady state for times long compared to L_n/v_t. However, we may also include "collisional" dissipation by including terms proportional to $-\nu \nabla_\perp^\alpha W$ on the RHS of each equation, where W is the moment being evolved, and $\alpha = 4$ for the density equation and 2 for the remainder to reduce the fluctuations at the grid scale. Results shown here use $\nu \leq 0.005$. Convergence studies have been carried out to make certain that the effect of these terms is in fact neglible.

Adiabatic Electrons An adiabatic electron response of the form $n_{e1} = n_0 e\Phi/T_e$ is often assumed for many types of tokamak instabilities. As we will explain, this form is usually incorrect for the $k_y = 0, k_z = 0$ component of the fluctuations, which can make a big difference in nonlinear simulations. The adiabatic electron response is derived from the linearized parallel force balance equation for electrons, which in the $k_\| v_{te} \gg \omega$ limit is $\nabla_\|(p_{\|1} - en_0\Phi) = 0$. Assuming that the electrons are isothermal (so that $p_{\|1} = n_{e1}T_{e0}$) yields

$$\nabla_\| n_{e1} = n_0 \nabla_\| \frac{e\Phi}{T_{e0}} . \qquad (11)$$

Thus, only the parallel gradient of the electron density is determined, and the constant of integration in

$$n_{e1} = C + n_0 \frac{e\Phi}{T_e} \qquad (12)$$

must be determined from some other constraint. In the usual case of electrostatic waves in a plasma with good magnetic flux surfaces, there is no net radial transport of particles if the electrons are exactly adiabatic (the radial particle flux $n_e v_{Ex} \propto n_e \partial\Phi/\partial y$ vanishes when averaged over a flux surface). This means that the number of electrons on each flux surface must be constant, thus determining the constant of integration in Eq. (12),

$$n_{e1} = n_0 \frac{e}{T_e} (\Phi - \langle\langle\Phi\rangle\rangle) , \qquad (13)$$

so that flux-surface-average $\langle\langle n_{e1} \rangle\rangle = 0$. For convenience, we will also use the notation $\langle\langle\Phi\rangle\rangle = \delta_m \delta_n \Phi$, where $\delta_m \delta_n$ is an operator in Fourier wave-number space which is 1 for the $m = 0, n = 0$ component and 0 for all other Fourier components. To demonstrate the importance of this, consider the solution of Eq. (10) for Ψ,

$$\Psi = \frac{\bar{n}_i}{\tau(1 - \delta_m \delta_n) + 1 - \Gamma_0} , \qquad (14)$$

where $\bar{n}_i$ is the non-polarization part of the ion density, i.e., the first term on the right hand side of Eq. (10). Expanding the Bessel function in the long wavelength limit, this reduces to

$$\Psi = \frac{\bar{n}_i}{\tau(1 - \delta_m \delta_n) + k_\perp^2 \rho_i^2} . \qquad (15)$$

For $m \neq 0$ or $n \neq 0$, this gives the familiar form $\Psi = \bar{n}_i/(\tau + k_\perp^2 \rho_i^2)$. But for the $m = n = 0$ component (the part representing poloidal flows $\propto \partial \Psi_{0,0}/\partial x$) this gives

$$\Psi_{0,0} = \frac{\bar{n}_{i0,0}}{k_x^2 \rho_i^2} , \qquad (16)$$

which at long wavelengths, $k_x^2 \rho_i^2 \ll 1$, gives a very large enhancement in the poloidal flow over the usual formula which ignores the $\delta_m \delta_n$ term. In physical terms, the usual $n_{e1} = n_0 e\Phi/T_e$ formula allows electrons to artificially move radially across flux surfaces and short out the radial electric field responsible for the poloidal flow, while the expression $n_{e1} = n_0 e(\Phi - \langle\langle\Phi\rangle\rangle)/T_e$ prevents this radial current. [Actually, it should be pointed out that the usual form may be acceptable if the magnetic field lines are fully chaotic and space-filling rather than forming good flux surfaces. Radial electric fields will then have a component parallel to this chaotic magnetic field, which allows the electrons to flow radially as well. However, in the more conventional case where good flux surfaces are assumed, one should include the $\langle\langle\phi\rangle\rangle$ term so that the electrons do not respond to a potential which is constant along a field line.] Finally, we should point out that this effect will continue to be important even if weak non-adiabatic effects (such as collisions or trapped electrons) are included. As long as the electron response is still close to adiabatic, the adiabatic component should not respond to a potential which is constant along a field line.

Equations (6–9) along with Eq. 10 are used to advance the fields $(n, u_\parallel, T_\parallel, T_\perp)$ in time. This completes the "3+1" gyrofluid model.

5.3.3 GENERIC SIMULATION ISSUES

Simulation Domain

ITG modes are naturally localized around a rational surface in this geometry because they are damped for $k_\parallel v_t \gg \omega_*$. Thus, if the driving temperature gradient relaxes nonlinearly in the vicinity of the rational surface, the instability is effectively stabilized. Because two-dimensional sheared-slab simulations have only a single rational surface, this stabilization mechanism plays an especially important role in single-helicity simulations, as noted previously [9, 15].

In a realistic plasma, the relaxation of the temperature gradient around one rational surface leads to a steeper gradient at another nearby rational surface. It is important to include this effect in three-dimensional simulations. That is, if the rational surfaces corresponding to unstable modes occupy a small region of the simulation domain, or if they are spaced too far apart, the transport rate predicted may be significantly misleading.

Consider the simulation in Fig. 5.3.3. The parameters are $L_x = 40\rho_i$, $L_y = 25\rho_i$, $L_z = 52 L_n$, $\eta_i = 4$, $L_s/L_n = 40$, and $T_i = T_e$. The nonlinear terms were evaluated on a grid with $(x, y, z) = (64, 16, 16)$ grid points. Because we use a de-aliased pseudo-spectral algorithm to solve Eqs. (6–10), the linear terms are advanced in (x, k_y, k_z) space with lower resolution in k_y and k_z. They are calculated with $k_y = 2\pi m/L_y$, $m = 0, \pm 1, \ldots, \pm M$, $M = 4$ and $k_z = 2\pi n/L_z$, $n = 0, \pm 1, \ldots, \pm N$, $N = 4$. Radially periodic boundary conditions were used, as described in Sec. 5.3.3.5.3.3.

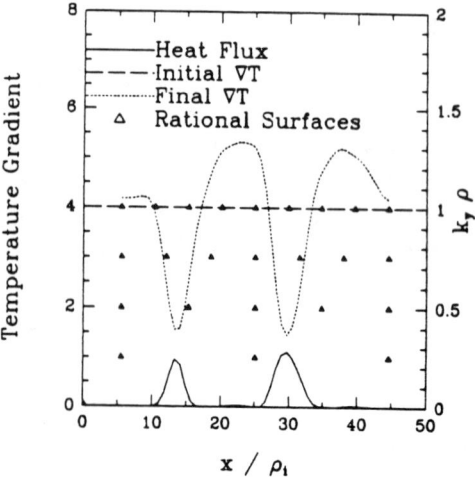

Fig. 5.1 Heat flux *vs.* x for a simulation with rational surfaces too thinly spread.

A fair amount of information is included in this figure. First, the flux-surface averaged, time averaged heat flux

$$\langle\langle \overline{Q} \rangle\rangle \equiv \frac{1}{TL_y L_z} \int dy\, dz\, dt\, Q$$

is measured by the scale on the left axis and is shown as a solid line at the bottom of the figure. Next, the imposed background temperature gradient (normalized to the density gradient) is shown as a dashed line at a constant value of $\eta_i = 4$. The dotted line is the temperature gradient averaged over the nonlinear phase of the simulation. Finally, the triangles represent the positions of the rational surfaces located within the simulation domain, with the corresponding $k_y \rho_i$ on the right axis.

The paucity of rational surfaces (and hence unstable modes) in this example allows the temperature gradient to flatten until $\eta_i \simeq 2$ at two points in the box without causing significant steepening at the neighboring rational surfaces. A volume averaged χ would be misleading in this case, since the thermal flux is actually confined to a small region of the simulation domain, and the driving temperature gradient varies widely over the box. In this example, a significant amount of perpendicular (y) rotation is occuring as well (not shown). Few conclusions can be drawn from such a simulation, since the resolution is simply inadequate. The lesson to be learned is that even though the average temperature gradient across the entire box remains fixed at $\eta_i = 4$, as it should, the flux is suppressed because of local modifications of the temperature gradient. Hence, a prediction of χ based on such a low-resolution simulation should not be trusted. Nothing prevents this behavior from also occurring in low-resolution particle simulations (even if the multiple scale expansion [13] is used).

The same problem occurs in the simulation shown in Fig. 5.1 for different reasons. In part, the problem is again the spacing of the rational surfaces, as the outer regions of the simulation domain have no rational surfaces. This follows directly from the choice of L_z, the size of which causes the rational surfaces to be confined to the center of the box. Because this case is also an example of the effect of the boundary conditions, it is discussed further in the next section.

Periodic Boundary Conditions

The usual sheared-slab simulation domain is periodic in y and z. The radial (x) boundary condition has been treated in several ways in the past. The most common approach is to require all quantities ($n, u_\parallel, \Phi$, etc.) to be zero at the edges of the box (with reflecting boundary conditions for particles in kinetic simulations). This approach is followed in kinetic and gyrokinetic simulations if the entire distribution function is being evolved, with strong gradients ($\rho/L_T \sim 1/10$) across the radial direction. A multiple-scale expansion technique [13] is employed to separate the evolution of the background gradient from the simulation scales. With the advent of "partially linearized" [16] or δf techniques [17, 18], it is possible to separate the background and fluctuating scales asymptotically in particle simulations, obviating the need for the multiple-scale expansion. Fluid simulations have long made this separation by using scaled fluid quantities [9] which make use of the ordering $\rho \ll L$.

One must treat the ($k_y = 0, k_z = 0$) mode with care in any case. Because the gyrokinetic ordering allows $\nabla_\perp F_1 \sim \nabla_\perp F_0$, the fluctuating quantities generate gradients in the ($k_y = 0, k_z = 0$) mode which can effectively cancel the driving background gradients and help to saturate the turbulence. This is a physical saturation mechanism, which should be included in a microturbulence simulation; however, if the ($k_y = 0, k_z = 0$) mode is handled carelessly (as in the previous example simulation), this mechanism can completely dominate the dynamics, which is probably not physical. Conversely, if the ($k_y = 0, k_z = 0$) modes are simply removed from the system, an important source of stabilization may be lost.

In particular, (F_1, Φ) = 0 radial boundary conditions do not adequately model the effects of heat sources and sinks which give rise to the overall gradient. Heat is not allowed to cross the radial boundaries of the box, and the background temperature gradient is gradually steepened near the edges of the box where the drive is intrinsically reduced by the boundary conditions. As in a low-resolution simulation, the gradients in the bulk of the simulation domain can relax, reducing the predicted thermal flux non-physically. Figure 5.2 shows the heat flux vs. x and the background temperature gradients for a nonlinear run which suffers from these deficiencies.

The parameters for this run were chosen to compare with the weak-shear case of Ref. [19]. Hence, $L_x = 64\rho_i$, $L_y = 126\rho_i$, $L_z = 1257 L_n$, $\eta_i = 4$, $L_s/L_n = 20$, and $T_i = T_e$. The nonlinear terms were evaluated on a grid with $(x, y, z) = (64, 128, 32)$ grid points. In this run, the nonlinear phase-mixing terms were not included and $\nu = 0.005\, v_t/L_n$. A gaussian filter $\exp\{-\beta^2 k_\perp^2 \rho^2\}$ with $\beta = 1.2$ was included in Poisson's equation to simulate the effect of finite-sized particles. Finally, in accord with a common practice in gyrokinetic simulations, the ($k_y = 0, k_z = 0$) component of the potential was removed at every time step.

The curves are the time and flux-surface averaged quantities obtained with our simulation. Several features can be noted. First, even though many rational surfaces are included, their distribution is uneven. In addition, the role of the boundary conditions is evident; the temperature gradients are steepening near the boundaries where the

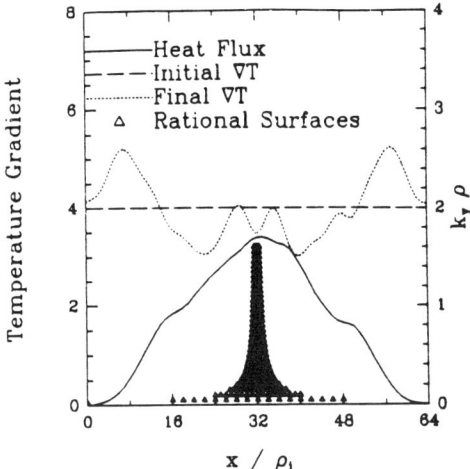

Fig. 5.2. Heat flux, ∇T_0, and rational surfaces *vs.* x using $\Phi = 0$ radial boundary conditions.

flux is forced to zero. The χ predicted from our simulation is difficult to interpret on these grounds, and an improved gyrofluid simulation is shown after a discussion of radial periodicity conditions. However, if the peak value from the central region of the box were to be believed, this simulation would predict $\chi_i \sim 2 \times 10^3$ cm^2/sec for $T_i \sim 250\,\text{eV}$, and $B = 2\,\text{T}$, which is much lower than the χ_i calculated in Ref. [19]. We do not understand this discrepancy; comparisons with other particle simulations reveal both gyrofluid and gyrokinetic simulations tend to recover the mixing-length χ_i in sheared-slab simulations [20, 21], in sharp contrast to the results of Ref. [19].

A better boundary condition is to make the box periodic in the radial direction, so as to maintain the average temperature gradient everywhere in the simulation domain. Heat which flows out of the cold side re-enters on the warm side, mimicking a heat source. Because the simulation domain is typically very much smaller than the background gradients, the actual temperature and density differences between the

356   Nonlinear Gyrofluid Model

edges of the domain are negligible. The difficulty lies in how to handle the sheared magnetic field lines.

A direct consequence of the ordering $k_\parallel \ll k_\perp$ is the tendency of the turbulent structures to be aligned with the pitch of the local magnetic field. It is not sufficient simply to enforce periodicity by requiring any variable W to satisfy

$$W(x + L_x, y, z) = W(x, y, z)$$

as a result. For example, Fig. 5.3(a) shows the electrostatic potential from a typical simulation in the linear phase in the $y - z$ plane at $\tilde{x} = 0$. The effect of the sheared magnetic field is clearly evident, especially when compared to Fig. 5.3(b), which shows the potential at the other side of the box (after periodicity has been enforced). The periodicity constraint used here is first to untwist the magnetic field so as to take into account the geometric effect, and then to enforce periodicity.

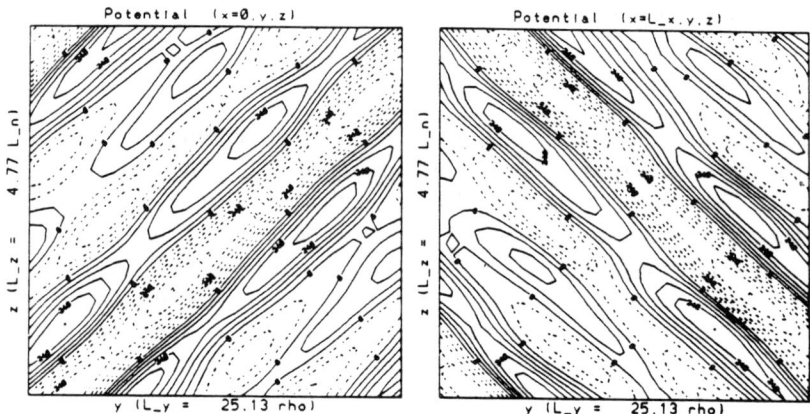

Fig. 5.3 Φ in the $y - z$ plane at (a) $\tilde{x} = 0$ and (b) $\tilde{x} = L_x$. The vertical axis is y in units of (ρ_i), and the horizontal axis is z in units of L_n.

Mathematically, one transforms the periodic variables to a new coordinate system (x', y', z') which is aligned with the magnetic field. These coordinates [22, 23] are given by

$$x' = x$$

$$y' = y - sxz$$

$$z' = z ,$$

so that (x', y') labels a field line and z' measures the distance along a field line. In the new coordinates, the periodicity constraint is stated as

$$W'(x' + L_x, y', z') = W'(x', y', z') . \qquad (17)$$

One then maps Eq. (17) back into the untwisted coordinates (using $W'(x', y', z') = W(x', y' + sx'z', z')$, and then dropping the primes) to find

$$W(x + L_x, y + s(x + L_x)z, z) = W(x, y + sxz, z).$$

Fourier transforming in the y and z directions according to

$$W(x, y, z) = \sum_{m,n} W_{mn}(x) e^{ik_y y + ik_z z}$$

leads to (using the definitions of k_y and k_z from above):

$$\sum_{mn} W_{mn}(x + L_x) \exp\left\{ i2\pi \left[\frac{m}{L_y}(y + s(x + L_x)z) + z\frac{n}{L_z} \right] \right\}$$

$$= \sum_{mn} W_{mn}(x) \exp\left\{ i2\pi \left[\frac{m}{L_y}(y + sxz) + z\frac{n}{L_z} \right] \right\} .$$

Shifting and matching the indices leads to the twisting periodicity condition [11]:

$$W_{m,n+2n_{rat}}(x + L_x) = W_{m,n}(x) \qquad (18)$$

where

$$n_{rat}(m) \equiv \frac{m}{2} \frac{L_x L_z}{L_y} \frac{L_n}{L_s}$$

is constrained to be an integer. One of the dimensions of the simulation domain is quantized by the magnitude of the magnetic shear and the number of twists enclosed in the box. Also, many rational surfaces are forced to lie at the edge of the simulation domain to allow their

358 Nonlinear Gyrofluid Model

periodic continuation elsewhere. For example, the position $x_{rat}(m,n)$ of the rational surface of the (m,n) mode lies where $\hat{\mathbf{b}} \cdot \nabla = 0$. This may be written as

$$x_{rat}(m,n) = \left(1 - \frac{n}{n_{rat}}\right) \frac{L_x}{2} \rho_i$$

(using Eq. (2)). The rational surfaces of modes with $n = \pm n_{rat}$ lie exactly at the edge of the box.

Returning to the simulations, we now show an example with the same physical parameters as the simulation of Fig. 5.4, but with periodic boundary conditions, the nonlinear phase-mixing terms included and the correct adiabatic electron constraint. The nonlinear resolution is given by (64,32,32). The artificial dissipation parameter $\nu = 0.005$.

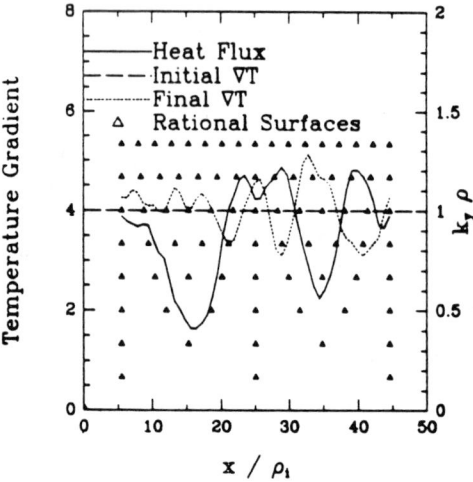

Fig. 5.4 Heat flux vs. x for a simulation with periodic boundary conditions and reasonable resolution.

As a result of the periodic boundary conditions, the temperature gradients are now roughly constant across the box. While there remains a region of reduced thermal flux, there is no clear correlation of this region with the edge or with a region of low resolution. The predicted χ_i from this simulation is $0.7\rho_i^2 v_t/L_n = 9 \times 10^2 \text{cm}^2/\text{sec}$. This is again lower than the experimental value by at least an order of magnitude, but in reasonable agreement with the mixing-length estimate $\chi_{ML} \sim (\gamma/k_\perp^2)_{\max} \sim 0.2\rho_i^2 v_t/L_n$. The addition of toroidal drive would increase the heat flux, perhaps to experimentally observed values.

5.3.4 NONLINEARLY GENERATED PERPENDICULAR SHEAR FLOW

The parameters for the run shown in Figs. 5–6 are $L_x = 40\rho_i$, $L_y = 25\rho_i$, $L_z = 64L_n$, nonlinear resolution $(x, y, z) = (64, 16, 32)$, $L_n/L_s = 0.025$, $\eta_i = 4.0$, $T_i = T_e$, $\nu = 0.1$, no nonlinear phase-mixing, and twisting periodicity in the radial direction. Figure 5.5 shows χ_i vs. tv_t/L_n. The initial peak is quickly quenched by the nonlinearly generated sheared perpendicular flow whose time-averaged radial profile is shown in Fig. 5.6.

Figure 5.7 suggests that the relevant wavelength may be the box-size. While we observe modes with larger $k_x \rho$ with regularity, the flow is characterized by long radial wavelengths. We are still investigating the possibility of box-size scaling of the flow. In Fig. 5.7, the kinetic energy $\langle\langle u_\parallel^2 + (\nabla_\perp \Phi)^2 \rangle\rangle$ for each (k_y, k_z) pair is plotted vs. tv_t/L_n. The mode which is dominant over most of the time of the simulation is the $(k_y = 0, k_z = 0)$ mode. This mode has no linear drive, and hence is a product of $(k_x, k_y, k_z) + (k_x', -k_y, -k_z)$ types of terms from the $\mathbf{E} \times \mathbf{B}$ nonlinearities. The only damping it sees sees in this simulation is $\propto \nu \nabla_\perp^4$, since the nonlinear phase-mixing terms are not included. Since $\nu \sim 0.1$, and the dominant wavelength $(\lambda \sim 10)$ implies $k_x \rho_i \sim 0.4$, we would expect a damping rate $\gamma \sim 2.5 \times 10^{-3}$. This is consistent with the kinetic energy, which falls roughly 3.5e-foldings between $tv_t/L_n = 200$ and 800.

When the $(k_y = 0, k_z = 0)$ mode is sufficiently damped, the remaining modes begin to grow again, driving the perpendicular rotation which damps all competing modes. This recurring process shows no sign of diminishing over very long times. If ν is taken to be small, the $(k_y = 0, k_z = 0)$ mode does not decay significantly, and the turbulent

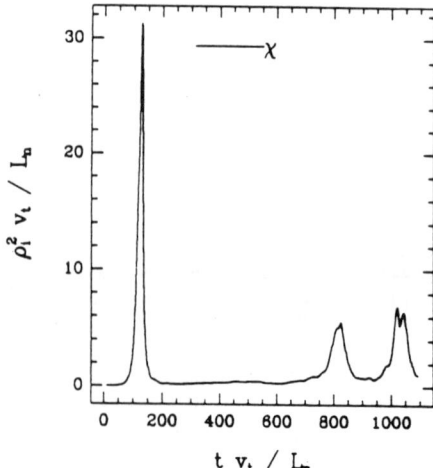

Fig. 5.5 Heat flux vs. tv_t/L_n for a simulation with proper adiabatic electron response.

heat flux is effectively suppressed to very low levels.

5.3.5 WEAK TURBULENCE ANALYSIS

We have begun a weak turbulence analysis of a highly simplified set of gyrofluid equations in an unsheared slab. In this model, nonlinear FLR effects are poorly modeled; though the approximations are well-behaved for $k_\perp \rho$ large, the equations are not even second-order accurate in $k_\perp \rho$ nonlinearly. Here, however, we are concerned primarily with the Landau damping terms. Correcting the FLR effects should not significantly alter these results as long as $k_\perp \rho$ remains small. Furthermore, we present only the "bare" term contributions to the nonlinear ion Compton scattering process. A more complete presentation of this calculation (including the "shielded" contributions) will be reported elsewhere.

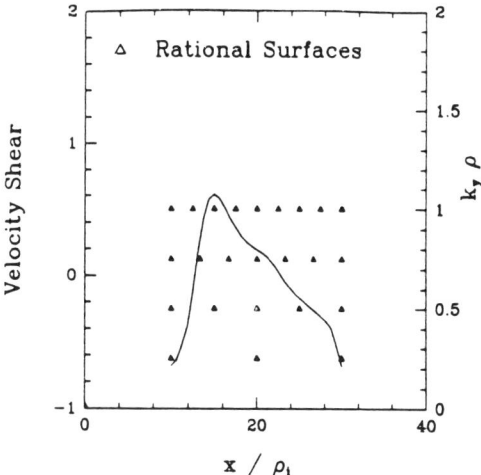

Fig. 5.6 Time-averaged perpendicular velocity shear.

In this section, $\psi \equiv \langle J_0^2 \rangle \Phi = \Gamma_0(b)\Phi$, and consequently

$$\widehat{\nabla}_\perp^2 \Gamma_0 \equiv b \frac{\partial \Gamma_0}{\partial b} \, .$$

Other normalizations and operators are unchanged. The equations we solve are:

$$\frac{\partial n}{\partial t} + \nabla_\parallel u_\parallel + (1 + \eta_{i\perp} \widehat{\nabla}_\perp^2) \frac{\partial \psi}{\partial y} = -\mathbf{v}_\psi \cdot \nabla n \qquad (19)$$

$$\frac{\partial u_\parallel}{\partial t} + \nabla_\parallel (n + T_\parallel) + \nabla_\parallel \psi = -\mathbf{v}_\psi \cdot \nabla u_\parallel \qquad (20)$$

$$\frac{\partial T_\parallel}{\partial t} + \nabla_\parallel 2u_\parallel + \widehat{\chi} |k_\parallel| T_\parallel + \eta_{i\parallel} \frac{\partial \psi}{\partial y} = -\mathbf{v}_\psi \cdot \nabla T_\parallel \qquad (21)$$

Physically, the form of Eqs. (19–21) suggests very strongly that the present approach should be adequate, especially in a weak turbulence regime, in which the linear dynamics are dominant. After all, these equations express the nonlinear conservation of density, momentum and energy, possess a reasonably accurate approximation to the kinetic

362 Nonlinear Gyrofluid Model

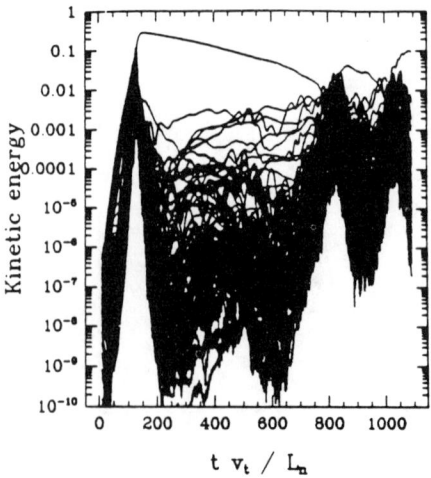

Fig. 5.7 Kinetic energy for each (k_y, k_z) pair vs. tv_t/L_n. The dominant mode is $(k_y = 0, k_z = 0)$.

linear propagator, and have the same driving terms as in the gyrokinetic description. Moreover, there is a physical distribution function for which this closure is linearly exact, implying that this model does not suffer from realizability difficulties.

The simplified Poisson's equation is

$$\tau\psi = \Gamma_0 n + (\Gamma_0 - 1)\psi . \qquad (22)$$

Equations (19–22) yield the local, linear dispersion relation

$$1+\tau+\Gamma_0\left\{\left[\zeta - \zeta_*\left(1 - \eta_i b(1 - \frac{I_1}{I_0})\right)\right] Z_3(\zeta) - \zeta_*\eta_i\left[\zeta + (\zeta^2 - \frac{1}{2})Z_3(\zeta)\right]\right\} = 0$$

which is an excellent approximation to the gyrokinetic linear dispersion relation; the only modification is that the plasma dispersion function Z is replaced by the three-pole approximation Z_3, defined in Ref. [14]. The normalized frequencies are defined as $\zeta \equiv \omega/(\sqrt{2}|k_\parallel|v_t), \zeta_* \equiv \omega_*/(\sqrt{2}|k_\parallel|v_t)$.

For $\gamma \ll \omega$, a weak turbulence expansion may be carried out [24]. We expand the equations in the small parameter γ/ω to third order, and solve for the self-consistent density response. The nonlinear coupling in each of Eqs. (19-21) allows weakly unstable linear modes characterized by $(\omega_\mathbf{k}, \mathbf{k})$ and $(\omega_{\mathbf{k}'}, \mathbf{k}')$ to interact, producing a beat wave with $(\omega_\mathbf{k} - \omega_{\mathbf{k}'}, \mathbf{k} - \mathbf{k}')$. We denote the beat wave quantities with a double prime, remembering that $(\omega'', \mathbf{k}'')$ is not required to be a solution of the linear dispersion relation.

In the drift wave limit, with $\eta_i = 0$ and $\zeta, \zeta' \gg 1$, the gyrofluid density response is

$$n_\mathbf{k}^{(3)} = \sum_{\mathbf{k}=\mathbf{k}'+\mathbf{k}''} \frac{1}{\omega}\left(\frac{\omega_*}{\omega} - \frac{\omega'_*}{\omega'}\right) \frac{\zeta''}{\omega''} Z_3(\zeta'')(\mathbf{k} \times \mathbf{k}' \cdot \hat{b})^2 \mathcal{G}(b, b')|\Phi_{\mathbf{k}'}^{(1)}|^2 \Phi_\mathbf{k}^{(1)} \tag{23}$$

in which

$$\mathcal{G}(b, b') = \Gamma_0(b'_*)\Gamma_0(b) .$$

This result compares favorably with the usual kinetic theory [24],

$$n_\mathbf{k}^{(3)} = \sum_{\mathbf{k}=\mathbf{k}'+\mathbf{k}''} \frac{1}{\omega}\left(\frac{\omega_*}{\omega} - \frac{\omega'_*}{\omega'}\right) \frac{\zeta''}{\omega''} Z(\zeta'')(\mathbf{k} \times \mathbf{k}' \cdot \hat{b})^2 G(b, b')|\Phi_{\mathbf{k}'}^{(1)}|^2 \Phi_\mathbf{k}^{(1)} \tag{24}$$

where the FLR effects are contained in the function

$$G(b, b') = \int_0^\infty dv_\perp \, v_\perp e^{-v_\perp^2} J_0^2(\sqrt{2}k_\perp \rho v_\perp) J_0^2(\sqrt{2}k'_\perp \rho v_\perp) f_M(v_\perp) .$$

The regime of validity of the weak turbulence expansion appears to be quite small for ITG modes [25, 26], but it is still useful in providing another analytic nonlinear test of the gyrofluid approach. Mattor recently concluded that, in this narrow ITG weak-turbulence regime $\zeta, \zeta' \ll 1 \ll \zeta_*, \zeta'_*$, kinetic effects become more pronounced and may not be adequately represented by the 3-moment gyrofluid equations he considered. We cannot yet draw any firm conclusions on this particular topic, as we are still investigating this ITG weak turbulence limit ourselves. In particular, Mattor neglected the induced potential and considered only the "bare" response. It is well-known, however, that in the long-wavelength limit significant cancellation occurs when the complete response in considered [27, 28]. Thus, a more detailed analysis may reveal better agreement between the fluid and kinetic calculations.

On fundamental grounds, we believe that the gyrofluid approach must converge to the proper answer if enough fluid moments are kept. This is because keeping more fluid moments is essentially equivalent to keeping more terms in a Hermite-polynomial expansion of the distribution function. It is known that the Hermite polynomials provide a complete basis set, and that the expansion converges. So the relevant question becomes not *whether* the fluid approach works but *how many* moments are required for it to work. Indeed, Smith and Hammett [29] have recently shown that even nonlinear plasma echoes may be reproduced with fluid equations, though this can require a large number of moments (~ 50, even for an echo which occurs fairly soon). However, we believe that the fine scale structure of velocity space is not important for most types of turbulence relevant to anomalous transport in tokamak experiments, and one probably needs to keep just the first few fluid moments (such as the "3+1" gyrofluid model of Eqs. (6-10)).

5.3.6 CONCLUSIONS

The nonlinear gyrofluid model is probably adequate to model the drift-type ITG turbulence which is a candidate to explain anomalous transport in modern magnetic fusion experiments. The early nonlinear results presented here show the importance of several issues which affect both fluid and kinetic simulations.

The most significant physics issue we discuss which is generic to microturbulence simulation is the correct adiabatic electron response. One must take care to prevent non-physical radial electron transport. In light of this, we showed that if the magnetic shear is weak, "bursty" behavior of the heat flux is observed, with long periods of very low transport when the simulation is dominated by the nonlinearly excited velocity shear, and intermittent bursts of high levels of transport characterized by large numbers of modes interacting. In this sheared-slab geometry, the only physics limiting the ($k_y = 0, k_z = 0$) mode is collisionality, modelled here with crude viscosity and hyperviscosity terms.

We showed that the box size should be chosen so as to spread the rational surfaces evenly throughout the domain, (even near the edges) and that the resolution should be high enough that profile flattening around a few low-$k_y\rho_i$ rational surfaces does not dominate the nonlinear dynamics. In general, of course, one should check one's results by doing convergence studies in the non-physical parameters, such as the geometrical concerns discussed here. Only when the predicted transport is insensitive to the number of (modes, particles, etc.) can a given simulation technique hope to have any predictive power.

We discussed a way to implement periodic radial boundary conditions in a sheared magnetic field. This allows the heat flux leaving the box on the cold side to re-enter the simulation domain on the warm side, mimicking the effect of a heat source and sink. The benefit of this method over simply zeroing out the ($k_y = 0, k_z = 0$) modes was discussed, as we pointed out that these modes may play an important role in the actual physical dynamics, and thus should not be completely neglected.

We presented the "bare" contribution to the ion Compton scattering process from a standard weak turbulence calculation using a highly simplified version of our gyrofluid equations. The result is very similar to the kinetic result, differing only slightly in the FLR effects and in the nonlinear Landau damping rate. The approximation to the resonant contribution to the ion Compton scattering rate was shown to be precisely as accurate as the linear propagator, previously shown to be a reasonable approximation to the full plasma dispersion function. We commented that while Mattor recently discussed a "kinetic" limit in which our model may fail, we are still investigating the contribution of the shielded response to the calculation which he neglected.

Overall, we have found good agreement with some published gyrokinetic particle simulation results, though more work certainly needs to be done on this point. Our method is frugal, since it involves only a few moments of the velocity distribution, and thus may offer a promising route to further investigations of anomalous transport in more realistic geometries.

Bibliography

[1] M. C. Zarnstorff, in *This Conference*.

[2] S. D. Scott, in *This Conference*.

[3] D. Brower, in *This Conference*.

[4] W. Dorland and G. W. Hammett, Phys. Fluids B **5**, 812 (1992).

[5] R. Linsker, Phys. Fluids **24**, 1485 (1981).

[6] W. Dorland, G. W. Hammett, and M. A. Beer, submitted to Phys. Fluids B.

[7] M. A. Beer, G. W. Hammett, W. Dorland, and S. C. Cowley, Bull. Am. Phys. Soc. **37**, 1478 (1992).

[8] M. A. Beer, private communication, 1992.

[9] W. Horton, R. D. Estes, and D. Biskamp, Plasma Physics **22**, 663 (1980).

[10] S. Hamaguchi and W. Horton, Phys. Fluids B **2**, 1833 (1990).

[11] M. Kotschenreuther, Bull. Am. Phys. Soc. **36**, 2435 (1991).

[12] E. A. Frieman and L. Chen, Phys. Fluids **25**, 502 (1982).

[13] W. W. Lee, Phys. Fluids **26**, 556 (1983).

[14] G. W. Hammett and F. W. Perkins, Phys. Rev. Lett. **64**, 3019 (1990).

[15] A. Dimits, J. F. Drake, P. N. Guzdar, and A. B. Hassam, Phys. Fluids B **3**, 620 (1991).

[16] A. M. Dimits and W. W. Lee, submitted to J. Comput. Phys.

[17] M. Kotschenreuther, Bull. Am. Phys. Soc. **33** (1988).

[18] S. E. Parker and W. W. Lee, Phys. Fluids B **5**, 77 (1992).

[19] R. D. Sydora, T. S. Hahm, W. W. Lee, and J. M. Dawson, Phys. Rev. Lett. **64**, 2015 (1990).

[20] M. Kotschenreuther, H. L. Berk, R. Denton, S. Hamaguchi, W. Horton, C.-B. Kim, M. LeBrun, P. Lyster, S. Mahajan, W. H. Miner, P. J. Morrison, D. Ross, T. Tajima, J. B. Taylor, P. M. Valanju, H. V. Wong, S. Y. Xiao, and Y.-Z. Zhang, in *Plasma Physics and Controlled Nuclear Fusion Research, 1990*, volume 2, p. 361, International Atomic Energy Agency, Vienna, 1991.

[21] B. Cohen, private communication, 1993.

[22] K. V. Roberts and J. B. Taylor, Phys. Fluids **8**, 315 (1965).

[23] S. C. Cowley, R. M. Kulsrud, and R. Sudan, Phys. Fluids B **3**, 2767 (1991).

[24] R. Z. Sagdeev and A. A. Galeev, Nonlinear plasma theory, Benjamin, New York, 1969, T. M. O'Neil and D. L. Book, Eds.

[25] N. Mattor and P. H. Diamond, Phys. Fluids B **1**, 1980 (1989).

[26] N. Mattor, Phys. Fluids B **4**, 3952 (1992).

[27] L. Chen, R. L. Berger, J. G. Lominadze, M. N. Rosenbluth, and P. H. Rutherford, Phys. Rev. Lett. **39**, 754 (1977).

[28] J. A. Krommes, Phys. Fluids **23**, 736 (1980).

[29] S. A. Smith, private communication, 1992.

Chapter 6
STABILITY AND TRANSPORT THEORY

6.1 KINETIC TOROIDAL ANALYSIS OF TRANSPORT TRENDS IN TFTR PLASMAS

G. Rewoldt and W.M. Tang
Plasma Physics Laboratory, Princeton University, Princeton, New Jersey 08543

Abstract. Low frequency drift-type microinstabilities driven by trapped-electron and ion temperature gradient dynamics continue to be among the more strongly supported candidates to account for the anomalous transport observed in tokamak plasmas. The possible relevance of these instabilities is addressed by applications of realistic kinetic stability calculations in toroidal geometry using parameters specific to TFTR confinement experiments of interest. With regard to examining the basic question of how sensitive transport might be to spatial gradient variations, representative L-mode cases have been analyzed by artificially varying the local density and temperature gradients for each species.

6.1.1 INTRODUCTION

Given the continuing possible relevance of drift-type microinstabilities for tokamak-type confinement, it is useful to apply a kinetic stability analysis to the interpretation of confinement trends, using a realistic TFTR case as a starting point. The principal theoretical tool employed here is a comprehensive kinetic toroidal microinstability code[1, 2] for high-n (toroidal mode number) tokamak instabilities. Using physical parameters for density, density gradient, electron temperature, electron temperature gradient, ion temperature, and ion temperature gradient specific to TFTR confinement experiments of interest, the linear eigenfrequencies and spectral properties, along with the corresponding quasilinear particle and energy fluxes, are systematically calculated for toroidal electrostatic drift-type modes destabilized by the combined effects of all of these gradients, in the presence of magnetically trapped particles. The results are used to estimate confinement trends as these parameters are varied individually.

The comprehensive toroidal kinetic code employed here for computing linear and quasilinear properties of high-n toroidal microinstabilities has been described already in detail in Refs. [1] and [2]. This calculation employs the 'ballooning representation' to obtain eigenmode solutions including the effects of all relevant resonances, including transit frequency resonances for untrapped particles of each species, bounce frequency resonances for trapped particles, and magnetic curvature and gradient precession drift resonances. It also includes finite Larmor radius effects to all orders and finite banana orbital dynamics. It includes finite β effects in the MHD equilibrium and can include electromagnetic effects for the perturbations, though the results presented here are obtained in the electrostatic limit (the electromagnetic corrections are rather small for this case).

The results presented here are obtained only to lowest order in $1/n$, within the context of the ballooning representation, which means that the analysis is local to a single, chosen magnetic surface. The code is interfaced with realistic, numerically calculated MHD equilibria. It can employ non-Maxwellian equilibrium distribution functions for energetic particle species when appropriate, but, for the results presented here, Maxwellian distributions are chosen for all species, including the hot beam ion species.

The case considered here employs data from the TFTR L-mode shot 42990 at $t = 4.4$ sec as the starting point. The magnetic surface chosen for the instability calculation is at $r = 50$ cm; the gradient and parameter values are varied individually just on this one magnetic surface, so that the MHD equilibrium can remain fixed. The local starting values for the parameters on this surface are: $T_e = 2.1396$ keV, $T_i = 2.2885$ keV $n_e = 3.01253 \times 10^{13}$ cm^{-3}, $q = 1.7218$, $\nu_e^* = 0.09643$, $\eta_e = 2.5965$, and $\eta_i^e \equiv (d \ln T_i/dr)/(d \ln n_e/dr) = 2.2804$. Electrons, deuterium ions, and sometimes carbon impurity ions are included as background species, and also sometimes a hot deuterium beam species is included.

For these results, the quasilinear particle and energy fluxes are calculated with the arbitrary choice of a mixing-length-type saturation level for the perturbed electrostatic potential, $e|\tilde{\phi}|/T_e = 1.0/k_\perp r_p$ (or else $1.0/k_\theta r_p$). Any variation of the saturation level with the local equilibrium parameters can cause additional variation of these particle and energy fluxes. However, ratios of fluxes will not change.

6.1.2 RESULTS

The variation of the linear growth rate and real frequency of the toroidal drift mode, maximized over n or $k_\theta \rho_i$, with the ion temperature gradient, normalized to the experimental value, is shown in Fig. 1. The growth rate is seen to increase monotonically, and almost linearly, with the ion temperature gradient, at fixed density gradient and electron temperature gradient. The corresponding real frequency is in the ion diamagnetic direction, and increases in magnitude similar to the growth rate. This is the expected behavior when the ion temperature gradient, or η_i, instability mechanism is dominant, for η_i^e sufficiently large, as was seen previously in Ref. 3. The associated particle fluxes, shown in Fig. 2, have a somewhat more complicated dependence. The electron particle flux Γ_e in fact changes sign from inward to outward as η_i^e increases. The energy fluxes are shown in Fig. 3; the ion energy flux is outward and dominant for $\eta_i^e \gtrsim 4$, but is inward and small for $\eta_i^e < 3$. This behavior is also consistent with that seen in Ref. 3.

The variation of the growth rate (maximized over n or $k_\theta \rho_i$) and real frequency with the electron temperature gradient, normalized to the experimental value, is shown in Fig. 4, for fixed density gradient and ion temperature gradient. The growth rate is seen to depend rather weakly on the electron temperature gradient, as would be expected in this case with $\eta_i^e > 2$, though the real frequency is somewhat more sensitive to it. The electron and background deuterium ion particle fluxes, shown in Fig. 5, appear to be more strongly affected by changes in η_e, with the latter again changing sign. The corresponding energy fluxes, shown in Fig. 6, are not so strongly affected by the changes in η_e.

The variation of the growth rate (maximized over n or $k_\theta \rho_i$) and real frequency with the density gradient, normalized to the experimental value, is shown in Fig. 7, for fixed electron and ion temperature gradients. The growth rate is seen to exhibit some variation with density gradient, though it is not monotonic, while the real frequency increases almost linearly with density gradient. It also changes sign from the ion to the electron diamagnetic direction as the density gradient increases, so that η_i^e decreases, in a familiar fashion.[3] The particle fluxes display an interesting variation in Fig. 8, with the electron and the background deuterium ion fluxes both changing sign as the density gradient increases. The energy fluxes, shown in Fig. 9, also show more variation than the growth rate, and the background deuterium ion energy flux changes sign.

The variation of the linear growth rate and real frequency with temperature is shown in Fig. 10. Here, the carbon impurity and hot beam species are omitted, and $T_i/T_e = 1.07$ and $k_\theta \rho_i = 0.47$ are held fixed. Even though T_i is varying proportional to T_e in this figure, the variation of the eigenfrequency and the fluxes dominantly come from T_e here. The growth rate shows standard regimes, varying as $T_e^{3/2}$ at low T_e (high collisionality) for the dissipative trapped-electron mode regime and as $T_e^{1/2}$ at high T_e (low collisionality) for the collisionless trapped-electron mode regime. The associated particle and energy fluxes are shown in Fig. 11. The particle flux is outward at low T_e where the dominant collisional dissipation is positive, while it becomes inward at high T_e where the dominant resonant dissipation can be negative. However, the energy fluxes for both electrons and ions are outward. With the assumed form for the saturation level, at high T_e, Q_e and Q_i both vary as $T_e^{5/2}$ and $\Gamma_e = \Gamma_i$ varies as $T_e^{3/2}$. At low T_e, Q_e varies as $T_e^{9/2}$ and Q_i varies as $T_e^{7/2}$, while $\Gamma_e = \Gamma_i$ has a more complicated dependence. Also, Q_i is larger than Q_e over the entire range of T_e for this case.

The variation of the growth rate and real frequency with T_i is shown in Fig. 12 for a different TFTR shot, namely the supershot 35782. Here, $T_e = 6.38\,\text{keV}$ and $k_\theta \rho_i = 0.641$ are kept fixed, though this value of $k_\theta \rho_i$ is always close to that which maximizes the growth rate over this range of T_i. The interesting result here is that increasing ion temperature decreases the growth rate monotonically, contrary to the result in Fig. 10 for increasing electron temperature. This sort of stabilization with increasing T_i may help explain the experimentaly observed decrease in χ_i with increasing T_i in TFTR supershots.[4] Part of this stabilizing effect comes from the corresponding increase in the plasma β, and thus in the Shafranov shift, which decreases the amount of 'bad' curvature available to drive the more, as can be seen by comparing the curves labeled "Varying MHD Equilibrium" and "Frozen MHD Equilibrium" in Fig. 12. Additionally, increasing T_i increases the ion transit frequency and the ion magnetic drift frequency, which affect both the resonant (Landau damping) and nonresonant ion responses. The real frequency is changed from the electron diamagnetic direction to the ion diamagnetic direction by increasing ion temperature.

The effect of varying the plasma density, which in this parameterization affects only the dimensionless collisionality parameter $\nu_e^* \equiv$ effective trapped-electron collision frequency/average trapped-electron bounce frequency $\propto n_j$, is shown in Fig. 13 for the TFTR L-mode shot 42990. Here, for $\nu_e^* < 0.1$ the collisionless trapped electron mode instability mechanism, which is the trapped-electron orbit-average magnetic drift (toroidal precession) frequency resonance, is dominant, and for $\nu_e^* > 0.1$ the dissipative trapped electron mode (trapped-electron collision) mechanism is dominant. In this case there are separate maxima in the growth rate curve for these two mechanisms, while more typically there is a monotonic stabilization with increasing n_j blending together both mechanisms. The mode becomes completely stabilized in the plateau regime, where $\nu_e^* > 1$. The real frequency is moved from the ion diamagnetic direction to the electron diamagnetic direction by increasing density and collisionality in this case.

6.1.3 CONCLUSIONS

The linear growth rate of the toroidal drift mode, which is destabilized by the combined effects of the temperature and density gradients and of the trapped electrons, is most sensitive to the ion temperature gradient (almost linear), but much less sensitive to the electron temperature gradient and the density gradient (no secular trend), as is expected when the ion temperature gradient instability mechanism is dominant. The quasilinear particle and energy fluxes have a more complicated dependence on the gradients. For instance, the *sign* of Γ_e can be changed by changing *any* of the three gradients. Increasing electron temperature is destabilizing, while increasing ion temperature is stabilizing. Increasing density (and thus collisionality) is generally stabilizing. These results for this trapped-electron-η_i mode are calculated for a particular

TFTR L-mode case. In general, the results, especially for the fluxes, are relatively sensitive to the input parameter and gradient values in a complicated way, so that the numerical calculation really needs to be done over again for each new case (*i.e.* shot). Simple generalizations are not supported by the results obtained so far for this and other cases.

ACKNOWLEDGMENTS

The authors would like to thank Drs. P. Efthimion and M. Zarnstorff for providing the TFTR experimental data used as input data for these calculations. This work was supported by United States Department of Energy Contract No. DE-AC02-76-CHO-3073.

Bibliography

[1] G. Rewoldt, W.M. Tang, and M.S. Chance, Phys. Fluids **25**, 480 (1982).

[2] G. Rewoldt, W.M. Tang, and R.J. Hastie, Phys. Fluids **30**, 807 (1987).

[3] G. Rewoldt and W.M. Tang, Phys. Fluids B **2**, 318 (1990).

[4] D.M. Meade, V. Arunasalam, C.W. Barnes, M.G. Bell, R. Bell, et al., in *Proceedings of the Thirteenth International Conference on Plasma Physics and Controlled Nuclear Fusion Research* (International Atomic Energy Agency, Vienna, 1991), Vol. 1, p. 9.

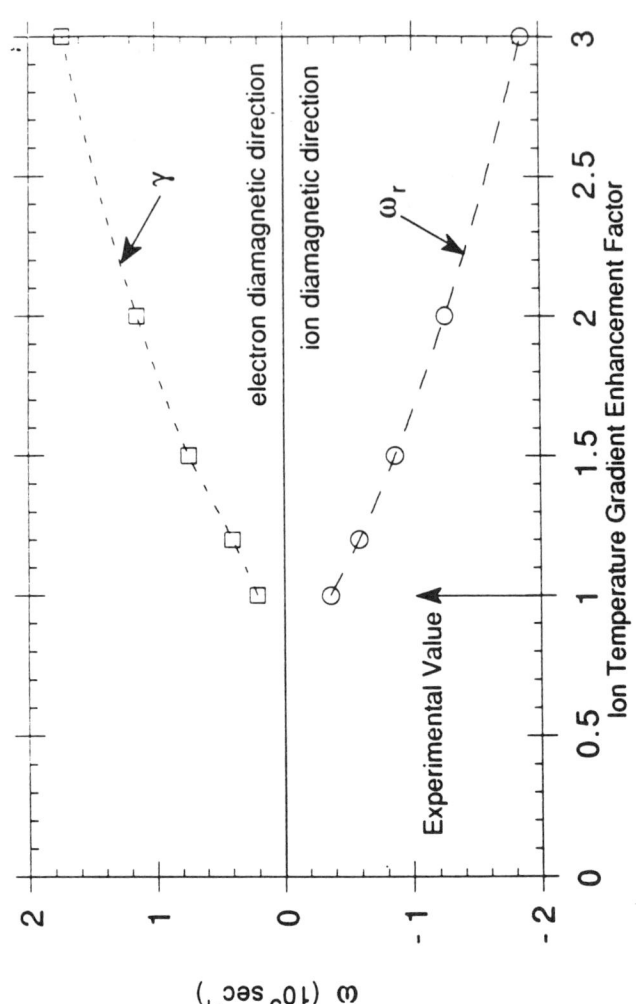

Fig. 1. Variation of growth rate (maximized over n or $k_\theta \rho_i$) and real frequency of toroidal drift mode with ion temperature gradient, based on TFTR L-mode shot 42990 at r = 50 cm and t = 4.4 sec, including carbon and Maxwellian beam, with MHD equilibrium held fixed, and density and electron temperature gradients held fixed.

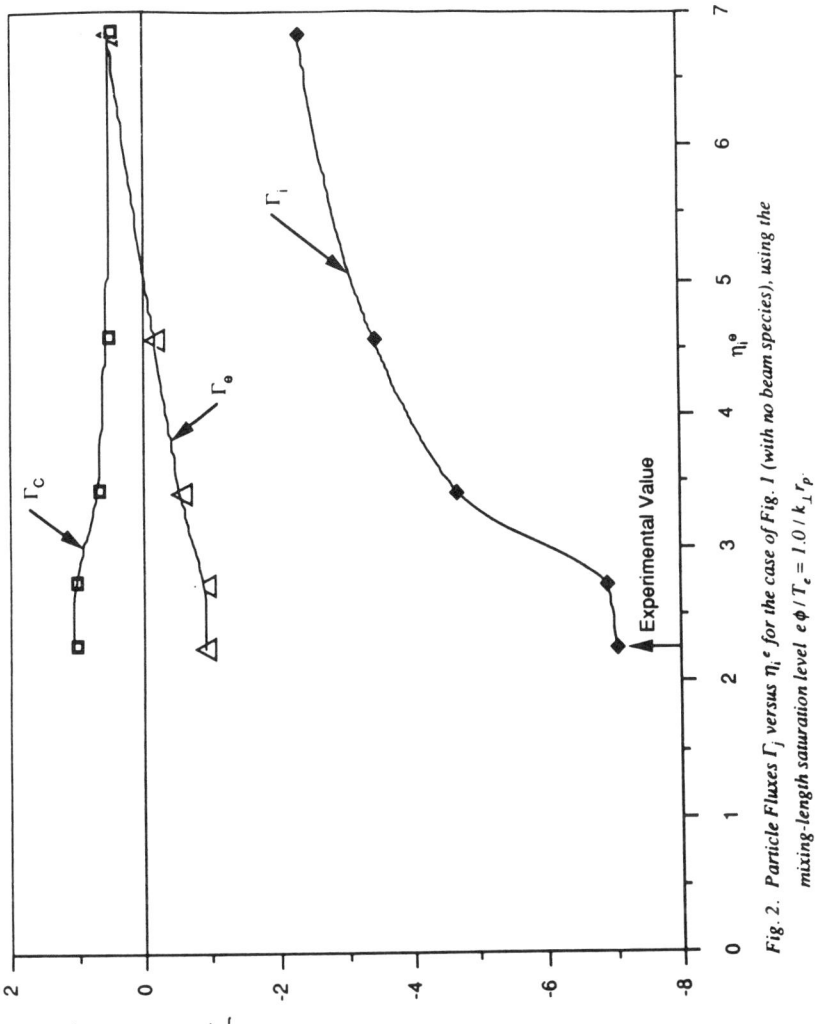

Fig. 2. Particle Fluxes Γ_j versus η_i^e for the case of Fig. 1 (with no beam species), using the mixing-length saturation level $e\phi/T_e = 1.0/k_\perp r_p$.

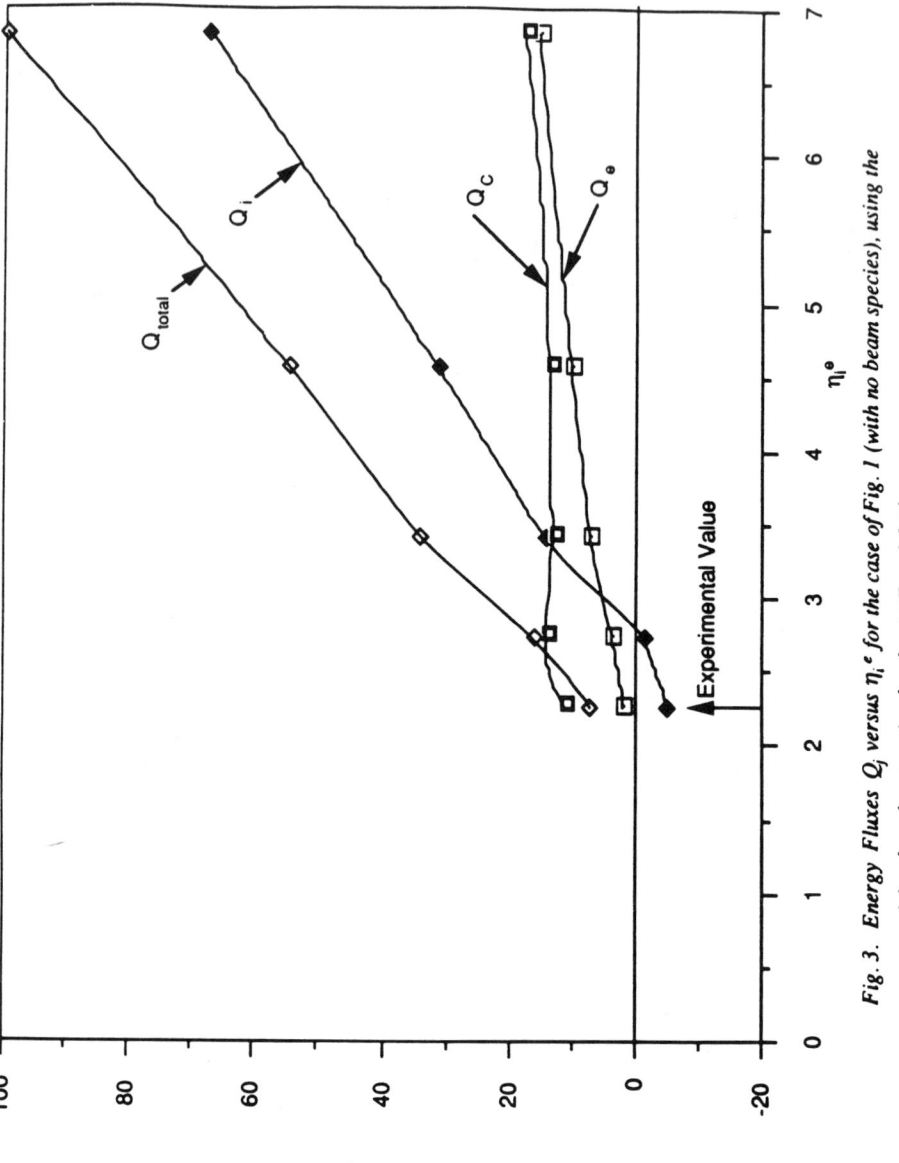

Fig. 3. Energy Fluxes Q_j versus η_i^e for the case of Fig. 1 (with no beam species), using the mixing-length saturation level $e\phi/T_e = 1.0/k_\perp r_p$.

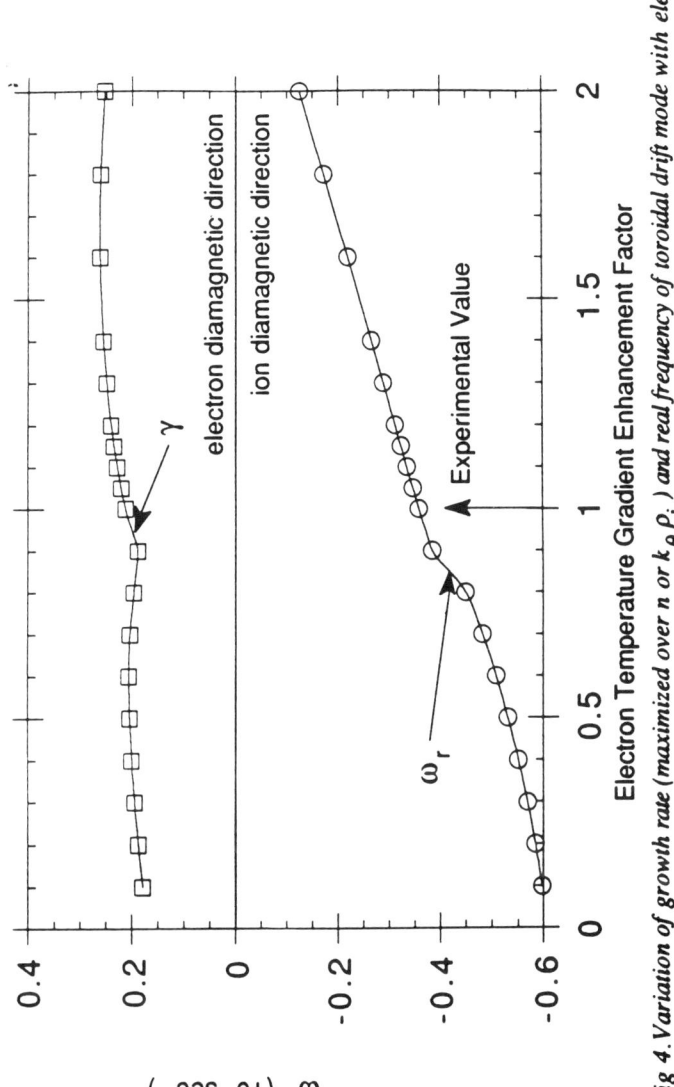

Fig 4. Variation of growth rate (maximized over n or $k_\theta \rho_i$) and real frequency of toroidal drift mode with electron temperature gradient, based on TFTR L-mode shot 42990 at r = 50 cm and t = 4.4 sec, including carbon and Maxwellian beam, with MHD equilibrium held fixed, and density and ion temperature gradients held fixed

382 Kinetic Toroidal Analysis

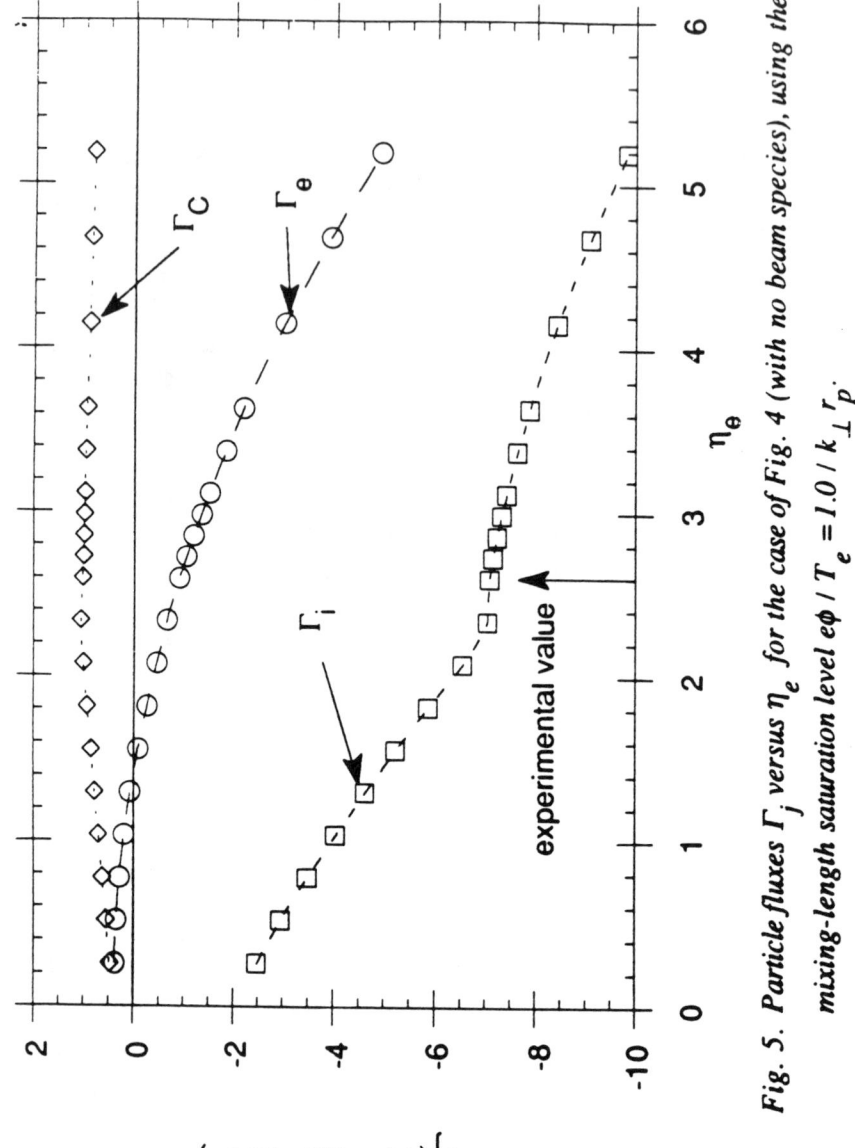

Fig. 5. Particle fluxes Γ_j versus η_e for the case of Fig. 4 (with no beam species), using the mixing-length saturation level $e\phi/T_e = 1.0/k_\perp r_p$.

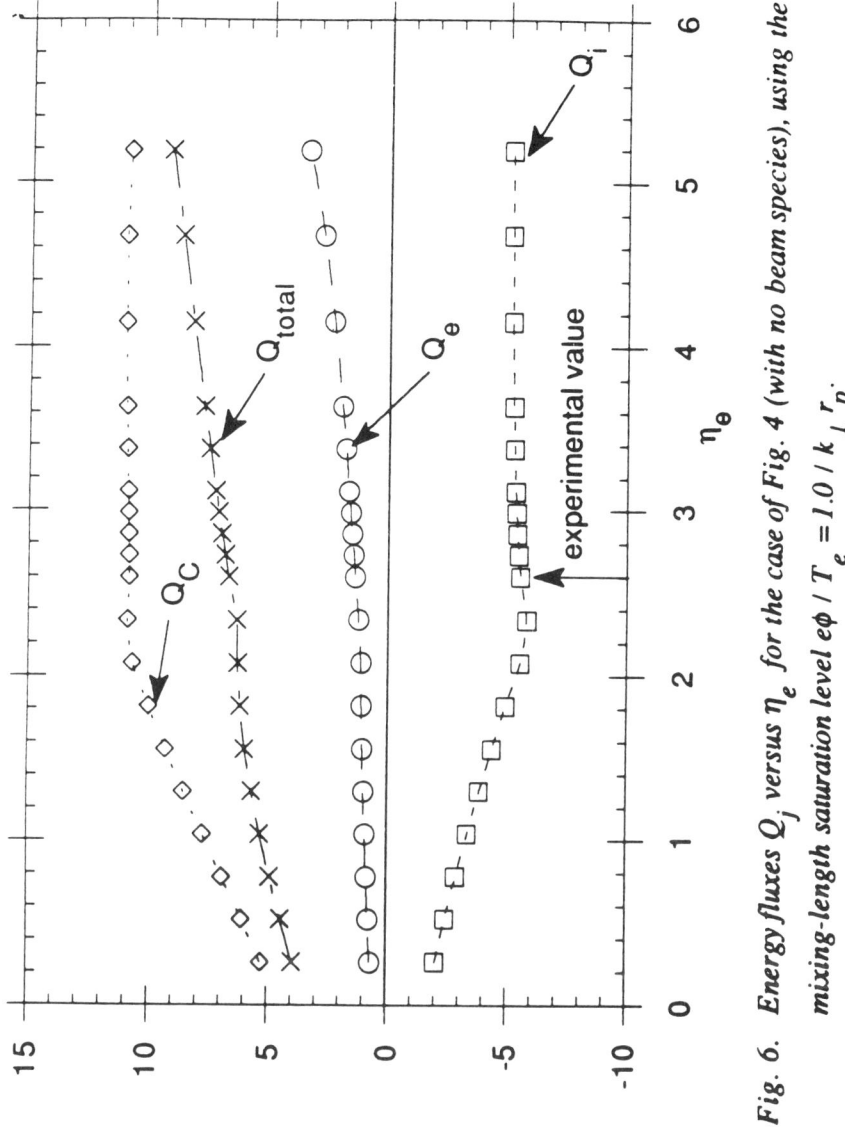

Fig. 6. *Energy fluxes Q_j versus η_e for the case of Fig. 4 (with no beam species), using the mixing-length saturation level $e\phi/T_e = 1.0/k_\perp r_p$.*

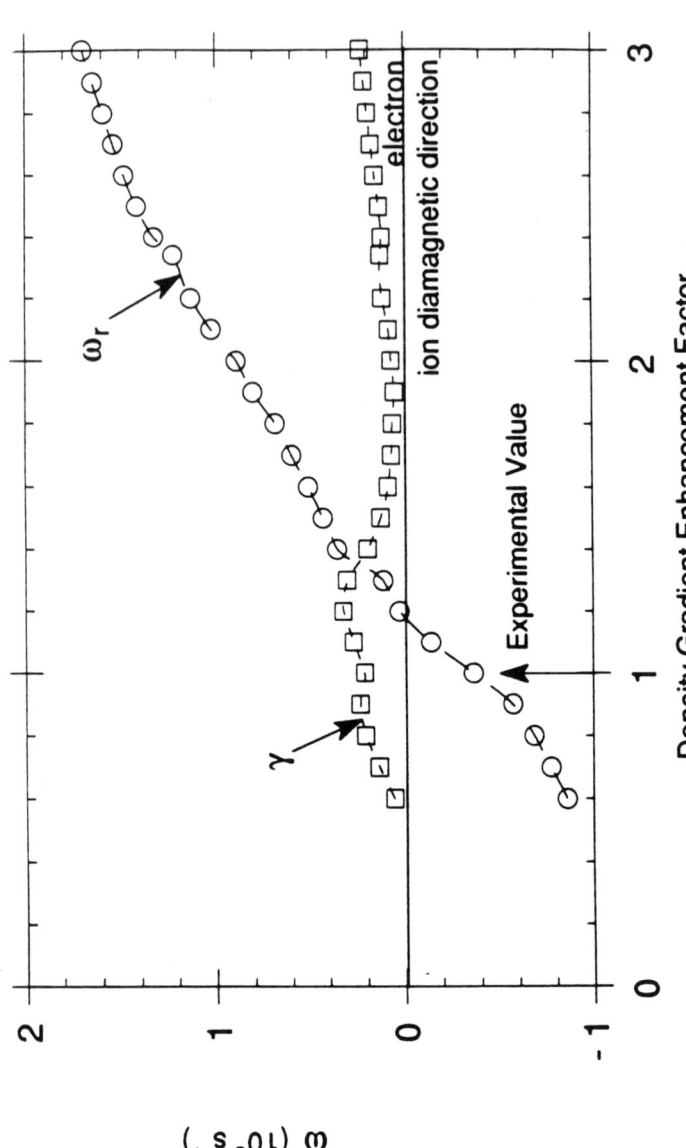

Fig. 7. Variation of growth rate (maximized over n or $k_\theta \rho_i$) and real frequency of toroidal drift mode with density gradient, based on TFTR L-mode shot 42990 at r = 50 cm and t = 4.4 sec, including carbon and Maxwellian beam, with MHD equilibrium held fixed, and temperature gradients held fixed.

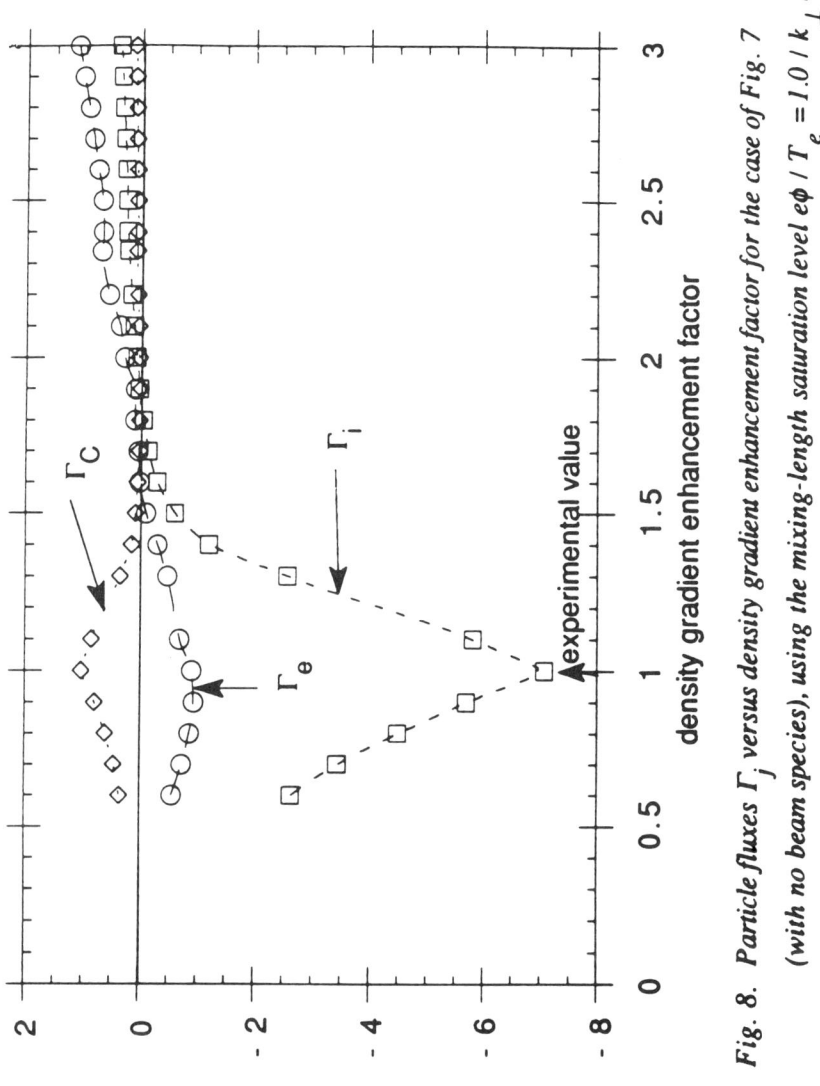

Fig. 8. Particle fluxes Γ_j versus density gradient enhancement factor for the case of Fig. 7 (with no beam species), using the mixing-length saturation level $e\phi/T_e = 1.0/k_\perp r_p$.

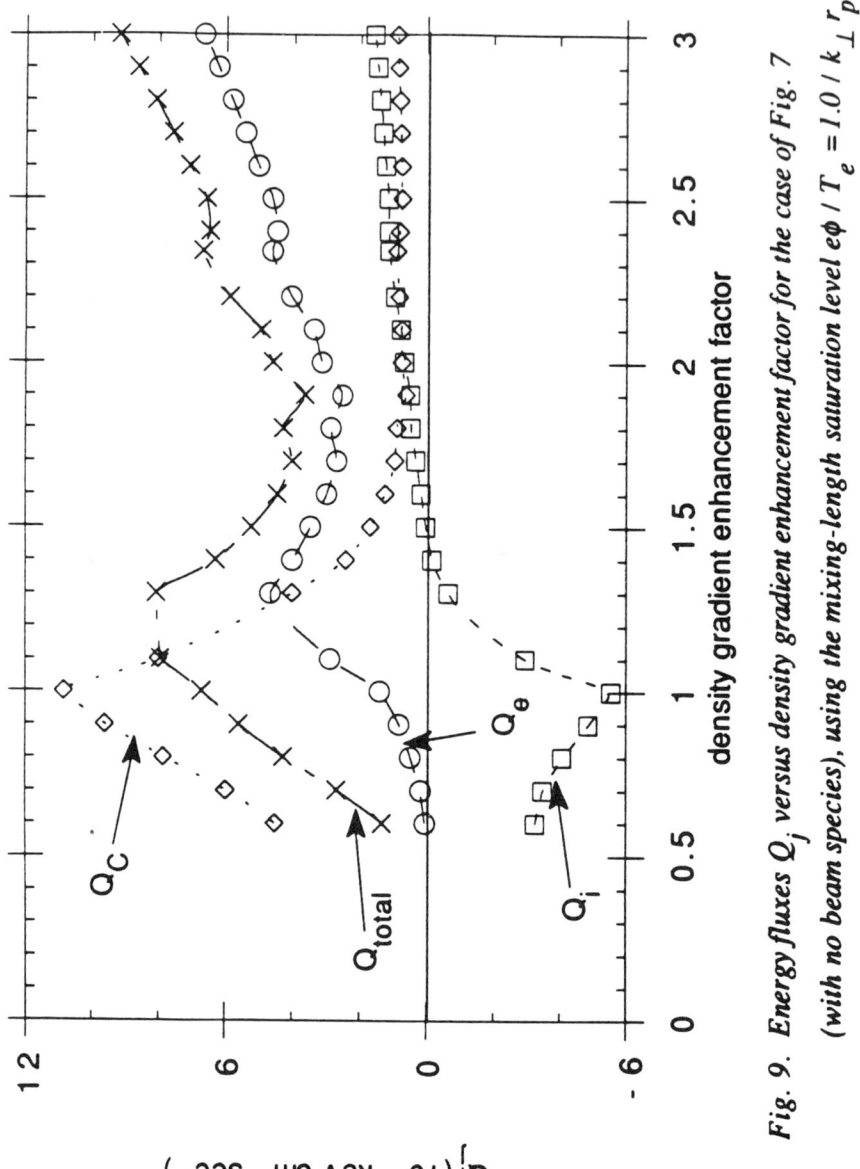

Fig. 9. Energy fluxes Q_j versus density gradient enhancement factor for the case of Fig. 7 (with no beam species), using the mixing-length saturation level $e\phi/T_e = 1.0/k_\perp r_p$.

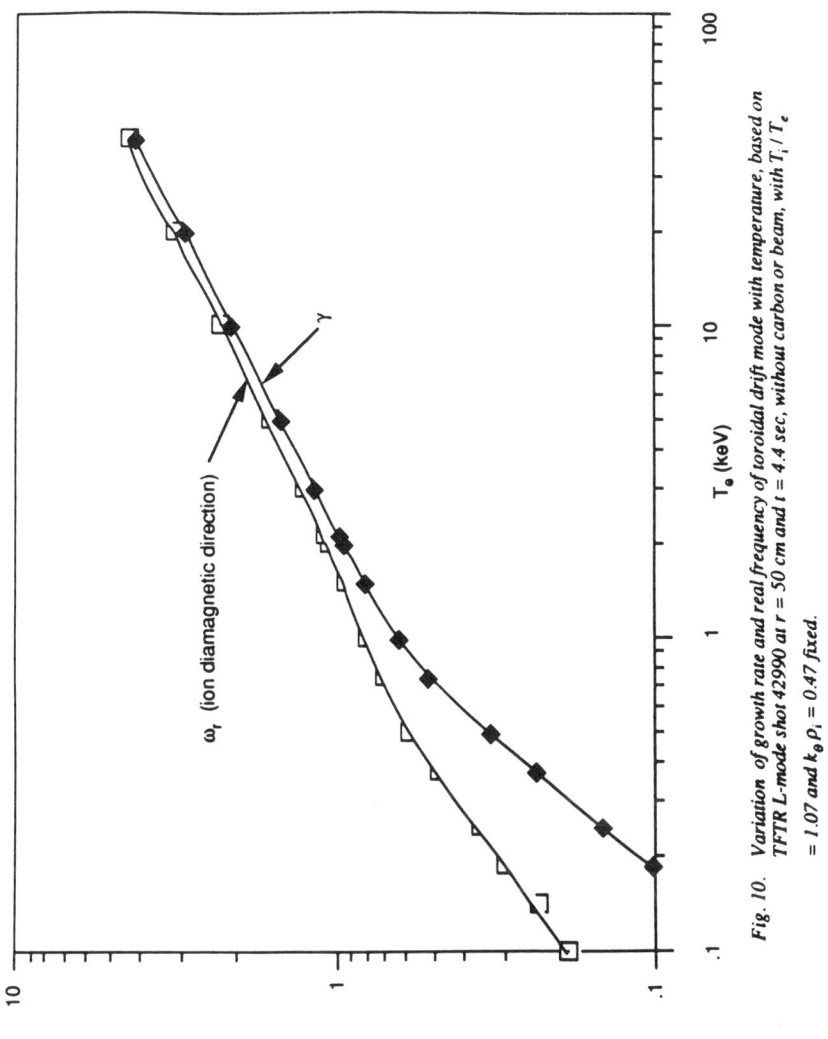

Fig. 10. Variation of growth rate and real frequency of toroidal drift mode with temperature, based on TFTR L-mode shot 42990 at $r = 50$ cm and $t = 4.4$ sec, without carbon or beam, with $T_i/T_e = 1.07$ and $k_\theta \rho_i = 0.47$ fixed.

388   Kinetic Toroidal Analysis

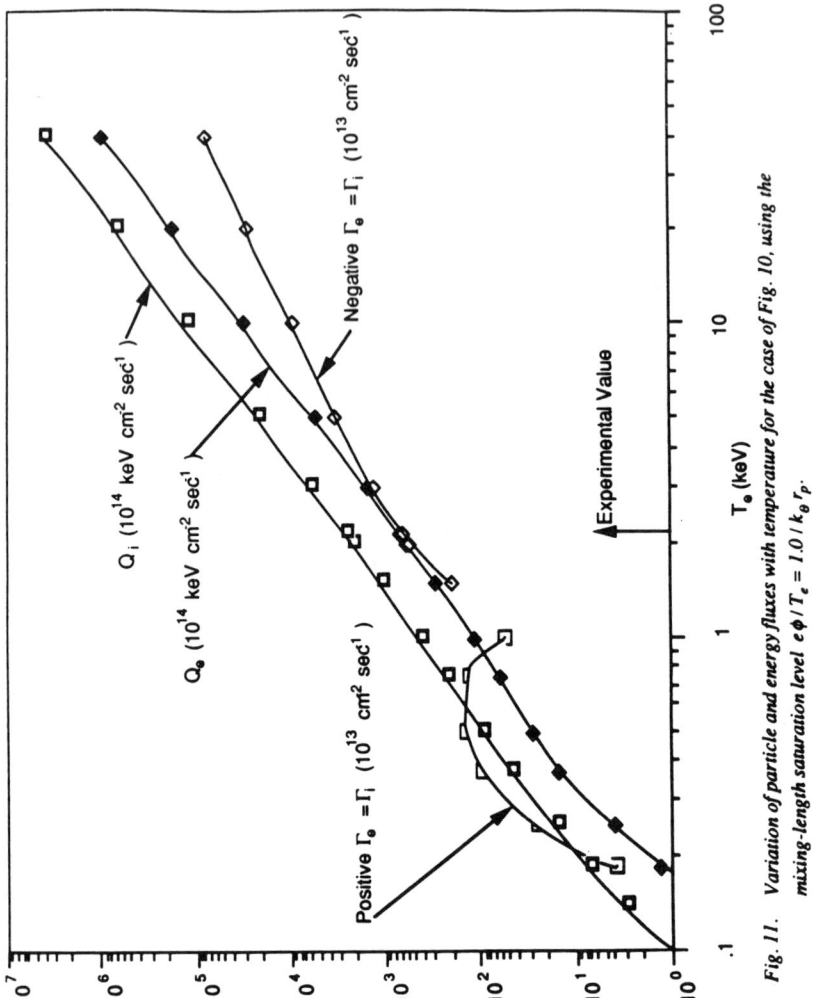

Fig. 11. *Variation of particle and energy fluxes with temperature for the case of Fig. 10, using the mixing-length saturation level* $e\phi/T_e = 1.0/k_\theta r_p$.

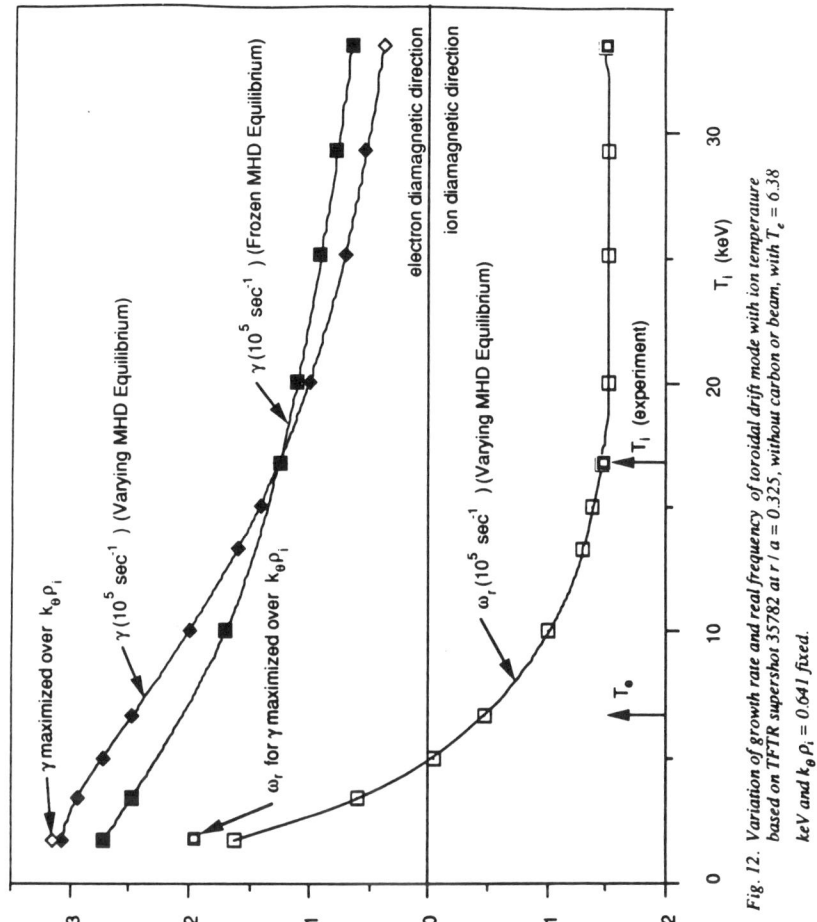

Fig. 12. Variation of growth rate and real frequency of toroidal drift mode with ion temperature based on TFTR supershot 35782 at $r/a = 0.325$, without carbon or beam, with $T_e = 6.38$ keV and $k_\theta \rho_i = 0.641$ fixed.

390 Kinetic Toroidal Analysis

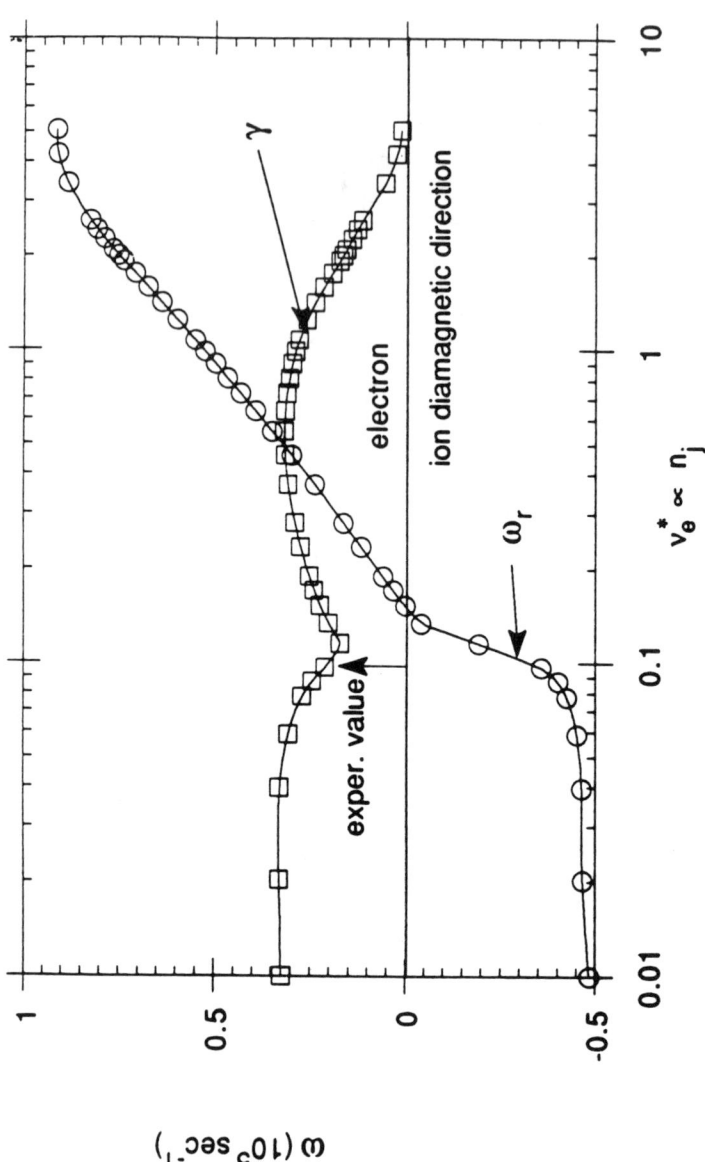

Fig. 13. *Variation of growth rate (maximized over n or $k_\theta \rho_i$) and real frequency of toroidal drift mode with density, based on TFTR L-mode shot 42990 at r = 50 cm and t = 4.4 sec, including carbon and Maxwellian beam, with MHD equilibrium held fixed.*

6.2 PROGRESS IN THE UNDERSTANDING OF THE ION TEMPERATURE GRADIENT DRIVEN TRANSPORT

F. Romanelli
Associazione EURATOM-ENEA sulla Fusione,
C.R.E. Frascati, C.P. 65
00044, Frascati, Roma, Italy

A b s t r a c t. *An analysis of the radial structure of the ion temperature gradient driven mode is presented and the dependence of the radial correlation length L_r on parameters such as magnetic shear is discussed. It is found that L_r decreases algebraically with increasing shear for moderate shear values, and it decreases exponentially with decreasing shear for low shear values. These results seem in qualitative agreement with several experiment which observe strong reduction of the transport coefficients close to the magnetic axis.*

6.2.1 INTRODUCTION

Over the past few years the ion temperature gradient driven mode (η_i-mode $\eta_i = d\ln T_i/d\ln n$) dynamics has been investigated in different parameter ranges, as low and moderate magnetic shear ($s = rq'/q$ with q being the safety factor), short and long wavelengths, peaked and flat density profiles [1, 2, 3, 4]. Even though the theoretical understanding has significantly improved, a model for the energy transport capable of accounting for all the parameter dependances has not yet been derived. Among the various problems which arise in the comparison between experiment and mixing-length estimates of the diffusion coefficients, the most apparent discrepancy is in the radial dependence of the thermal conductivity χ. Indeed, while a monotonic increase toward the edge is observed in the experiments, the theoretical models predict a decrease in the outer part of the discharge. Such a disagreement can be traced back to the estimate given for the characteristic step size of the diffusion process, which can be identified with the mode radial correlation length L_r. Such a quantity is usually estimated as the perpendicular wavelength of the mode which is of the order of the ion Larmor radius ρ_i, and therefore decreases very rapidly toward the edge.

In order to understand whether such an estimate is correct, it is worthwhile recalling here that in a tokamak the perturbed electrostatic potential can be written as a superposition of different poloidal harmonics coupled by the effect of toroidicity. The estimate for L_r discussed above corresponds to identify the radial correlation length with the width of each harmonic. However such an estimate is incorrect, and, in a toroidal system, L_r should instead be identified as the radial range of correlation between different harmonics. This has two major consequences. First, L_r is expected to be much larger than ρ_i, and indeed there is growing experimental evidence in favour of radially elongated turbulence [5]. Second, magnetic shear can be expected to play a major role in determining L_r. It is interesting to note that several experiments have focused the attention on the magnetic shear dependence of the energy transport. To be specific, the TFTR current ramp experiment [6] has shown that the ion thermal diffusivity decreases with increasing shear in the outer part of a discharge where the magnetic shear is larger than unity. On the other hand, the JET pellet enhanced performance [7] has been obtained only in the presence of a very low (or inverted) shear region. Furthermore, many experiments observe a strong reduction in χ close to the magnetic axis, where s is small, if the effect of sawteeth is suppressed [8, 9].

In order to determine the radial correlation length, the solution of the eigenmode problem at the first order in the ballooning representation is required [10]. It should be noted that, with very few exceptions [11, 12, 13], nobody has addressed this issue in the past. The primary aim of this paper is to determine the radial structure of the η_i-mode eigenfunctions both for low and moderate shear values.

This paper is organized as follows: in Sec. 2 the general formalism is presented. In Sec. 3 the radial problem is solved for the short wavelength, moderate shear case. In Sec. 4 the same problem is solved for the long wavelength, low shear case. Concluding remarks are presented in Sec. 5.

6.2.2 GENERAL FORMALISM

In a two-dimensional system a generic perturbation can be expressed as a superposition of different poloidal harmonics.

$$\phi = \sum_p \phi_p(x) e^{i[-(m_o+p)\theta + n\zeta - \omega t]} \tag{1}$$

where ϕ is, for example, the electrostatic potential, n the toroidal mode number, ζ (θ) the toroidal (poloidal) angle, ω the mode frequency and $x = r - r_o$ the radial variable with r_o being a reference radius, m_o being defined by $q(r_o) = m_o/n$, and q the safety factor.

Within the context of the ballooning mode representation [10] it is assumed that neighboring poloidal harmonics ϕ_p have similar shape, simply shifted by an amount $1/nq'$ which corresponds to the distance between neighboring rational surfaces. Such a structure, formed by a superposition of harmonics with similar shape, is modulated by a slowly varying envelope function $A(x)$ which determines the range of the toroidicity-induced coupling

$$\phi_p(x) = A(x)\phi_o(x - p/nq') . \tag{2}$$

It is generally assumed that the radial variation of $A(x)$, even though slower than the radial variation of ϕ_o, remains faster than the variation of the equilibrium quantities. Thus an eikonal representation for $A(x)$ can be employed

$$A(x) = e^{i\int k_x(x')dx'} = e^{inq'\int \theta_o(x')dx'} \tag{3}$$

where the radial wave vector k_x has been normalized to nq'.

The solution of the global eigenmode problem is thus accomplished in two steps. At the lowest order in the ballooning mode problem the shape of the function ϕ_o is determined, whilst at the first order the envelop $A(x)$ is obtained. It is convenient to define the Fourier transform of the function ϕ_o as $\phi_o(x) = \int \hat{\phi}_o(\theta)\exp(-i\theta x)d\theta$. On substituting Eqs. (1) to (3) into the quasineutrality equation in the fluid approximation the following differential equation is obtained

$$\frac{\omega_{ti}^2}{\omega^2}\frac{\partial^2 \hat{\phi}_o}{\partial \theta^2} + \left[\frac{\frac{1}{\tau} + \frac{\omega_{*i}}{\omega}}{1 - \frac{\omega_{*pi}}{\omega}} + (k_\theta \rho_i)^2(1 + s^2(\theta - \theta_o)^2) - \frac{\omega_D}{\omega}(\cos\theta + s(\theta - \theta_o)\sin\theta)\right]\hat{\phi}_o = 0 \tag{4}$$

with $\omega_D = 2ck_\theta T_i/(eBR)$, $\tau = T_e/T_i$, $\omega_{*pi} = \omega_{*i}(1+\eta_i)$, $\omega_{ti} = v_{ti}/2^{1/2}qR$, $\omega_{*i} = -2ck_\theta T_i/(eBL_n)$, $L_n^{-1} = -\nabla n_o/n_o$, $\eta_i = d\ln T_i/d\ln n$, B is the equilibrium magnetic field, R is the major radius, $v_{ti} = (2T_i/m_i)^{1/2}$ is the thermal velocity and $\rho_i = v_{ti}/2^{1/2}\Omega_i$ is the ion Larmor radius. The usual low-β equilibrium with circular magnetic surfaces is employed.

With the boundary condition of exponentially vanishing solution for $|\theta| \to \infty$, the above equation determines a one dimensional eigenvalue problem, with an associated dispersion relation of the form

$$F(\omega, x, \theta_o)A(x) = 0 \ . \tag{5}$$

In most of the analysis of the η_i-mode x and θ_o are set to zero and a local eigenvalue is determined. A proper analysis [15, 16] should instead solve the pseudo-differential problem obtained from Eq. 5 with the substitution $\theta_o = -i(1/nq')d/dx$. This corresponds to the second step in the ballooning mode problem. It can be shown [15] that a WKB analysis leads to the following global dispersion relation

$$nq' \int x(\theta_o)d\theta_o = \pi[l + \beta] \tag{6}$$

with $x(\theta_o)$ being determined from Eq. 5, l being an integer and β a numerical constant which depends on the structure of the phase space trajectories. The position of the turning points $\theta_o = \theta_T$ is given by the condition $(\partial F/\partial \theta_o)_{\theta_o=\theta_T} = 0$. For further details on this problem we refer to Ref. [15]. In the following Eq. 4 will be solved in the two different limits of short wavelengths, moderate shear and long wavelengths or low shear.

6.2.3 SHORT WAVELENGTHS, MODERATE SHEAR LIMIT

The standard short wavelength ordering corresponds to balancing parallel ion compressibility, curvature, inertia and adiabatic electron response on the connection length scale $\theta \approx 1$, yielding [1, 2] $|\omega| \approx |\omega_{*pi}\omega_D|^{1/2}$, $k_\theta\rho_i \approx \epsilon_T^{1/4}$, $\epsilon_T = L_T/R$ and $|\theta - \theta_o| \approx 1$. In this limit the eigenfunction has a moderate ballooning structure which can be analytically determined by using the strong coupling approximation [14], yielding

$$F = \frac{\frac{1}{\tau} + \frac{\omega_{*i}}{\omega}}{1 - \frac{\omega_{*pi}}{\omega}} + (k_\theta \rho_i)^2 - \frac{\omega_D}{\omega}\cos\theta_o + i\frac{\omega_{ti}}{\omega}k_\theta \rho_i s = 0 \qquad (7)$$

For the sake of simplicity it is convenient to discuss Eq. 7 in the flat density limit

$$F = -\frac{\omega_D}{2\omega}\left[\Omega^2 f(x) + i\frac{s}{q} + 2\cos\theta_o\right] = 0 \qquad (8)$$

where $\Omega = \omega/\omega_o$

$$\omega_o = \frac{nc(T_{io}T_{eo})^{1/2}}{eBa(aR)^{1/2}}, \qquad f(x) = -\frac{r^2}{a^2q^2\widehat{T}_e\widehat{T}_i'}, \qquad (9)$$

$\widehat{T}_j = T_j/T_{jo}$, with T_{jo} being the central temperature of the species j, and a being the minor radius. It is apparent from the above equation that the mode localization depends on the shape of the temperature profile through the function $f(x)$. For the sake of simplicity it is convenient to consider two limiting situations. Assume first that the temperature profiles are flat near the magnetic axis, $T \approx 1 + \lambda(r/a)^\beta$ with $\beta > 3$, as e.g. the profiles after a sawtooth crash. In such a case the function f has a minimum around a point which can be arbitrarily assumed to be $r = r_o(x = 0)$. Upon expanding f as $f \approx f(0)(1 + x^2/L^2)$, the resulting phase space trajectories in the x, θ_o plane, are similar to those of a pendulum. Therefore it is possible to distinguish between closed and open orbits. The closed orbits are characterized by turning points located at $\theta_o = 0$. Since along the orbits $\theta_o \ll 1$ it is possible to expand the $\cos\theta_o$ term as $1 - \theta_o^2/2$, and, after making the substitution $\theta_o = -i(1/nq')d/dx$, a Weber equation is obtained. The resulting eigenvalue is given by

$$\Omega^2 f(0) + i\frac{s}{q} + 2 = \frac{[l+1/2]\pi}{nq'L}. \qquad (10)$$

The turning points are located at $x = \pm x_T$ with

$$x_T \approx L\left(\frac{l+1/2}{nq'L}\right)^{1/2}. \qquad (11)$$

Upon taking the lowest radial eigenmode $l = 1$ it is possible to see that its radial extension corresponds to a coupling of a number $(nq'L)^{1/2}$ of mode rational surfaces. The radial correlation length is thus much larger than the distance between mode rational surfaces which is of the order of the ion Larmor radius. It is interesting to express the radial correlation length in terms of ρ_i yielding

$$x_T = \frac{(L\rho_i/s)^{1/2}}{(k_\theta \rho_i)^{1/2}} \qquad (12)$$

with $k_\theta \rho_i \approx \epsilon_T^{1/4}$.

Note that the assumption $x_T/L \ll 1$ is satisfied for $l \ll nq'L$. It is important to stress that the shear dependence of the radial correlation length in the short wavelengths, moderate shear limit is algebraic $x_T \approx s^{-1/2}$.

For larger radial mode numbers the radial width increases and it is no longer possible to assume the local expansion for f.

For typical temperature profiles the function f is monotonic with radius and no points exist in the plasma at which $f'(x) = 0$. In order to discuss such a case it is convenient to assume $f(x) = f(0)(1 + x/L_1)$. Furthermore it is necessary to keep in Eq. 7 also the first order correction in the quantity $\omega/\omega_{*pi} = -\Omega g(x)$, where $g(x) = O(\epsilon_T^{1/2})$ is assumed to have a radial variation as $g = g(0)(1 + x/L_2)$.

The phase space trajectories in this case are similar to the open trajectories of a pendulum. The turning points are located at $\theta_o = 0, \pi$ i.e. at x_T given by

$$x_T = \pm 2L_1 \frac{1 + \Omega g(0)(1 + L_1/L_2)}{\Omega^2 f(0)} . \qquad (13)$$

Note that the above expression is independent of the toroidal mode number n, and corresponds to a mode localized on a macroscopic space scale. In order to determine the expression for the eigenvalue it is convenient to use the WKB condition Eq. 6 which, together with Eq. 7, yields

$$\Omega \approx -f(0)^{-1/2} e^{-i\pi/4} \left(\frac{s}{q}\right)^{1/2} \left[1 - \frac{g(0)e^{-i\pi/4}}{2f(0)^{1/2}} \left(\frac{s}{q}\right)^{1/2}\right] . \qquad (14)$$

After the substitution of Eq. 14 into Eq. 13 it is possible to see that the turning points are located in the x plane at positions approximately symmetric with respect to the real axis. The projection of the turning points on the real axis, which can be used as an estimate of the radial correlation length, turns out to be

$$\mathrm{Re}(x_T) \approx \pm 2^{3/2} \frac{L_1^2}{L_2} \frac{g(0)}{f(0)^{1/2}} \left(\frac{q}{s}\right)^{1/2}. \tag{15}$$

Again the shear dependence is algebraic $x_T \approx s^{-1/2}$. Note that, since a local expansion has been used for the equilibrium profiles, the above result is valid for $s/q < 1$ ($|x_T| \ll L_1, L_2$). Furthermore, such a result is in agreement with recent numerical findings of particle simulation codes [17] or 2-D linear eigenvalue codes [13] which indeed observe the formation of radially elongated structures.

6.2.4 LONG WAVELENGTHS, LOW SHEAR LIMIT

In the long wavelength limit ($k_\theta \rho_i \ll \epsilon_T^{1/4}$) or for low shear values, the eigenfunction becomes broad and the strong coupling approximation can no longer be applied. In this case, two branches exist, a toroidal and slablike branch [3]. The eigenmode considered in this paper in the short wavelength limit is connected with the toroidal branch in the long wavelength limit. The eigenfunctions belonging to such a branch exhibit a fast variation, along the equilibrium field, over the connection length scale with a superimposed slow variation over a secular scale. The other branch, which exists also in slab geometry, is characterized by a variation along the equilibrium magnetic field dominated by the secular scale, with superimposed small oscillations on the connection length scale. For the sake of simplicity the latter branch will be ignored here.

The optimal ordering for the toroidal eigenmodes, as given in Ref. [3], is obtained by balancing parallel compressibility and adiabatic electron response on the scale $\theta \approx 1$, yielding $\omega \approx (\omega_{*pi}\omega_{ti}^2)^{1/3}$. The characteristic scale for the slow variation is obtained by balancing parallel compressibility and geodesic curvature on the secular scale, yielding $\theta \approx s^{-1/2}[\epsilon_T^{1/2}/(k_\theta \rho_i)^2]^{1/3}$.

In order to determine the radial mode width, the 1-D eigenvalue problem must be solved to find the dependence on the quantity θ_o. This, however, is not straightforward as in the moderate shear limit, because a regular perturbation theory fails to give any θ_o dependence of the dispersion relation at all the orders in s. The reason is that, since each poloidal harmonic tends to become localized with respect to the distance between mode rational surfaces, the toroidal coupling tends to vanish faster than any power in s.

It is convenient to solve Eq. 4 using a variational method. Upon defining the functional L as

$$L = \int_{-\infty}^{+\infty} \left[-(\partial_\theta \phi)^2 + \left(\frac{1}{4} + \lambda + \epsilon(\theta - \theta_o) \sin\theta \right) \phi^2 \right] \quad (16)$$

where the normal curvature and inertia terms have been neglected and

$$\epsilon = -\frac{\omega \omega_D}{\omega_{ti}^2} s \, , \quad \lambda = \frac{\frac{1}{\tau} + \frac{\omega_{*i}}{\omega}}{1 - \frac{\omega_{*pi}}{\omega}} \frac{\omega^2}{\omega_{ti}^2} - \frac{1}{4} \quad (17)$$

and using as a trial function the eigenfunctions determined in Ref. [3] using a matched asymptotic expansion method,

$$\phi = \left[\cos\theta/2 - \frac{2\sigma + \epsilon/2}{\lambda}(\theta - \theta_o)\sin\theta \right] e^{-\sigma(\theta - \theta_o)^2} \quad (18)$$

the variation over the parameter σ yields, at the lowest order the result of Ref. [3] ($\sigma = \epsilon/4$). After the substitution of Eq. 18 into the expression of the functional and keeping the lowest significant term in s, the following dispersion relation is obtained which generalizes that of Ref. [3]

$$F = \frac{\lambda}{2} - \frac{\epsilon}{2\lambda} - \frac{3e^{-1/2\epsilon}}{8\epsilon} \cos\theta_o = 0 \quad (19)$$

for modes propagating in the electron diamagnetic direction and

$$F = \frac{\lambda}{2} - \frac{3e^{-1/2\epsilon}}{8} \cos\theta_o = 0 \quad (20)$$

for modes propagating in the ion direction. It is now possible to determine the radial mode structure following the analysis of Sec. 3. Only the expression for the modes propagating in the ion diamagnetic direction will be reported here. For flat temperature profiles around the magnetic axis the mode localization turns out to be

$$x_T \approx \pm \left(\frac{6^{1/2} L}{(nq'L)^{1/2}} \right)^{1/2} e^{-1/8|\epsilon|} \qquad (21)$$

while for monotonically increasing profiles the mode is localized at

$$x_T \approx \pm 3 L_1 e^{-1/2|\epsilon|} . \qquad (22)$$

In addition to the change in the eigenvalue scaling, the most important result is that the radial eigenmode width turns out to be exponentially decreasing as $\exp(-c/s)$, with $c \approx \epsilon_T^{1/3} (qk_\theta \rho_i)^{-4/3} \tau^{-1/3}$ being of order unity for $k_\theta \rho_i \approx \epsilon_T^{1/4}$. Therefore, if a mixing length estimate is employed for the thermal conductivity, a strong decrease in χ_i is expected toward the magnetic axis. It is important to stress that the result of the 1-D analysis is opposite. It always yields a monotonic increase of χ_i with decreasing shear since the radial width of each poloidal harmonics scales with shear as $s^{-1/2}$.

6.2.5 CONCLUSIONS

The most important result of the present paper is the determination of the radial eigenmode structure of the ion temperature gradient driven mode for the case of short wavelengths, moderate shear and long wavelengths, low shear. It has been found that the radial mode localization is determined by the equilibrium profiles. For typical temperature profiles the radial correlation length is independent of the toroidal mode number, and thus of the ion Larmor radius. For moderate shear values the radial correlation length has an inverse algebraic dependence on s, $L_r \approx s^{-1/2}$. For low shear values the radial correlation length decreases as $\exp(-c/s)$, with $c \approx \epsilon_T^{1/3} (qk_\theta \rho_i)^{-4/3} \tau^{-1/3}$ being of order unity. Such a strong dependence is due to the rapid decrease of the toroidal coupling which is associated to the sharp localization of each poloidal harmonics, compared with the distance between mode rational surfaces. For very low shear values the number of coupled poloidal harmonics becomes of order unity and the conventional sheared slab limit can be employed.

On applying a mixing length estimate the above results predict a strong reduction of transport for $s < 1$ and, in particular, close to the magnetic axis. This is in agreement with several experimental findings which seem to indicate very low level of transport close to the magnetic axis [8] or when sawtooth activity is stabilized using RF current drive [9].

Furthermore it should be noted that the scaling of the radial correlation length with the ion Larmor radius indicates that conventional gyroBohm estimate of drift wave transport might be not necessarily appropriate. In order to draw definitive conclusion on this issue, a deep understanding of the nonlinear evolution of the instability is needed. It has been speculated that elongated structures, such those described in the present paper, could be destroyed by secondary instabilities [18]. If this hypothesis turns out to be correct, the Larmor radius dependence of the radial correlation length might be restored.

ACKNOWLEDGMENTS

The author is indebted for discussion about the radial eigenmode problem with Dr. F. Zonca.

Bibliography

[1] W. Horton, D.I. Choi, W.M. Tang, Phys. Fluids **24**, 1077 (1981).

[2] F. Romanelli, Phys. Fluids B **1**, 1018 (1989).

[3] Liu Chen, S. Briguglio and F. Romanelli, Phys. Fluids B **3**, 611 (1991).

[4] F. Romanelli, Liu Chen, and S. Briguglio, Phys. Fluids B **3**, 2496 (1991).

[5] N. Bretz, E. Mazzuccato, R. Nazikian, S. Paul, R. Fonck, R. Durst in *Plasma Physics and Controlled Nuclear Fusion Research 1992*, paper A-3-37.

[6] M. Zarnstorff these proceedings.

[7] A. Taroni, Ch. Sack, E. Springmann, F. Tibone in *Controlled Fusion and Plasma Physics*, Proc. 18th European Conference, Berlin (1991) Vol. 15C, Part. I, 181.

[8] D. Pasini, R. Giannella, L. Lauro-Taroni, G. Magyar, M. Mattioli proc. 1992 International Conference in Plasma Physics Vol. I, p. 283.

[9] G.V. Pereverzev, F.X. Soldner, R. Bartiromo, F. Leuterer, V.V. Parail, Nucl. Fusion **32**, 1023 (1992).

[10] J.W. Connor, R.J. Hastie, and J.B. Taylor, Proc. Roy. Soc. London Ser. A **365**, 1 (1979).

[11] R. Marchand, W.M. Tang, and G. Rewoldt Phys. Fluids **23**, 1164 (1980).

[12] S. Briguglio, F. Romanelli, C.M. Bishop, J.W. Connor, and R.J. Hastie Phys. Fluids B **1**, 1449 (1989).

[13] W.M. Tang, to appear in Phys. Fluids.

[14] J.B. Taylor in *Plasma Physics and Controlled Nuclear Fusion Research 1976* (IAEA, Vienna, 1977), Vol. 2, p. 323.

[15] F. Zonca, Ph.D. Thesis, Princeton University (January,1993); F. Zonca and Liu Chen, International Sherwood Fusion Theory Conference, Santa Fe, N.M., (April, 1992).

[16] J.W. Connor, J.B. Taylor, and H.R. Wilson, AEA FUS Report (August 1992).

[17] W.W. Lee, these proceedings.

[18] S. Cowley, R.M. Kulsrud, and R. Sudan, Phys. Fluids B **3**, 2767 (1991).

6.3 MODELING OF ANOMALOUS ION THERMAL TRANSPORT IN HOT PLASMA

J.Y. Kim and W. Horton
Institute for Fusion Studies, The University of Texas at Austin,
Austin, Texas 78712

Abstract. *The radial profile problem of the anomalous ion thermal conductivity χ_i which is observed to increase with the minor radius, unlike the expectation from the usual drift wave turbulence transport model, is reconsidered in terms of the toroidal ion temperature gradient mode transport model. First, we discuss several stabilizing factors which may be closely related to the decrease of χ_i in the core region. Secondly, we present some possible models to explain the large χ_i near the edge region.*

6.3.1 INTRODUCTION

Recently many experimental results [1, 2] in high ion temperature plasma show a general profile of the ion thermal conductivity $\chi_i(r)$ which increases with the minor radius. In terms of usual drift wave turbulence transport models, like the η_i transport model, this profile dependence of $\chi_i(r)$ appears to be difficult to understand at present. Here, we discuss various possible factors which may be important to resolve this radial profile problem in terms of the toroidal η_i mode turbulence model. In the first part, we consider various stabilizing factors which may be related to the decrease of χ_i in the core region. In the second part, we consider various models which may explain the large χ_i near the edge region.

Before we consider these problems, it is useful to review briefly toroidal η_i mode stability property in the simple limit. Let us consider the electrostatic local kinetic dispersion relation

$$1 + \frac{T_i}{T_e} - \int \frac{\omega - \omega_{*i}^T}{\omega - \omega_{Di} - k_\parallel v_\parallel} J_o^2(k_\perp v_\perp) f_M d^3v = 0 , \qquad (1)$$

where

$$\omega_{*i}^T = \omega_{*i}(1 + \eta_i(v^2/2v_{ti}^2 - 3/2)),$$

$$\omega_{Di} = \varepsilon_n \omega_{*i}(v_\parallel^2 + v_\perp^2/2)/v_{ti}^2.$$

and ω_{*i} is the ion diamagnetic drift frequency, $\eta_i = d\ln T_i/d\ln n$, $\varepsilon_n = -(R\,d\ln n/dr)^{-1}$, $v_{ti} = (T_i/m_i)^{1/2}$, and f_{Mi} is the Maxwellian distribution of ion. Taking the approximation $k_\parallel = 0$, and $(v_\parallel^2 + v_\perp^2/2) \to \frac{2}{3}v^2$ in ω_{Di} (*constant energy model*), we can obtain the following marginal stability conditions [3, 4, 5],

$$\omega_r/\omega_{Di} = (\eta_c - \frac{2}{3})/(\eta_c - \frac{4}{3}\varepsilon_n), \qquad (2)$$

$$\eta_c \simeq \frac{4}{3}\varepsilon_n\left(1 + \frac{T_i}{T_e}\right)\Gamma_0^{-1}(b_i), \qquad (3)$$

when

$$\omega_r/\omega_{Di} > 0 \quad \text{or} \quad 2\varepsilon_n\left(1 + \frac{T_i}{T_e}\right)\Gamma_0^{-1}(b_i) > 1 \quad \text{(resonant limit)}.$$

On the other hand, in the opposite non-resonant limit of $\omega_r/\omega_D \leq 0$, we can obtain a similar stability condition

$$\eta_c \simeq 1 - \frac{2}{3}\varepsilon_n\left(1 + \frac{T_i}{T_e}\right)\Gamma_0^{-1}(b_i). \qquad (4)$$

Here, we have approximately included the finite Larmor radius (FLR) effect using the Γ_0 function. The simple marginal stability conditions (2)–(4) are useful to understand various important properties of the toroidal η_i mode stability. We note that $(\eta_c)_{\min} = 2/3$ and which occurs when $2\varepsilon_n\left(1 + \frac{T_i}{T_e}\right)\Gamma_0^{-1}(b_i) = 1$. Figure 1 shows the dependence of the threshold value η_c on the three important parameters of ε_n, T_i/T_e, and k_y, obtained from Eqs. (2)–(4).

6.3.2 VARIOUS STABILIZING FACTORS IN THE CORE REGION OF HOT ION PLASMA

The small χ_i in the core region with high ion temperature suggests that in the core region the plasma may be near the marginal state with $\eta_i \sim$

η_c. However, the calculation by many previous approximate models yields the η_c which is significantly smaller than the measured η_i in the core region, and thus failing to explain the observed small χ_i. This situation suggests that we should try to calculate more accurately the η_c in the core region. In particular, we should be careful in considering several stabilizing factors, which are possible in the core region, such as the large T_i/T_e ratio, finite impurity and beam ions fraction, finite $k_\parallel v_i$, and electromagnetic effect. Here, we discuss the effects of these stabilizing factors on the toroidal η_i mode.

Large T_i/T_e Effect

As shown in Fig. 1(b) and also in many works [3, 6, 7] the large T_i/T_e can give a significant stabilization when the density profile is not so peaked. Thus, in the core region of hot ion or supershot plasma this effect is expected as an important stabilizing factor of the ion temperature gradient mode.

Hot Beam Ions Effects

The data files from TFTR supershot discharges indicate that there exist a non-negligible fraction of high energy beam ion in hot ion plasma heated by NBI. Thus, when we analyze this plasma, it appears to be important to consider the effect of the high energy beam ions.

Assuming the Maxwellian distribution for the ion and beam, the local kinetic dispersion relation takes the form of

$$\frac{T_i}{T_e} + (1-\alpha_b)\left[1 - \int \frac{\omega - \omega_{*i}^T}{\omega - \omega_{Di} - k_\parallel v_\parallel} J_0^2(k_\perp \rho_i v_\perp/v_{ti}) f_{Mi} d^3v\right]$$

$$+ \alpha_b \frac{T_i}{T_b}\left[1 - \int \frac{\omega - \omega_{*b}^T}{\omega - \omega_{Db} - k_\parallel v_\parallel} J_0^2(k_\perp \rho_b v_\perp/v_{tb}) f_{Mb} d^3v\right] = 0 \quad (5)$$

where $\alpha_b = n_b/n_e$. Noting $\alpha_b < 1$, $T_i/T_b \ll 1$, and thus $J_0^2(k_\perp \rho_b) \ll J_0^2(k_\perp \rho_i)$, it is easy to see that the adiabatic and non-adiabatic parts of beam which is the last term in Eq. (5) is almost negligible. We can also see that in the $\alpha_b \to 1$ limit, the possibility that the beam driven η_b mode will be generated is very small due to the strong stabilization by the large T_b/T_e term. Thus, the main effect of beam ions appears to dilute the main ion density through the term $(1-\alpha_b)$. This dilution of main ion density is effectively equivalent to increase T_i/T_e by $T_i/T_e \to (T_i/T_e)/(1-\alpha_b)$, so that its effect is expected to give a similar stabilizing effect as the large T_i/T_e effect which considered in previous subsection. In Fig. 2 we show the actual growth rate as a function of α_b, calculated from Eq. (5). We can see that the beam ions give a strong stabilization and the inclusion of the non-adiabatic part of beam ions makes the stabilization a little stronger. Thus, high energy beam ions appear to be an important stabilizing factor in the NBI heated plasma and this suggests that the α particles in the burning plasma will contribute to the stabilization of the η_i mode.

Impurity Ions Effect

The data files from TFTR discharges also show that there is a nonnegligible fraction of impurity ions in the present high temperature plasmas. While it is desirable ultimately to make the plasma clean without the impurity, to analyze the present-day experiments, it appears to be important to consider the impurity effect. There are many previous works [8, 9, 10] which have considered the impurity effect on the toroidal η_i mode. Here, we present a brief summary on the effect in the full kinetic limit including the FLR effect and all resonance effects. We start from the local kinetic dispersion relation

$$\frac{T_i}{T_e} + (1 - Z\alpha_z)\left[1 - \int \frac{\omega - \omega_{*i}^T}{\omega - \omega_{Di} - k_\| v_\|} J_0^2(k_\perp \rho_i v_\perp / v_{ti}) f_{Mi} d^3v\right]$$
$$+ Z^2 \alpha_z \frac{T_i}{T_z}\left[1 - \int \frac{\omega - \omega_{*z}^T}{\omega - \omega_{Dz} - k_\| v_\|} J_0^2(k_\perp \rho_z v_\perp / v_{tz}) f_{Mz} d^3v\right] = 0 ,$$
(6)

where $\alpha_z = n_z/n_e$ and Z is the impurity charge number. First, assuming $k_\| = 0$, $\varepsilon_{nz} = \varepsilon_{ni} = \varepsilon_{ne}$, $\eta_i = \eta_z$ and using the constant energy model of ω_D, we can obtain following marginal stability condition in the resonance limit of $\omega_r/\omega_{Di} > 0$, similarly with Eq. (2),

$$(1 - Z\alpha_z)(\widehat{\omega}_0 - \widehat{\omega}_r)\Gamma_0(b_i)e^{-\frac{3}{2}\widehat{\omega}_r} + Z^2\alpha_z\frac{T_i}{T_z}(\widehat{\omega}_0 - Z\widehat{\omega}_r)\Gamma_0(b_z)e^{-\frac{3}{2}Z\widehat{\omega}_r} = 0 ,$$
(7)

where $\hat{\omega}_r = \omega_r/\omega_{Di}$ and $\hat{\omega}_0 = (\eta_c - \frac{2}{3})/(\eta_c - \frac{4}{3}\varepsilon_n)$. ¿From Eq. (7), it is easy to see that there is just one single root of ω_r for arbitrary α_z between 0 to $1/Z$, and this means that the ion temperature gradient mode makes a transition from the working gas drive to the impurity mode continuously if $\hat{\omega}_r > 0$. We can also see that the marginal real frequency reduces from ω_i to ω_i/Z as $Z\alpha_z$ increases. In the pure ion or impurity limits, we can obtain following threshold values, similarly with Eq. (3),

$$\eta_{ci} = \frac{4}{3}\left(1 + \frac{T_i}{T_e}\right)(1 + b_i) \quad \text{for} \quad Z\alpha_z = 0$$

$$\eta_{cz} = \frac{4}{3}\left(1 + \frac{1}{Z}\frac{T_z}{T_e}\right)(1 + b_z) \quad \text{for} \quad Z\alpha_z = 1.$$

These show that the pure impurity mode has the lower threshold value η_c. An another characteristic of the impurity mode is that the maximum growth rate occurs for the mode with the shorter wavelength than for the main ion mode, since the impurity receives less stabilization from the FLR effect due to its smaller gyroradius ($\rho_z = (m_z/m_i)^{1/2}(Z_i/Z)\rho_i$ for $T_z = T_i$).

It is useful to consider the fluid limit which is possible in the case of $Z\alpha_z \ll 1$, that is, the small impurity fraction limit, since here $\omega \sim \omega_{Di} \gg \omega_{Dz}$. Assuming $T_i = T_z$, $\varepsilon_{ni} = \varepsilon_{nz}$, $\eta_i = \eta_z$, in the fluid limit Eq. (6) reduces to

$$\omega^2 - \omega_{*e}\omega - (1 - (Z-1)\alpha_z)\omega_{De}\omega_{*pi} = 0. \tag{8}$$

¿From this equation, we can see that the impurity reduces the driving force coefficient from 1 to $1 - (Z-1)\alpha_z$, giving the stabilization. This fluid approximation fails as $Z\alpha_z$ increases, since the real frequency ω_r decreases with the increasing $Z\alpha_z$, becoming $\omega_r \sim \omega_{Dz}$ in the $Z\alpha_z = 1$ limit.

Now, we present the full kinetic results obtained from Eq. (6). Figure 3 shows that the impurity effect is stabilizing initially but destabilizing later when $Z\alpha_z$ increases. The maximum stabilization thus occurs when the mixture of the two ion species is almost equal in terms of the charge density fraction. It is also shown that the real frequency decreases significantly as the fraction of impurity increases. Finally, we have considered the effect of the impurity density profile, and find that more stabilization occurs as the density profile of impurity becomes steeper when $L_{nz}L_{ni} > 0$, while flatter when $L_{nz}L_{ni} < 0$.

Finite $k_\parallel$ effect

For the toroidal mode, a finite $k_\parallel$ occurs from the ballooning structure of the mode, which is inversely proportional to the connection length qR (which is different from that occurring by the magnetic shear). The local kinetic analysis [11] showed that the effect gives a significant stabilization when

$$\frac{k_\parallel v_i}{\omega_{Di}} \sim \frac{c_o}{2qk_y} \geq 1, \qquad (9)$$

where $c_o \leq 1$. The numerical value of c_o depends on the ballooning degree of the eigenfunction. Thus, the $k_\parallel$ effect is expected to be stronger at the core region with small q and for the long wavelength mode, even though the non-local linear stability analysis shows that for small q or the long wavelength mode, $k_\parallel$ becomes smaller, by taking weaker ballooning structure. However, the long wavelength modes may be also excited by the inverse cascade in the turbulent regime from the shorter wavelength modes which are more unstable and have a stronger ballooning structure. In this case, the driven long wavelength mode may have a strong ballooning structure, taking a larger $k_\parallel$ value. On the other hand, as will be shown in the next subsection, the stabilizing effect of the electromagnetic term becomes stronger as $k_\parallel$ decreases, so that the unstable toroidal mode should have a finite $k_\parallel$ value in the finite β plasma.

Electromagnetic Effect

Finally, we consider the electromagnetic effect. Numerous previous works [12, 13, 14, 15, 16] have reported studies of the electromagnetic effect in various limits and found that increasing plasma β gives a stabilization. Here, we study the effect in the full kinetic limit using local and nonlocal methods. First, let us consider the simple local limits to obtain some physical understanding. Electromagnetic local kinetic dispersion relation is given by

$$1 + \tau - P_0 = \frac{[\tau(\omega - \omega_{*e}) - k_\| P_1]^2}{\frac{2k_\|^2 k_\perp^2}{\beta_i} + k_\|^2 P_2 - \tau\omega(\omega - \omega_{*e}) - \omega_{Di}(\omega - \omega_{*pe})}, \quad (10)$$

where

$$P_m = \int \left(\frac{v_\|}{v_{ti}}\right)^m \frac{\omega - \omega_{*i}^T}{\omega - \omega_{Di} - k_\| v_\|} J_0^2 f_M i \, d^3 v \ .$$

Taking the fluid limit, Eq. (5) reduces further to

$$\tau + \frac{\omega_{*i}}{\omega} - \frac{k_\|^2}{\omega^2}\left(1 - \frac{\omega_{*pi}}{\omega}\right) = -\left(k_\perp^2 - \frac{\omega_{Di}}{\omega}\right)\left(1 - \frac{\omega_{*pi}}{\omega}\right)$$

$$\cdot \left[1 - \frac{\tau\omega(\omega - \omega_{*e}) - k_\|^2\left(1 - \frac{\omega_{*pi}}{\omega}\right)}{\frac{2k_\|^2 k_\perp^2}{\beta_i} - \omega_{Di}(\omega - \omega_{*pe})}\right]. \quad (11)$$

Here, we note that the last term in the right-hand side represents the electromagnetic effect. From Eq. (11) it is easy to see that the coefficient of the destabilizing force term $\omega_{Di}\omega_{*pi}$ of the toroidal η_i mode changes its sign at about

$$\beta_i \simeq \beta_{it} = \frac{2k_\|^2 k_\perp^2}{\omega_{Di}(\omega_{*i} - \omega_{*pe})}, \quad (\text{note } \omega \sim \omega_{Di} \ll \omega_{*pe}) . \quad (12)$$

This means that for $\beta_i \geq \beta_{it}$, the η_i mode is completely stabilized. Note that, here, β_{it} has a similar form as the threshold value β_{ic} for the ideal ballooning instability, which is given by

$$\beta_{ic} = \frac{2k_\|^2 k_\perp^2}{\omega_{Di}(\omega_{*pi} - \omega_{*pe})} . \quad (13)$$

This means that the stabilization of the toroidal η_i mode occurs almost simultaneously with the excitation of the ideal ballooning mode. In fact, the stabilization of the electrostatic η_i mode comes from the coupling of the mode with the electromagnetic shear Alfven wave. The coupling makes the driving force of the η_i mode almost null at the point where the ideal ballooning mode becomes excited. Figure 4 shows the growth rates of the η_i mode and the ideal ballooning mode as a function of β_i, calculated from the local kinetic equation (10). We observe that the stabilization of the electrostatic η_i mode is in fact closely related to the excitation of the ideal ballooning mode. An another important feature shown in Fig. 4 is that the kinetic effects reduce further β_{it} and β_{ic} from the fluid theory value in Eqs. (12) and (13). This means that the electromagnetic effect can be more important in the actual low β plasma. It is interesting to note that β_{it} is independent of k_y, and proportional to $k_\parallel^2$ and $(1+\eta_i)^{-1}$. In fact, numerical results shown in Figs. 5 and 6, indicate that at a given β_i the electromagnetic stabilizing effect increases with the decreasing $k_\parallel$ and increasing η_i. Thus, in the electromagnetic regime, the most unstable mode occurs when it has a finite $k_\parallel$ value. Also, the dependence of the growth rate on η_i becomes significantly weaker than the η_i dependence of the purely electrostatic growth rate. In Fig. 7, we show the dependence of the growth rate on the shear s and the safety factor q, calculated from a nonlocal code [17] which solves the gyrokinetic equation using the integral equation method. We find that at a given β_i the electromagnetic effect increases as q increases or s decreases.

There are some experimental results which seem to support the importance of the electromagnetic effect. An actual analysis of a TFTR supershot discharge shows a significant effect of electromagnetic term, as given in Fig. 8. Also, a JET experiment by Taroni, et al. [2] indicates that χ_i in JET is substantially smaller than in TFTR, in almost the same parameter regime except β_i which is three times larger for JET (due to the smaller B). A DIII-D experiment by Burrell, et al. [2] also show a significant decrease of χ_i after L to H transition during which the density ($\propto \beta$) increases substantially. Finally, a recent JT-60 experiment by Hirayama, et al. [18] shows a significant decrease of the χ_i with the increase of the density. These experimental results suggest that the electromagnetic effect may be playing an important stabilizing role in the high ion temperature plasma system.

6.3.3 SOME MODELS TO EXPLAIN LARGE χ_i NEAR EDGE REGION

Most experimental results show that the χ_i increases with the minor radius. The small χ_i in the core region may be explained as due to that the plasma system is near the marginal stability state by various stabilizing effects, discussed in Sec. 2. However, near the edge region most experimental χ_i's appear still to be much larger than the value estimated using the mixing length formula. Here, we discuss three possible models which may provide a resolution of this radial problem near the edge region.

The radial mode width of the dominant mode may be larger than that estimated by the simple mixing length formula

This possibility occurs from the inverse cascade phenomena which was shown in many numerical simulations of η_i mode in the slab limit. The inverse cascade suggests that the dominant radial mode width in the turbulent state will be that of the longest relevant wavelength mode, rather than that of the fastest growing mode. If we assume that the saturation level of the dominant mode is still proportional to the maximum growth rate, and isotropic spectrum, the inverse cascade suggests a new mixing length formula

$$\chi_i = (\gamma)_{\max}/(k_y)_{\min}^2 . \tag{14}$$

Here, we note that the long wavelength toroidal modes receive the strong stabilization by the finite $k_\parallel v_i$ effect, and the cut-off wavelength may be obtained approximately by the condition $\omega_D \simeq k_\parallel v_i$ or

$$(k_y)_{\min}\rho_i \simeq k_\parallel L_n/2\varepsilon_n .$$

The long wavelength modes driven by the inverse cascade (not by linearly) are expected to take a strong ballooning structure like the driving short wavelength modes, so that we may approximate as $k_\parallel \simeq 1/qR$, and then

$$(k_y)_{\min} \sim 1/2q\rho_i . \tag{15}$$

Now, the new transport formula Eq. (14) becomes finally

$$\chi_i = 4(\gamma)_{\max}\rho_i^2 q^2 . \qquad (16)$$

We note that this new formula (16) is different from the conventional mixing length formula by the factor of q^2. This q^2 dependence can provide a large enhancement of χ_i in the edge region (here, note that $\gamma_{\max}$ also depends on q and this can weaken a little the strong q^2 dependence). By the strong q dependence we can also obtain the scaling law of $\chi_i \propto B^0/I^\alpha$, with $\alpha \sim 1-2$, which agrees better to the experimental results than the simple mixing length formula. The χ_i values calculated from the new formula for the two TFTR Supershot discharges 44669A and 51025 also show a reasonable agreement with the experimental results, as given in Fig. 9. Finally, we note that our new formula is based on following assumptions,

1. inverse cascade by strong turbulence to the long wavelength side.

2. a cutoff wavelength $(k_y)_{\min}$ is determined by the parallel ion transit effect.

3. isotropic fluctuation spectrum $k_x \sim k_y$ in steady state (the anisotropic structure like $k_x < k_y$ can be an another possible enhancement factor of χ_i. For example, a recent simulation [19] of the toroidal η_i mode shows an anisotropic spectrum structure where the fluctuations form the streamer structure along the radial direction).

These three assumptions should be checked by more advanced numerical simulations including the kinetic and toroidal effects.

The fluctuation amplitude near the edge region may be larger than the mixing length estimate by the nonlocal transport from the core region

The fluctuation amplitude $\tilde{\phi}(r)$ near the edge region might not be determined instantaneously by the local stability parameters, as discussed in the previous subsection using the mixing length argument, but built up over the time period required for the nonlocal effects to propagate from the core region. Under this model the temperature T_i, the radial heat flux $q_r = -n_i \chi_i r \frac{\partial T_i}{\partial r}$ (neglecting the convection term), and the fluctuation amplitude $\tilde{\phi}(r = r_o)$ near the edge region are expected to take the following spatial and temporal evolution, as shown in Fig. 10.

1. Initially, the $\tilde{\phi}$, χ_i, and q_r at a point r is locally determined by the local stability parameters like η_i and T_i at r ($t = t_1$, $\tilde{\phi} = \tilde{\phi}_1$). If we assume the simple mixing length formula, the initial χ_i and q_r will have the profiles shown in Fig. 10.

2. The temperature profile will then become steeper in the outer region, while flatter in the core region, to satisfy the steady state heat balance condition

$$\frac{1}{r}\frac{\partial}{\partial r}(rq_r) = 0, \quad (t = t_2),$$

making the $\tilde{\phi}, \chi_i$, and q_r increase in the outer region, while decrease in the core region.

3. By the reduced heat flux from the core region, the temperature profile will soon come back to that similar with the initial profile ($t = t_3$).

4. At this stage, however, the fluctuation amplitude does not decay to zero, but reaches a finite value by decaying more slowly (for example, by developing the self-organized structure shown recently by Kishimoto, et al. [19]) ($\tilde{\phi}_3 \sim \tilde{\phi}_2$).

5. Now, the temperature profile will not change so much, since the steady state heat balance condition will be satisfied even with a small change of the temperature profile by the finite $\tilde{\phi}(r)$ ($t = t_4$).

In this relaxation model, the edge transport is closely connected to the core transport. Thus, if the core becomes more stable, the edge also becomes more stable. Also, a heat pulse which can be generated by an abrupt increase of η_i from η_c due to a strong local heating in the core region can increase the fluctuation amplitude in the outer region. Under this model, the overall transport scaling is expected to be determined by the confinement property in the core region rather than in the outer edge region. This model may provide a possible explanation of a recent

Zarnstorff, *et al.*'s current profile change experiment [2] which seems to suggest that a nonlocal transport effect is present, at least, during the change of the profiles.

The Effects of Trapped Particles

The trapped electrons are known to give the destabilizing effect on the toroidal η_i model (even though how much the destabilization is significant is still questionable since the trapped particles behave similarly with the passing particles in the outer region). Since the fraction of trapped electrons increases with the minor radius, the trapped electron effects can be an enhancement factor of χ_i near the edge region. The trapped particles are also less sensitive to the parallel ion sound wave coupling effect, or the shear damping effect, so that they can increase the radial mode width near the edge region. The trapped particles can also generate by themselves an unstable mode, like the trapped electron or trapped ion modes which can have quite a different frequency and the mode width from the toroidal η_i mode. The overall relevance of these trapped particle effects to the radial profile problem of χ_i is still unclear and should be investigated more.

6.3.4 SUMMARY

Here, we have considered the radial profile problem of the ion thermal conductivity which increases with the minor radius. In a attempt to explain this problem in terms of the toroidal η_i mode transport model, we have first discussed several stabilizing factors which may provide an explanation of the decrease of χ_i in the core region of hot ion plasma. It seems to be important to consider all these effects when we calculate η_c since in the core region a small difference in η_c can give a significantly different χ_i due to core's high ion temperature. Secondly, we have discussed three possible models to explain the large χ_i near the edge region. The actual relevance of these various models to the experiment should be investigated more thoroughly with more advanced numerical simulations including the full kinetic effect in the 3D toroidal geometry, and of course we also need to obtain more exact experimental data for the plasma system, such as the turbulence spectrum in the wavenumber and frequency, the saturation amplitudes of various fields, and T_i, T_e, and χ_i profiles.

ACKNOWLEDGMENTS

We would like to thank Prof. B. Coppi and Dr. J.Q. Dong for useful discussions. This work is supported by the U.S. Department of Energy under contract #DE-FG05-80ET-53088.

Bibliography

[1] S.D. Scott, et al., Phys. Rev. Lett. **64**, 531 (1990); S.D. Scott et al., Phys. Fluids B **2**, 1300 (1990).

[2] M.C. Zarnstorff, et al., in *Plasma Physics and Controlled Nuclear Fusion Research*, 1990, Proceedings of the 12th International Conference, Washington (IAEA, Vienna, 1991), Vol. I, p. 109; A. Taroni, et al., ibid, Vol. I, p. 93; K.H. Burrell, et al., ibid, Vol. 1, p. 123.

[3] F. Romanelli, Phys. Fluids B **1**, 1018 (1989).

[4] R.R. Dominguez and R.E. Waltz, Phys. Fluids **31**, 3147 (1988).

[5] W. Horton, B.G. Hong, and W.M. Tang, Phys. Fluids **31**, 2971 (1988).

[6] T.S. Hahm and W.M. Tang, Phys. Fluids B **1**, 1185 (1989).

[7] J.Q. Dong, W. Horton, and J.Y. Kim, Phys. Fluids B **4**, 1867 (1992).

[8] P. Terry, W. Anderson, and W. Horton, Nucl. Fusion **22**, 487 (1982).

[9] R.R. Dominguez and M.N. Rosenbluth, Nucl. Fusion **29**, 844 (1989).

[10] R. Peccagnella, F. Romanelli, and S. Briguglio, Nucl. Fusion **30**, 545 (1990).

[11] J.Y. Kim and W. Horton, Phys. Fluids B **3**, 1167 (1991).

[12] C.Z. Cheng, Nucl. Fusion **22**, 773 (1982).

[13] W. Horton, J.E. Sedlak, D.I. Choi, and B.G. Hong, Phys. Fluids **28**, 3050 (1985).

[14] R.R. Dominguez and R.W. Moore, Nucl. Fusion **26**, 85 (1986).

[15] A. Jarmen, P. Anderson, and J. Weiland, Nucl. Fusion **27**, 841 (1987).

[16] B.G. Hong, W. Horton, and D.-I. Choi, Plasma Phys. Controlled Fusion **31**, 1291 (1989); B.G. Hong, W. Horton, and D.-I. Choi, Phys. Fluids B **1**, 1589 (1989).

[17] J.Y. Kim, W. Horton, and J.Q. Dong, IFS# 598, 1993.

[18] T. Hirayama, H. Shirai, M. Yagi, K. Shimizu, Y. Koide, M. Kikuchi, and M. Azumi, Nucl. Fusion **32**, 89 (1992).

[19] Y. Kishimoto, T. Tajima, M.J. Lebrun, M.G. Gray, J.Y. Kim, and W. Horton, IFSR #589 (1993).

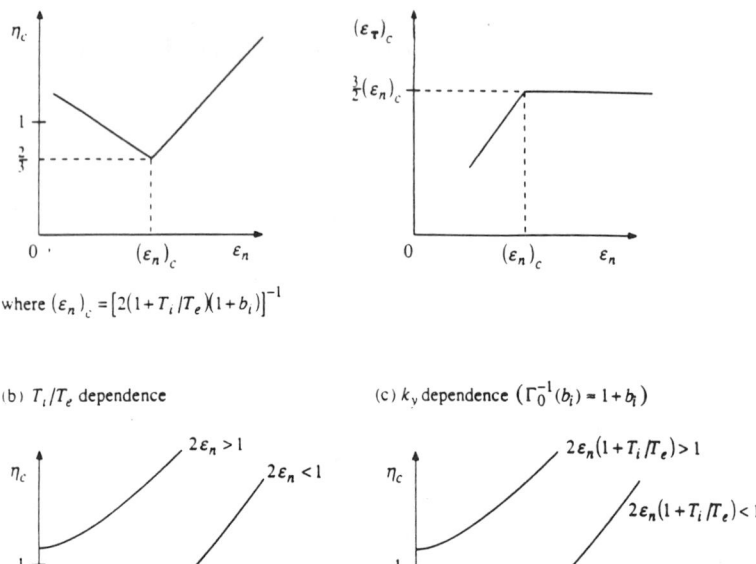

Figure 1. The threshold value η_c as function of (a) ε_n, (b) T_i/T_e, and (c) $k_y\rho_i$, obtained from the marginal stability conditions (2)–(4).

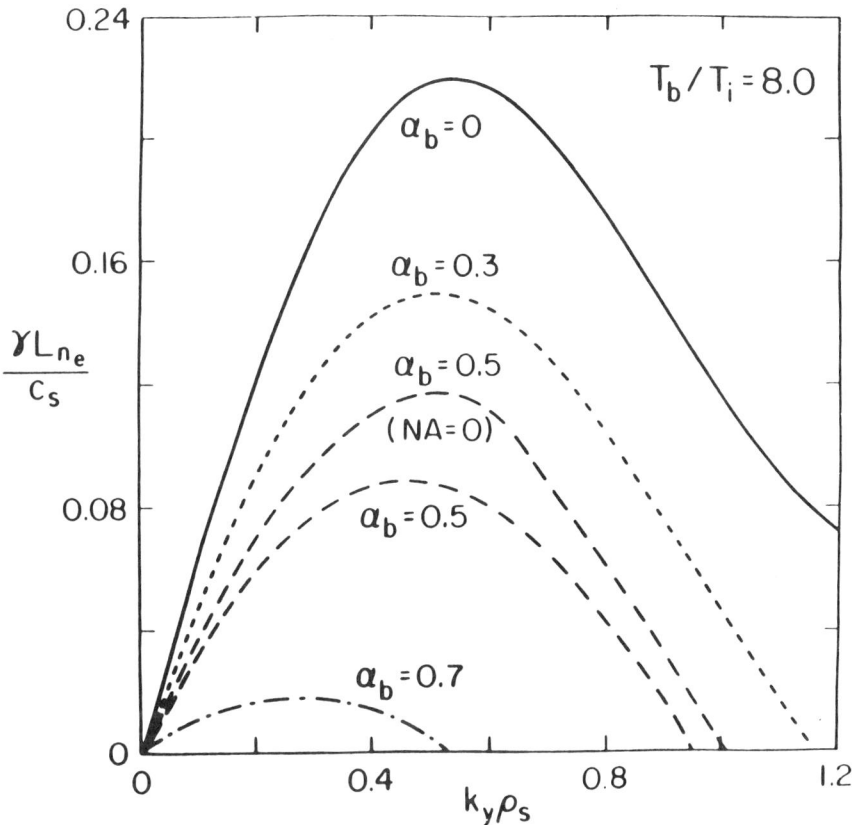

Figure 2. The normalized growth rate as a function of wavenumber for various values of $\alpha_b \equiv n_b/n_e$ when $\eta_i = \eta_b = 2.5$, $\varepsilon_{ni} = \varepsilon_{nb} = \varepsilon_{ne} = 0.2$, $k_\parallel = 0$, $T_e/T_i = 1$, and $T_b/T_e = 8$.

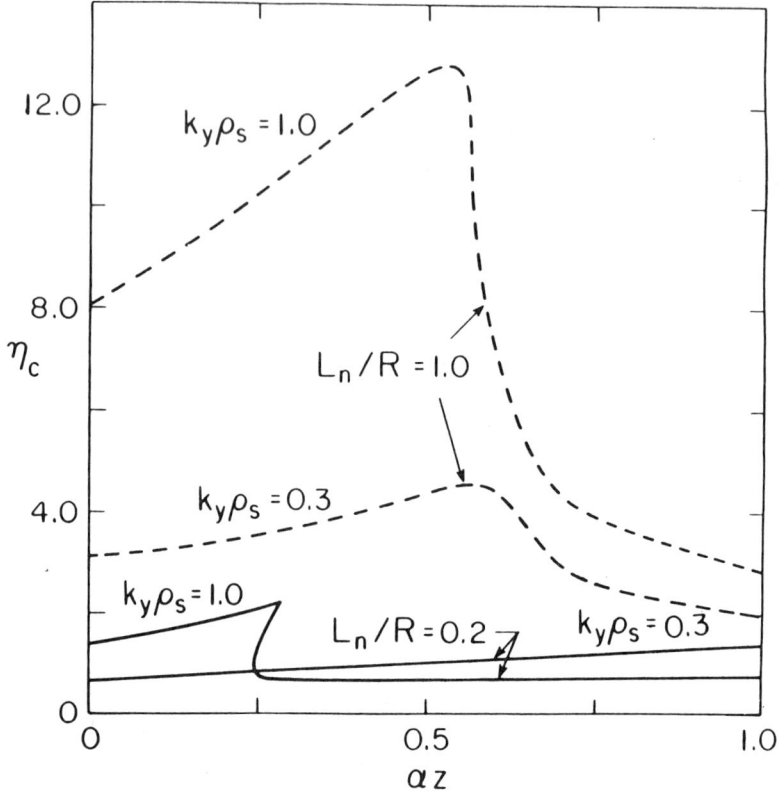

Figure 3. The threshold η_c as a function of the impurity charge density fraction $Z\alpha_z$ for various cases when $\eta_i = \eta_z$, $\varepsilon_{ni} = \varepsilon_{nz} = \varepsilon_{ne}$, $k_\parallel = 0$, and $T_i = T_z = T_e$.

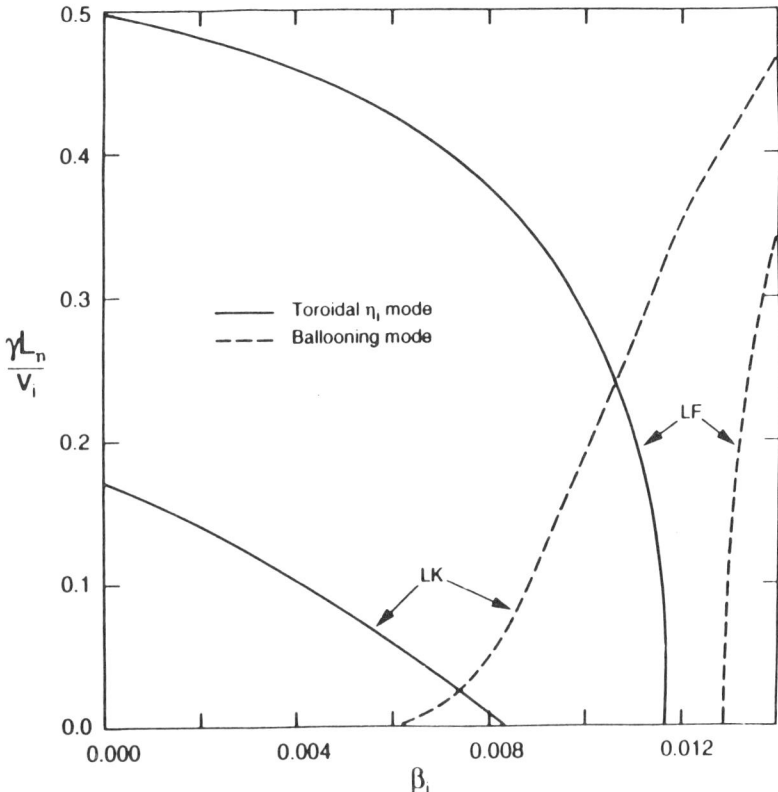

Figure 4. The normalized growth rate as a function of β_i in the local kinetic (LK) and local fluid (LF) limits when $\eta_i = 2.5$, $\eta_e = 2$, $k_y\rho_i = 0.5$, $k_\parallel = 0.1$, $\varepsilon_n = 0.18$, and $T_i/T_e = 1.5$.

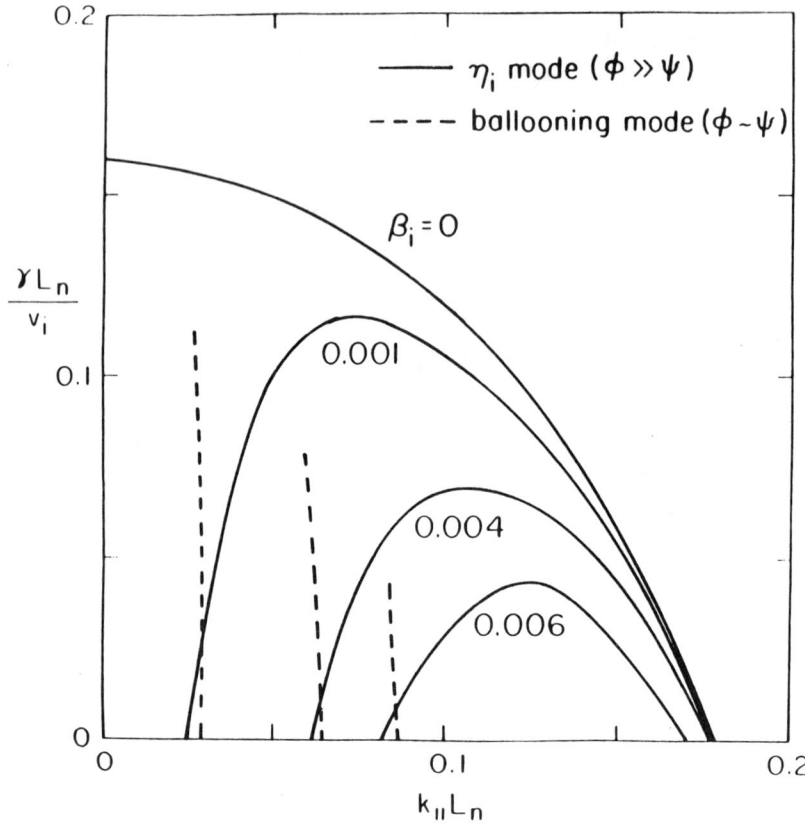

Figure 5. The normalized growth rate as a function of $k_\|$ for various β_i values with other parameters same as Fig. 4.

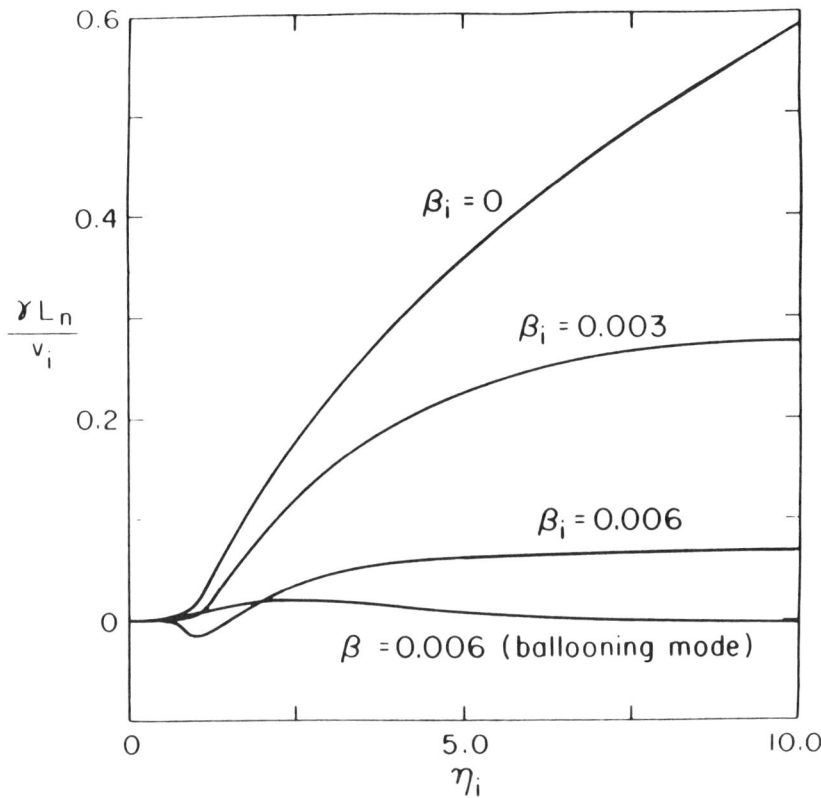

Figure 6. The normalized growth rate as a function of η_i for various β_i values with other parameters as Fig. 4.

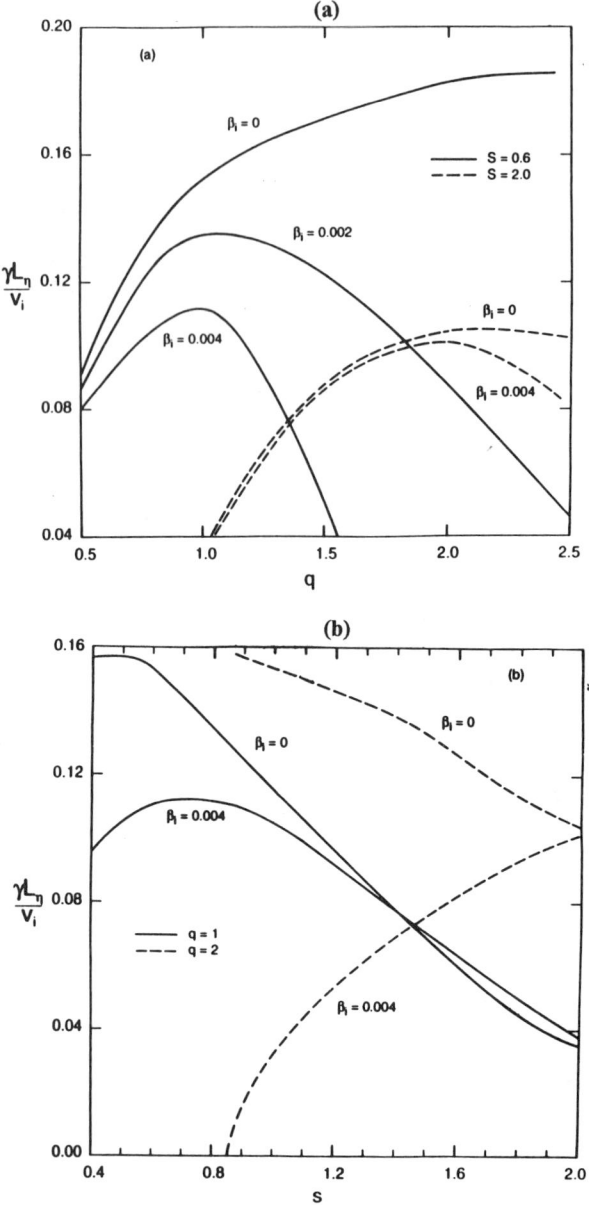

Figure 7. The normalized growth rate of the toroidal η_i mode for various β_i values; (a) as a function of q for two s values, (b) as a function of s for two q values, when $\eta_i = 2.5, \eta_e = 2, k_y\rho_i = 0.5, \varepsilon_n = 0.18$, and $T_i/T_e = 1$.

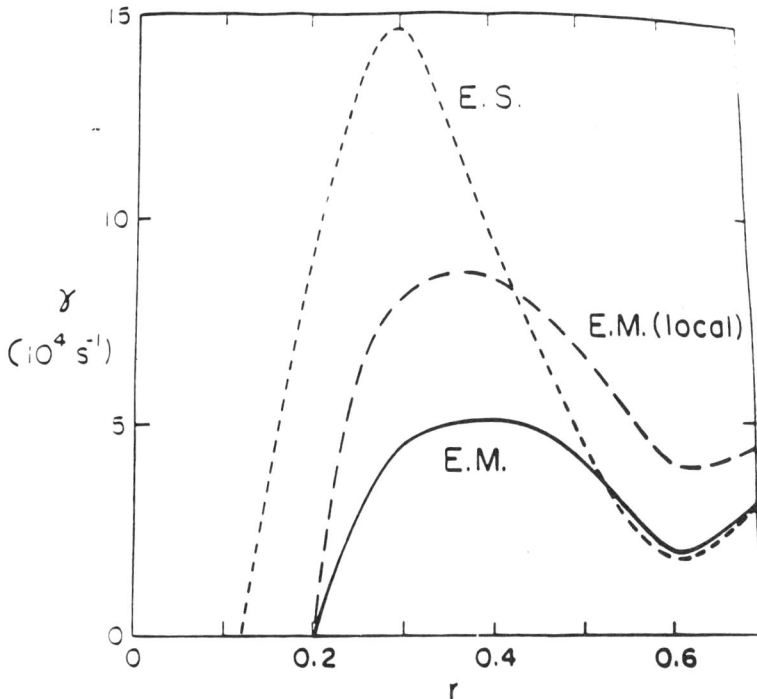

Figure 8. The maximum growth rate as a function of minor radius in the electrostatic (nonlocal) and electromagnetic (local and nonlocal) cases for the TFTR 44669A supershot discharge.

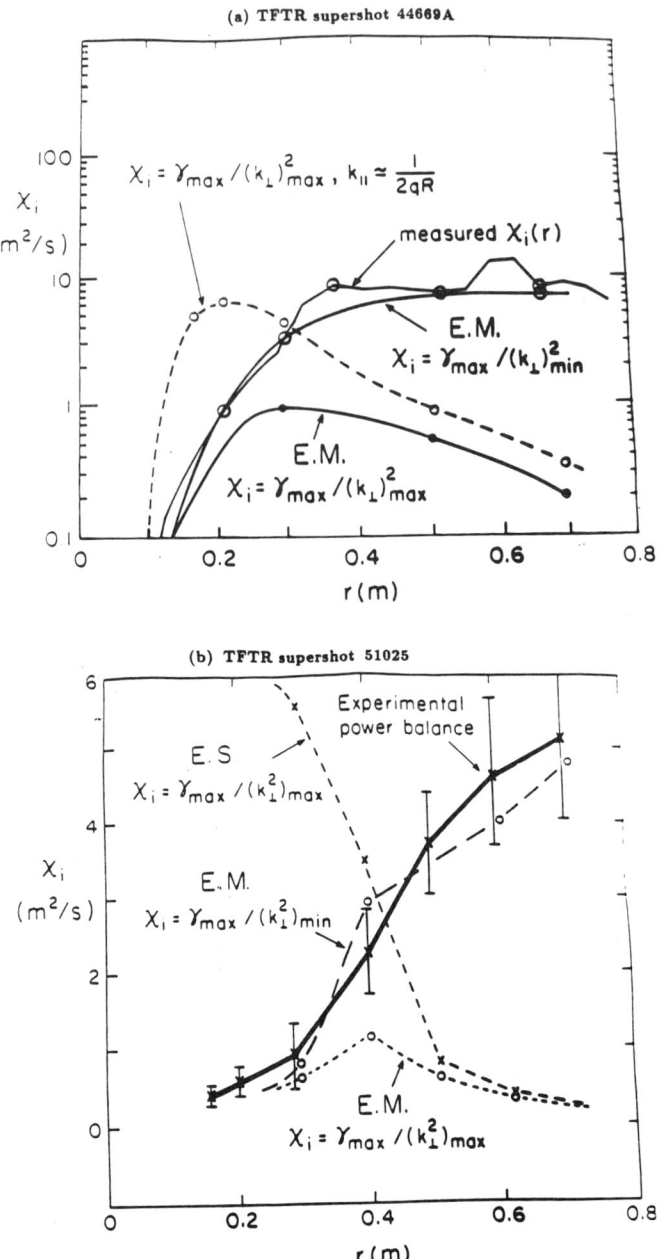

Figure 9. Comparison of two theoretical models of ion thermal conductivity with experimental result in the electromagnetic regime for the two TFTR supershot discharges (a) 44669A, (b) 51025.

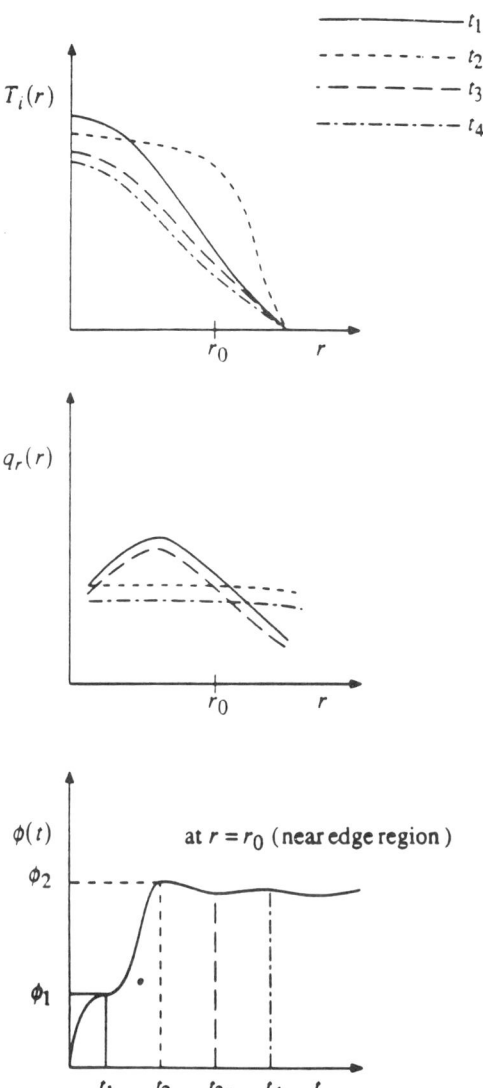

Figure 10. A model of the time evolution of $T_i(r)$, $q_r(r)$, and $\phi(t)$.

6.4 CROSS-FIELD ENERGY FLUX DUE TO ION TEMPERATURE GRADIENT MODE IN A TOKAMAK

Tomejiro Yamagishi
Fukui Institute of Technology
Gakuen Fukui 910, Japan

Abstract. Cross-field energy flux which is consistently related to the dispersion relation has been evaluated in a tokamak. Resultant ion heat conductivity profile shows similar behavior as the simple one derived under many assumptions. The effect of continuous eigenvalue spectrum given by the toroidal ion drift resonance on the thermal flux, which may be considered as the free streaming ion effect, is also evaluated. It is found that the continuum contribution to heat conductivity increases sharply near the edge, which may have correlation with experimentally observed behavior.

6.4.1 INTRODUCTION

Ion heat conductivity observed in Tokamaks discharges is larger than the one estimated by the neoclassical theory by more than one order of magnitude. Numerous efforts to interpret this anomalous heat conductivity by the ion temperature gradient drift instability (η_i mode) have been made [1, 2, 3, 4]. The ion diffusivity may be evaluated by the simple mixing length formula $D_\perp = \gamma/k_\perp^2$ with γ and $k_\perp$ being the growth rate and perpendicular wavenumber, respectively. The heat conductivity χ_i may also be evaluated by calculating the cross-field flux $\langle \tilde{v}_x \tilde{f} \rangle$ where $\tilde{v}_x$ is the radial component of $\mathbf{E} \times \mathbf{B}$ drift motion and $\tilde{f}$ is the perturbed distribution function, respectively [2, 5].

Since $\tilde{v}_x$ is imaginary, the imaginary component of $\tilde{f}$ which is induced by the imaginary part of the discrete eigenvalue, the growth rate, makes the real flux. By making use of the flux, a simple formula has been derived for convenience under several assumptions [2]. This formula for the ion thermal conductivity induced by the η_i mode has been used to examine the correlation with experimental observations [6].

One of the purposes of this paper is to examine these approximations used to derive the simple formula by numerically calculating the flux introducing the discrete η_i mode eigenvalue obtained from a dispersion relation.

The gyrokinetic equation has the continuous eigenvalue on the real axis induced from the continuous wave-particle resonance condition. This continuous eigenvalue is one of the most important characteristics of the Vlasov system [7, 8]. The existence of the discrete η_i mode eigenvalue depends on the ion temperature and density profiles or parameters $\varepsilon_n = L_n/R$ and $\eta_i = d\ell n\, T_i/d\ell n\, n$ where $1/L_n = d\ell n/dr$, R is the major radius and T_i and n represent ion temperature and density, respectively. On the other hand, the continuous eigenvalue always exists independently of these parameters. On the continuous eigenvalue spectrum, the distribution function has the characteristic imaginary part. The combination of this imaginary part with $\tilde{v}_x$ may make real cross-field flux. To examine the effect of the continuous eigenvalue spectrum on the quasilinear energy flux and ion heat conductivity is another purpose of this paper.

6.4.2 LOCAL DISPERSION RELATION

We start with the gyrokinetic equation

$$i \frac{v_\parallel}{Rq} \frac{\partial h}{\partial \theta} + (\omega - \omega_D)h = (\omega - \omega_{*T})J_0(\alpha)F_M\widehat{\varphi}(\theta) \qquad (1)$$

where all notations are standard as in Ref. 2; $\omega_D = \widehat{\omega}_D(v_\perp^2/2 + v_\parallel^2)$, $\omega_{*T} = \omega_*\{1 + \eta_i(v_\perp^2 + v_\parallel^2 - 3/2)\}$, $\widehat{\omega}_D = 2\varepsilon_n\omega_*(\cos\theta + s\theta\sin\theta)$, $\omega_* = -k_\theta\, cT/eB|L_n|$, $\alpha = (2b)^{1/2}v_\perp$, $L_n = |\nabla n/n|^{-1}$, $s = rq'/q$ (q; safety factor), $F_M = (\pi v_{th}^2)^{-3/2}\exp(-E/T)$ and $\widehat{\varphi} = e\phi/T_i$. The first-order differential equation (1) is easily solved with respect to h. Substitution of the solution h into the neutrality condition

$$\left(1 + \frac{1}{\tau}\right)\varphi(\theta) = \int d^3 v\, J_0(\alpha)h \qquad (2)$$

yields the integral equation with respect to ϕ [2, 3], where $\tau = T_e/T_i$.

We consider a localized mode in the outside region $\theta \cong 0$ in the torus, i.e. $\omega_D = 2\varepsilon_n \omega_*$ and $\varphi(\theta) \cong \varphi(0)$, and approximate the first term in Eq. (1) by $-\omega_t h$ with $\omega_t = k_\parallel v_\parallel$. In this case, the solution of Eq. (1) can be written in the form

$$h = \frac{\omega - \omega_{*T}}{\omega - \omega_D - \omega_t} J_0(\alpha) F_M \widehat{\varphi} . \tag{3}$$

Introducing Eq. (3) into Eq. (2), we have the local dispersion relation

$$D_{es} \equiv 1 + \frac{1}{\tau} - \int d^3v \, \frac{\omega - \omega_{*T}}{\omega - \omega_D - \omega_t} J_0^2(\alpha) F_M = 0 . \tag{4}$$

In this case, the localized dispersion relation can be written in the form of double integral:

$$D_{es}(\omega) = 1 + \frac{1}{\tau} - \frac{2}{\sqrt{\pi}} \int_0^\infty dx\, x\, e^{-x^2} J_0^2(\alpha)$$

$$\times \int_{-\infty}^\infty dy \, \frac{\omega - \omega_*(1 + \eta_i(x^2 + y^2 - 3/2))}{\omega - \widehat{\omega}_D(x^2/2 + y^2) - \omega_{ti} y} e^{-y^2} = 0 . \tag{5}$$

The toroidal η_i mode is induced by the toroidal drift frequency ω_D, in which the energy dependence of ω_D is essential. If we use an averaged constant for ω_D, Eq. (5) does not give the correct η_i mode.

The dispersion relation (5), therefore, is numerically solved by a conformal mapping method. The double integral was carried out by making use of the subroutine TWODQ in the IMSL system. Without Larmor radius effect ($\alpha = 0$), numerical results are presented for various values of ε_n and η_i in Fig. 1, in which both real frequency ω_r and growth rate ω_i are normalized by the electron drift frequency ω_*. The growth rate ω_i is larger for greater η_i. As ε_n increases, however, it tends to zero at which the discrete η_i mode disappears. This marginal condition $\omega_i = 0$ gives the critical value η_c above which the η_i mode becomes stable. This critical η_c for each ε_n gives a boundary curve in the (ε_n, η_i) plane as shown in Fig. 3. For smaller ε_n, which may correspond to the position near the edge or larger major radii, the real frequency ω_r tends to positive region and approaches to the electron drift branch. This may be reasonable because larger major radii, $R \to \infty$, means the slab limit.

In the derivation of the local dispersion relation (5), the uniform eigenfunction was assumed. If the eigenfunction is approximated by $\phi = (1 + \cos\theta)/(3\pi)^{1/2}$ like the strong ballooning mode, Eq. (5) is applicable, in the same form, to the nonlocal problems by replacing $k_\perp^2, k_\parallel^2$ and ω_D respectively by the averaged formulas: $\langle k_\perp^2 \rangle = k_\theta^2(1 + (\pi^2/3 - 5/2)s^2)$, $\langle k_\parallel^2 \rangle = 1/3(Rq)^2$, and $\langle \omega_D \rangle = 2\varepsilon_n \omega_*(2/3 + 5s/9)$. The eigenvalue calculated from this semi-local approximation agrees well [9] with the eigenvalue obtained from the nonlocal integral equation [4].

We have seen the behavior of the discrete eigenvalue ω_0 given by $D_{es}(\omega_0) = 0$ in Fig. 1. We also have the continuous eigenvalue spectrum given by the toroidal drift resonance condition $\omega = \omega_{Di}$. Since ω_{Di} is a continuous function of particle velocity, the continuous eigenvalue spectrum consists of the negative real axis $C = \{\omega_r | \omega_i = 0, -\infty < \omega_r < 0\}$. The existence of this continuous eigenvalue spectrum is one of the characteristics of Vlasov equations [7, 8]. We will discuss the effect of this continuum contribution to the energy flux in Sec. 4.

The double numerical integration in Eq. (5) is time consuming, particularly near the resonance condition $\omega = \omega_{Di}$, at which the integrand becomes singular. We approximate this double integral by a single integral for the case without Larmor radius effect $\alpha = 0$, by replacing the toroidal drift frequency: $\hat{\omega}_D(x^2/2 + y^2) \cong \hat{\omega}_D E$ where $E = m_i v^2/2T_i$. In this case, the dispersion relation can be written in the form

$$D_{es}(\omega) = 1 + \frac{1}{\tau} - \frac{2}{\sqrt{\pi}} \int_0^\infty \frac{\omega - \omega_*(1 + \eta_i(E - 3/2))}{\omega - \omega_D E} e^{-E} E^{1/2} dE = 0 . \tag{6}$$

Variation of discrete η_i mode eigenvalue ω_0 given by Eq. (6) are presented in Fig. 2 for various values of ε_n and η_i. As seen in Figs. 1 and 2, both results show similar behavior although the frequency difference is not so small. The stability boundary curves for both cases are also compared in Fig. 3. From these results, the essential characteristics of the η_i mode may be kept in the simple dispersion relation (6). Since the numerical integration in Eq. (6) is much faster, we will adopt the latter model in what follows.

6.4.3 CROSS-FIELD FLUX

We now evaluate the cross-field thermal flux due to the η_i mode making use of the dispersion relation obtained in the previous section. We calculate the energy flux induced by the $\mathbf{E} \times \mathbf{B}$ drift motion

$$Q_i = \int d^3v \left\langle E\tilde{v}_x \tilde{f}\right\rangle \equiv -\chi_i n \frac{dT_i}{dr} \tag{7}$$

where E is the particle kinetic energy, $\tilde{v}_x = -ik_y\phi/B$ is the radial component of $\mathbf{E} \times \mathbf{B}$ drift velocity and the angular brackets mean an ensemble average. When the perturbation is Fourier analyzed in the form $\tilde{f} = \sum \hat{f} e^{-i\omega t} e^{ikr}$, and applying the random phase approximation, the flux may be written as

$$Q_i = \sum_{k,\omega} \int d^3v\, E\hat{v}_x^* \hat{f}. \tag{8}$$

Since $\hat{v}_x$ is imaginary, the imaginary part of $\hat{f}$ gives real flux. The sum with respect to the wavenumber k in Eq. (8) is replaced by the average over k-space which is approximated by the peak value of the k-spectrum around $k\rho_i = k_y\rho_i \cong 0.1$ [2, 3].

When we consider the discrete η_i mode the sum with respect to the frequency ω may be replaced by the single term at $\omega = \omega_0$, which can be written as [2]

$$Q_i = \frac{5}{3}\left|\frac{e\varphi}{T_i}\right|^2 \frac{k_y T_i}{eB} nT_i \hat{q}_i^0 \tag{9}$$

where the normalized flux $\hat{q}_i^0$ due to the discrete eigenvalue ω_0 is defined by

$$\hat{q}_i^0 \equiv -\operatorname{Im} \int d^3v\, \frac{\omega_0 - \omega_{*T}}{\omega_0 - \omega_D} F_M \frac{E}{T_i} J_0^2(\alpha) \tag{10}$$

which is induced by the imaginary part of the discrete eigenvalue ω_0.

To derive a simple form of ion heat conductivity χ_i, Romanelli [2] assumed that $\operatorname{Re}\omega_0 \ll \operatorname{Im}\omega_0$, and away from the marginal state $\eta \gg \eta_c$, the growth rate is of the form $\operatorname{Im}\omega_0 = (\omega_* \omega_D \eta_i)^{1/2} \gg \omega_D$, and obtained

$$\chi_i = 5 \frac{v_i \rho_i^2}{L_n} (\varepsilon_n(\eta_i - \eta_c))^{1/2}. \tag{11}$$

This formula together with other formulas derived from different models is applied to JT-60 and TFTR tokamaks [6] in order to examine the correlation between the heat conductivity induced by the η_i mode and that observed in these large tokamak devices. As seen in Figs. 1 and 2, these assumptions may be valid when ε_n is small. In actual situations, however, the real frequency Re ω_0 is not small, and the approximation Im $\omega_0 = (\omega_* \omega_D \eta_i)^{1/2}$ is also broken down, because η_i is close to η_c and even becomes less than η_c in certain radial intervals.

Although the simple form of χ_i given by Eq. (11) is applied for convenience, it is important to examine whether the flux without these assumptions actually has similar radial profile as the simple formula. To examine the radial profile of heat conductivity, we calculate numerically the flux (10) without these assumptions making use of the discrete eigenvalue ω_0 obtained in Sec. 2.

We assume the density and ion temperature profile, respectively in the forms: $n(x) = n_0((1-x^2)^{\alpha_n} + \alpha_{n0})$, $T_i(x) = T_0((1-x^2)^{\alpha_T} + \alpha_{T0})$, where $x = r/a$ with a being the plasma radius. For these profiles, at the plasma center $x = 0$, $\eta_i = \alpha_t \alpha_{n0}/\alpha_n \alpha_{T0}$. Away from the center, η_i tends to zero when $\alpha_{T0} = 0$. When $\alpha_{T0} = 0$ and $\alpha_{n0} = 0$, on the other hand, η_i increases sharply at the edge $x = 1$.

In JT-60 discharges [6], the density changes sharply near the edge, while the ion temperature profile is steeper in the central region. For these profiles, we set $\alpha_n = 0.5$, $\alpha_T = 2$. As we move from the center $x = 0$ to the edge $x = 1$, the parameter (ε_n, η_i) varies as shown by the dotted broken line in Fig. 3. Since the discrete eigenvalue ω_0 is completely determined when (ε_n, η_i) is given, the corresponding discrete eigenvalue ω_0 varies as shown by the dotted broken curve in Fig. 2. We calculate the radial dependence of flux (10) by introducing the discrete eigenvalue ω_0 along with the trajectory in Fig. 2.

Numerical results of flux q_i^0 is presented in Fig. 4 for $\alpha_n = 0.5$ and $\alpha_T = 2.0$. We express, from Eq. (9), the heat conductivity in the form

$$\chi_i^0(x) = \left(\frac{v_i \rho_i^2}{R}\right)_0 f(x) . \tag{12}$$

We also write Eq. (11) in the form

$$\chi_i^R(x) = \left(\frac{v_i \rho_i^2}{R}\right)_0 f_R(x) \qquad (13)$$

where ( )$_0$ indicates the quantity within the brackets at the plasma center $x = 0$, $f(x)$ and $f_R(x)$ are profile functions. Variations of present model χ_i^0 and χ_i^R are shown in Fig. 4. The magnitude of χ_i^0 is smaller by about factor 2 and the profile is peaked slightly outside as compared with χ_i^R, although overall behaviors are similar. This difference may be induced by the effect of frequency ω_r.

Due to the factor $T^{3/2}$, both χ_i^0 and χ_i^R sharply decreases in the edge region, while experimentally observed χ_i in JT-60 and TFTR sharply increases near the edge [6]. This is the major discrepancy of the heat conductivity induced from the η_i mode and that of experimental observations. In the edge region, the dissipative trapped electron mode or some other MHD activities and/or the continuum contribution which will be discussed in Sec. 4 may dominate the transport process.

6.4.4 EFFECT OF CONTINUOUS EIGENVALUE SPECTRUM

We have seen that as ε_n increases the discrete η_i mode eigenvalue ω_0 tends to the negative real axis in the complex ω-plane; i.e., the discrete eigenvalue ω_0 attains the continuous eigenvalue spectrum C and vanishes. When the discrete mode disappears the energy flux $\hat{q}_i^0$ induced by the discrete mode also vanishes. This may happen in actual experimental situation in JT-60 and TFTR [6]. However, the continuous eigenvalue spectrum C, which is the continuous set of frequency with the resonance condition $\omega = \omega_{Di}(E)$ for $0 < E < \infty$, does not change as ε_n and η_i vary.

As we approach the continuum C, the distribution function given by Eq. (2) has the characteristic imaginary part from the formula: $x^{-1} = Px^{-1} - i\pi\delta(x)$ where P means the Cauchy principal value. We have, therefore, a contribution from the continuous eigenvalue spectrum C for the energy flux independently of the existence of the discrete eigenvalue. In this case, the sum over frequencies in Eq. (9) may be replaced by integral over the continuum C.

Let us evaluate this continuum contribution to the flux, which may be written in the form

$$\widehat{q}_i^c = \int_c d\omega\, S(\omega) \left[-\mathrm{Im} \int d^3v (\omega - \omega_{*T}) F_M \left(P \frac{1}{\omega - \omega_D} - i\pi\delta(\omega - \omega_D) \right) \right] \tag{14}$$

The principal part is real and no real contribution to the flux. If we make an assumption: $\omega_D = \widehat{\omega}_D E$ employed in Sec. 2, we have

$$\widehat{q}_i^c = \frac{\sqrt{\pi}}{\varepsilon_n} \int_{-\infty}^0 d\omega\, S(\omega) u_\omega^{3/2} e^{-u_\omega} \left((\eta_i - 2\varepsilon_n) u_\omega + 1 - \frac{3}{2}\eta_i \right) \tag{15}$$

where $u = \omega\tau/2\omega_* \varepsilon_n$.

If the width of frequency spectrum, ω_s, is about the diamagnetic drift frequency, $\omega_s = \omega_*$, we have

$$\widehat{q}_i^c = \frac{3}{4} \pi (1 + \eta_i - 5\varepsilon_n) \,. \tag{16}$$

In the case of the Gaussian frequency power spectrum function S, the same results as Eqs. (16) and (17) are also obtained with different numerical coefficients. As seen in Eqs. (16) and (17), depending on the frequency width ω_s, the ε_n-dependence of the continuum contribution entirely changes. In the latter case $\omega_s = \omega^*$, however, ω_s sharply increases near the edge ($L_n \to 0$), which may not likely happen in experimental situations. The normalized heat flux $\widehat{q}_i^c$ given by Eq. (16) is also plotted in Fig. 4. The flux $\widehat{q}_i^c$ becomes negative in the central region, which is due to the third term in Eqs. (16) and (17). This inward transport term comes from the first frequency term in $(\omega - \omega_{*T})$ in Eq. (14).

The third inward transport term in Eqs. (16) and (17) has no correspondence to the force term dT/dr as in Eq. (7). We delete this term, therefore, to define the heat conductivity χ_i^c corresponding to the continuum contribution. The heat conductivity χ_i^c thus defined for Eq. (16) may be written in the form

$$\chi_i^c = \lambda \frac{\rho_i^2 v_i}{R} \cdot \frac{\eta_i + 1}{\varepsilon_n} \,. \tag{17}$$

The variation of coefficient χ_i^c for various values of the profile parameter α_T is presented by a surface graphics in Fig. 5. The sharp increase of χ_i^c near the edge is due to the first term in Eq. (16). As seen in Fig. 5, for $\alpha_T > 1$, χ_i^c is peaked in the intermediate radial region, which may correspond to the high-T_i discharge case [6]. While for $\alpha_T < 1$, χ_i^c increases sharply toward the periphery, which may correspond to the L-mode discharges [6]. To compare more detail with experimental observations both χ_i^0 and χ_i^c together with the neoclassical result should be considered by applying more realistic temperature and density profiles.

6.4.5 SUMMARY AND DISCUSSION

The toroidal η_i mode is induced by the toroidal ion drift motion, in which the energy dependence of the drift frequency is very important. An averaged constant drift frequency does not give the correct η_i mode. Numerical evaluation for the toroidal η_i mode dispersion relation was, therefore, necessary. The local electrostatic dispersion relation has been evaluated by the double integral and single integral methods.

The latter approximation was employed to evaluate the cross-field thermal energy which was converted to ion heat conductivity. The radial profile of this heat conductivity was similar to the simple one derived under many assumptions.

The effect of the continuous eigenvalue spectrum induced from the ion toroidal drift resonance on the thermal energy flux has also been evaluated in the same case without Larmor gyromotion and parallel motion. The flux induced by the discrete η_i mode may vanish when the discrete eigenvalue ω_0 vanishes, which may occur in actual situations. The flux and corresponding heat conductivity due to the continuum contribution, on the other hand, always exists independently of the existence of the discrete eigenvalue. The heat conductivity due to the continuum contribution, which may be considered as the free streaming ion effect, χ_i^c can be larger in the edge region, and may have correlation with the experimentally observed behavior in large tokamaks.

ACKNOWLEDGMENTS

The author would like to thank Prof. A. Hirose, Dr. A. Smolyakov and members of the Plasma Physics Laboratory, University of Saskatchewan for valuable discussions and their hospitality. This study was initiated during my stay at the Plasma Physics Laboratory, which was supported by the Natural Sciences and Engineering Research Council of Canada. The numerical results in Fig. 1 was obtained with the help of Dr. Smolyakov by using the computer system at the University of Saskatchewan. Acknowledgment is also due to Prof. M. Wakatani and Dr. M. Azumi for providing valuable information.

The author is grateful to Prof. W. Horton for taking interest in this study, and to Mrs. Suzy Mitchell for typing this manuscript. This study was a joint research effort with the National Institute for Fusion Science.

Bibliography

[1] H. Nordman and J. Wieland, Nuclear Fusion **29**, 251 (1989).

[2] F. Romanelli, Phys. Fluids B **1**, 1018 (1989).

[3] F. Romanelli and Briguglio, Phys. Fluids B **2**, 754 (1990).

[4] J.Q. Dong, W. Horton, and J-Y. Kim, Phys. Fluids B **4**, 1867 (1992).

[5] T. Antonsen, B. Coppi, and R. Englade, Nuclear Fusion **19**, 641 (1979).

[6] H. Shirai, T. Hirayama, Y. Koide, M. Azumi, D.R. Mikelsen, and S.D. Scott, Ion Temperature Profile Simulation of JT-60 and TFTR Plasmas with Ion Temperature Gradient Mode, submitted to Nucl. Fusion; also in this conference.

[7] N.G. Van Kampen and B.V. Felderhof, Theoretical Methods in Plasma Physics (translated into Japanese by M. Nishida) Kinokuniya, Tokyo, p. 153, 1973.

[8] T. Yamagishi, Transport Theory and Statistical Physics **3**, 107 (1973).

[9] A. Hirose, Phys. Fluids B **5**, 281 (1993).

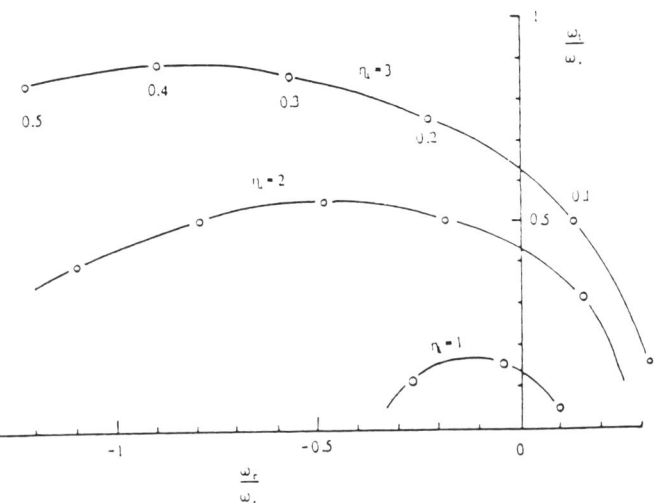

Figure 1. Variation of discrete η_i mode eigenvalue of the double integral dispersion relation for $\tau = 1$ and $b = 0$, and various values of ε_n and η_i.

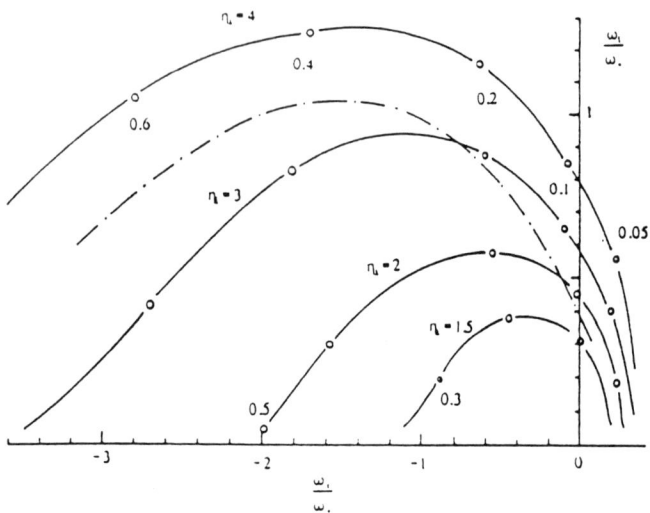

Figure 2. Variation of discrete η_i mode eigenvalue of the single integral dispersion relation for $\tau = 1$ and $b = 0$, and various values of ε_n and η_i. The dotted broken curve is the eigenvalue trajectory corresponding to the trajectory in Fig. 3.

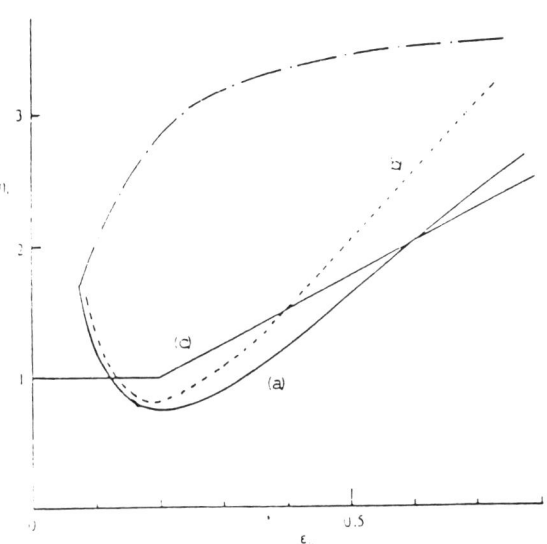

Figure 3. Stability boundary of η_i mode in (ε_n, η_i)-plane for (a) dispersion relation (5), (b) dispersion relation (6), (c) Romanelli [2]. The dotted broken curve represents variation of (ε_n, η_i) from center (right) to the edge (left) for $\alpha_n = 0.5$, $\alpha_T = 2.0$ and $\alpha_{T0} = 0.1$.

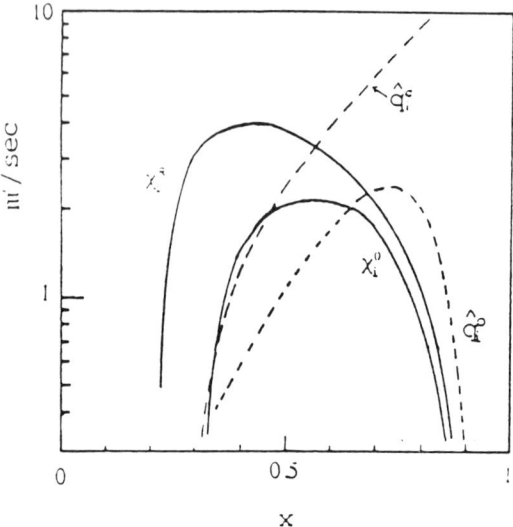

Figure 4. Variations of thermal flux $\hat{q}_i^0$ due to discrete eigenvalue and heat conductivities χ_i^0 and χ_i^c for $T_0 = 5.7\,\text{keV}$, $B = 4.5\,\text{T}$, $a = 0.9\,\text{m}$, $R = 2.9\,\text{m}$ and the same profile parameters as in Fig. 3. The normalized flux $\hat{q}_i^0$ has no dimension.

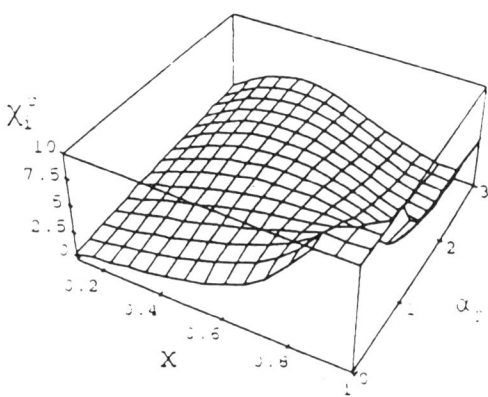

Figure 5. Surface graphics of heat conductivity χ_i^c given by Eq. (18) for $\alpha_n = 0.5$, $\alpha_{n0} = 0$, $\alpha_{T0} = 0.1$, and $\rho_i^2 v_i/R [\text{m}^2/\text{sec}]$ at the center $x = 0$.

6.5 ANOMALOUS PARTICLE AND ELECTRON ENERGY PINCHES

R.R. Dominguez
General Atomics, San Diego, California 92186-9784

Abstract. Interest in plasma pinches has recently increased due to the observation in electron cyclotron heated (ECH) plasmas in the Doublet III-D (DIII-D) tokamak [Plasma Physics and Controlled Nuclear Fusion Research 1986 (International Atomic Energy Agency, Vienna, 1987), Vol. I, p. 159] of an electron energy pinch. Previous work on drift wave-induced particle and electron energy pinches concluded that they were only possible in strictly collisionless plasmas, $\nu_* = 0$, an assumption not applicable to electron cyclotron heating experiments. In the present work it is shown that a more detailed quasilinear analysis predicts a particle pinch at moderate collisionality, $\nu_* = 0.2$ to 1.0, when the electron temperature gradient parameter η_e is in the range $\eta_e = 2$ to 3. Further, the quasilinear analysis is expanded to include the effects of turbulent resonance broadening on the circulating electrons. It is then found that, in addition to a particle pinch (qualitatively the same as the quasilinear prediction), a simultaneous particle and electron energy pinch at moderate collisionality can occur at $\eta_e \simeq 1$, which is lower than in ECH experiments where $\eta_e = 2$ to 3.

6.5.1 INTRODUCTION

One of the several unresolved issues for the drift wave anomalous transport model of tokamaks is an explanation for the existence of a plasma particle and/or energy pinch. Moderate to high density tokamak discharges would seem to require a plasma particle pinch to maintain observed peaked density profiles [1, 2] in the central plasma region where there are no significant plasma sources. It has also been concluded from recent off-axis electron cyclotron heating (ECH) experiments [3] in DIII-D that an electron energy pinch is operative in the central plasma region. A possible mechanism to obtain plasma pinches at moderate values of the electron temperature gradient parameter η_e ($\eta_e \simeq 2$), based on the trapped electron response to ion temperature gradient (ITG) mode turbulence, has been recently suggested by several authors [4, 5, 6]. However, these calculations concluded that plasma pinches were only possible at very low collisionality $\nu_* \ll 1$, a condition not typically found in present tokamaks.

In the present work, the ITG mode turbulent pinch mechanism is reconsidered in a sheared slab model, first by performing a more detailed analysis using a standard quasilinear formulation of anomalous transport flows [7]. This is then extended by incorporating nonlinear corrections into the circulating electron response [8, 9] for both the stability analysis and calculation of the plasma flows.

The results obtained from revisiting standard quasilinear theory are significantly different from previous work [4, 5, 6] in several ways. First, it is found that a particle pinch can result at experimentally relevant values of the electron collisionality parameter ν_*, in the range $\nu_* = 0.2$ to 1.0, for $\eta_e = 2$ to 3 ($\eta_e = \partial \ln T_e / \partial \ln n_e$). This differs from the conclusions of Refs. 4 and 5 primarily because self-consistent normal mode solutions were not used to evaluate the quasilinear fluxes; in Ref. 6 the authors considered the toroidal ITG mode but only for the case $\nu_* = 0$. A second, surprising result is that the role of circulating electrons in determining the quasilinear particle and electron heat fluxes driven by ITG mode turbulence is generally larger than that of the trapped electrons. The effect of circulating electrons is usually neglected in ITG mode calculations; indeed, the electrons are often treated as adiabatic even though this is only justified for $\nu_* > 1$ [10]. This neglect is motivated by the belief that trapped electrons dominate the low $v_\parallel$ region of velocity space and therefore greatly reduce the accessible region for the circulating electron resonance $v_\parallel = \omega/k_\parallel$ [7]. However, in a sheared magnetic field, the spatial variation in $k_\parallel$, as well as the weighting due to spatial dependence of the eigenfunctions found in the quasilinear fluxes must be considered in determining the net effect of the circulating electrons. For the ITG mode, the heuristic picture of the effect of the trapped electron subtraction [7] is greatly modified by the spatial weighting, while for the dissipative trapped electron (DTE) mode the heuristic picture is qualitatively correct. It seems that this can be understood by noting that the ITG mode is quite narrow, with typically normal mode widths [11] $\Delta x < 2\rho_s$ ($\rho_s = c_s/\Omega_i$ with c_s the sound speed and Ω_i the ion cyclotron frequency); conversely, the DTE mode has mode widths $\Delta x \sim 4$ to $5\rho_s$. Thus the spatial region over which the trapped electron subtraction is dominant is much smaller for the ITG mode than for the DTE mode.

The extension of the stability and quasilinear transport analysis incorporates nonlinear effects into the circulating electron response. The nonlinearity of interest is the turbulent modification of the circulating electron response in a sheared slab geometry, originally considered by Hirschman and Molvig [7] and later analyzed in more detail by Diamond and Rosenbluth [8]. The circulating orbit modification due to turbulence in a sheared magnetic field produces resonance broadening which is fundamentally different than in the shear-free case, a difference which turns out to be crucial to the final results. Resonance broadening does not change the qualitative features of the particle pinch predicted from simple quasilinear theory. However, it predicts a new feature not found in simple quasilinear theory, a simultaneous particle and electron heat pinch at moderate ν_* for $\eta_e \simeq 1$. This prediction seems to be qualitatively consistent with ECH experiments [3]. but the experimental η_e is larger, $\eta_e = 2$ to 3. This latter situation can occur when the magnetic shear-induced coupling between $\mathbf{E} \times \mathbf{B}$ and parallel electron motion dominates the ambient $\mathbf{E} \times \mathbf{B}$ turbulent motion in the circulating electron response.

The organization of the paper is that the fundamental eigenmode stability and quasilinear transport relations are discussed in Sec. 2; these complicated results have been given in a number of previous works [7, 10] and are not repeated here. Section 3 presents the numerical results based on the material of Sec. 4. In Sec. 4, the circulating electron resonance broadening effect is discussed and the way this is incorporated into the formulation of Sec. 2 is noted. Numerical results from this expanded model are also presented here. The final section summarizes the conclusions from the present calculation.

6.5.2 BASIC EQUATIONS

The formulation of the basic electrostatic eigenmode stability analysis and associated quasilinear flux calculations that are employed here are well known [7, 10]. It is sufficient to note the main assumptions without reproducing the lengthy results here. The perturbed density responses to an electrostatic perturbation are calculated in a sheared slab geometry. The ion and trapped electron responses (which are fully kinetic) are taken to be linear throughout. The circulating electron response is initially assumed linear, but later modified to account for turbulent resonance broadening. The eigenmode stability differential equation for the perturbed potential $\delta\Phi$ is a second order differential equation $d^2\delta\Phi/d\bar{x}^2 + b_x(\bar{x})\delta\Phi = 0$ with $\bar{x} = x/\rho_s$ and $b_x(\bar{x}) = -(\tau Q_i + Q_e)/(\partial Q_i/\partial b_{\theta i})$ [7, 10].

The response functions $Q_{i,e}$ are given by

$$Q_i = 1 + \left\{\left(1 - \frac{\omega_i^*}{\omega}\right)\Gamma_0 + \frac{\eta_i}{2}\frac{\omega_i^*}{\omega}\left[\Gamma_0 - 2b_{\theta_i}(\Gamma_1 - \Gamma_0)\right]\right\}\lambda_i Z_0(\lambda_i)$$

$$- \eta_i \frac{\omega_i^*}{\omega} \Gamma_0 \lambda_i Z_2(\lambda_i) \ , \tag{1}$$

and

$$Q_e = 1 + \left\{\left[1 - \frac{\omega_e^*}{\omega'}\left(1 - \frac{\eta_e}{2}\right)\right]\lambda_e Z_0^\dagger(\bar{\lambda}_e) - \eta_e \frac{\omega_e^*}{\omega}\lambda_e Z_2^\dagger(\bar{\lambda}_e)\right.$$

$$\left. + \left(1 - \frac{\omega_e^*}{\omega}\right)\epsilon^{1/2}A_2 - \eta_e \frac{\omega_e^*}{\omega}\epsilon^{1/2}A_1\right\} \ , \tag{2}$$

where ω_e^* is the electron (ion) diamagnetic drift frequency, $\lambda_i = \omega/|k_\| v_{T_i}|$, $\lambda_e = \omega/|k_\| v_{T_e}|$, $\bar{\lambda}_e = \lambda_e$ (a different $\bar{\lambda}_e$ is defined in Sec. 4),

$$A_{\{^1_2\}} = -2\pi^{-1/2}\int_{\nu_*^{1/2}}^\infty \frac{t^{1/2}\exp(-t)dt}{1 - (\omega_{de}/\omega)t + (i\nu_e/\epsilon\omega)t^{-3/2}}\left\{^{t-3/2}_{1}\right\} \ ,$$

$$Z_n^\dagger(\bar{\lambda}_e) = Z_n(\bar{\lambda}_e) - \left[Z_n(\bar{\lambda}_e') - \bar{Z}_n(\bar{\lambda}_e', \nu_*^{1/4})\right] \ , \tag{3}$$

$$\bar{Z}_n(\lambda, u) = \pi^{-1/2}\int_{-\mu}^{\mu} dt \frac{[t^n\exp(-t^2) - \mu^n\exp(-\mu^2)]}{t - \lambda} \ , \tag{4}$$

$$(\text{Im }\lambda > 0) \tag{5}$$

with $\bar{\lambda}'_e = \bar{\lambda}_e/\epsilon^{1/2}$ and $\omega_{de} = (L_{n_e}/R)\omega_e^*$. The eigenmode differential equation is solved using a "shooting" code and in the present work only the lowest order even ITG eigenmode is considered.

The plasma flows associated with the ITG mode instability are of primary interest here. The flows are evaluated in the quasilinear approximation [7, 10, 12]. The quasilinear transport equations are

$$\frac{\partial n_s}{\partial t} + \frac{\partial}{\partial x} \Gamma_x^{(s)} = 0 , \qquad (6a)$$

$$\frac{3}{2} \frac{\partial}{\partial t} (n_s T_s) + \frac{\partial}{\partial x} Q_x^{(s)} = \Delta_s , \qquad (6b)$$

where $s = i, e$. The radial energy flow is decomposed into conduction and convection, $Q_x^{(s)} = q_x^{(s)} + (5/2)T_s\Gamma_x^{(s)}$. The anomalous exchange terms Δ_s in Eq. (6b) are generally small for ITG mode turbulence [10] and will be neglected in the following discussion. The expressions for particle flux $\Gamma_x^{(e)}$ and the conduction heat flows $q_x^{(s)}$ are [7]

$$\Gamma_x^{(e)} = n_e L_{n_e} \int \frac{dk_\theta}{2\pi} \omega_e^* W_{k_\theta} \mathcal{H}_e^{(1)} , \qquad (7a)$$

$$q_x^{(e,i)} = n_{e,i} T_{e,i} L_{n_e} \int \frac{dk_\theta}{2\pi} \omega_e^* W_{k_\theta} \mathcal{H}_{e,i}^{(2)} , \qquad (7b)$$

with the amplitude factor

$$W_{k_\theta} = \int_{-\infty}^{+\infty} dx \left| \frac{e\delta\Phi}{T_e} \right|^2 \qquad (8)$$

and $\mathcal{H}_e^{(1)}$, $\mathcal{H}_{e,i}^{(2)}$ are weighting functions involving the linear eigenfrequency ω, the Larmor radius $b_{\theta_s} = (k_\theta \rho_s^2 = \tau b_{\theta_i}$ and the plasma profile parameters. The expressions for these weighting functions are very complicated and may be found in Refs. 7 and 10.

It is conventional and instructive to partition the particle and heat fluxes into contributions driven by density and temperature gradients. Thus the particle and electron heat fluxes can be expressed as $\Gamma_x^{(e)} = -D^{(e)}\partial n_e/\partial x + D_T^{(e)}(n_e/T_e)\partial T_e/\partial x$ and $q_x^{(e)} = -\kappa_T^{(e)}\partial T_e/\partial x - \kappa_n^{(e)}(T_e/n_e)\partial n_e/\partial x$. The weightings $\mathcal{H}_e^{(1)}$ and $H_{e,i}^{(2)}$ are divided as $\mathcal{H}_{s,n}^{(k)} + \eta_s \mathcal{H}_{s,T}^{(k)}$ ($s = e, i$,

$k = 1, 2$), so that $D_T^{(e)}/D_n^{(e)} = -\mathcal{H}_{e,T}^{(1)}/\mathcal{H}_{e,n}^{(1)}$ and $\kappa_T^{(e)}/\kappa_n^{(e)} = \mathcal{H}_{e,T}^{(2)}/\mathcal{H}_{e,n}^{(2)}$ (of course $D_n^{(e)}$, $D_T^{(e)}$, $\kappa_T^{(e)}$, and $\kappa_n^{(e)}$ generally depend on the density and temperature gradients as well). A particle pinch requires that $D_T^{(e)}/D_n^{(e)} > 0$ in order for the density gradient-induced plasma flow to balance the temperature gradient-induced plasma flow. Similarly, an electron heat pinch requires $\kappa_T^{(e)}/\kappa_n^{(e)} < 0$ in order for the density gradient-induced and temperature gradient-induced electron heat flows to balance.

6.5.3 QUASILINEAR FLUXES; NO RESONANCE BROADENING

The first topic considered is the existence of a particle pinch in the framework of strict quasilinear theory. A typical set of bulk plasma parameters is chosen, $\epsilon = 0.2$ ($\epsilon = r/R$), $\tau = 1$ ($\tau = T_e/T_i$), $L_{n_e}/L_s = 0.1$, $\eta_i = 2$, and $(k_\theta \rho_s)^2 = 0.2$. It is assumed that ITG mode turbulence generates an inverse cascade to long wavelength [13] so that the amplitude factor Wk_θ [Eq. (7)] is peaked at $b_{\theta_*} \ll 1$ and it is sufficient to examine the weighting factors $\mathcal{H}_e^{(1)}$ and $\mathcal{H}_{e,i}^{(2)}$ in Eq. (7) at $b_{\theta_*} \ll 1$. The threshold for the onset of a particle pinch driven by ITG mode turbulence is determined by varying the toroidicity parameter $\epsilon_n = L_{n_e}/R$ for a given η_e until $\mathcal{H}_e^{(1)} = 0$. The associated spectrum of linear normal mode eigenfrequencies for the ITG mode with $\mathcal{H}_e^{(1)} = 0$ is plotted versus η_e in Fig. 1 for several ν_*. The real and imaginary parts of the eigenfrequencies (normalized to c_s/L_{n_e}) are plotted in Fig. 1(a) and (b), respectively. The numerical fiduciaries in Fig. 1(b) are the threshold ϵ_n values at a given η_e; note that ν_* varies between the curves. It is clear that there is a large variation with ν_* in Fig. 1; in Ref. 4, ω was fixed while ν_* was varied. In Fig. 2 the threshold values of ϵ_n for the particle pinch are again plotted versus η_e for $\nu_* = 0.05, 0.2,$ and 1.0. As indicated in the figure, the region above each curve corresponds to inward particle flow while the region below corresponds to outward flow. The equivalent values of $\nu_{\text{eff}}/\omega_e^*$ are indicated along each of the curves. The vertical line at $\eta_e = 2.55$ in Fig. 2 is the threshold value for $\nu_* = 1.0$ when trapped electron effects are neglected. Comparing the threshold ϵ_n at $\nu_* = 1.0$ to the $\eta_e = 2.55$ vertical line, it is seen that trapped electron effects lower the critical η_e for $\epsilon_n > 0.46$ (the value of ϵ_n at the intersection of the two curves) but raise it for $\epsilon_n < 0.46$. Thus, for example, at small $\epsilon_n = 0.25$, the critical $\eta_e = 4$ versus $\eta_e = 2.55$ neglecting trapped electron effects; conversely, at larger $\epsilon_n = 0.6$, the critical $\eta_e = 2.3$ versus $\eta_e = 2.55$. Finally, the nonmonotonic behavior with ν_* evident in Fig. 2 is worth noting. The $\nu_* = 0.2$ curve obtains the maximum ϵ_n while for very small ν_*, $\nu_* < 0.01$, the threshold $\epsilon_n \to 0$. Thus in the collisionless limit $\nu_* \to 0$, a particle pinch obtains for all $\epsilon_n > 0$ when $\eta_e \geq 2$. A similar conclusion is obtained in Ref. 6, though for somewhat different parameters; a particle pinch occurs when $\epsilon_n > 0.2$ for $\eta_e \geq 2$.

The fact that the critical η_e obtained neglecting trapped electron effects, shown in Fig. 2, does not differ substantially from the full result suggests that trapped electron effects do not play a dominant role in obtaining a particle pinch. To investigate this conjecture in more detail, the contributions to the particle pinch condition $\mathcal{H}_e^{(1)} = 0$ of Fig. 2 from circulating and trapped electrons are determined separately. Recall that $\mathcal{H}_e^{(1)} = \mathcal{H}_{e,n}^{(1)} + \eta_e \mathcal{H}_{e,T}^{(1)}$. Each of the weighting coefficients $\mathcal{H}_{e,n}^{(1)}$ and $\mathcal{H}_{e,T}^{(1)}$ is the sum of a circulating and trapped electron contribution, $\mathcal{H}_{e,n}^{(1)} = \mathcal{H}_{e,n}^{(1,c)} + \mathcal{H}_{e,n}^{(1,t)}$ and $\mathcal{H}_{e,T}^{(1)} = \mathcal{H}_{e,T}^{(1,c)} + \mathcal{H}_{e,T}^{(1,t)}$. In Fig. 3(a) through (c) for $\nu_* = 1.0$, 0.2, and 0.05, respectively, the circulating and trapped electron weighting coefficients for density and temperature gradient-driven flows, $\mathcal{H}_{e,n}^{(1,c)}$, $\mathcal{H}_{e,T}^{(1,c)}$, $\mathcal{H}_{e,n}^{(1,t)}$, and $\mathcal{H}_{e,T}^{(1,t)}$, are plotted versus the critical ϵ_n at $\mathcal{H}_e^{(1)} = 0$. Note that the corresponding critical $\eta_e = -\mathcal{H}_{e,n}^{(1)}/\mathcal{H}_{e,T}^{(1)}$. Figure 3(a) and (b) ($\nu_* = 1.0$ and 0.2) are qualitatively the same and clearly show that $\mathcal{H}_{e,n}^{(1,c)}$ is substantially larger in magnitude than $\mathcal{H}_{e,n}^{(1,t)}$ and that $\mathcal{H}_{e,n}^{(1,t)} < 0$, i.e., the portion of the density gradient-driven flow from the trapped electrons is inward. It is also seen that $\mathcal{H}_{e,T}^{(1,c)}$ is larger in magnitude than $\mathcal{H}_{e,T}^{(1,t)}$, and again of opposite sign but $\mathcal{H}_{e,T}^{(1,t)} > 0$ so that the portion of the temperature gradient-driven flow from the trapped electrons is outward. In Fig. 3(c) ($\nu_* = 0.05$), the circulating electron weighting coefficients are qualitatively the same as at larger ν_*. The trapped electron-weighting coefficients behave differently, both changing sign at $\epsilon_n \simeq 0.2$. For $\epsilon_n > 0.2$, the behavior of the trapped electron-weighting coefficients is qualitatively the same as in Fig. 3(a) and (b). For $\epsilon_n < 0.2$, the direction of the density and temperature gradient-driven flows due to the trapped electrons reverses sign. However, the magnitude of the trapped electron flows is still less than the corresponding circulating electron flows. The conclusion from Fig. 3 is that over a broad range in ν_* the role of trapped electrons in establishing a particle pinch is always subordinate to the circulating electrons.

The dependence of the critical ϵ_n for the particle pinch on magnetic shear, L_{n_e}/L_s, is shown in Fig. 4 for $\eta_e = 2$ and 3. The other parameters for the figure are $\epsilon = 0.2$, $\tau = 1$, $\eta_i = 2$, $\nu_* = 1.0$, and $b_{\theta_s} = 0.2$. There is substantial variation in ϵ_n over the range of L_{n_e}/L_s in Fig. 4 for both $\eta_e = 2$ and 3. The trend of decreasing ϵ_n with decreasing L_{n_e}/L_s is consistent with the requirement of establishing a particle pinch in the central plasma region since the magnetic shear becomes progressively weaker in the central region.

The other topic of interest is the electron heat pinch. Again, previous theoretical work [4, 6] has shown the possibility of an electron heat pinch, but only for $\eta_e \ll 1$ in collisionless plasmas. Recent two-dimensional (no shear) fluid simulations [14] of coupled ITG and DTE mode turbulence concluded that an electron heat pinch was possible at moderate electron collisionality, $\nu_* = O(1)$, for $\eta_e = 2$, although no particle pinch was reported. On the basis of the strict quasilinear theory of Sec. 2 for the sheared slab model, results similar to Ref. 6 are obtained for the occurrence of an electron heat pinch. This is shown in Fig. 5 where the threshold η_e for the electron heat pinch, $\mathcal{H}_e^{(2)} = 0$ though $\mathcal{H}_e^{(1)} > 0$, is plotted versus ν_*; note that in the figure ϵ_n is fixed, $\epsilon_n = 0.6$ which is typical of the central plasma region where electron heat pinches are observed [3]. Additional parameters in the figure are $\epsilon = 0.2$, $\tau = 1$, $\eta_i = 2$, and $b_{\theta_s} = 0.2$. In the figure the pinch occurs in the region below the curve. The decomposition of the heat pinch condition $\mathcal{H}_e^{(2)} = 0$ of Fig. 5 again shows that the circulating electron contribution is dominant in both the density and temperature gradient driven heat flow contributions. At $\nu_* = 0$ the critical $\eta_e = 0.7$. This differs somewhat from Ref. 6 where at $\eta_e = 0.7$, ϵ_n was constrained to be smaller, $\epsilon_n \simeq 0.25$. It should also be noted that unlike Ref. 6, there is no coincidental ion heat pinch since $\mathcal{H}_i^{(2)} > 0$ for all cases considered. The small critical η_e at $\nu_* > 0.1$, along with the fact that there is a substantial outward particle flux, indicates that this does not offer a credible explanation for the heat pinch seen in ECH experiments [3].

6.5.4 QUASILINEAR FLUXES; RESONANCE BROADENING

In this section the circulating electron response is modified to include the effect of turbulent resonance broadening [8, 9]. After discussing the qualitative features of the resonance broadening effect, a simple model is employed to incorporate these features into both the stability analysis and quasilinear flux calculations summarized in Sec. 2.

The modification of the circulating electron response due to turbulent resonance broadening is manifest through orbit integrals of the type

$$I\left(\omega, k_\| v_\|\right) = i\omega \int_0^\infty d\tau \, e^{i(\omega - k_\| v_\| + i\tau_\theta^{-1})\tau - (\tau/\tau_s)^3 (v_\|/v_{T_e})^2} , \quad (9)$$

where $k_\| = k_\|' x$, $k_\|' = k_\theta/L_s$, L_s is the magnetic shear length, $v_{T_e} = (2T_e/m_e)^{1/2}$, $\tau_\theta^{-1} = \langle k_\theta^2 \rangle D_{\theta\theta}$, $\tau_s^{-1} = [\langle (k_\|' v_{T_e})^2 \rangle D_{rr}/3]^{1/3}$, and $\langle k_\theta^2 \rangle$ is the spectral-averaged k_θ^2. The diffusivities in τ_θ^{-1} and τ_s^{-1} denote the poloidal and radial particle diffusivities, respectively. Note that the test particle diffusivities are always positive, $D_{rr} \propto \langle \delta r^2 \rangle$, $D_{\theta\theta} \propto r^2 \langle \delta \theta^2 \rangle$ [8, 9]. It will be assumed for simplicity that the turbulence is isotropic $D_{\theta\theta} = D_{rr} = D_\perp$. Note that the ways in which τ_θ^{-1} and τ_s^{-1} enter Eq. (9) are quite different. τ_θ^{-1} derives from the turbulent $\mathbf{E} \times \mathbf{B}$ motion, which is decoupled from the parallel electric motion. Magnetic shear causes the turbulence to couple radial $\mathbf{E} \times \mathbf{B}$ motion with parallel motion in a qualitatively different way, as indicated by the τ_s^{-1} term in Eq. (9). Typically, the latter mechanism is more effective, $\tau_s^{-1} \gg \tau_\theta^{-1}$. The ions and trapped electrons would also be subject to the turbulent $\mathbf{E} \times \mathbf{B}$ motion (neglected here); the shear-induced turbulent resonance broadening of the ion response is negligible because of the slower velocity of the ions along field lines.

The integral I in Eq. (9) may be evaluated in terms of a special function e_0 [15], $e_0(\alpha) = \int_0^\infty dx \, exp - (\alpha x + x^3/3)$, which is related to the Airy function. The result is

$$I = \frac{i\omega\tau_s}{3^{1/3}(v_\parallel/v_{T_e})^{2/3}} e_0(\alpha), \qquad (10)$$

where $\alpha = \tau_s[\tau_\theta^{-1} - i(\omega - k_\parallel v_\parallel)]/3^{1/3}(v_\parallel/v_{T_e})^{2/3}$. for $\tau_s^{-1}, \tau_\theta^{-1} \to 0$, the usual linear result is recovered, $I = -\omega/(\omega - k_\parallel v_\parallel)$. To illustrate the behavior of the function I consider the case of real ω with $\tau_\theta^{-1} = 0$. Equation (10) can then be expressed as $I = i\Delta(\bar{v}_\parallel)^{-2/3}e_0[i\Delta(\bar{v}_\parallel - 1)\bar{v}_\parallel^{-2/3}]$ where $\bar{v}_\parallel = v_\parallel/(v_{T_e}\lambda_e)$, $\lambda_e = \omega/k_\parallel v_{T_e}$, and $\Delta = \omega\tau_s/(3^{1/3}\lambda_e^{2/3})$. In Fig. 6, the real and imaginary parts of I are plotted versus $\bar{v}_\parallel$ for several Δ. For $\Delta = 2$ (very strong turbulence, $\omega\tau_s \ll 1$), the resonance behavior at $\bar{v}_\parallel = 1$ is essentially eliminated, while at larger Δ (moderate turbulence, $\omega\tau_s \sim 1$), $\Delta = 10$, the rapid variation of I near $\bar{v}_\parallel = 1$ is apparent. For present purposes, the important point from the figure is that for large Δ, $Im\,I < 0$ outside of a narrow band centered at $\bar{v}_\parallel = 1$. Contrast this to the delta function at $\bar{v}_\parallel = 1$ of linear theory. The $\bar{v}_\parallel > 1$ range is most relevant for circulating electrons since the low velocity $\bar{v}_\parallel < 1$ region is likely dominated by the trapped electrons.

The qualitative features of Fig. 6 at large Δ (for $\bar{v}_\parallel \neq 1$) can be obtained from the large argument limit [15] of $e_0(\alpha) \sim \alpha^{-1} - 2\alpha^{-4} + \dots$ and Eq. (10) reduces to

$$I = \frac{-\omega}{\omega - k_\parallel v_\parallel + i\tau_\theta^{-1} - \dfrac{6\,i(v_\parallel/v_{T_e})^2}{\tau_s^3\left(\omega - k_\parallel v_\parallel + i\tau_\theta^{-1}\right)^2}}. \qquad (11)$$

Observe that in the shear-free case ($\tau_s^{-1} = 0$), Eq. (11) is exact. The resonance broadening from τ_θ^{-1} has the effect one would intuitively expect, which is that turbulence causes decorrelation (nonlinear damping). The effect of τ_s^{-1} is quite different as is apparent from Eq. (11) for $\tau_\theta^{-1} = 0$. In this case, the turbulence inhibits decorrelation (nonlinear driving) and $I \simeq -\omega[\omega - k_\parallel v_\parallel - 6\,i\tau_s^{-3}(k_\parallel v_{T_e})^{-2}]^{-1}$. For finite $\tau_\theta^{-1}, \tau_s^{-1}$ at moderate turbulence levels, $\omega\tau_s \simeq 1$, the turbulent modification to the circulating electron propagator cannot be characterized as a simple decorrelation process, although this assumption has been previously invoked [8]. Instead, there is a competition between the processes and the net effect may be manifest as either a nonlinear damping or driving.

To incorporate the qualitative features of resonance broadening into the framework of the previous formalism, a frequency shift is added to the circulating electron propagator $\omega \to \omega + i\omega_s$ where ω_s is real but either positive or negative. The parametric dependence of the nonlinear frequency shift represented by ω_s is quite complicated. To deduce a simple expression, Eq. (11) is approximated by $I = -\omega[\omega - k_\| v_\| + i\tau_\theta^{-1} - 6i\tau_s^{-3}(k_\| v_{T_e})^{-2}]^{-1}$ so that $\omega_s = [\tau_s/\tau_\theta - 6\tau_s^{-2}(k_\| v_{T_e})^{-2}]\tau_s^{-1}$. The complicated interplay between $\mathbf{E} \times \mathbf{B}$ diffusion and the shear-induced coupling of $\mathbf{E} \times \mathbf{B}$ and parallel motion is simply combined into a constant (positive or negative, depending on which process dominates) and ω_s is taken proportional to τ_s^{-1}. In units of c_s/L_{n_e}, $\omega_s = c_0 b_{\theta_s}^{1/3}(L_{n_e}/L_s)^{2/3}$ with c_0 either positive or negative but $|c_0| \lesssim 1$ to be consistent with the moderate turbulence condition $\omega\tau_s \sim 1$. Making the indicated change in the circulating electron response, the definition $\bar{\lambda}_e$ following Eq. (2) is changed to $\bar{\lambda}_e = (\omega + i\omega_s)/|k_\| v_{T_e}|$. The same change is made to electron weighting coefficients $\mathcal{H}_e^{(1)}$ and $\mathcal{H}_e^{(2)}$ in Eq. (7).

Typically, the effect of the resonance broadening correction on the normal mode eigenfrequencies is not large. To illustrate this point, Fig. 7 shows the real and imaginary parts of the eigenfrequencies (normalized to c_s/L_{n_e}) plotted versus η_e in the range $-2 \leq \eta_e \leq 2$ for several choices of c_0, $c_0 = -0.5, 0, 1.0$. Additional parameters in the figure are $\epsilon_n = 0.6$, $\epsilon = 0.2$, $\tau = 1$, $L_{n_e}/L_s = 0.1$, $\eta_i = 2$, $\nu_* = 1.0$, and $b_{\theta_s} = 0.2$. For $\eta_e < 0$, the nonlinear frequency shift $\omega_s < 0$ ($c_0 = -0.5$) has the largest effect and $\omega_s \simeq \text{Re}\,\omega$. For $\omega_s > 0$ ($c_0 = 1.0$), even large frequency shifts $\omega_s > |\text{Re}\,\omega|$ have a small effect. Compared to the strictly linear theory, the nonadiabatic electron response Q_e is virtually unchanged for $\omega_s > 0$ ($c_0 = 1.0$) while $\omega_s < 0$ ($c_0 = -0.5$) does have some effect on the ITG mode eigenfrequency.

The resonance broadening correction has a much larger effect on the particle and electron heat pinch boundaries. In Fig. 8, the critical ϵ_n for a particle pinch $[\mathcal{H}_e^{(1)} = 0]$ is plotted versus η_e for the same c_0 as in Fig. 7, $c_0 = -0.5, 0,$ and 1.0. In the figure $\nu_* = 1$ while the other parameters are the same as in Fig. 2. Along the $c_0 = -0.5$ curve the ratio of the real eigenmode frequency to ω_s is $\text{Re}\,\omega/\omega_s \simeq 0.7$ while for the $c_0 = 1.0$ curve $|\text{Re}\,\omega|/\omega_s \simeq 0.35$. Clearly the pinch boundary for $c_0 = -0.5$ has been dramatically lowered in ϵ_n from the strictly quasilinear case $c_0 = 0$. It is also important to note that the electron heat flux $\mathcal{H}_e^{(2)}$ along the particle pinch boundaries changes qualitatively between the $c_0 = 0$ and 1.0 curves and the $c_0 = -0.5$ curve. For the $c_0 = 0$ and 1.0 curves $\mathcal{H}_e^{(2)} > 0$, while along the $c_0 = -0.5$ curve $\mathcal{H}_e^{(2)}$ changes sign at $\eta_e \simeq 0.75$. This is indicated in Fig. 8 by the vertical line showing where $\mathcal{H}_e^{(2)} = 0$. The qualitatively new feature not found in strict quasilinear theory is that a particle and electron heat pinch can occur simultaneously. In reality the magnitude and sign of the nonlinear frequency shift modeled by ω_s is undoubtedly a complicated function of the plasma parameters (L_{n_e}/L_s, η_e, ..., etc.) and of the turbulence level. Figure 8 suggests that a particle pinch may sometimes be accompanied by outward electron heat flow ($c_0 = 0$ and 1.0) or in other cases by an inward heat flow ($c_0 = -0.5$), depending on the details that determine ω_s.

The electron heat pinch $\mathcal{H}_e^{(2)} < 0$ indicated in Fig. 8 occurs for $\eta_e < 1$. In order to establish the conditions for the occurrence of the electron heat pinch more fully, the critical η_e for the onset of the electron heat and particle pinches is plotted versus b_{θ_s} in Fig. 9. The parameters for the figure are $\epsilon_n = 0.6$, $\epsilon = 0.2$, $\eta_i = 2$, $\tau = 1$, $L_{n_e}/L_s = 0.1$, and $\nu_* = 1.0$. In Fig. 9(a) the boundaries for $\mathcal{H}_e^{(1)}$, $\mathcal{H}_e^{(2)} = 0$ for $c_0 = -0.5$ are shown with the shaded region indicating $\mathcal{H}_e^{(1)}$ and $\mathcal{H}_e^{(2)} < 0$. In Fig. 9(b) the boundaries for $\mathcal{H}_e^{(1)}$, $\mathcal{H}_e^{(2)} = 0$ for strict quasilinear theory ($c_0 = 0$) are shown; as previously noted, there is no overlap where both $\mathcal{H}_e^{(1)}$ and $\mathcal{H}_e^{(2)} < 0$. In the interval $0.1 \leq b_{\theta_s} \leq 0.4$ shown in Fig. 9, the critical η_e for $\mathcal{H}_e^{(2)} = 0$ increases rapidly in Fig. 9(a) while it is essentially constant in (b). Assuming that the amplitude weighting factor W_{k_θ} in Eq. (7) is peaked at $b_{\theta_s} \ll 1$, it follows from Fig. 9(b) that $q_x^{(e)} < 0$ (with $\Gamma_x^{(e)} \simeq 0$) for $\eta_e \simeq 0.5$ to 1.0. It must be noted though that this range of η_e, where there is a significant inward heat flow with small particle flow, is substantially less than reported from experiment [3] which obtains $\eta_e \simeq 2$ to 3.

6.5.5 SUMMARY

The present calculation has shown that a particle pinch is contained in the framework of a simple quasilinear theory for typical plasma parameters. However, an electron heat pinch is not found in this case. The addition of a model nonlinear frequency shift, representing the effect of resonance broadening on the circulating electrons, can change the conclusions substantially. A particle pinch can occur and be accompanied by either an outward or inward electron heat flux. The direction of the heat flux is determined by a competition between nonlinear processes that is modeled very simply in the present work. The predicted electron heat pinch occurs at values of η_e, $\eta_e \simeq 0.5$ to 1.0, that are lower than in the ECH experiments [3] where $\eta_e \simeq 2$ to 3 in the region of strong electron heat pinch.

The present work illustrates the sensitivity of anomalous electron fluxes to additional physics contained in the nonadiabatic electron response, which is often modeled very crudely using the so-called "$i\delta$" model. Certainly other effects besides resonance broadening in the circulating electron response may be important. One would not expect trapped electron nonlinearities to be important because ν_{eff} is generally larger than any other frequency of interest, including nonlinear frequency shifts. It also seems plausible that for typical turbulence levels, the lower frequency ITG mode would be more strongly affected than the higher frequency DTE mode (at least at moderate plasma beta). Perhaps the most interesting area for exploration on the subject of pinches is the role of magnetic fluctuations, particularly in regard to electron heat transport. Simulations of electromagnetic ITG mode turbulence [16] demonstrated that magnetic fluctuations drove an inward flow of electron heat, though the electrostatic fluctuations contributed a much larger outward flow.

A point that also should be emphasized is the nature of the anomalous particle and energy pinches discussed here; they occur in a plasma with normal density and temperature profiles (∇n_e, $\nabla T_e < 0$) because the particle and thermal diffusivities become negative. This is different from a critical gradient model or the situation where a diffusive loss mechanism generating an outward flow is offset by a nondiffusive inward flow mechanism, for example, $\Gamma^{(e)} = -D\nabla n_e - \mathbf{V} n_e$. An example of the latter is the neoclassical Ware pinch [17] that obtains $\Gamma_x^{(e)}$ and $Q_x^{(e)} < 0$, though the magnitude of the pinch velocity ($\lesssim 1$ m/sec) is too small to explain experimental results. The effective inward pinch velocity obtained from the present work is of order $\langle |e\delta\Phi/T_e|^2 \rangle c_s$, ($c_s$ is the sound speed), which is typically in excess of 10 m/sec. Finally, a conjecture about the difference between ECH heated and neubral beam heated (NBI) [18] plasmas in DIII–D is that the density profile is the crucial element. In the higher density, NBI heated H–mode plasmas, the electron density profiles are broad [19] while they are much more peaked in lower density ECH heated plasmas [3]. Even though the T_e profile is more peaked in the latter case, the net effect is that η_e is in the range $2 \leq \eta_e \leq 3$ for ECH heated plasmas, while $\eta_e \geq 3$ for the NBI heated plasmas. The smaller η_e characteristic of the ECH plasmas could correlate with the appearance of an energy pinch, although the experimental range is higher than the theoretical prediction. A corollary to this hypothesis is that peaking the density profile of an off-axis NBI heated plasma could also produce an electron energy pinch.

ACKNOWLEDGMENTS

This is a report of research sponsored by the U.S. Department of Energy under Grant No. DE-FG03-92ER54150; such financial support does not constitute an endorsement by DOE of the views expressed herein.

Bibliography

[1] M. Gaudreau, A. Gondhalekar, M. Hughes, D. Overskei, D. Pallas, R. Packer, S. Wolfe, E. Apgar, H. Helava, I. Hutchinson, E. Marmar, and K. Molvig, Phys. Rev. Lett. **39**, 1266 (1977).

[2] M. Greenwald and R. Parker, Bull. Amer. Phys. Soc. **22**, 1154 (1983).

[3] T.C. Luce, C.C. Petty, and J.C.M. de Haas, Phys. Rev. Lett. **68**, 52 (1992).

[4] R.E. Waltz and R.R. Dominguez, Phys. Fluids B **1**, 1935 (1989).

[5] P.W. Terry, Phys. Fluids B **1**, 1932 (1989).

[6] H. Nordman, J. Wieland, and A. Jarmèn, Nucl. Fusion **30**, 983 (1990).

[7] R.E. Waltz, W. Pfeiffer, and R.R. Dominguez, Nucl. Fusion **20**, 43 (1980).

[8] S.P. Hirschman and K. Molvig, Phys. Rev. Lett. **42**, 648 (1979).

[9] P.H. Diamond and M.N. Rosenbluth, Phys. Fluids **24**, 1641 (1981).

[10] R.R. Dominguez and G.M. Staebler, *Impurity Effects on Drift Wave Stability and Impurity Transport* (1992), to be published in Nucl. Fusion.

[11] R.R. Dominguez, Nucl. Fusion **31**, 2063 (1991).

[12] G.M. Staebler and R.R. Dominguez, Nucl. Fusion **31**, 1891 (1991).

[13] S. Hamaguchi and W. Horton, Phys. Fluids B **2**, 1883 (1990).

[14] O.T. Kingsbury and R.E. Waltz, Phys. Fluids B **3**, 3539 (1991).

[15] L.N. Nosova and S.A. Tumarkin, *Tables of Generalized Airy Functions*, Pergamon Press (1965).

[16] R.E. Waltz, Phys. Fluids **29**, 3684 (1986).

[17] F.L. Hinton and R.D. Hazeltine, Rev. Mod. Phys. **48**, 239 (1976).

[18] J. Luxon, P. Anderson, F. Batty, C. Baxi, G. Bramson, N. Brooks, B. Brown, B. Burley, K.H. Burrell, R. Callis, G. Campbell, T. Carlstrom, A. Colleraine, J. Cummings, L. Davis, J. DeBoo, S. Ejima, R. Evanko, H. Fukumoko, R. Gallix, J. Gilleland, T. Glad, P. Gohil, A. Gootgeld, R.J. Groebner, S. Hanai, J. Haskovec, E. Hackman, M. Heilberger, F.J. Helton, P. Henline, D. Hill, D. Hoffman, E. Hoffmann, R. Hong, N. Nosogane, C. Hsieh, G.L. Jackson, G. Jahns, G. Janeschitz, E. Johnson, A. Kellman, J.S. Kim, J. Kohli, A. Langhorn, L. Lao, P. Lee, S. Lightner, J. Lohr, M. Mahdavi, M. Mayberry, M. McHarg, T. McKelvey, R. Miller, C.P. Moeller, D. Moore, A. Nerem, P. Noll, T. Ohkawa, N. Ohyabu, T. Osborne, D. Overskei, P. Petersen, T. Petrie, J. Phillips, R. Prater, J. Rawls, E. Reis, D. Remsen, P. Riedy, P. Rock, K. Schaubel, D. Schissel, J. Scoville, R. Seraydarian, M. Shimada, T. Shoji, B. Sleaforth, J. Smith, Jr., P. Smith, T. Smith, R.T. Snider, R.D. Stambaugh, R. Stav, H. St. John, R. Stockdale, E.J. Strait, R. Street, T.S. Taylor, J. Tooker, M. Tupper, S.K. Wong, and S. Yamaguchi, in *Plasma Physics and Controlled Nuclear Research 1986* (International Atomic Energy Agency, Vienna, 1978), Vol. I, p. 159.

[19] D.P. Schissel, T.H. Osborne, J.C. DeBoo, J.R. Ferron, E.A. Lazarus, and T.S. Taylor, Nucl. Fusion **32**, 689 (1992).

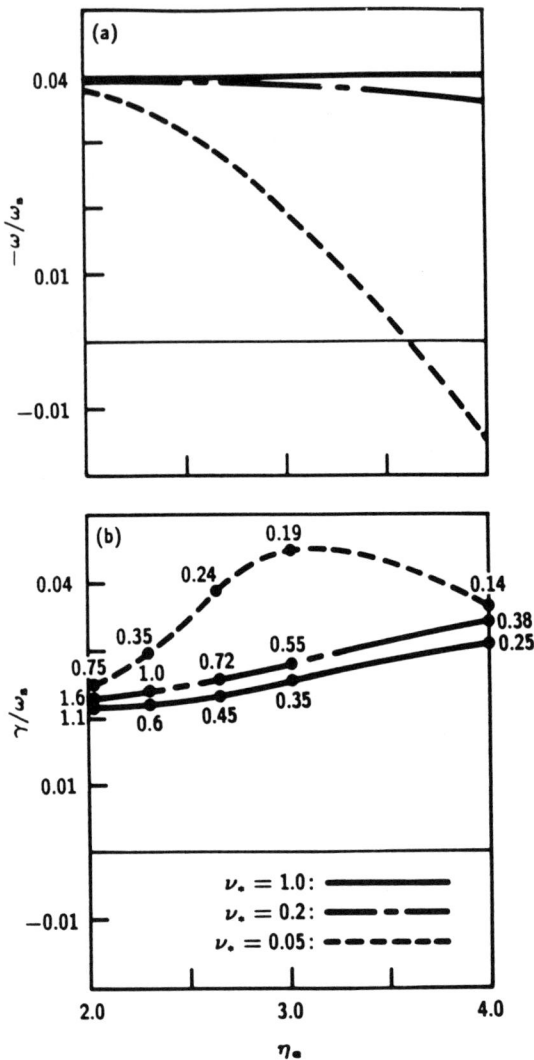

Figure 1. Real frequency (a) and growth rate (b) for ITG modes (normalized to c_s/L_{n_e}) versus η_e for $\nu_* = 0.05$, 0.2, and 1.0. Additional parameters are $b_{\theta_s} = 0.2$, $L_{n_e}/L_s = 0.1$, $\tau = 1$, $\epsilon = 0.2$, $\eta_i = 2$. No resonance broadening effects included.

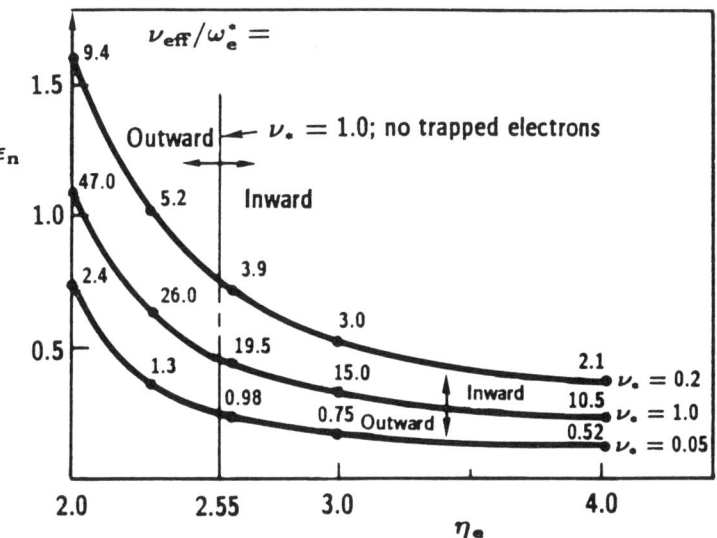

Figure 2. Critical ϵ_n for a particle pinch versus η_e for $\nu_* = 0.05$, 0.2, and 1.0. Other parameters are the same as in Fig. 1.

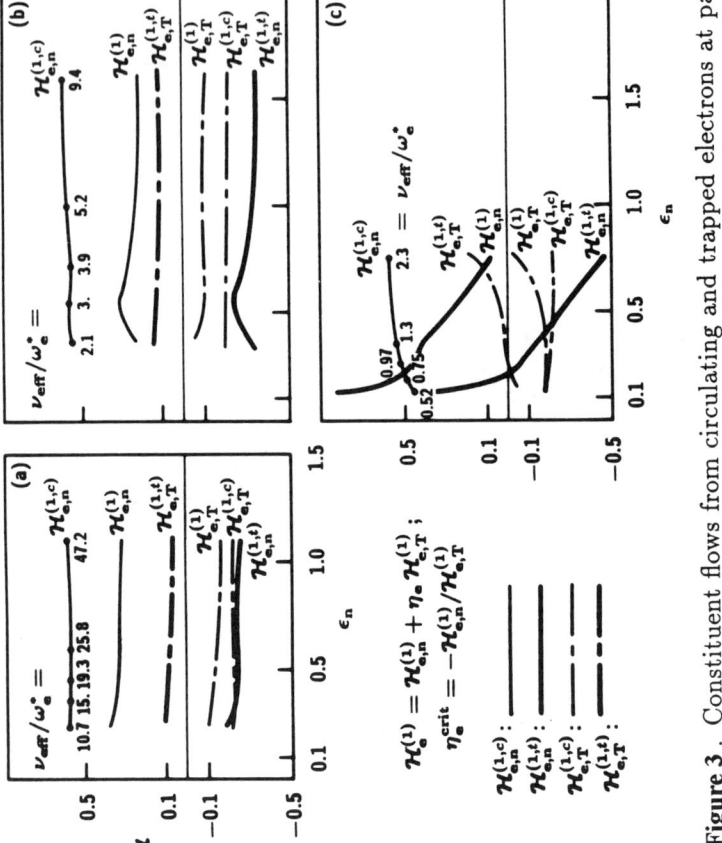

Figure 3. Constituent flows from circulating and trapped electrons at particle pinch boundary of Fig. 2 for (a) $\nu_* = 1.0$, (b) 0.2, and (c) 0.05.

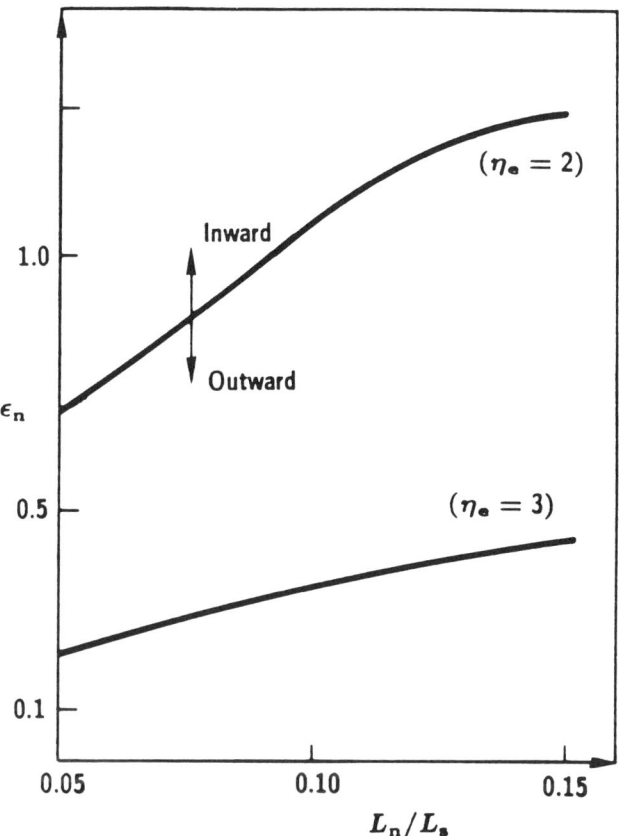

Figure 4. Critical ϵ_n for a particle pinch versus L_{n_e}/L_s for $\eta_e = 2$ and 3. Additional parameters are $b_{\theta_s} = 0.2$, $\tau = 1$, $\epsilon = 0.2$, $\eta_i = 2$, and $\nu_* = 1.0$. No resonance broadening effects included.

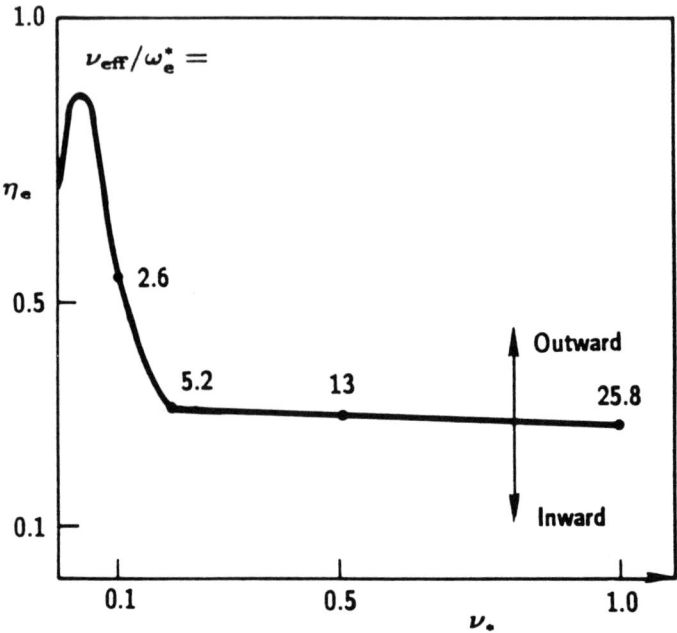

Figure 5. Critical η_e for an electron heat pinch versus ν_*. Additional parameters are $b_{\theta_s} = 0.2$, $L_{n_e}/L_s = 0.1$, $\tau = 1$, $\epsilon_n = 0.6$, $\epsilon = 0.2$, and $\eta_i = 2$. No resonance broadening effects included.

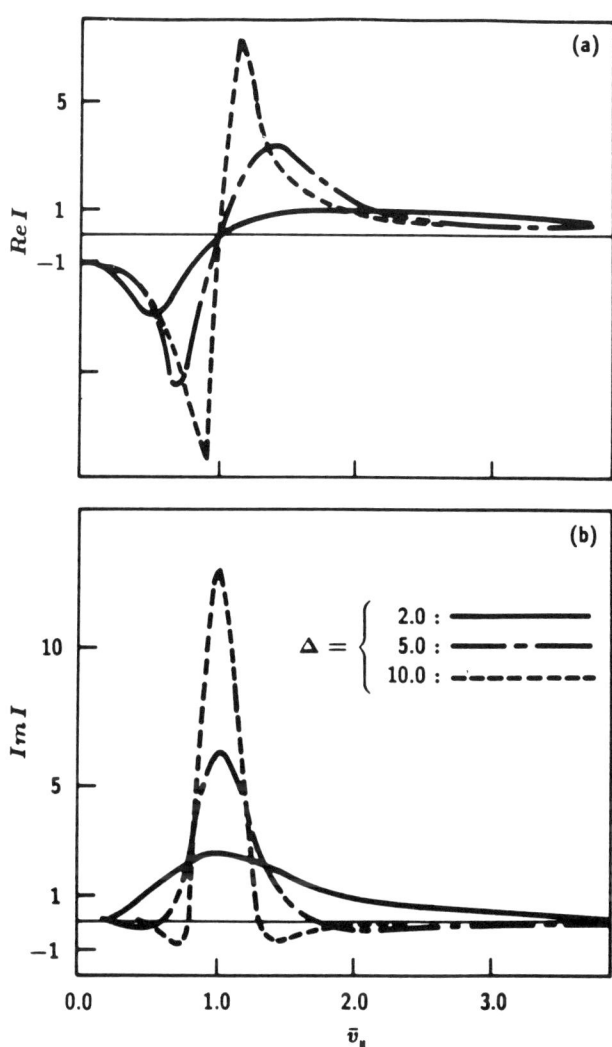

Figure 6. Real (a) and imaginary (b) parts of orbit integral I, Eq. (9), plotted versus $\bar{v}_\parallel$ for real ω and $\tau_\theta^{-1} = 0$ for $\Delta = 2$, 5, and 10.

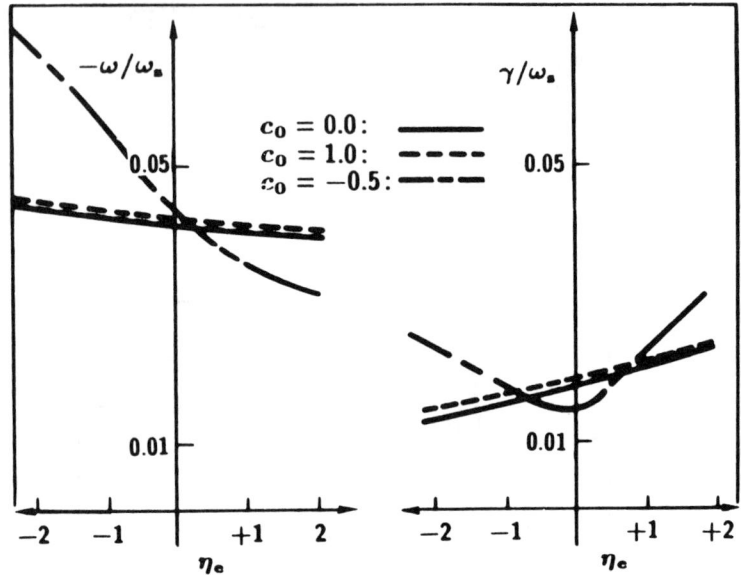

Figure 7. Real frequency (a) and growth rate (b) for ITG modes (normalized to c_s/L_{n_e}) versus η_e including effect of resonance broadening, $c_0 = -0.5$, 0, and 1.0. Additional parameters are $b_{\theta_s} = 0.2$, $L_{n_e}/L_s = 0.1$, $\tau = 1$, $\epsilon_n = 0.6$, $\epsilon = 0.2$, $\eta_i = 2$, and $\nu_* = 1.0$.

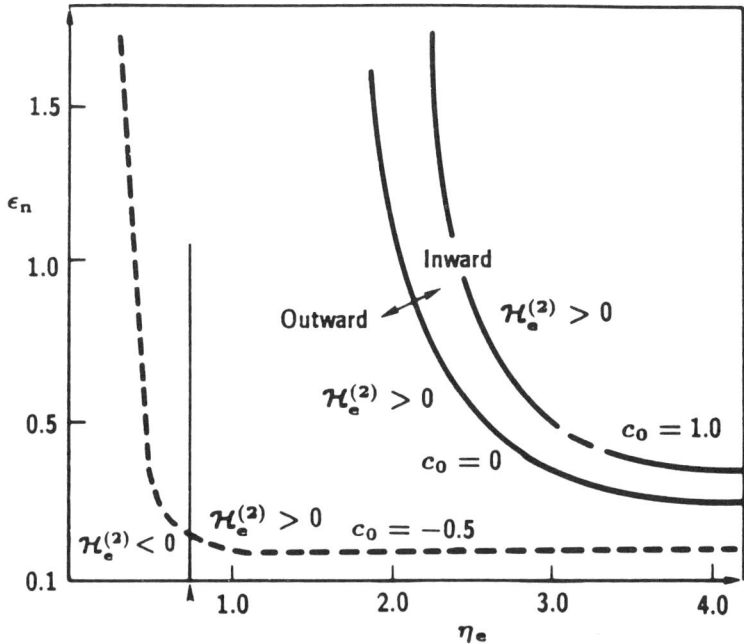

Figure 8. Critical ϵ_n for a particle pinch versus η_e including effect of resonance broadening, $c_0 = -0.5$, 0, and 1.0. Additional parameters are $b_{\theta_s} = 0.2$, $L_{n_e}/L_s = 0.1$, $\tau = 1$, $\epsilon = 0.2$, $\eta_i = 2$, and $\nu_* = 1.0$.

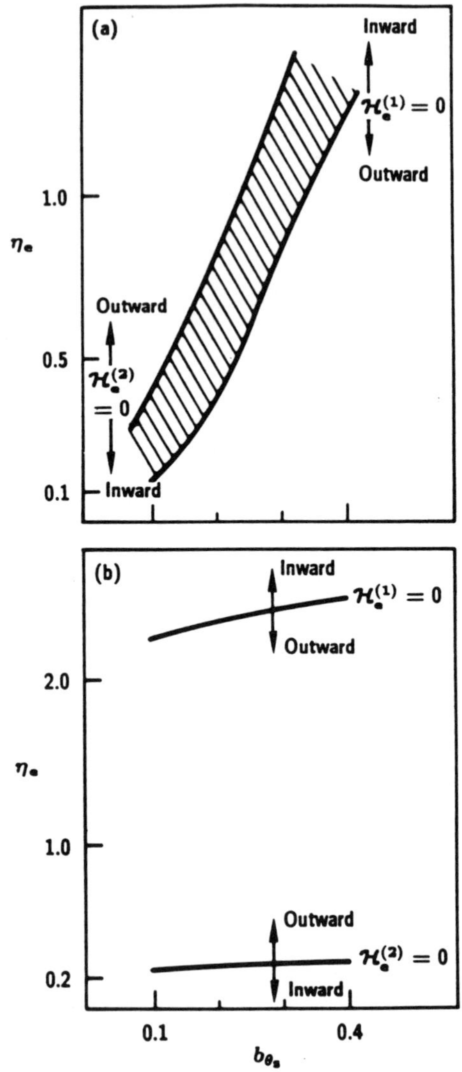

Figure 9. Critical η_e for a particle pinch and electron heat pinch versus b_{θ_s} for (a) $c_0 = -0.5$ and (b) 0. Shaded region is overlap where both $\mathcal{H}_e^{(1)}$ and $\mathcal{H}_e^{(2)} < 0$. Additional parameters are $L_{n_e}/L_s = 0.1$, $\tau = 1$, $\epsilon_n = 0.6$, $\epsilon = 0.2$, $\eta_i = 2$, and $\nu_* = 1$.

6.6 ION TEMPERATURE GRADIENT DRIVEN IMPURITY MODES

S. Migliuolo
Research Laboratory of Electronics, Massachusetts Institute of Technology, Cambridge, Massachusetts 02139-4307

Abstract. *The combined effects of collisions, an ion temperature gradient, and the presence of an impurity are studied in the context of a sheared slab magnetic configuration. Unstable electrostatic modes, that propagate in the electron diamagnetic direction, are found to exist for a wide range of parameters and to be unstable for both strong and weak magnetic shear. The linear eigenfunctions have extended structure in the radial direction.*

Theoretical understanding of transport phenomena in high temperature plasmas has long been a goal of the magnetic fusion program, since the early observation that particle and energy losses from the plasma are much faster than predicted by classical and neoclassical theories (cf. the review articles by Tang [1] and Liewer [2] as well as the transport analyses and simulations carried out by the Texas [3] and GA [4, 5] groups). With respect to this "anomalous" energy loss, theoretical modelling was given impetus by experiments [6, 7, 8], which indicated that peaked density profiles were beneficial to the containment of plasma energy, and the observation [9, 10] of fluctuations propagating in the ion diamagnetic direction, under conditions appropriate for the excitation of the ion temperature gradient instability [11, 12, 13, 14]. Since these early papers, a great deal of analysis has been published on the theory of these ion temperature gradient (ITG) modes, dealing with their linear and nonlinear properties. The drawback of models that depend on ITG modes (or, for that matter, any drift wave type of instability) for the anomalous transport of energy is that predicted diffusivities always scale with $v_{Ti}\rho_i^2/L_p \propto T_i^{3/2}$ and thus decrease sharply at the plasma edge, while experimentally inferred diffusivities peak at the edge (here v_{Ti}, ρ_i are the ion thermal velocity and gyro-radius, while L_p is a typical pressure gradient scale length). Though this "defect"

© 1994 American Institute of Physics

has been partially overcome by recent models that take into account the toroidal coupling of linear harmonics centered at different mode rational surfaces (thereby creating a radially extended linear eigenmode) [15, 16], this can only account for the region $r \leq 0.75a$, leaving the edge "uncovered." One possibility is that modes that are particular to the plasma edge can take up the slack and "carry" the energy over the last interval to the outside. One such candidate is the impurity drift mode, that is driven unstable by the temperature gradient of the plasma fuel (the primary ion population, which we assume to be hydrogenic, $Z_i = 1$), but which appears only in conjunction with the presence of a second ion species (the impurity). This mode has been studied previously, primarily in the local approximation [17, 18, 19, 20, 21, 22] (Tang, White, and Guzdar [23] analyzed this mode in collisionless plasmas and sheared magnetic fields, for reversed profiles $\eta_i \equiv d\ln T_i/d\ln n_i < 0$).

Note that the appearance of a new instability is one of (at least) three consequences that impurities have on the linear stability of plasmas. First, impurities "dilute" the concentration of the primary species: $n_i = n_e - Z_I n_I$ (here, i, I denote the primary ion and the impurity, respectively). This effect is beneficial in that the source of free energy for the ITG instability is diminished (local analyses [19, 20, 21, 22] have found a large increase in the critical value of η_i in the presence of impurities). The second is the possibility of triggering an impurity ITG mode (driven by η_I); this is likely to be a weaker instability, since ITG modes have growth rates that scale (not necessarily linearly) with the thermal velocity of the species driving the instability (far smaller for heavy impurities). Third, the new mode appears; we denote this the impurity drift wave, partly because it generally propagates in the electron diamagnetic direction, and partly to distinguish it from the impurity ITG mode discussed above.

In the present work, we consider these impurity drift modes anew, in a sheared slab magnetic geometry, $\mathbf{B} = B_0(\hat{e}_z + \hat{e}_y x/L_s)$, for parameters appropriate to the plasma edge. In particular, the effect of finite ion parallel thermal conductivity is examined, as it represents the "prototypical" dissipation required for the linear destabilization of these modes. Note that we make no attempt to analyse the extreme edge region (near the scrape-off layer), where spatially varying equilibrium electric fields are known to occur and to influence varying microinstabilities (including the ITG mode, cf. [24, 25] for early analyses of the linear stability theory). Since the parallel phase velocity of the perturbations is much smaller than the electron thermal speed, the common approximation of an adiabatic electron response, $\tilde{n}_e = en_e\tilde{\phi}/T_e$, is adequate (note: $\tilde{\phi}$ is the perturbed electrostatic potential; perturbations may be taken to be electrostatic at the plasma edge). Fluid theory is used to describe the ions. For reasonably small impurity density concentrations, the equations for conservation of parallel momentum and energy of the impurity ion species are dominated by the impurity–primary collisional friction term (cf. Ref. [18], we will return to this point in the Appendix, where the opposite limit is briefly discussed). As a consequence, we set $\tilde{u}_{\|I} = \tilde{u}_{\|i}$ and $\tilde{T}_I = \tilde{T}_i$ (where i, I refer to the primary and impurity ion species, respectively).

The remaining equations (cf. [26]) are the two continuity equations ($j = i, I$):

$$\omega \frac{\tilde{n}_j}{n_j} + \sigma_j \omega_* \tilde{\psi} - k_\| \tilde{u}_{\|i} = \frac{1}{2}\left[\omega - \frac{\sigma_j}{\tau Z_j}\omega_*(1+\eta_j)\right]\rho_{sj}^2 \nabla_\perp^2 \tilde{\psi} \quad (1)$$

($\tilde{\psi} \equiv e\tilde{\phi}/T_e$ denotes the normalized perturbed potential, $\sigma_j = d\ln n_j/d\ln n_i$ measures the relative peaking of the density profiles, $\rho_{sj} \equiv \sqrt{m_j T_e/Z_j c}$ is the sound Larmor radius (relative to species mass), $\tau \equiv T_e/T_i$, and $\omega_* \equiv k_y cT_e/eB_0 L_n$ is the drift frequency, defined with the electron temperature and the density scale length, L_n, of the primary ion, and $k_\| \equiv k_y x/L_s$, $\nabla_\perp^2 = d^2/dx^2 - k_y^2$), the momentum conservation equation (primary ions):

$$\left(\omega + i\frac{\mu}{\tau}\frac{k_\|^2 c_s^2}{\nu_{ii}}\right)\tilde{u}_{\|i} = k_\| c_s^2\left(\frac{1}{\tau}\frac{\tilde{n}_i}{n_i} + \frac{\tilde{T}_i}{T_e} + \tilde{\psi}\right) \quad (2)$$

and the energy conservation equation (primary ions):

$$\left(\omega + i\frac{2\chi}{3\tau}\frac{k_\|^2 c_s^2}{\nu_{ii}}\right)\frac{\tilde{T}_i}{T_e} = -\frac{1}{\tau}\eta_i\omega_*\tilde{\psi} + \frac{2}{3\tau}k_\|\tilde{u}_{\|i} + \frac{1}{3\tau}\left[\omega - \frac{1}{\tau}\omega_*\left(1+\frac{7}{2}\eta_i\right)\right]\rho_{si}^2\nabla_\perp^2\tilde{\psi} \quad (3)$$

where μ and χ are coefficients of order unity that characterize the strength of the parallel transport coefficients (perpendicular transport coefficients are neglected; we shall return to them at the end of this work), and $c_s = \sqrt{T_e/m_i}$ is the sound speed.

In order to illustrate the instability of the impurity drift wave, we consider the limit in which $\mu = 0$, $\omega \ll \omega_*$, $\nu_{ii}\omega \ll k_\parallel^2 c_s^2$, and $Z_I n_I/n_e \ll 1$. Simple algebra leads to:

$$\frac{5i}{6}\frac{\omega_*}{\omega_\kappa}\eta_i\rho_{si}^2\nabla_\perp^2\tilde{\psi} + \left\{\left(\frac{x}{\rho_{si}}\right)^2\left[1+\tau+\frac{Z_I n_I}{n_e}(\sigma_I - 1)\frac{\omega_*}{\omega}\right] - i\tau\frac{\omega_*}{\omega_\kappa}\left(\frac{2}{3}-\eta_i\right)\right\}\tilde{\psi} \quad (4)$$

where $\omega_\kappa \equiv (2/3)(k_y\rho_{si}c_s)^2/\nu_{ii}L_s^2$. This Weber equation is easily solved, leading to a solution in terms of Hermite functions:

$$\tilde{\psi} = \tilde{\psi}_0 H_N(\sqrt{\alpha}x/\rho_{si})\exp\left[-\frac{\alpha}{2}\left(\frac{x}{\rho_{si}}\right)^2\right] \quad (5)$$

where

$$\alpha \equiv (6\tau/5\eta_i)(\eta_i - 2/3)(2N+1)^{-1}$$
$$= \sqrt{\frac{6i}{5\eta_i}\frac{\omega_\kappa}{\omega_*}\left[1+\tau+\frac{Z_I n_I}{n_e}(\sigma_I - 1)\frac{\omega_*}{\omega}\right]} \quad (6)$$

This dispersion relation (6) yields the eigenfrequency:

$$\omega = -\frac{Z_I n_I}{n_e}(\sigma_I - 1)\omega_*\left\{1+\tau+\frac{6}{5}i\eta_i\frac{\omega_*}{\omega_\kappa}\left[\tau\frac{\eta_i - 2/3}{2N+1}\right]^2\right\}^{-1} \quad (7)$$

which shows a mode propagating in the electron diamagnetic direction, driven unstable by the primary ion temperature gradient, with the usual $\eta_i \geq 2/3$ fluid threshold. Note that, like standard ITG modes (cf. [27]), higher harmonics ($N > 1$) are more extended in radius. Clearly, optimal conditions for instability are those where the three characteristic frequencies of the system, ω_κ (denoting dissipation), $\omega_{*T} = \eta_i\omega_*$ (denoting the ITG frequency), $(n_I/n_e)\sigma_I\omega_*$ (the impurity drift wave frequency, weighted by the impurity fraction) are commensurate. Each frequency represents a "mode" coupling for the system: the first yields the collisional coupling of the drift wave to the sound wave, the second couples the parallel ion dynamics to the temperature perturbation, the third connects the natural "drift wave" carried by the impurities to the primary ion dynamics. When one of the three frequencies is much higher than the rest, the three "modes" decouple and the instability is turned off (this criterion is "elastic" with respect to ω_{*T}, this frequency can be an order of magnitude greater than the other two and still cause an instability; this is because η_i represents the free energy for the mode). Also when dissipation (thermal conductivity) is either absent or infinite (isothermal primary ions), the instability disappears. If the ion density gradients are equal, $\sigma_I = 1$, a weak residual instability may be provided by the difference in FLR contributions to the two continuity equations; like any other drift-type mode, this instability is sensitively dependent to charge separation between the non-adiabatic particle responses.

In order to study trends, Eqs. (1)–(3) are solved numerically with the help of a shooting code. For purposes of illustration, we adopt parameters appropriate to the edge region of, e.g. JET [28], choose oxygen ($A_I = 16$, $Z_I = 6$; we consider temperatures in the $100 - 200\,\text{eV}$ range) as the impurity ion, take $\sigma_I = 2$ as an example and vary the rest of the parameters (note that we assume equal temperatures for all species, thus $\eta_I = \eta_i/\sigma_I$). As expected (cf. Eq. (7)), charge concentration $\hat{\rho}_I \equiv Z_I n_I/n_e$, rather than density concentration (n_I/n_e), plays the important role in determining the growth rate. An example is shown in Fig. 1: the residual changes for the case where charge concentration is kept constant are due to the FLR impurity terms. The charge of the impurity tends to mostly influence the real part of the frequency (as Z_I decreases, the mode frequency also decreases in absolute value). Parallel ion viscosity (μ) has a small stabilizing effect, notable only near marginal stability. Pure FLR (in the sense of the $k_y \rho_{si} < 1$ contribution to the operator part of Eqs. (1) and (3)) has a negligible effect on the growth rate. The growth rate follows a nonlinear trend with any one parameter (L_s/L_n, η_i, $\hat{\rho}_i$, ω_κ). Typically, it increases from zero as the parameter exceeds a threshold, reaches a maximum (where the three "modes" couple most strongly) and then decays asymptotically as the parameter is taken to very large values. Note that there never is re-stabilization (i.e., γ remains positive) in the absence of parallel viscosity ($\mu = 0$). Restabilization may also occur~in the presence of a second source of dissipation (i.e., an energy sink), that may be provided by breaking the strong coupling approximation, $\tilde{T}_I = \tilde{T}_i$, thereby introducing the thermal conductivity of the impurities (an example showing such a trend is discussed in the Appendix). Fixing $k_y \rho_{si} = 0.01$, $\tau = 1$, $\sigma_I = 2$, $A_I = 16$, and $Z_I = 6$ in all that follows, we show the trend against magnetic shear (L_s/L_n) in Fig. 2. Clearly strong shear, $L_s = L_n$, is unable to stabilize the mode. Weak shear $(L_s \gg L_n)$ makes for low thermal conductivity and poor coupling to the temperature perturbation. Figure 3 shows the trends with respect to the parallel thermal conductivity parameter; again we have instability over a considerable range of values. Figure 4 shows the eigenfrequency as a function of the ion temperature gradient parameter; note that maximum growth (for this case) occurs in the experimentally meaningful range $1 \leq \eta_i \leq 2$ (for the standard ITG mode, the maximum is somewhat broader and occurs at $\eta_i \sim 2-5$).

Figure 5 shows the parameters for maximum growth (in terms of impurity charge concentration and ion collisionality, for $L_s = 6L_n$) indicating that the mode has appreciable growth (a finite fraction of an inverse transit time for the acoustic wave) over a large region of parameter space. Note that there is remarkably little variation in this curve when we change charge states from $Z_I = 6$ to fully ionized oxygen. A typical eigenmode will extend over several ion gyral radii: its wavelength of oscillation will often be $\geq 3\ \rho_{si}$, while the envelope, $|\tilde{\psi}|$ will span 10-15 ρ_{si}; a typical case is shown in Fig. 6.

Perpendicular transport coefficients can also be considered, in the limit of small parallel thermal conductivity and parallel viscosity, leading to a second order differential equation in conjugate (i.e., Fourier) space (cf. [26]). An exhaustive search has failed to find any unstable impurity drift eigenmodes for parameters that can reasonably be applied to the edge of the plasma (the standard ITG root, mildly stabilized by $\mu_\perp$, was easily identified). In conclusion, the instability of the "impurity drift wave," destabilized by the ion temperature gradient, has been shown to survive in the presence of magnetic shear and parallel thermal conduction, and to be a good candidate for the anomalous transport of energy over the outer portion of the plasma column. A nonlinear analysis, in the same spirit as that carried out for the ITG mode [29], will begin shortly and will, hopefully, elucidate the properties of the nonlinear convective cells that are ultimately responsible for transport.

This work was sponsored in part by the U.S. Department of Energy, under contract DE-FG02-91ER-54109. The author wishes to thank B. Coppi for encouraging this research.

APPENDIX

Throughout this work, we have used the simplifying assumption that friction forces between the primary and impurity ions dominate the equations of conservation of momentum and energy for the impurity species. Referring, for instance, to the energy conservation equation in the limit where $T_I = T_e$:

$$\left(\omega + i\frac{2\chi}{3}\frac{k_\parallel^2 T_I}{m_I \nu_{II}}\right)\frac{\tilde{T}_I}{T_e} = -\sigma_I \eta_I \omega_* \tilde{\psi} + \frac{2}{3}k_\parallel \tilde{u}_{\parallel I} + \frac{A_I}{3}\left[\omega - \frac{\sigma_I}{Z_I}\omega_*\left(1 + \frac{7}{2}\eta_I\right)\right]\rho_s^2 \nabla_\perp^2 \tilde{\psi}$$

$$+ 2i\frac{n_i m_i}{n_I m_I}\nu_{iI}\left(\frac{\tilde{T}_i}{T_e} - \frac{\tilde{T}_I}{T_e}\right) \quad (A1)$$

we see that the ratio of the energy exchange term (due to friction) to the parallel thermal conduction term is:

$$\left(\frac{n_i m_i}{n_I m_I}\nu_{iI}\right) \bigg/ \left(\frac{k_\parallel^2 T_I}{m_I \nu_{II}}\right) \approx \frac{Z_I^6}{\sqrt{A_I}}\frac{n_I}{n_i}\left(\frac{qR\nu_{ii}}{c_s}\right)^2 \quad (A2)$$

a number of order 100 for 5% density concentration of O^{+6} and typical parameters. There may exist cases (e.g., singly ionized impurities), where this ratio becomes less than unity and frictional effects are negligible. We now study this situation, using some simplifying assumptions, analytically. The momentum conservation equation for the impurities is (we take $\mu = 0$ for all species):

$$\omega \tilde{u}_{\parallel I} = \frac{1}{A_I}k_\parallel c_s^2\left(\frac{\tilde{n}_I}{n_I} + \frac{\tilde{T}_I}{T_e} + \tilde{\psi}\right) \quad (A3)$$

We solve Eqs. (1)–(3) and (A1), (A3) in the usual low frequency limit, $(\omega \nu_{ii}, A_I \omega^2) \ll k_\parallel^2 c_s^2$. The resulting dispersion equation can be written as:

$$\rho_{si}^2 \nabla_\perp^2 \tilde{\psi} + \left\{\frac{\delta\eta}{\eta} + \frac{i}{\eta}\frac{\omega}{\nu_{ii}}\left[1 + \frac{Z_I n_I}{n_e}(\sigma_I A_I - 1)\right] - 2\frac{i}{\eta}\frac{k_\parallel^2 c_s^2}{\nu_{ii}\omega_*}\right\}\tilde{\psi} \quad (A4)$$

leading to the usual Hermite function solutions and eigenfrequency:

$$\omega = i\nu_{ii}\bar{\eta}\frac{\delta\eta/\bar{\eta} - k_y^2\rho_{si}^2 - (2N+1)(1+i)\sqrt{(2/\bar{\eta})(k_y^2\rho_{si}^2 c_s^2)/(L_s^2\nu_{ii}\omega_*)}}{1 + (Z_I n_I/n_e)(\sigma_I A_I - 1)}. \tag{A5}$$

The following definitions were used:

$$\delta\eta \equiv \frac{1}{\chi_i}\left(1 - \frac{Z_I n_I}{n_e}\right)\left(\frac{3}{2}\eta_i - 1\right) + \frac{1}{\chi_I}\frac{Z_I n_I}{n_e}\left(\frac{3}{2}\eta_I - \sigma_I\right) \tag{A6}$$

$$\bar{\eta} \equiv \frac{5}{4}\left(\frac{1}{\chi_i}\eta_i + \frac{1}{\chi_I}\frac{A_I\sigma_I}{Z_I}\eta_I\right) \tag{A7}$$

where $\chi_I \approx (n_i/n_I)\chi_i/Z_I^4 A_I^{3/2}$. Note that an unstable mode still exists, driven by the thermal gradient and parallel thermal conductivity of the primary ions, though its growth rate is now reduced by the dissipative response of the impurities (i.e., their thermal conductivity): for equal temperatures and peaked impurity density profiles ($\sigma_I > 1$, $\eta_I = \eta_i/\sigma_I$), the impurity contribution to $\delta\eta$ is negative.

Bibliography

[1] W.M. Tang, Nucl. Fusion **18**, 1089 (1978).

[2] P.C. Liewer, Nucl. Fusion **25**, 543 (1985).

[3] W. Horton and R.D. Estes, Nucl. Fusion **19**, 203 (1979).

[4] R.J. Groebner, W. Pfeiffer, F.P. Blau, K.H. Burrell, E.S. Fairbanks, R.P. Seraydarian, H. St. John, and R.E. Stockdale, Nucl. Fusion **26**, 543 (1986).

[5] R.R. Dominguez and R.E. Waltz, Nucl. Fusion **27**, 65 (1987).

[6] S.M. Wolfe, M. Greenwald, R. Gandy, R. Granetz, C. Gomez, D. Gwinn, B. Lipschultz, S. McCool, E. Marmar, J. Parker, R.R. Parker, and J. Rice, Nucl. Fusion **26**, 329 (1986).

[7] O. Gehre, O. Gruber, H.D. Murmann, D.E. Roberts, F. Wagner, B. Bomba, A. Eberhagen, H.U. Fahrbach, G. Fussmann, J. Gernhardt, K. Hubner, G. Janeschitz, K. Lackner, E.R. Muller, H. Niedermeyer, H. Rohr, G. Staudenmeier, K.H. Steuer, and O. Vollmer, Phys. Rev. Lett. **60**, 1502 (1988).

[8] F.X. Soldner, E.R. Muller, F. Wagner, H.S. Bosch, A. Eberhagen, H.U. Fahrbach, G. Fussmann, O. Gehre, K. Gentle, J. Gernhardt, O. Gruber, W. Herrmann, G. Janeschitz, M. Kornherr, H.M. Mayer, K. McCormick, H.D. Murmann, J. Neuhauser, R. Nolte, W. Poschenrieder, H. Rohr, K.H. Steuer, U. Stroth, N. Tsois, and H. Verbech, Phys. Rev. Lett. **61**, 1105 (1988).

[9] D.L. Brower, W.A. Peebles, S.K. Kim, N.C. Luhmann, Jr., W.M. Tang, and P.E. Phillips, Phys. Rev. Lett. **59**, 48 (1987).

[10] D.L. Brower, M.H. Redi, W.M. Tang, R.V. Bravenec, R. D. Durst, S.P. Fan, Y.X. He, S.K. Kim, N.C. Luhmann, Jr., S.C. Mc Cool, A.G. Meigs, M. Nagatsu, A. Ouroua, W.A. Peebles, P.E. Phillips, T.L. Rhodes, B. Richards, C.P. Ritz, W.L. Rowan, and A.J. Wooton, Nucl. Fusion **29**, 1247 (1989).

[11] L.I. Rudakov and R.Z. Sagdeev, Sov. Phys. Dokl. **6**, 415 (1961).

[12] B. Coppi, M.N. Rosenbluth, and R.Z. Sagdeev, Phys. Fluids **10**, 582 (1967).

[13] M. Porkolab, Nucl. Fusion **8**, 29 (1968).

[14] T. Antonsen, B. Coppi, R. Englade, Nucl. Fusion **19**, 641 (1979).

[15] M.G. Gray, M.J. LeBrun, T. Tajima, G. Furnish, and W. Horton, Bull. Am. Phys. Soc. **37**, 1433 (1992).

[16] W.M. Tang, Bull. Am. Phys. Soc. **37**, 1527 (1992).

[17] B. Coppi, H.P. Furth, M.N. Rosenbluth, and R.Z. Sagdeev, Phys. Rev. Lett. **17**, 377 (1966).

[18] B. Coppi, G. Rewoldt, and T. Schep, Phys. Fluids **19**, 1144 (1976).

[19] R. Paccagnella, F. Romanelli, and S. Briguglio, Nucl. Fusion **30**, 545 (1990).

[20] M. Fröjdh, M. Liljeström, and H. Nordman, Nucl. Fusion **32**, 419 (1992).

[21] S. Migliuolo, Nucl. Fusion **32**, 1331 (1992).

[22] W.M. Tang, R.B. White, and P.N. Guzdar, Phys. Fluids **23**, 167 (1980).

[23] S. Migliuolo and A.K. Sen, Phys. Fluids B **2**, 3047 (1990).

[24] G.R. Staebler and R.R. Dominguez, Nucl. Fusion **31**, 1891 (1991).

[25] A.B. Hassam, T.M. Antonsen, Jr., J.F. Drake, and P.N. Guzdar, Phys. Fluids B **2**, 1822 (1990).

[26] P.W. Terry, J.-N. LeBoeuf, P.H. Diamond, D.R. Thayer, J.E. Sedlak, and G.S. Lee, Phys. Fluids **31**, 2920 (1988).

[27] L. DeKock, P.E. Stott, S.K. Erents, P.J. Harbour, M. Laux, G.M. McCracken, C.S. Pitcher, M.F. Stamp, P.C. Strangeby, D.R.R. Summers, and A.J. Tagle, in *Plasma Physics and Controlled Nuclear Fusion Research 1988* (IAEA, Vienna, 1989), vol. 1, p. 467.

[28] J.F. Drake, P.N. Guzdar, and A. Dimits, Phys. Fluids B **3**, 1937 (1991).

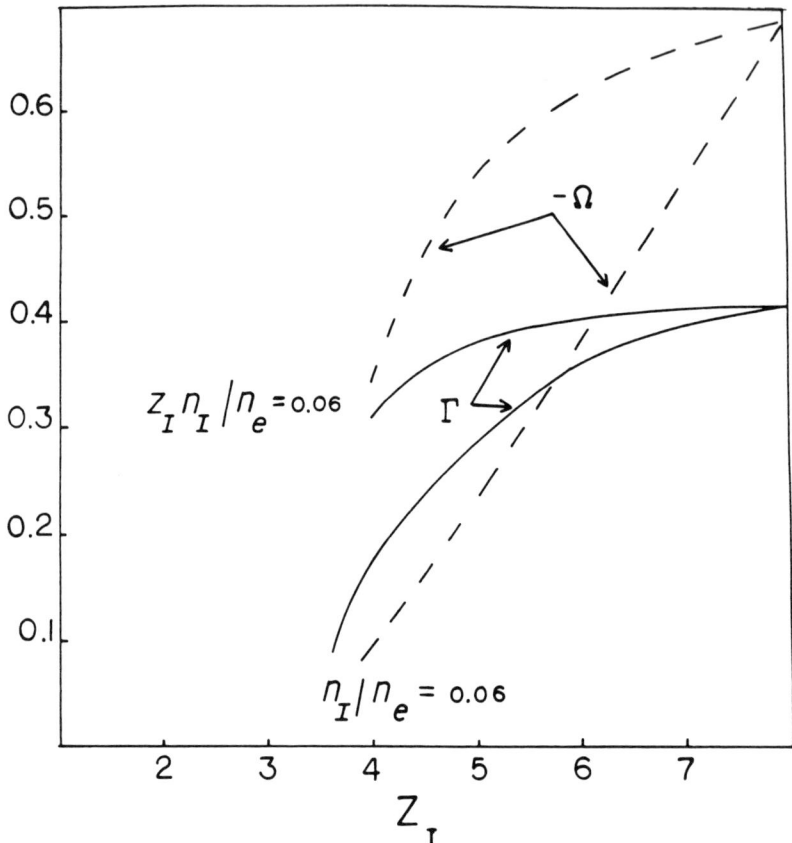

Figure 1. Eigenfrequency, $\Omega + i\Gamma \equiv \omega L_s/k_y \rho_{si} c_s$, as a function of impurity ionisation state. Parameters: $T_e = T_i = T_I$, $\eta_i = 4$, $L_s/L_n = 6$, $\mu = 0$, $k_y \rho_{si} = 0.01$, $k_y \rho_{si} c_s/\nu_{ii} L_s = 0.015$, $\sigma_I = 2$, and $A_I = 16$ (oxygen).

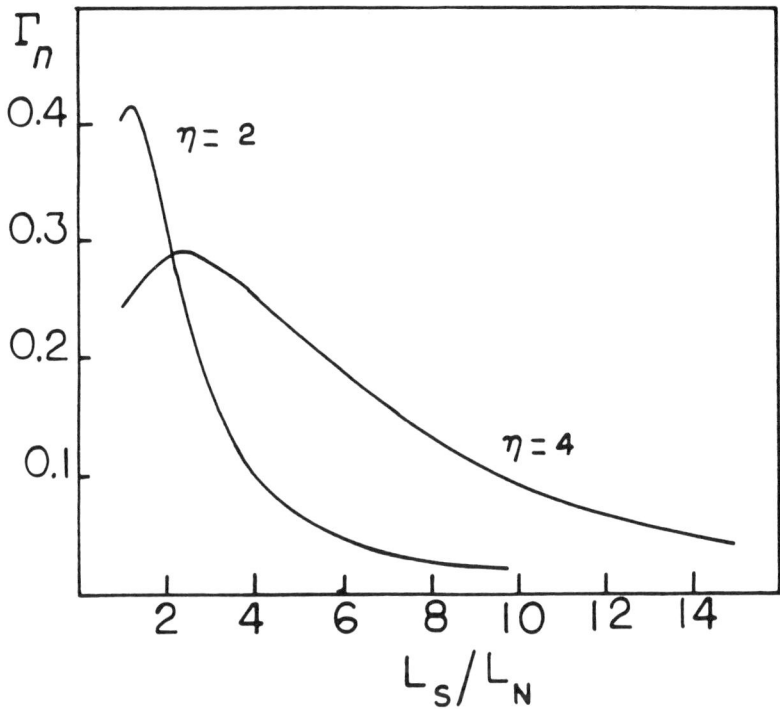

Figure 2. Growth rate $\Gamma_n \equiv \omega L_n/k_y \rho_{si} c_s$ vs. the magnetic shear parameter L_n/L_s. Other parameters are: $k_y \rho_{si} c_s / \nu_{ii} L_n = 0.015$, $n_I/n_e = 0.05$, $\eta_i = 2, 4$.

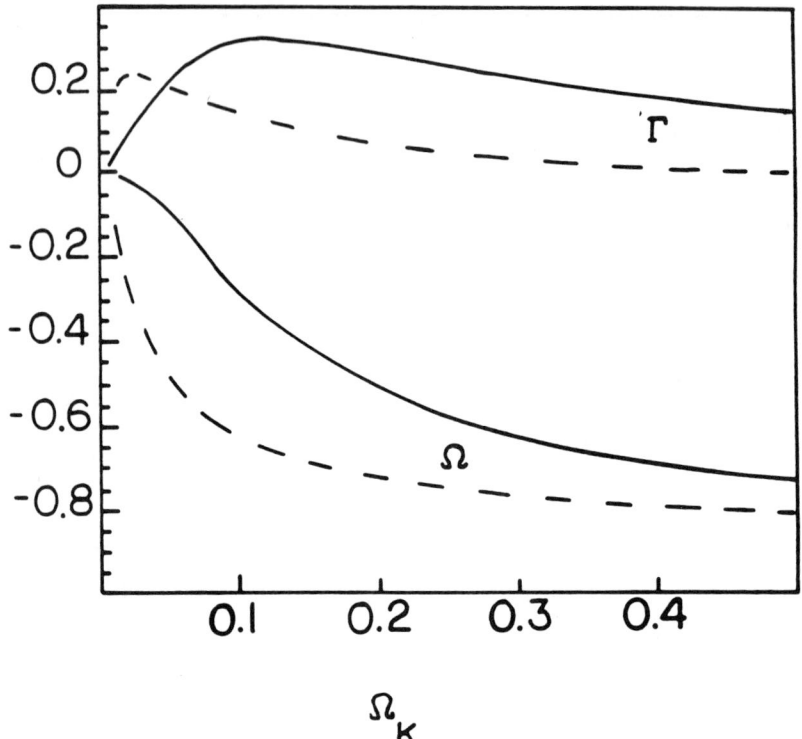

Figure 3. Eigenfrequency, $\Omega + i\Gamma \equiv \omega L_s/k_y\rho_{si}c_s$, as a function of the parallel thermal conductivity parameter, $\Omega_\kappa = 2k_y\rho_{si}c_s/3\nu_{ii}L_s$. Parameters: $L_s/L_n = 6$, $n_I/n_e = 0.05$, (dashed) solid lines show $\eta_i = (2)4$

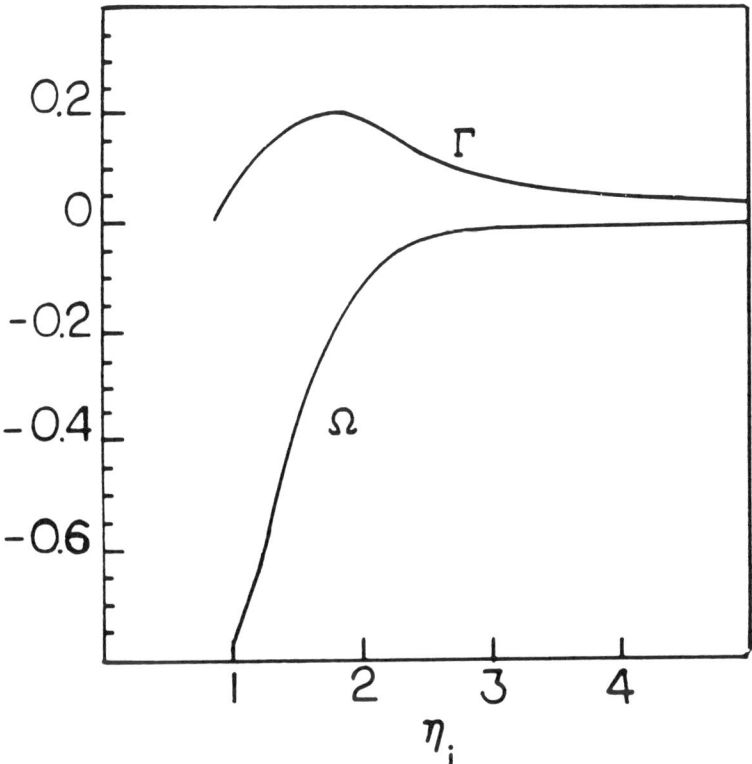

Figure 4. Eigenfrequency, $\Omega + i\Gamma \equiv \omega L_s/k_y\rho_{si}c_s$, as a function of the temperature gradient parameter. Other parameters are: $L_s/L_n = 6$, $k_y\rho_{si}c_s/\nu_{ii}L_s = 0.015$, $n_I/n_e = 0.05$.

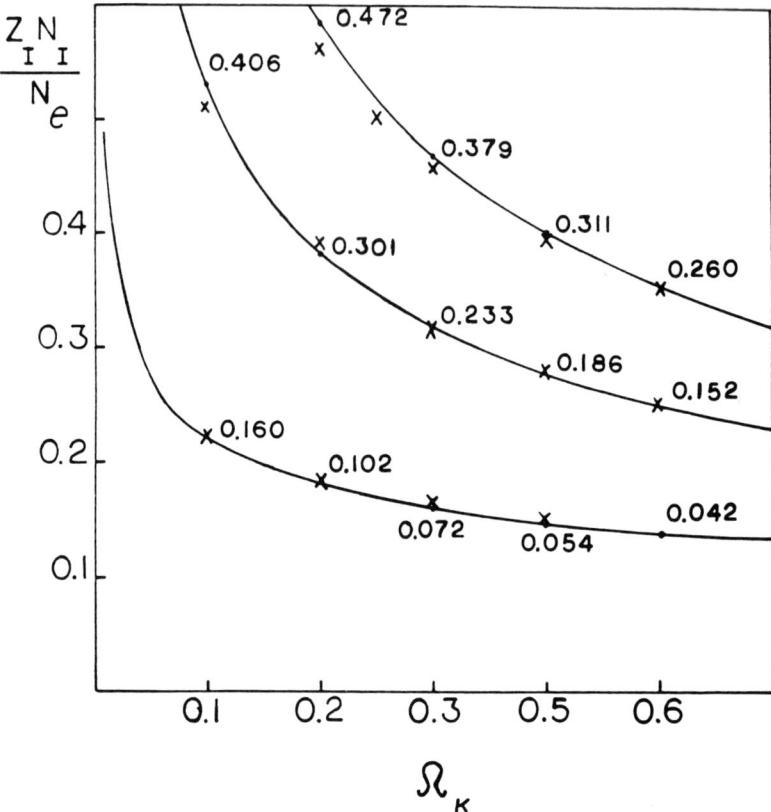

Figure 5. Loci of maximum growth rate, in $\Omega_\kappa - \hat{\rho}_I$ space ($\hat{\rho}_I \equiv Z_I n_i / n_e$ is the impurity charge concentration, while $\Omega_\kappa \equiv \omega_\kappa L_s / k_y \rho_{si} c_s = 2 k_y \rho_{si} c_s / 3 \nu_{ii} L_s$). Values of the normalised growth rate, $\gamma L_s / k_y \rho_{si} c_s$ are indicated at sample points (the real part of the frequency, normalised in the same manner, is of order -0.4, -0.7, -0.8 for $\eta_i = 2$, 4, 6 on these curves). Other parameters are $L_s/L_n = 6$, $Z_I = 6$ (oxygen). The three curves (in ascending order, top to bottom) correspond to $\eta_i = 2, 4, 6$. Crosses near each of the curves indicate sample points for fully ionized ($Z_I = 8$) oxygen.

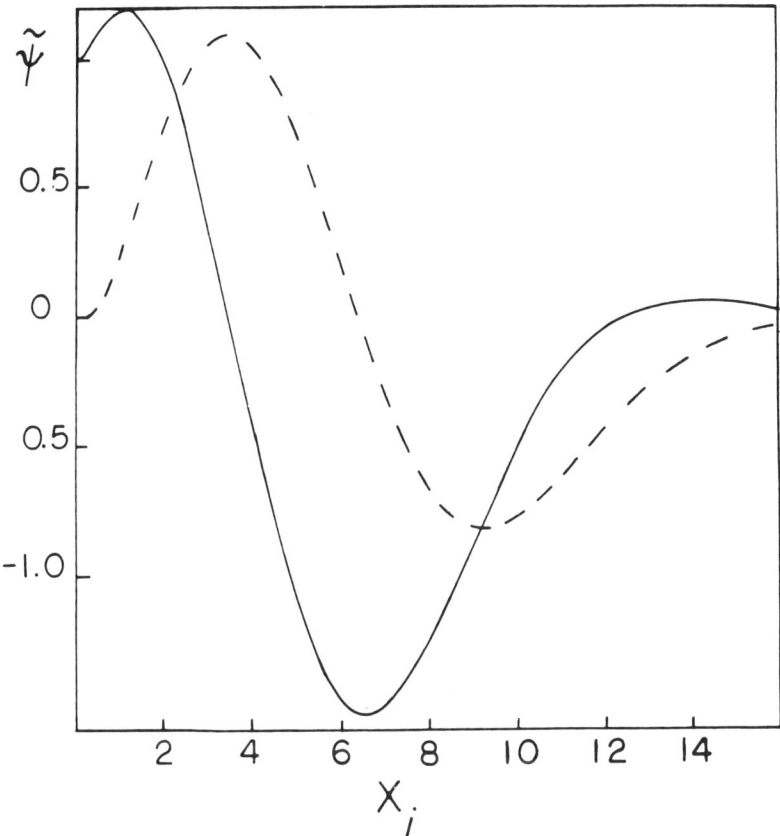

Figure 6. Typical eigenfunction (near marginal stability, for $\eta_i = 2$); shown is the normalised potential $\tilde{\psi} = e\tilde{\phi}/T_e$ as a function of the dimensionless coordinate $x_i = x/\rho_{si}$. Solid (dashed) lines denote the real (imaginary) part.

6.7 IMPURITY EFFECT ON KINETIC η_i-MODE IN TOKAMAK PLASMA

J.Q. Dong,[1] W. Horton, and X.N. Su
Institute for Fusion Studies, The University of Texas at Austin,
Austin, Texas 78712

A b s t r a c t. The impurity effects on the ion temperature gradient driven mode, or η_i-mode, in a tokamak plasma are studied in shearless slab, sheared slab and toroidal magnetic confinement configurations. The stability analysis shows that the impurity ions have a stabilizing effect on the mode in high temperature tokamak plasmas and such effects depend on (i) the impurity concentration in the plasma (the dilution effect), (ii) the density profile of the impurity ions (when $T_z = T_i$ is assumed with T_i and T_z being the temperature of the hydrogen and impurity ions, respectively), (iii) the dynamics of the impurity ions. The analysis also shows the importance of taking impurity parameters into account when experimental results are analyzed with η_i-mode theory. In the limit of adiabatic electron dynamics the ambipolar of the total particle flux is achieved by the balance of the outward working gas (hydrogenic) particle flux with the inward impurity flux.

6.7.1 INTRODUCTION

Based on the fact that fusion plasmas are not pure single element plasmas composed of electrons and one kind of ions only, say hydrogen, rather, they are usually composed of mixture of ionic elements we find it is important to develop ion temperature gradient driven mode theory for plasmas of two ion species. When the second ion species is subdominant this is called the impurity effects which were studied in 60's and 80's [1, 2]. In Refs. [1] and [2] it is found that impurity ions drive so-called impurity drift mode unstable in the plasma where the impurity density profiles are outwardly peaked while the main ion and electron densities are inwardly peaked. On the other hand if the impurity ions

---

[1]Permanent address: Southwestern Institute of Physics, Leshan, China

are inwardly peaked as the main ions and the electrons are, the instabilities do not appear. Moreover, for such a situation the impurities tend to exert a strong stabilizing effect on the ITG mode. Shearless or sheared slab magnetic models and long radial wavelength ($k_x \rho_i < 1$) approximation are used and the impurity response is considered to be fluid–like in those studies. Impurity effect on the η_i-mode is later studied in toroidal geometry with the impurity ions being adiabatic and the stabilizing effect coming from the dilution of the main ions is found [3] It is pointed out, recently, that the impurity drift mode can be unstable for some restricted parameter regimes even when the impurity and main ion profiles are both inwardly peaked if the full impurity ion dynamics is considered [4].

On the other hand, the impurity effects on the plasma confinement and the fluctuation in the periphery of tokamak plasma are widely studied in experiment [5, 6, 7, 8, 9, 10]. So-called Z-mode discharge which is a plasma having an increased low-Z impurity content and exhibiting an enhancement confinement relative to the L-mode value has been found and extensively studied. Several physical mechanisms such as radiation, profile control and recycling are studied in those experiments. At the same time there is evidence that the enhancement of the confinement may relate to the suppression of the ITG driven turbulence due to profile evolution [8].

Although the comparison of the ITG mode theory with the experiment is in reasonably good agreement [11, 12, 13], more realistic models for the tokamak plasma are required in order either to compare the theoretical results on the thermal conductivity $\chi_i(\eta_i, \epsilon_n, s, \tau...)$ with the experimental values in conventional tokamak plasma (L-, H-mode or OH discharge) or to further understand the Z-mode discharge plasma. In addition, the including of the impurity species into the ITG mode study will introduce a new mechanism into the analysis of ITG mode driven particle diffusion in the plasmas of adiabatic electrons.

The impurity effects on the ITG driven mode in tokamak plasmas are studied in the shearless slab, the sheared slab, and the toroidal magnetic configurations in this work. Comparison of the stability properties between these three geometries aids in understanding the dynamical processes producing instabilities. In most of the studies carried out here the full dynamics of the impurity ions are considered that leads to the pure impurity ITG driven mode which is neglected in previous studies. That it is the present treatment takes into account the fact that as the impurity element dominants the plasma (and thus the term "impurity" is no longer appropriate) the ITG mode appears for the plasma with ions of charge Ze and mass m_z. In addition the stability analysis shows that the stabilizing influence of the impurity ions on the ITG driven mode is due not only to the dilution of the main ions but also to the density profile and dynamics of the impurity ions as well.

The low frequency impurity drift mode mentioned above [1, 2, 4] is not concerned in this work since that mode seems less important in the core of tokamak plasmas.

The remainder of this work is organized as follows. In Sec. 2 the ITG mode is studied in a shearless geometry with the dynamics of the impurity ions included. The eigenvalue equation in sheared slab is given and solved under fluid approximation in Sec. 3. The integral dispersion equation in a toroidal magnetic configuration is presented and numerically solved in Sec. 4 while Sec. 5 is devoted to concluding remarks and discussions.

6.7.2 Shearless Slab Approximation

The equilibrium distribution for both hydrogen and impurity ions are Maxwellian,

$$f_{0j} = \frac{n_{0j}}{\pi^{\frac{3}{2}} v_{tj}^3} \exp(-\frac{v^2}{v_{tj}^2}) . \qquad (1)$$

where v_{tj} ($j = i, z$) is the thermal speed of the hydrogen and impurity ions, respectively, and $T_i = T_z$ is assumed. The dispersion equation can easily be obtained with standard linearization procedure from Vlasov equation as follows:

$$1 + (1 - f_z)\left\{\tau + \eta_i \zeta_i \frac{\omega_{*e}}{k_\| v_{ti}} L_{ei} \Gamma_0(b_i) \right.$$

$$\left. + \Gamma_0(b_i) \zeta_i Z(\zeta_i) \left[\tau + \frac{\omega_{*e}}{k_\| v_{ti}} L_{ei} - \eta_i \frac{\omega_{*e}}{k_\| v_{ti}} L_{ei} \left(\frac{1}{2} + b_i \delta_i - \zeta_i^2\right)\right]\right\}$$

$$+Zf_z\left\{\tau + \eta_z\zeta_z\frac{\omega_{*e}}{k_{\parallel}v_{tz}}L_{ez}\Gamma_0(b_z)\right.$$

$$\left.+\Gamma_0(b_z)\zeta_z Z(\zeta_z)\left[\tau + \frac{\omega_{*e}}{k_{\parallel}v_{tz}}L_{ez} - \eta_z\frac{\omega_{*e}}{k_{\parallel}v_{tz}}L_{ez}\left(\frac{1}{2}+b_z\delta_z - \zeta_z^2\right)\right]\right\} = 0 \tag{2}$$

where

$$L_{kj} = \frac{L_{nk}}{L_{nj}}, \ (k,j=e,\ i,\ z)\,, \qquad f_z = \frac{Zn_{0z}}{n_{0e}}, \qquad \tau = \frac{T_e}{T_i}, \eta_i = \frac{L_{ni}}{L_{Ti}},$$

$$b_j = \frac{T_j}{m_j\Omega_j^2}\,, \qquad \zeta_j = \frac{\omega}{k_{\parallel}v_{tj}}\,, \qquad \Gamma_0(b_j) = I_0(b_j)e^{-b_j}\,, \qquad \omega_{*e} = \frac{k_y c T_e}{eBL_{ne}}$$

$$L_{ei} = \frac{1-f_z L_{ez}}{1-f_z}, \qquad \eta_z = \frac{\eta_i(\frac{L_{nz}}{L_{ne}}-f_z)}{1-f_z}, \qquad \delta_j = 1-\frac{I_1(b_j)}{I_0(b_j)}\,. \tag{3}$$

Equation (2) has been studied by Migliuolo for impurity drift mode [4]. In the fluid limit $\frac{\omega}{k_{\parallel}v_{tj}} \gg 1$, the dispersion equation reduces to

$$a_0 + a_1\frac{\omega_{*e}}{\omega} + a_2\left(\frac{\omega_{*e}}{\omega}\right)^2 + a_3\left(\frac{\omega_{*e}}{\omega}\right)^3 = 0\,, \tag{4}$$

where

$$a_0 = 1 + \tau(1-f_z)(1-\Gamma_0(b_i)) + Z\tau f_z(1-\Gamma_0(b_z))\,,$$

$$a_1 = -[(1-f_z)\Gamma_0(b_i)L_{ei}(1-\eta_i b_i\delta_i) + f_z\Gamma_0(b_z)L_{ez}(1-\eta_z b_z\delta_z)]\,,$$

$$a_2 = -\frac{1}{2\omega_{*e}^2}[(1-f_z)\tau\Gamma(b_i)k_{\parallel}^2 v_{ti}^2 + f_z\tau\Gamma(b_z)Zk_{\parallel}^2 v_{tz}^2]\,,$$

$$a_3 = -\frac{1}{2\omega_{*e}^2}\{(1-f_z)L_{ei}\Gamma_0(b_i)k_{\parallel}^2 v_{ti}^2[1+\eta_i(1-b_i\delta_i)]$$

$$+ f_z L_{ez}\Gamma_0(b_z)k_{\parallel}^2 v_{tz}^2[1+\eta_z(1-b_z\delta_z)]\}\,. \tag{5}$$

It is obvious that only one unstable mode is possible in this limit. If $b_i \ll 1$, $b_z \ll 1$, and $|\frac{\omega_{*e}}{\omega}| \gg 1$ then

$$\omega \simeq i\frac{1}{\sqrt{2}}\{k_\parallel^2 v_{ti}^2(1+\eta_i) - f_z L_{ez}[k_\parallel^2 v_{ti}^2(1+\eta_i) - k_\parallel^2 v_{tz}^2(1+\eta_z)]\}^{\frac{1}{2}} . \quad (6)$$

In this case the impurity stabilization effect is only due to its larger ion mass thus lower thermal velocity $v_{ti} > v_{tz}$. On the other hand, if $b_i \ll 1$, $b_z \ll 1$, $\omega_{*e} \to 0$ and $\frac{\omega_{*i}(1+\eta_i)}{\omega} \gg 1$, then

$$\omega^3 = -\frac{\{k_\parallel^2 v_{ti}^2(1+\eta_i) - f_z L_{ez}[k_\parallel^2 v_{ti}^2(1+\eta_i) - k_\parallel^2 v_{tz}^2(1+\eta_z)]\}}{2[1+\tau+f_z(Z-1)\tau]}\omega_{*e} . \quad (7)$$

It is easy to notice that impurity effect stems from its larger mass, as in the above Eq. (6) case, as well as the higher electric charge $Z > 1$. In addition, Eq. (7) indicates the destabilization effect of the impurity ions when the impurity ion density is peaked to the opposite direction against the electron density profile ($L_{ez} < 0$). Equation (2) needs to be solved numerically for more general cases and the results are presented in Figures 1–3. Generally speaking, as illustrated by the numerical results, the impurity effect depends on the concentration, density profile, inertia and charge of the impurity ions.

In Fig. 1 the normalized mode growth rate γ/ω_{*e} is presented as a function of f_z for helium, beryllium, carbon and neon with $\eta_i = 2.5$, $k_\perp \rho_i = 0.75$, $\omega_{*e}/k_\parallel v_{ti} = 10$ and $L_{ez} = 1$. Those typical parameters will be used unless otherwise declared in this Section. The mode growth rate decreases with f_z almost linearly for helium. For heavy impurities such as neon the decreasing of the growth rate exhibits two different regimes. The growth rate decreases rapidly with the increasing of f_z in the regime $f_z < 0.3$ and slowly in the regime $f_z > 0.6$. The same is shown in Fig. 2 for neon impurities of different ionization states. It is easy to be noticed that the impurity stabilizing effect increases with the electric charge of the impurity ions with same mass. Comparison the line Ne^{+6} in Fig. 2 with the line carbon in Fig. 1 shows that heavier impurities have stronger effects than the lighter ions with the same electric charge do. Figure 3 shows the density profile influence. Helium is used and η_z is calculated from Eq. (3) so that the effect of f_z is from both the direct change of f_z and the change through η_z in Eq. (2). It is clearly shown that more inwardly peaked impurity profiles ($L_{ez} = 2$) has stronger suppression effect on the mode (compared with the helium line in Fig. 1) and that outwardly peaked profiles ($L_{ez} < 0$) enhance the mode growth.

6.7.3 SHEARED SLAB GEOMETRY

Sheared slab magnetic field, $\mathbf{B} = B_0(\hat{\mathbf{z}} + \frac{x}{L_s}\hat{\mathbf{y}})$ and the long radial wavelength approximation $k_\perp \rho_j \ll 1$ are used in this section. It is straightforward to get the dispersion equation [2] from Eq. (2),

$$\frac{\partial^2 \phi}{\partial x^2} + \frac{A(x,\omega)}{B(x,\omega)}\phi = 0 \qquad (8)$$

where

$$A(x,\omega) = 1 + (1 - f_z)\{\tau + (\tau + \frac{\omega_{*e}}{\omega}L_{ei} - \frac{\eta_i}{2}\frac{\omega_{*e}}{\omega}L_{ei})\zeta_i Z(\zeta_i) + \frac{\omega_{*e}}{\omega}L_{ei}\eta_i(1 + \zeta_i Z(\zeta_i))\zeta_i^2\}$$

$$+ f_z\{\tau Z + (\tau Z + \frac{\omega_{*e}}{\omega}L_{ez} - \frac{\eta_z}{2}\frac{\omega_{*e}}{\omega}L_{ez})\zeta_z Z(\zeta_z) + \frac{\omega_{*e}}{\omega}L_{ez}\eta_z(1 + \zeta_z Z(\zeta_z))\zeta_z^2\},$$

$$B(x,\omega) = (1 - f_z)\{(\tau + \frac{\omega_{*e}}{\omega}L_{ei} + \frac{\eta_i}{2}\frac{\omega_{*e}}{\omega}L_{ei})\zeta_i Z(\zeta_i) + \frac{\omega_{*e}}{\omega}L_{ei}\eta_i(1 + \zeta_i Z(\zeta_i))\zeta_i^2\}$$

$$+ \frac{\mu}{Z^2}f_z\{(\tau Z + \frac{\omega_{*e}}{\omega}L_{ez} + \frac{\eta_z}{2}\frac{\omega_{*e}}{\omega}L_{ez})\zeta_z Z(\zeta_z) + \frac{\omega_{*e}}{\omega}L_{ez}\eta_z(1 + \zeta_z Z(\zeta_z))\zeta_z^2\}, \qquad (9)$$

with $\mu = m_z/m_i$.

This equation has been numerically studied by Tang et al. [2] and Dominguez [3] with fluid-like impurity ion response. If both the hydrogen and impurity ions have fluid-like response, then

$$\frac{A(x,\omega)}{B(x,\omega)} \simeq \frac{A_1}{B_1} + \frac{A_2}{B_1}\frac{k_\parallel^2 v_{ti}^2}{2\omega^2}, \qquad (10)$$

where

$$A_1 = L_{ie}\hat{\omega} - (1 - f_z) - f_z L_{iz},$$

$$A_2 = -(1 - f_z)(\tau\hat{\omega}L_{ie} + 1 + \eta_i) - f_z\frac{L_{iz}}{\mu}(\tau Z L_{ze}\hat{\omega} + 1 + \eta_z),$$

$$\bar{B}_1 = -[(1-f_z)(\tau\hat{\omega}L_{ie}+1+\eta_i) + f_z\frac{\mu}{Z^2}(\tau Z L_{ie}\hat{\omega}+L_{iz}+L_{iz}\eta_i)]$$

$$\hat{\omega} = \frac{\omega}{\omega_{*e}}.$$

The dispersion equation becomes a Weber equation and the eigenvalue equation is

$$\pm i\frac{(2n+1)}{\hat{\omega}}\frac{L_{ne}}{L_s} \simeq \frac{\tau A_1}{A_2}. \tag{11}$$

One approximation is

$$\pm i\frac{(2n+1)}{\hat{\omega}}\frac{L_{ne}}{L_s} \simeq \frac{\tau[\frac{(\hat{\omega}-f_z L_{ez})}{(1-f_z)} - L_{ei}]}{\hat{\omega}\tau + L_{ei}(1+\eta_i)} \tag{12}$$

and another approximate solution is

$$\hat{\omega} \simeq i(2n+1)\frac{L_{ne}}{L_s}\tau\left[1+\eta_i - f_z(1+\eta_i - \frac{1}{\mu}(1+\eta_z))\right] \tag{13}$$

Eq. (12) clearly shows the profile dependence of the impurity effect as pointed out in Ref. [2]. If the impurities are more inwardly peaked than the electrons are ($L_{ez} > 1$) then from Eq. (3) $L_{ei} < 1$ and the driving term $(1+\eta_i)$ in Eq. (12) is reduced. On the other hand, if the impurities are outwardly peaked ($L_{ez} < 0$) then $L_{ei} > 1$ and the driving term is enhanced.

From Eq. (13) we see that only the mass ratio enters the expression for the mode growth rate. Thus for a pure impurity plasma $\gamma_z = \gamma_i/\mu$.

6.7.4 TOROIDAL MAGNETIC CONFIGURATION

The gyrokinetic integral equation [14] for the study of low frequency drift modes, such as ITG mode, is extended to include impurity species in this section. The curvature and magnetic gradient effects $\omega_D(v_\perp^2, v_\parallel^2, \theta)$ of both hydrogen and impurity ions are included. The ballooning representation is used so that the mode coupling due to toroidal magnetic configuration of tokamak device is taken into account. The full ion transit $k_\| v_\|$ and finite Larmor radius effects are retained while the ion bouncing is neglected. Electron response is assumed to be adiabatic for simplicity. The integral dispersion equation derived in Ref. [14] is easily extended to include the second species of ions,

$$[1 + \tau_i(1 - f_z) + \tau_z Z f_z]\hat{\phi}(k) = \int_{-\infty}^{+\infty} \frac{dk'}{\sqrt{2\pi}} K(k, k')\hat{\phi}(k'), \quad (14)$$

where

$$K(k, k') = -i \int_{-\infty}^{0} \omega_{*e} d\tau \sqrt{2} e^{-i\omega\tau} \left[(1 - f_z) \frac{\exp\left[-\frac{(k'-k)^2}{4\lambda}\right]}{\sqrt{a}(1+a)\sqrt{\lambda}} \right.$$

$$\times \left\{\frac{\omega}{\omega_{*e}} \tau_i + L_{ei} - \frac{3}{2}\eta_i L_{ei} + \frac{2\eta_i L_{ei}}{(1+a)}\right.$$

$$\times \left[1 - \frac{k_\perp^2 + k'^2}{2(1+a)\tau_i} + \frac{k_\perp k'_\perp}{(1+a)\tau_i} \frac{I_1}{I_0}\right] + \frac{\eta_i L_{ei}(k-k')^2}{4a\lambda}\right\} \Gamma_0(k_\perp, k'_\perp) +$$

$$+ f_z \frac{\exp\left[-\frac{(k'-k)^2}{4\lambda_z}\right]}{\sqrt{a_z}(1+a_z)\sqrt{\lambda_z}} \left\{\frac{\omega}{\omega_{*e}} Z\tau_z + L_{ez} - \frac{3}{2}\eta_z L_{ez} + \frac{2\eta_z L_{ez}}{(1+a_z)}\right.$$

$$\left.\times \left[1 - \frac{(k_\perp^2 + k'^2)\mu}{2(1+a_z)Z^2\tau_z} + \frac{k_\perp k'_\perp \mu}{(1+a_z)Z^2\tau_z} \frac{I_{1z}}{I_{0z}}\right] + \frac{\eta_z L_{ez}(k-k')^2}{4a_z\lambda_z}\right\} \Gamma_{0z}(k_\perp, k'_\perp)\right] \quad (15)$$

with

$$\lambda = \frac{\tau^2 \omega_{*e}^2}{\tau_i a} \left(\frac{\hat{s}}{q}\epsilon_n\right)^2, \quad \lambda_z = \frac{\tau^2 \omega_{*e}^2}{\tau_z a_z \mu} \left(\frac{\hat{s}}{q}\epsilon_n\right)^2,$$

$$a = 1 + \frac{i2\epsilon_n}{\tau_i}\omega_{*e}\tau\left(\frac{(\hat{s}+1)(\sin\theta - \sin\theta') - \hat{s}(\theta\cos\theta - \theta'\cos\theta')}{(\theta - \theta')}\right),$$

$$a_z = 1 + \frac{i2\epsilon_n}{\tau_z Z}\omega_{*e}\tau\left(\frac{(\hat{s}+1)(\sin\theta - \sin\theta') - \hat{s}(\theta\cos\theta - \theta'\cos\theta')}{(\theta - \theta')}\right),$$

$$\theta = \frac{k}{\hat{s}k_\theta}, \quad \theta' = \frac{k'}{\hat{s}k_\theta},$$

$$\Gamma_0 = I_0\left(\frac{k_\perp k'_\perp}{(1+a)\tau_i}\right) \exp\left[-(k_\perp^2 + k'^2_\perp)/2\tau_i(1+a)\right]$$

$$\Gamma_{0z} = I_0 \left(\frac{k_\perp k'_\perp \mu}{(1+a_z)\tau_z Z^2} \right) \exp\left[-(k_\perp^2 + k'^2_\perp)\mu/2Z^2\tau_z(1+a_z) \right] ,$$

$$k_\perp^2 = k_\theta^2 + k^2 , \qquad k'^2_\perp = k_\theta^2 + k'^2 ,$$

$$\epsilon_n = \frac{L_{ne}}{R}, \quad \eta_i = \frac{L_{ni}}{L_{Ti}}, \quad \eta_z = \frac{L_{nz}}{L_{Tz}}, \quad \tau_i = \frac{T_e}{T_i}, \quad \tau_z = \frac{T_e}{T_z},$$

$$f_z = \frac{Zn_{0z}}{n_{0e}}, \quad \mu = \frac{m_z}{m_i}, \quad L_{ei} = \frac{L_{ne}}{L_{ni}}, \quad L_{ez} = \frac{L_{ne}}{L_{nz}},$$

k, k' and k_θ are normalized to $\rho_i^{-1} = \sqrt{\tau_i}\Omega_i/v_{ti}$ and x to ρ, and $I_j(j = 0,1)$ is the modified Bessel function of order j.

Equation (14) has to be solved numerically. The typical parameters used here are $\eta_i = 3.9$, $\hat{s} = 0.83$, $q = 1.5$, $\epsilon_n = 0.56$, $\tau = 1$, $\eta_z = \eta_i$, $k_\theta \rho_i = 0.42$, $L_{ei} = L_{ez} = 1$, $T_z = T_i$ unless otherwise stated. Figure 4(a) shows the normalized mode growth rate as function of f_z for deuterium, helium, tritium, carbon, neon and iron impurities. For deuterium and tritium the growth rate decreases with f_z almost linearly. This is the isotopic stabilization effect first discussed by Dominguez [15]. For the other impurities except of iron the mode growth rate first decreases rather rapidly just like in Fig. 1 for $f_z < 0.3$ then increases slowly for $f_z > 0.6$ unlike what happens in Fig. 1 where it keeps decreasing slowly, when f_z increases from 0 (pure hydrogen plasma) to 1 (pure "impurity" plasma). We note that the pure neon, carbon, or typical plasma will have clearly its own collective thermal transport driven by its own temperature gradient. When there are two elements, however, the collective motions that are most effective for one element are not the most effective for the other. Thus, the collective oscillations can be weakened due to the competition from the oscillations trying to accommodate the requirement of the temperature gradient in each plasma element. The computation shows up a region of smaller growth rate in Fig. 4(a). The dip in the growth rate is only $10 - 20\%$ for the mixtures of light elements D, T, He, or Li with hydrogen. For hydrogen mixed with the heavier elements such as Ne or C, however, there is a very substantial stabilization of the mode. The stabilization is the strongest when the mixtures are approximately 50-50% in terms of their charge contribution. It is seen that, as long as the growth rate is concerned, most the pure "impurity" plasmas studied here are more stable than the pure hydrogen plasma is. The mode real frequency is shown in Fig. 4(b). It is clear that the mode real frequency decreases with the ion mass and charge of the impurities. In Fig. 4(c) the eigenvalues presented in Figs. 4(a) and 4(b) are given in the complex plane. It is indicated that from pure hydrogen to pure "impurity" plasmas the eigenvalue of the mode changes smoothly. Fig. 5 shows the mode growth rate as function of η_i. We note that the threshold value of η_i changes substantially with the combination of the elements in the plasma. Similar results are obtained with the parallel ion dynamics neglected [16, 17]. As if the threshold value of η_i is concerned the pure "impurity" plasmas are the most unstable. Figs. 6 and 7 illustrate that the stabilizing effect of the impurities does not change with ϵ_n and k_θ. Mixed plasmas are more stable than the pure hydrogen plasmas are for all ϵ_n and k_θ values analyzed here.

6.7.5 CONCLUSIONS

The effects of the impurity ions on the ITG mode are studied in shearless slab, sheared slab and toroidal magnetic configurations. It is shown that such effects strongly depend on the peakness of the electron and impurity ion density profiles. The impurities destabilize the mode if the electron and impurity ion density profiles are peaked towards opposite directions. Impurity ions have stabilizing effect on ITG mode in high temperature tokamak plasmas if the electron and impurity ion density profiles are peaked in the same direction. Such stabilizing effect depends on

(i) the impurity concentration in the plasma (dilution effect),

(ii) the density scale length of the impurity ions (when $T_i = T_z$ is assumed), and

(iii) the dynamics of the impurity ions. Qualitatively speaking, the heavier the impurity ions and the larger the electric charge each impurity ion has the stronger the stabilization effect on the η_i mode. The plasmas of two ion species with 50-50% mixture in terms of their charge contribution is the most stable and the pure "impurity" plasmas are usually more stable than the pure hydrogen plasmas are. It is shown that impurity effects may change the threshold value of η_i significantly. The analysis also shows the importance of taking impurity parameters-into account when experimental results are analyzed with η_i-mode theory.

ACKNOWLEDGMENTS

This work was supported by the U.S. Department of Energy contract #DE-FG05-80ET-53088.

Bibliography

[1] B. Coppi, H.P. Furth, M.N. Rosenbluth, and R.Z. Sagdeev, Phys. Rev. Lett. **17**, 377 (1966).

[2] W.M. Tang, R.B. White, and P.N. Guzdar, Phys. Fluids **20**, 167 (1980).

[3] R.R. Domingues, Nucl. Fusion **31**, 2063 (1991).

[4] S. Migliuolo, Nucl. Fusion **32**, 1331 (1992).

[5] M. Bessenrodt, et al., Nucl. Fusion **31**, 155 (1991).

[6] C. Hidalgo, et al., Nucl. Fusion **31**, 1661 (1991).

[7] Y.Y. Karzhavin, et al., Bull. Am. Phys. Soc. (1991).

[8] V.V. Alikaev, et al., Plasma Phys. and Cont. Fusion **30**, 381 (1988).

[9] M. Murakami, et al., IAEA-CN-44/A-II-2.

[10] E.A. Lazarus, et al., Nucl. Fusion **25**, 135 (1985).

[11] W. Horton, D. Lindberg, J.Y. Kim, J.Q. Dong, G.W. Hammett, S.D. Scott, M.C. Zarnstorff, Phys. Fluids B **4**, 952 (1992).

[12] D.L. Brower, M.H. Redi, W.M. Tang, et al., Nucl. Fusion **29**, 1247 (1989).

[13] A.L.M. Rogister, G. Hasselberg, F.G. Waelbroeck, J. Weiland, Nucl. Fusion **28**, 1053 (1988).

[14] J.Q. Dong, W. Horton, and J.Y. Kim, Phys. Fluids B **4**, 1867 (1992).

[15] R.R. Domonguez, Nucl. Fusion **31**, 2063 (1991).

[16] M. Fröjdh, M. Liljeström, H. Nordman, Nucl. Fusion **32**, 419 (1992).

[17] R. Paccagnella, F. Romanelli, Nucl. Fusion **30**, 545 (1990).

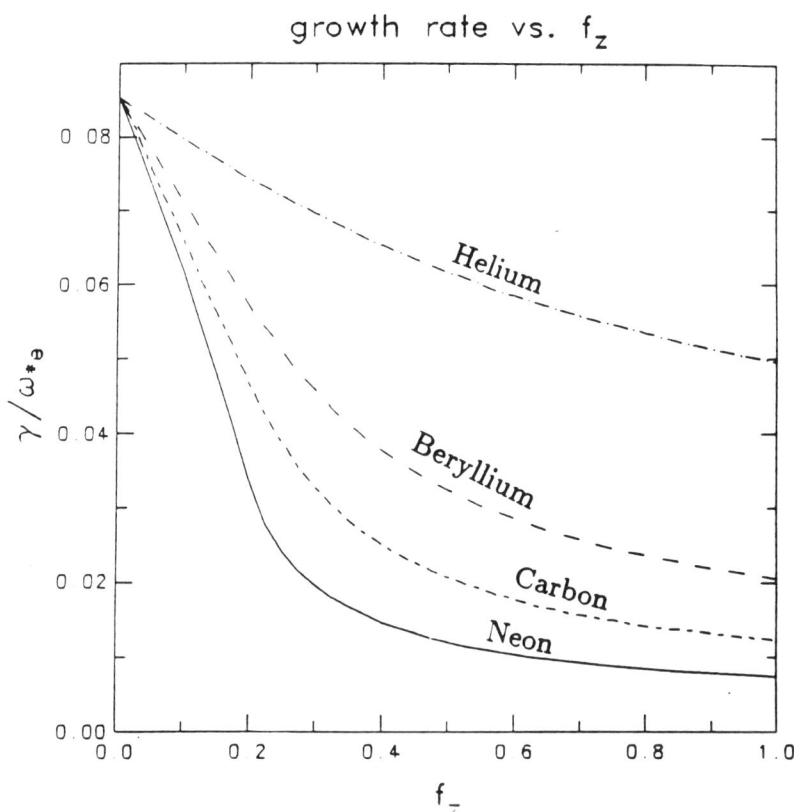

Figure 1. Normalized mode growth rate γ/ω_{*e} vs. the charge concentration of the impurities in the plasma for different impurity elements.

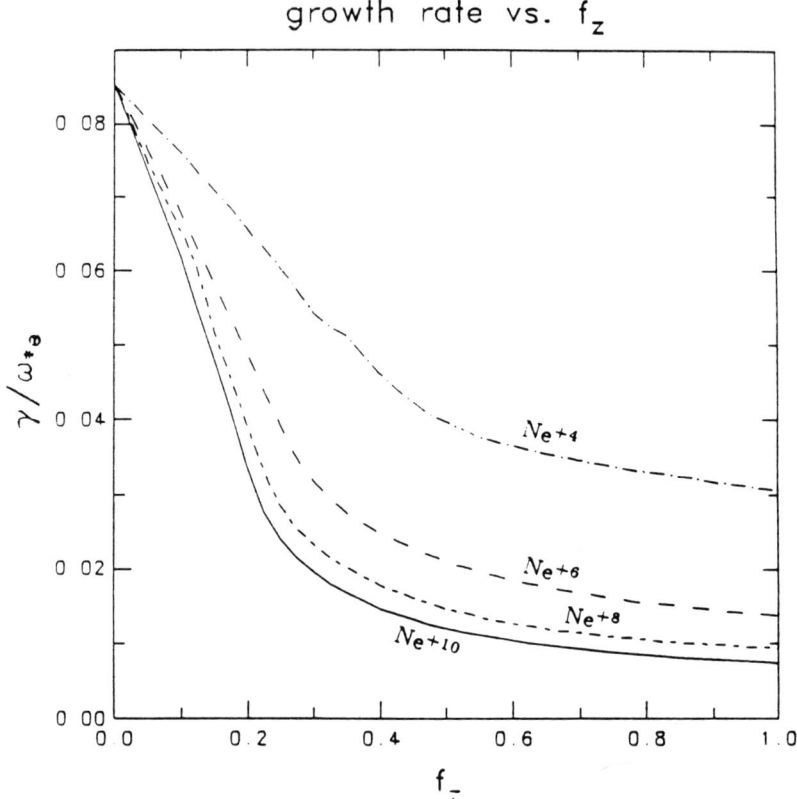

Figure 2. Normalized mode growth rate γ/ω_{*e} vs. the charge concentration of the impurities in the plasma for N_e impurities of different ionization staes.

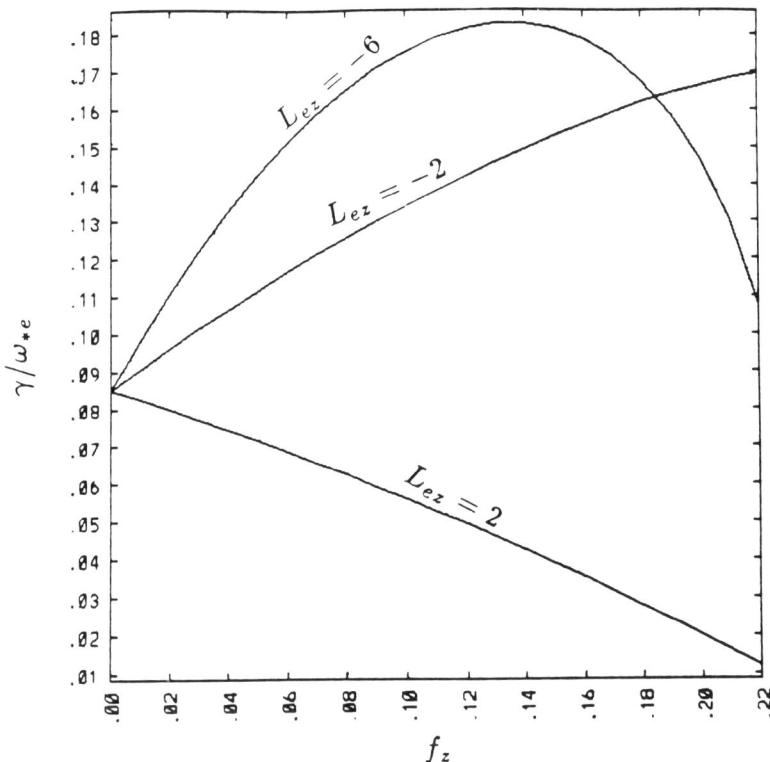

Figure 3. Normalized mode growth rate γ/ω_{*e} vs. the charge concentration of the He impurity in the plasma for different impurity density scale lengths.

502 Impurity Effect on Kinetic η_i-Mode

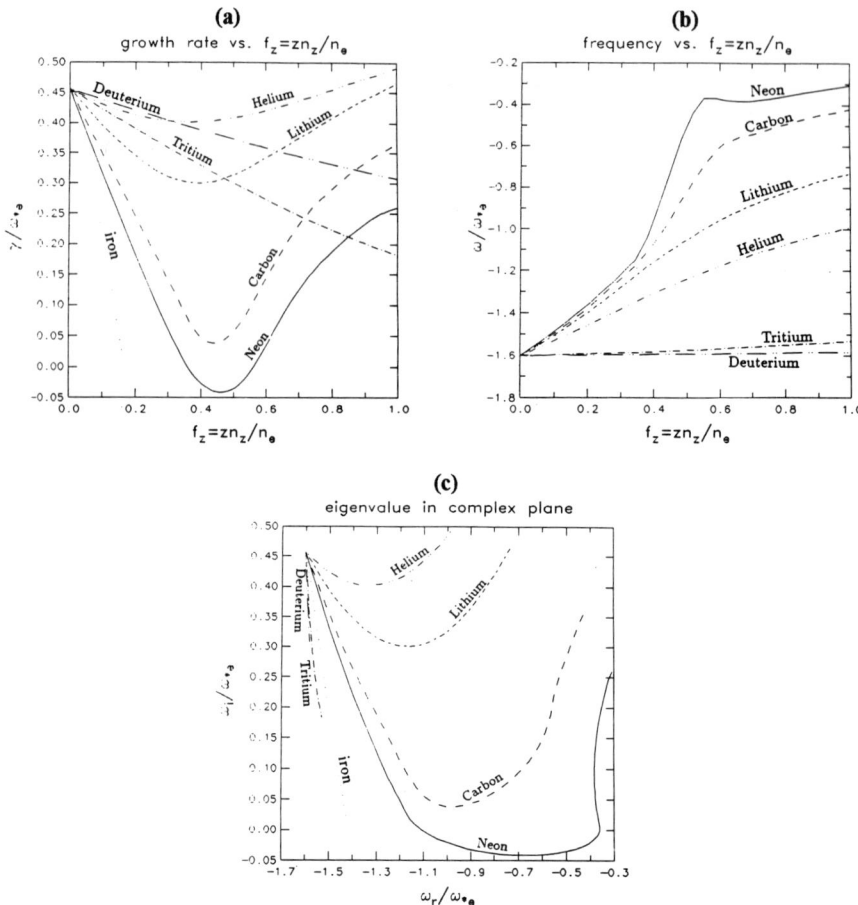

Figure 4. Normalized mode growth rate (a) γ/ω_{*e} and the real frequency ω_r/ω_{*e} vs. the charge concentration of the impurities in the plasma f_z for different impurity elements in toroidal geometry. (c) The eigenvalues $(\omega_r/\omega_{*e}, \gamma/\omega_{*e})$ in the complex plane.

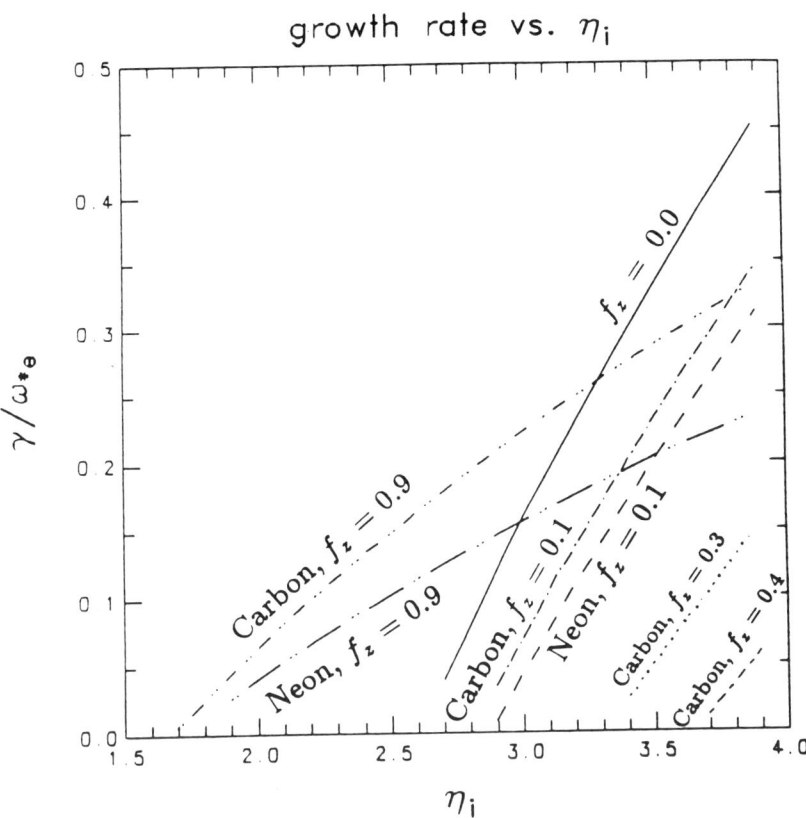

Figure 5. Normalized mode growth rate γ/ω_{*e} vs. η_i for plasmas of different composition.

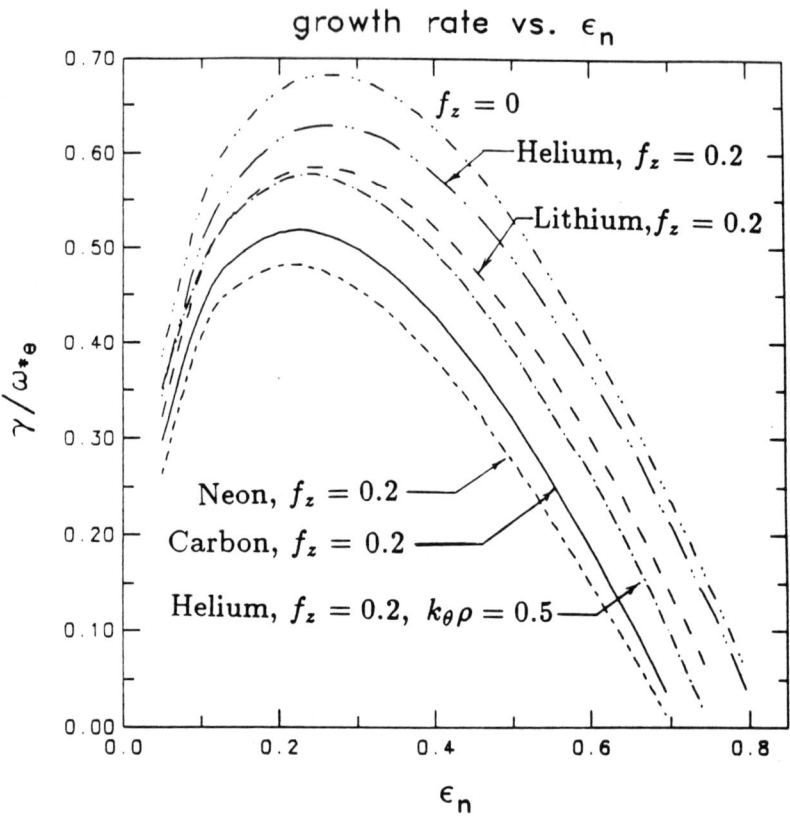

Figure 6. Normalized mode growth rate γ/ω_{*e} vs. ϵ_n for plasmas of different composition.

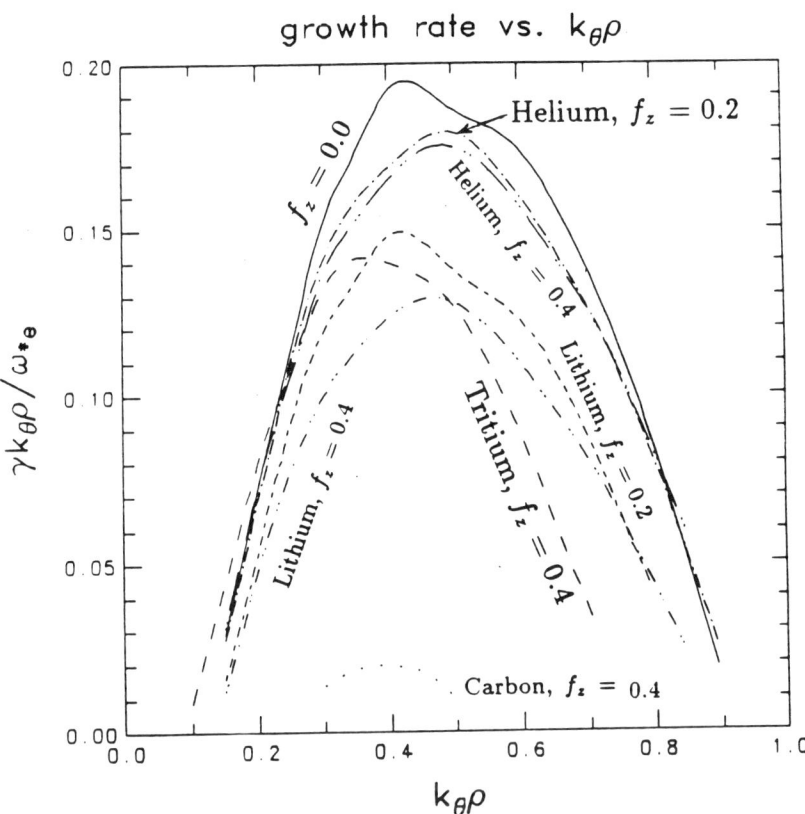

Figure 7. Normalized mode growth rate $\gamma/\omega_{*e}\, k_\theta \rho_i$ vs. $k_\theta \rho_i$ for plasmas of different composition.

Chapter 7
TURBULENCE THEORY

7.1 TRANSPORT ANALYSIS BASED ON K-ϵ ANOMALOUS TRANSPORT MODEL

H. Sugama and M. Okamoto
National Institute for Fusion Science, Chikusaku, Nagoya 464-01, Japan

M. Wakatani
Plasma Physics Laboratory, Kyoto University, Gokasho, Uji 611 Japan

Abstract. A K-ϵ anomalous transport model for resistive interchange turbulence is presented and applied to the transport analysis of ECH plasmas in Heliotron E. In this model, the turbulent kinetic energy $K \equiv \frac{1}{2}\langle \tilde{v}^2 \rangle$ and its viscous dissipation rate ϵ characterize the local turbulence and the anomalous transport coefficient is given by $D \sim K^2/\epsilon$, which has some nonlocal properties not included in the conventional expressions since their temporal and spatial variations are determined by taking into account the transport of the turbulent energy itself. In the case of the homogeneous turbulence where the anomalous transport may be described in terms of the local plasma parameters, the dimensional analysis applied to our model yields the two types of local parameter expressions of the anomalous diffusivity in the high and low collisional cases. We find a familiar diffusivity for the resistive interchange turbulence derived in the high collisional case and we have another one similar to the gyro-reduced Bohm (GRB) diffusivity in the low collisional case. However, it is shown from the transport simulation using our model that, in the region where the turbulence inhomogeneity is significant, the anomalous diffusivity deviates from the local parameter expression due to the transport terms in the K-ϵ equations. Our model explains the experimental results consistently in that it gives the GRB or LHD scaling for the energy confinement time and reproduces the experimentally obtained profile of the anomalous diffusivity which has large values in the peripheral region in contrast with the GRB model.

7.1.1 INTRODUCTION

Conventional treatments for anomalous transport have been based on the local transport coefficients (D or χ) which are expressed as functions of local plasma parameters such as local density n, temperature T, magnetic field B and a number of gradient scale length $L_n, L_T, L_s, \cdots$: D or $\chi = F(n, T, B, L_n, L_T, L_s, \cdots)$. These treatments assume that the mixing length l and the time scale τ of the turbulence responsible for the anomalous transport are determined by the local plasma parameters. However it is possible that the turbulence structure has nonlocal nature and the validity of the expression for the local transport coefficients as given above is limited.

Here we present a K-ϵ type model for the analysis of anomalous transport in the resistive interchange turbulence. A K-ϵ model was originally proposed for modeling the turbulent (or eddy) viscosity of the large Reynolds number turbulent shear flow [1]. In the K-ϵ model the turbulent kinetic energy $K \equiv \frac{1}{2}\langle \tilde{v}^2 \rangle$ and its viscous dissipation rate ϵ characterize the local turbulence spectral structure and their temporal and spatial variations are governed by transport equations. The turbulent transport coefficient is given by $D \sim K^2/\epsilon$, which has some nonlocal properties not included in the conventional expressions since the mixing length $l \sim K^{3/2}/\epsilon$, the turbulent time scale $\tau \sim K/\epsilon$ and the turbulent transport coefficient $D \sim l^2/\tau \sim K^2/\epsilon$ are determined not locally but globally by the solution of K-ϵ transport equations.

The resistive interchange turbulence has been extensively studied as a cause of anomalous transport in the peripheral region of stellarator plasmas [2, 3]. The K-ϵ equations for the resistive interchange turbulence have the turbulent energy production terms, which are given by the pressure gradient multiplied by the average magnetic curvature, the viscous and Joule dissipation terms, and the transport terms. In the case of the homogeneous turbulence, the transport terms vanish and the anomalous transport may be well described in terms of the local plasma parameters. However, when the turbulence is significantly inhomogeneous, the anomalous diffusivity deviates from the local parameter expression due to the transport terms in the K-ϵ equations.

7.1.2 TURBULENT TRANSPORT IN INHOMOGENEOUS PLASMA

Here we present a brief explanation of 'Two-Scale Direct-Interaction Approximation (TSDIA)' [4] which is useful for treating turbulent transport in inhomogeneous fluids. It is based on Kraichnan's direct-interaction approximation (DIA) [5, 6] and the two-scale expansion technique [7] utilizing the fact that the characteristic scale of the turbulent fluctuations is much smaller than that of the mean fields. A random field $f(\mathbf{x}, t)$ such as a turbulent velocity field $\mathbf{v}(\mathbf{x}, t)$ is written as

$$f(\mathbf{x}, t) = F(\mathbf{X}, T) + \tilde{f}(\mathbf{x}, \mathbf{X}; t, T), \qquad F = \langle f \rangle \qquad (1)$$

where $\mathbf{X} \equiv \delta \mathbf{x}$ and $T \equiv \delta t$ represent weak spatial and temporal dependence of the mean field and δ the small expansion parameter. Here $\langle \cdot \rangle$ denotes the ensemble average. The fluctuation part $\tilde{f}$ is Fourier transformed with respective to $\mathbf{x}$ as

$$\tilde{f}(\mathbf{x}, \mathbf{X}; t, T) = \int d^3 k\, f(\mathbf{k}, \mathbf{X}; t, T) \exp(i\mathbf{k} \cdot \mathbf{x}) \qquad (2)$$

where $\mathbf{X}$ is treated as a parameter. Then we expand $f(\mathbf{k}; t) \equiv f(\mathbf{k}, \mathbf{X}; t, T)$ in powers of δ as

$$f(\mathbf{k}; t) = f_0(\mathbf{k}; t) + \delta f_1(\mathbf{k}; t) + \delta^2 f_2(\mathbf{k}; t) + \cdots . \qquad (3)$$

As an example, let us consider the passive scalar $\theta = \Theta + \tilde{\theta}$ ($\Theta = \langle \theta \rangle$) which satisfies

$$\left(\frac{\partial}{\partial t} + \mathbf{v} \cdot \nabla \right) \theta = \kappa \nabla^2 \theta . \qquad (4)$$

Then, the equation for the fluctuation part $\tilde{\theta}$ is given by

$$\left(\frac{\partial}{\partial t} - \kappa \nabla^2 \right) \tilde{\theta} + \nabla \cdot \left(\tilde{\theta} \tilde{\mathbf{v}} - \langle \tilde{\theta} \tilde{\mathbf{v}} \rangle \right) = -\tilde{\mathbf{v}} \cdot \nabla \Theta . \qquad (5)$$

Applying the scale expansion to the above equation, we obtain $O(\delta^0)$ and $O(\delta^1)$ equations as

$$\left(\frac{\partial}{\partial t} + \kappa k^2 \right) \theta_0(\mathbf{k}) + i k^a \sum^{\triangle} v_0^a(\mathbf{p}) \theta_0(\mathbf{q}) = 0 \qquad (6)$$

$$\left(\frac{\partial}{\partial t} + \kappa k^2\right)\theta_1(\mathbf{k}) + ik^a \overset{\triangle}{\sum}[v_0^a(\mathbf{p})\theta_1(\mathbf{q}) + v_1^a(\mathbf{p})\theta_0(\mathbf{q})]$$

$$= -\frac{\partial \Theta}{\partial X^a}v_0^a(\mathbf{k}) - \left(\frac{\partial}{\partial T} - 2ik^a\frac{\partial}{\partial X^a}\right)\theta_0(\mathbf{k}) - \overset{\triangle}{\sum}\frac{\partial}{\partial X^a}[v_0^a(\mathbf{p})\theta_0(\mathbf{q})] \quad (7)$$

where $\overset{\triangle}{\sum} \equiv \int d^3p \int d^3q \delta(\mathbf{k}-\mathbf{p}-\mathbf{q})$. From these equations combined with the DIA propagator-renormalization, we can calculate the turbulent flux $\langle \tilde{\theta}\tilde{v}^a \rangle$ which is given up to $O(\delta)$ as

$$\langle \tilde{\theta}\tilde{v}^a \rangle = -\frac{\partial \Theta}{\partial x^a} \int d^3k \int_{-\infty}^{t} dt_1 G_\theta(\mathbf{k};t,t_1) Q^{aa}(\mathbf{k};t_1,t) \quad (8)$$

where the velocity correlation function $Q^{\alpha\beta}$ and the average propagator (or response function) G_θ are defined by

$$Q^{\alpha\beta}(\mathbf{k};t,t') \equiv \langle v_0^\alpha(\mathbf{k};t)v_0^\beta(-\mathbf{k};t') \rangle / \delta(0), \quad G_\theta(\mathbf{k};t,t') \equiv \langle \hat{G}_\theta(\mathbf{k};t,t') \rangle \quad (9)$$

$$(\partial_t + \kappa k^2)\hat{G}_\theta(\mathbf{k};t,t') + ik^a \overset{\triangle}{\sum} v_0^a(\mathbf{p};t)\hat{G}_\theta(\mathbf{q};t,t') = \delta(t-t'). \quad (10)$$

It is seen that the turbulent diffusion tensor is a functional of the wavenumber spectra of the velocity fluctuation and the propagator. These spectral functions include $\mathbf{X}$ as a parameter, which implies that they correspond to the local turbulence around the point $\mathbf{X}$. However direct calculation of these spectra on each local region is very complicated since the degree of freedom in them is extremely large. Thus it is reasonable to characterize the local turbulence by a small number of characteristic quantities in stead of full description of the wavenumber spectra. The turbulent energy K and its transfer rate ϵ to small wavelength regions are considered as such quantities to characterize the properties of the local turbulence. From dimensional analysis, we have the turbulent diffusivity as $D \sim K^2/\epsilon$. From the inertial-range theory for the wavenumber spectra of locally isotropic three-dimensional turbulence, we obtain

$$\langle \tilde{\theta}\tilde{v}^a \rangle = -D_\theta \frac{\partial \Theta}{\partial x^a}, \quad D_\theta = 0.135 \frac{K^2}{\epsilon}. \quad (11)$$

As mentioned earlier, in conventional models of plasma anomalous transport, anomalous diffusivities and characteristic quantities such as K and ϵ are assumed to be determined by local plasma parameters only. However that assumption is not valid when spatial transport of turbulent energy itself occurs due to strong inhomogeneities in a plasma. In order to investigate production, dissipation and transport of the turbulent energy, we consider the following magnetohydrodynamical (MHD) equations:

$$\left(\frac{\partial}{\partial t} + \mathbf{v} \cdot \nabla\right)\mathbf{v} = -\nabla p^* + \frac{1}{4\pi\rho_0}\mathbf{B}\cdot\nabla\mathbf{B} + \frac{\rho}{\rho_0}\mathbf{g} + \nu\nabla^2\mathbf{v} \quad (12)$$

$$\left(\frac{\partial}{\partial t} + \mathbf{v} \cdot \nabla\right)\rho = S \quad (13)$$

$$\left(\frac{\partial}{\partial t} + \mathbf{v} \cdot \nabla\right)\mathbf{B} = \mathbf{B}\cdot\nabla\mathbf{v} + \frac{c^2\eta}{4\pi}\nabla^2\mathbf{B} \quad (14)$$

where $\mathbf{v}$ denotes the plasma flow velocity, $\mathbf{B}$ the magnetic field, ρ the mass density, $\mathbf{g}$ the acceleration due to gravity, ν the kinematic viscosity, η the resistivity, c the light velocity in the vacuum and $p^* = (p + B^2/8\pi)/\rho_0$ the sum of the kinematic pressure p and the magnetic pressure $B^2/8\pi$ divided by the averaged mass $\rho_0 = \langle\rho\rangle$. In the above equations (12)–(14) we used the solenoidal condition for $\mathbf{B}$ and $\mathbf{v}$

$$\nabla\cdot\mathbf{B} = 0, \quad \nabla\cdot\mathbf{v} = 0 \quad (15)$$

in which the latter represents the incompressibility. In the momentum balance equation (12), the spatial and temporal variation of the mass density ρ are neglected except in the acceleration term $\rho\mathbf{g}$ according to the Boussinesq approximation. The source term S is included in the density equation (13). The turbulent quantities are divided into the average and fluctuating parts as

$$\mathbf{v} = \mathbf{V} + \tilde{\mathbf{v}}, \quad \mathbf{V} = \langle\mathbf{v}\rangle \quad (16)$$

$$\mathbf{B} = \mathbf{B}_0 + \tilde{\mathbf{b}}, \quad \mathbf{B}_0 = \langle\mathbf{B}\rangle \quad (17)$$

$$\rho = \rho_0 + \tilde{\rho}, \quad \rho_0 = \langle\rho\rangle \quad (18)$$

$$p^* = P + \tilde{p}, \quad P = \langle p^*\rangle. \quad (19)$$

From (12)–(19), we obtain the following equations for the average and turbulent parts of energy

$$\left(\frac{\partial}{\partial t} + \mathbf{V} \cdot \nabla\right)\left(\frac{1}{2}\rho_0 V^2 + \frac{1}{8\pi} B^2 + \rho_0 \Phi_g\right)$$

$$= \nabla \cdot \left[-\rho_0 P \mathbf{V} + \frac{1}{4\pi}(\mathbf{V} \cdot \mathbf{B}_0)\mathbf{B}_0 + \left(-\rho_0 \langle \tilde{\mathbf{v}}\tilde{\mathbf{v}} \rangle + \frac{1}{4\pi}\langle \tilde{\mathbf{b}}\tilde{\mathbf{b}} \rangle\right) \cdot \mathbf{V}\right.$$

$$\left. + \frac{1}{4\pi}(\langle \tilde{\mathbf{b}}\tilde{\mathbf{v}} \rangle - \langle \tilde{\mathbf{v}}\tilde{\mathbf{b}} \rangle) \cdot \mathbf{B}_0 - \Phi_g \langle \tilde{\rho}\tilde{\mathbf{v}} \rangle + \frac{\mu}{2}\nabla V^2 + \left(\frac{c}{4\pi}\right)^2 \frac{\eta}{2}\nabla B_0^2 \right]$$

$$+ \left(\rho_0 \langle \tilde{\mathbf{v}}\tilde{\mathbf{v}} \rangle - \frac{1}{4\pi}\langle \tilde{\mathbf{b}}\tilde{\mathbf{b}} \rangle\right) : \nabla \mathbf{V} + \frac{1}{4\pi}(\langle \tilde{\mathbf{b}}\tilde{\mathbf{v}} \rangle - \langle \tilde{\mathbf{v}}\tilde{\mathbf{b}} \rangle) : \nabla \mathbf{B}_0$$

$$+ S\Phi_g - \mu \left\langle \frac{\partial V^a}{\partial x^b} \frac{\partial V^a}{\partial x^b} \right\rangle - \left(\frac{c}{4\pi}\right)^2 \eta \left\langle \frac{\partial B_0^a}{\partial x^b} \frac{\partial B_0^a}{\partial x^b} \right\rangle - \langle \tilde{\rho}\tilde{\mathbf{v}} \rangle \cdot \mathbf{g} \quad (20)$$

$$\left(\frac{\partial}{\partial t} + \mathbf{V} \cdot \nabla\right)\left(\frac{1}{2}\rho_0 \langle \tilde{v}^2 \rangle + \frac{1}{8\pi}\langle \tilde{b}^2 \rangle\right)$$

$$= \nabla \cdot \left[-\frac{1}{2}\rho_0 \langle \tilde{v}^2 \tilde{\mathbf{v}} \rangle - \frac{1}{8\pi}\langle \tilde{b}^2 \tilde{\mathbf{v}} \rangle - \langle \tilde{p}\tilde{\mathbf{v}} \rangle + \frac{1}{4\pi}\langle \tilde{\mathbf{v}} \cdot \tilde{\mathbf{b}} \rangle \mathbf{B}_0 + \frac{1}{4\pi}\langle (\tilde{\mathbf{v}} \cdot \tilde{\mathbf{b}})\tilde{\mathbf{b}} \rangle \right.$$

$$\left. + \frac{\mu}{2}\nabla \langle \tilde{v}^2 \rangle + \left(\frac{c}{4\pi}\right)^2 \frac{\eta}{2}\nabla \langle \tilde{b}^2 \rangle \right]$$

$$+ \left(-\rho_0 \langle \tilde{\mathbf{v}}\tilde{\mathbf{v}} \rangle + \frac{1}{4\pi}\langle \tilde{\mathbf{b}}\tilde{\mathbf{b}} \rangle\right) : \nabla \mathbf{V} + \frac{1}{4\pi}(\langle \tilde{\mathbf{v}}\tilde{\mathbf{b}} \rangle - \langle \tilde{\mathbf{b}}\tilde{\mathbf{v}} \rangle) : \nabla \mathbf{B}_0$$

$$+ \langle \tilde{\rho}\tilde{\mathbf{v}} \rangle \cdot \mathbf{g} - \mu \left\langle \frac{\partial \tilde{v}^a}{\partial x^b} \frac{\partial \tilde{v}^a}{\partial x^b} \right\rangle - \left(\frac{c}{4\pi}\right)^2 \eta \left\langle \frac{\partial \tilde{b}^a}{\partial x^b} \frac{\partial \tilde{b}^a}{\partial x^b} \right\rangle \quad (21)$$

where we assumed $\mathbf{g}$ to be expressed in terms of a time-independent potential Φ_g as $\mathbf{g} = -\nabla \Phi_g$. Here we treated $\mu \equiv \rho_0 \nu$ and η as constants and the density source term S as a non-random quantity. We can see that the production of the turbulent energy due to the gradients of the mean velocity and magnetic field are transferred from the mean part of energy since the corresponding turbulent energy production terms appear in (20) and (21) with opposite signs. The turbulent energy production due to the gravity, which is essential to the resistive interchange turbulence, also comes from the loss of the mean part of energy as can be seen from the signs of $\langle \tilde{\rho}\tilde{\mathbf{v}}\rangle \cdot \mathbf{g}$ in (20) and (21). In addition to the energy supply due to the transport through the plasma boundary surface, the mean energy are produced by the density source term $S\Phi_g$ which represents the supply of particles with gravity potential. Both the mean and turbulent energy equations (20) and (21) contain the viscous and Ohmic dissipation terms, which are proportional to μ and η, respectively. In (20) and (21), terms in the form of $\nabla \cdot [\cdots]$ represent transport of energy. The flux terms such as $\nabla \cdot \left(\frac{1}{2}\rho_0 \langle \tilde{v}^2\tilde{\mathbf{v}}\rangle\right)$ represent turbulent (or anomalous) transport of turbulent energy itself similar to the anomalous transport of particles and thermal energies. When the effects of the turbulent energy transport terms are significant compared to the local production and dissipation terms, the turbulent energy should be determined not by the local plasma parameters only but by its transport equation.

7.1.3 K-ϵ ANOMALOUS TRANSPORT MODEL

The K-ϵ model for anomalous transport in resistive interchange turbulence was derived in Ref. [8] by applying two-scale direct-interaction approximation (TSDIA) [4] to the resistive magnetohydrodynamical (MHD) equations. Assuming that the mean velocity vanishes $\langle \mathbf{v}\rangle = 0$ and that there is the inhomogeneity of turbulence only in the minor radial direction, statistical analyses of the resistive MHD equations show that the equations governing the turbulent kinetic energy $K \equiv \frac{1}{2}\langle \tilde{v}^2\rangle$ and its dissipation rate ϵ are written as follows,

$$\frac{\partial K}{\partial t} = P_K - \epsilon - \epsilon_J + T_K \qquad (22)$$

$$\frac{\partial \epsilon}{\partial t} = C_{\epsilon 1}\frac{\epsilon}{K}P_K - C_{\epsilon 2}\frac{\epsilon^2}{K} - C_{\epsilon J}\frac{\epsilon\epsilon_J}{K} + T_\epsilon . \qquad (23)$$

where ϵ_J denotes the Joule dissipation term and the turbulent energy production term P_K, the transport terms T_K and T_ϵ are given by

$$P_K = \langle \tilde{p}\tilde{v}_r \rangle \frac{1}{\rho_m} \frac{d\Omega}{dr} \tag{24}$$

$$T_K = \frac{1}{r} \frac{\partial}{\partial r} \left(rC_K \frac{K^2}{\epsilon} \frac{\partial K}{\partial r} \right), \quad T_\epsilon = \frac{1}{r} \frac{\partial}{\partial r} \left(rC_\epsilon \frac{K^2}{\epsilon} \frac{\partial \epsilon}{\partial r} \right) \tag{25}$$

Here ρ_m denotes the mass density, c the light velocity in the vacuum, η the resistivity, B the magnitude of the magnetic field, and $d\Omega/dr$ the average magnetic curvature. As seen from (24), the turbulent energy production is in the form of the product of the flux and the centrifugal force due to the average magnetic curvature. The turbulent pressure flux $\langle \tilde{p}\tilde{\mathbf{v}} \rangle$ is expressed in terms of the mean pressure gradient dP/dr and the turbulent diffusivity D_p as

$$\langle \tilde{p}\tilde{\mathbf{v}} \rangle = -D_p \frac{dP}{dr} = -C_p \frac{K^2}{\epsilon} \frac{dP}{dr}. \tag{26}$$

In the same manner as in (26), the turbulent diffusivity for passive scalars in the turbulent velocity field is given by

$$D_\theta = C_\theta \frac{K^2}{\epsilon}. \tag{27}$$

C_K, C_ϵ, $C_{\epsilon 1}$, $C_{\epsilon 2}$, $C_{\epsilon J}$, C_p, and C_θ are non-dimensional constants, which are determined empirically or theoretically from TSDIA.

With the electrostatic approximation, the Joule dissipation term $\epsilon_J = \eta \tilde{J}^2/\rho_m$ can be expressed in terms of K and ϵ as follows. For high collision frequencies such that $L_s^2 m_e \nu_e T_e > \tau (\sim K/\epsilon)$, we have

$$\epsilon_J = C_J \frac{B_0^2}{\rho_m c^2 \eta L_s^2} \frac{K^4}{\epsilon^2} \quad \text{for} \quad L_s^2 \frac{m_e \nu_e}{T_e} > \tau \tag{28}$$

where L_s is a magnetic shear length and C_J a non-dimensional numerical constant. Here we used the Ohm's law $\tilde{J} \sim \eta^{-1} k_\parallel \tilde{\phi}$ and estimated that $K \sim \tilde{v}^2 \sim (ck_\perp \tilde{\phi}/B)^2$ and $k_\parallel/k_\perp \sim 1/L_s$. On the other hand, for low collision frequencies such that $L_s^2 m_e \nu_e/T_e < \tau$, we need to take account of the adiabaticity of electron response to potential fluctuations to evaluate ϵ_J. From the balance between the time scales of the electron parallel conduction and the potential fluctuation $k_\parallel^2 T_e/(m_e \nu_e) \sim \tau^{-1}$ with $k_\parallel \sim k_\perp \Delta/L_s$, the width of the non-adiabatic layer Δ is given by $k_\perp^2 \Delta^2 \sim L_s^2 (m_e \nu_e/T_e)\tau^{-1}$. Therefore, in this low collisional case, Δ becomes smaller than the turbulent mixing length $l \sim K^{3/2}/\epsilon$. Using the generalized Ohm's law $\tilde{J} \sim (T_e/e\eta)k_\parallel(\tilde{n}/n_0 - e\tilde{\phi}/T_e)$ and noting that we have the Boltzmann relation $\tilde{n}/n_0 \sim e\phi/T_e$ outside the non-adiabatic layer and $\tilde{n}/n_0 < e\phi/T_e$ inside it, we obtain $\tilde{J}^2 \sim \eta^{-2}(\Delta/L_s)^2 (k_\perp \tilde{\phi})^2$. From these relations, we have

$$\epsilon_J = C'_J \frac{1}{\rho_s^2} \frac{K^3}{\epsilon} \quad \text{for} \quad L_s^2 \frac{m_e \nu_e}{T_e} < \tau \tag{29}$$

where C'_J is a numerical constant and $\rho_s = c\sqrt{m_i T_e}/(eB_0)$ the ion Larmor radius at the electron temperature.

If the profiles of the mean pressure gradient dP/dr and the magnetic curvature $d\Omega/dr$ are given, we can obtain the turbulent diffusivity (27) by solving the K-ϵ transport equations (22) and (23) with proper boundary conditions imposed. The self-consistent model for the analysis of anomalous transport is obtained by combining the K-ϵ equations with the transport equations for the density and temperature. The profiles of the density and temperature affect the turbulence through the production and dissipation terms in the K-ϵ equations while the turbulence has dominant effects on the transport of the density and temperature through the anomalous diffusivities. The diffusion terms given by (25) describe the radial propagation of the turbulence which have not been taken into account by the conventional anomalous transport model.

7.1.4 SCALING IN TERMS OF LOCAL PARAMETERS

Here, we consider the stationary case in which the transport term in the turbulent energy can be neglected so that the energy production and dissipation terms are balanced with each other. Even then, the turbulent energy flux does not need to vanish although it should have a constant value. Then we obtain from (22), (24) and (26),

$$C_p \left(\frac{-1}{\rho_m} \frac{dP}{dr} \frac{d\Omega}{dr}\right) \frac{K^2}{\epsilon} - \epsilon - \epsilon_J = 0 \ . \tag{30}$$

In this case, as seen from (28)–(30), it can be assumed that the turbulence property is characterized locally by two parameters, which are for high collision frequencies ($L_s^2 m_e \nu_e / T_e > \tau$),

$$\frac{-1}{\rho_m} \frac{dP}{dr} \frac{d\Omega}{dr} \ , \quad \frac{\rho_m c^2 \eta L_s^2}{B_0^2} \tag{31}$$

and for low collision frequencies ($L_s^2 m_e \nu_e / T_e < \tau$),

$$\frac{-1}{\rho_m} \frac{dP}{dr} \frac{d\Omega}{dr} \ , \quad \rho_s \ . \tag{32}$$

In the case considered here, the turbulence is regarded as locally homogeneous and these parameters need to be nearly constant. When the variation of the parameters is large, the turbulence can be no longer homogeneous and the turbulent energy transport term becomes important so that the assumption given above is not valid. The first parameter of (31) or (32) is written as $-P'\Omega'/\rho_m \sim c_s^2/(L_p L_c)$ and has a dimension of square frequency, which gives the characteristic time scale $\sqrt{L_p L_c}/c_s$ of the turbulence driven by the pressure gradient and the magnetic curvature. Here $dP/dr = P' = -P/L_p$, $d\Omega/dr = \Omega' = 1/L_c$ and $c_s = \sqrt{T_e/m_i}$ are used and $T_e \geq T_i$ is assumed.

We have the following scaling in terms of the above local parameters from the dimensional analysis. First, in the high collisional case,

$$K \sim c^2 \eta \rho_m^{-1/2} (-P'\Omega')^{3/2} L_s^2 / B_0^2 \ , \quad l \sim K^{3/2}/\epsilon \sim c \eta^{1/2} (-\rho_m P'\Omega')^{1/4} L_s / B_0^2 \ ,$$
$$\epsilon \sim c^2 \eta \rho_m^{-1} (-P'\Omega')^2 L_s^2 / B_0^2 \ , \quad \tau \sim K/\epsilon \sim (-P'\Omega'/\rho_m)^{-1/2} \sim (c_s/\sqrt{L_p L_c})^{-1} \ . \tag{33}$$

Then we obtain the anomalous diffusivity D as

$$D \sim \frac{K^2}{\epsilon} \sim \frac{l^2}{\tau} \sim \frac{c^2 \eta (-P'\Omega') L_s^2}{B_0^2} \sim D_{cl} \frac{L_s^2}{L_p L_c} \tag{34}$$

where we used the classical diffusivity $D_{cl} = c^2 \eta P/B_0^2$. Equation (34) is the same expression as that of the anomalous diffusivity for the resistive interchange turbulence obtained from the reduced MHD equations using the dimensional analysis or the scale invariance technique. Using the time scale $\tau \sim \sqrt{L_p L_c}/c_s$, we can write the condition for the high collisional case as $L_s^2 m_e \nu_e / T_e > \sqrt{L_p L_c}/c_s$.

Next, in the same way as above, we obtain the scaling in the low collisional case as follows

$$K \sim \rho_s^2 c_s^2 / L_p L_c, \qquad l \sim K^{3/2}/\epsilon \sim \rho_s,$$
$$\epsilon \sim \rho_s^2 c_s^3 / (L_p L_c)^{3/2}, \quad \tau \sim K/\epsilon \sim (c_s/\sqrt{L_p L_c})^{-1} \qquad (35)$$

which gives the anomalous diffusivity D as

$$D \sim \frac{K^2}{\epsilon} \sim \frac{l^2}{\tau} \sim \frac{\rho_s^2 c_s}{\sqrt{L_p L_c}} \sim D_B \frac{\rho_s}{\sqrt{L_p L_c}} \qquad (36)$$

where we used the Bohm diffusivity $D_B = \rho_s c_s = cT_e/(eB_0)$. Equation (36) has the form of the gyro-reduced Bohm (GRB) diffusivity [9] which is the Bohm diffusivity multiplied by the factor $\rho_s/\sqrt{L_p L_c}$. The condition for the low collisional case is written as $L_s^2 m_e \nu_e / T_e < \sqrt{L_p L_c}/c_s$.

7.1.5 TRANSPORT SIMULATION OF ECH PLASMAS IN HELIOTRON E

We have combined the K-ϵ anomalous transport model with the transport code for stellarators [10] to simulate ECH plasmas in Heliotron E ($R = 2.2m$, $a = 0.2m$) [11, 12]. The anomalous particle diffusivity D is given by (27). We have assumed that both the electron and ion thermal diffusivities are given by the same expression $\chi_e = \chi_i = \frac{5}{2} D$. In our simulations, the averaged electron density $\bar{n}_e$, the ECH absorbed power P_{abs} and the magnetic field strength B were scanned in the following ranges: $1 \leq \bar{n}_e \leq 3 \times 10^{19} m^{-3}$, $96 \leq P_{abs} \leq 288\,\text{kW}$ and $B = 0.95, 1.9\,\text{T}$, respectively. We have done numerically the time integration of the electron and ion temperatures, T_e, T_i as well as the turbulent energy K and its dissipation rate ϵ while the electron density profile was fixed. We gave the profiles of the electron density and the absorbed power density as $n_e(r) = n_e(0)[0.95\,(1-(r/a)^6)+0.05]$ and $p_{abs}(r) = p_{abs}(0)\,(1-(r/a)^4)^2$ which fit the experimentally observed results. The average magnetic curvature $d\Omega/dr$ was given from the vacuum magnetic field configuration of Heliotron E since the beta values of the ECH plasmas simulated here are very low. The neoclassical diffusivities were included in our simulations although the effects of the radial electric field and the neutral particles are assumed to be negligible in order to clarify the effects of the anomalous transport. In the parameter regime of the ECH plasmas in Heliotron E, the whole plasma is considered to satisfy the low collision frequency condition and therefore we employed the Joule dissipation term given by (29). The numerical constants used here are $C_K = 0.09$, $C_\epsilon = 0.07$, $C_{\epsilon 1} = C_{\epsilon 2} = C_{\epsilon J} = 1.7$, $C_p = C_\theta = 0.135$ and $C'_J = 0.05$.

After adequate time steps, we obtained the stationary states in which the radial profiles of T_e, T_i, K and ϵ did not depend on the initial conditions. Figure 1 shows the radial profile of the anomalous thermal diffusivity obtained by K-ϵ model, $\chi_e^{K-\epsilon} = \frac{5}{2} C_\theta K^2/\epsilon$, in the stationary state for $\bar{n}_e = 1 \times 10^{19}, km^{-3}$, $P_{abs} = 192\,\mathrm{kW}$ and $B = 1.9\,\mathrm{T}$. In this case, the boundary conditions for K and ϵ were given such that the energy confinement time took the experimentally observed value. There also shown is the profile of the anomalous diffusivity expressed in terms of the local parameters as in (36), $\chi_e^{local} \equiv C\,(\rho_s/\sqrt{L_p L_c})\,c(T_e+T_i)/eB$, where the value of the numerical coefficient employed in Fig. 1 is $C = 0.57$. It can be seen that both of the diffusivities have the same radial dependence in the region $0.1 < r/a < 0.6$ while the discrepancy between their profiles becomes large in the other regions. Especially, in the peripheral region, $\chi_e^{K-\epsilon}$ increases in approaching the boundary whereas χ_e^{local} decreases. Figure 2 shows the radial profiles of the turbulent energy production, viscous and Joule dissipations in the same case as in Fig. 1. It is found that the production and dissipation occur mostly near the peripheral region where the average magnetic curvature becomes large. In this case, the inward transport of the turbulent energy appears. The ratio of the Joule dissipation to the viscous one increases in the peripheral region due to the decrease in the temperature, which is correlated with the deviation of $\chi_e^{K-\epsilon}$ from χ_e^{local} since the ratio between the production, viscous and Joule dissipations need to be homogeneous or constant in order to ensure the validity of the scaling by the local parameters. Thus the local parameter expression poorly predicts the anomalous transport coefficient in the peripheral region where the inhomogeneities of the local parameters in (32) are significant.

In Fig. 3, the experimentally obtained thermal diffusivity χ_e^{exp} in the case corresponding to Fig. 1 is compared with the numerically predicted total diffusivity $\chi_e^{\text{total}} = \chi_e^{K-\epsilon} + \chi_e^{\text{neo-ax}} + \chi_e^{\text{ripple}}$, where $\chi_e^{\text{neo-ax}}$ and χ_e^{ripple} denote the neoclassical axisymmetric and non-axisymmetric (ripple) thermal diffusivities, respectively. The disagreement between χ_e^{total} and χ_e^{exp} seems to be within the accuracy of the experimental results although the predicted diffusivity χ_e^{tot} may be relatively smaller than χ_e^{exp} in the inner region $r < 0.3a$ since there the magnetic curvature is quite small and other turbulence sources are not taken into account in our model. It is seen that the anomalous diffusivity is a dominant contribution to the whole plasma confinement although χ_e^{ripple} is comparable to $\chi_e^{K-\epsilon}$ at $0.3 < r/a < 0.5$ and $\chi_e^{\text{neo-ax}}$ is the largest at $r/a < 0.1$.

We have scanned the electron density, the absorbed power and the magnetic field strength in the ranges mentioned earlier. Since we have seen in Fig. 1 that the local parameter expression (36) is valid in the regions except for the peripheral and central regions, we adjusted the boundary conditions for K and ϵ in all the simulations in the above ranges such that the K-ϵ anomalous diffusivity coincide with the local expression at $r = a/2$: $\chi_e^{K-\epsilon}(a/2) = \chi_e^{\text{local}}(a/2)$. In Fig. 4, the energy confinement times $\tau_E^{K-\epsilon}$ obtained from the simulations are compared with the LHD scaling [13]. It is found that the simulation results are in good agreement with the LHD scaling. This seems to be natural since the thermal diffusivity imposed at $r = a/2$ obeys a type of GRB scaling which gives almost the same energy confinement time as the LHD scaling. Thus our model predicts the experimental results consistently in the two aspects: the first is that it gives the energy confinement time following the GRB or LHD scaling and the second is that it supplements the drawback of the GRB diffusivity, i.e., it reproduces the experimentally observed profile of the anomalous diffusivity which has large values in the peripheral region.

Bibliography

[1] BRADSHAW, P., CEBECI T., WHITELAW, J.H., Engineering Calculation Method for Turbulent Flow, Academic, London (1981) 37.

[2] CARRERAS, B.A., GARCIA, L., DIAMOND, P.H., Phys. Fluids **30**, (1987) 1388.

[3] SUGAMA, H., WAKATANI, M., J. Phys. Soc. Jpn. **57**, (1988) 2010.

[4] YOSHIZAWA, A., Phys. Fluids **27**, (1984) 1377.

[5] KRAICHNAN, R.H., J. Fluid Mech. **5**, (1959) 497.

[6] LESLIE, D.C., Developments in the Theory of Turbulence (Clarendon Press, Oxford, 1973).

[7] NAYFEH, A.H., Perturbation Methods (Wiley, New York, 1973), p. 243.

[8] SUGAMA, H., WAKATANI, M., J. Phys. Soc. Jpn. **61**, (1992) 3166.

[9] PERKINS, F.W., in *Heating in Toroidal Plasmas* (Proc. 4th Int. Symp. Rome, 1984), Vol. 2 (1984) 977.

[10] NAKAMURA, Y., WAKATANI, M., Transport Simulation of New Stellarator/Heliotron Devices Based on the Neoclassical Ripple Transport Associated with an Edge Turbulence, Rep. PPLK-R-24, Plasma Phys. Lab., Kyoto Univ., Uji (1988).

[11] ZUSHI, H., *et al.*, Nucl. Fusion **28**, (1988) 1801.

[12] SUDO, S., *et al.*, Nucl. Fusion **31**, (1991) 2349.

[13] SUDO, S., *et al.*, Nucl. Fusion **30**, (1990) 11.

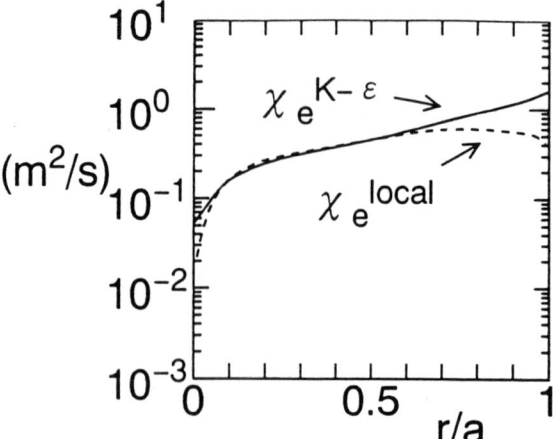

Figure 1. Comparison between the anomalous thermal diffusivity of the K-ϵ model, $\chi_e^{K-\epsilon}$, and that of the local parameter expression, χ_e^{local}.

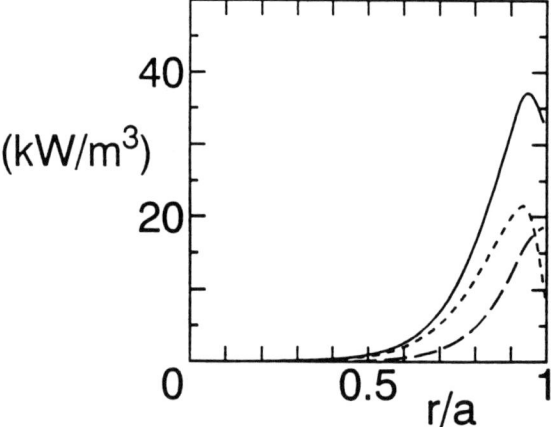

Figure 2. Radial profiles of the turbulent kinetic energy production (solid line), the viscous dissipation (dotted line) and the Joule dissipation (dashed line).

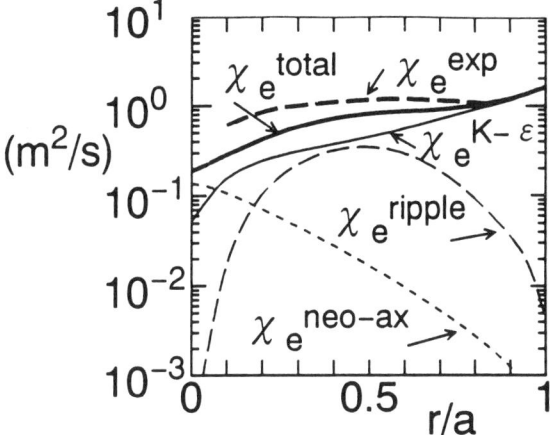

Figure 3. Radial profiles of the neoclassical axisymmetric thermal diffusivity $\chi_e^{\rm neo-ax}$, the neoclassical ripple diffusivity $\chi_e^{\rm ripple}$, the K-ϵ anomalous diffusivity $\chi_e^{K-\epsilon}$, the total diffusivity $\chi_e^{\rm total} = \chi_e^{K-\epsilon} + \chi_e^{\rm neo-ax} + \chi_e^{\rm ripple}$ and the experimentally obtained diffusivity $\chi_e^{\rm exp}$.

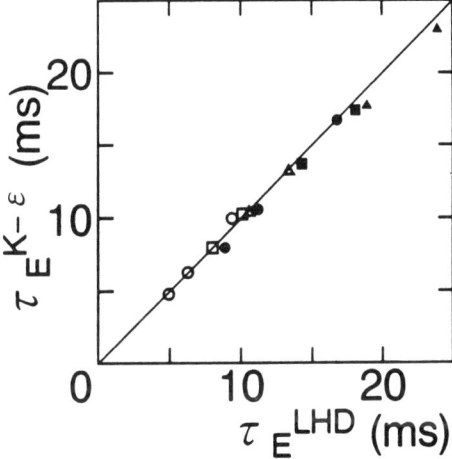

Figure 4. Comparison between the energy confinement times obtained from the simulations using the K-ϵ model, $\tau_E^{K-\epsilon}$, and those of the LHD scaling [13], $\tau_E^{\rm LHD}$.

7.2 RESISTIVE INTERCHANGE INSTABILITIES IN HELIOTRON E

M. Wakatani
Plasma Physics Laboratory, Kyoto University, Gokasho, Uji, Japan 611

Abstract. In Heliotron E resistive interchange (or g) modes are unstable even for sufficiently low beta plasmas, since magnetic hill governs a substantial region of the plasma column. From linear theory of g modes, both low and high-m modes are relevant to present experiments, where m is a poloidal mode number in the cylindrical plasma limit. Evidence of low-m g mode is clearly seen in the pressure driven sawtooth oscillations observed in Heliotron E. This phenomena are well explained by the nonlinear evolution of low-m g mode. Unstable high-m g modes may show turbulent behavior in Heliotron E. Nonlinear theory of g mode turbulence and predicted anomalous transport are briefly reviewed. Also theoretical models are extended to include coupling to resistive drift waves and ion temperature gradient (or η_i) modes, and to study poloidal shear flow effect.

Recent measurements of density and magnetic field fluctuations are compared with the theoretical predictions for the g mode turbulence qualitatively. The anomalous edge transport seems to be induced by the g mode turbulence.

7.2.1 INTRODUCTION

Since several stellarators such as Heliotron E [1], ATF [2] and CHS [3] can not exclude the magnetic hill in the edge plasma region, the resistive interchange (or g) modes [4] become crucial from the point of view of both the beta limit and the anomalous transport. In the context of the ideal MHD model, interchange modes give the theoretical beta limit of Heliotron E [5]. Necessary stability condition for interchange modes is Mercier criterion [6] in the toroidal geometry or Suydam criterion [7] in the cylindrical approximation. According to the recent numerical results the growth rate of the interchange mode increases very slowly [8, 9] when the above stability condition is violated and the beta is increased. It is also shown that the radial mode structure becomes extremely localized at beta just above the critical value [8]. Even for low beta plasmas below the Mercier (or Suydam) limit, g modes are always unstable in Heliotron E, although the growth rate γ depends on resistivity η as $\gamma \propto \eta^{1/3}$ [4].

In 1984 the sawtooth type relaxation oscillation was firstly observed in the soft-X ray measurement of Heliotron E [10]. After this experiment such a sawtooth type oscillation is always observed in the soft-X ray, density and electron cyclotron emission measurements for finite beta plasmas with $\beta(0) \gtrsim 1\%$ [11]. According to the comparison between the numerical studies of nonlinear g mode with $(m, n) = (1, 1)$ and the experimental results, it is concluded that the low-m g mode is the origin of the sawtooth type oscillation [12, 13, 14]. Since the toroidal plasma current is negligible except a small amount of bootstrap current less than $(1-2)$ kA in Heliotron E, the tearing mode [4] with $(m, n) = (1, 1)$ is not a candidate for the sawtooth type oscillation.

From the linear theory of g mode, when the low-m mode is unstable, it is highly probable that high-m modes are also destabilized in the same resistive plasma. In Heliotron E density and potential fluctuations were measured by the Langmuir probe [15] and the laser phase contrast method [16], and they showed turbulent properties similar to these observed in ATF [17]. There are many results of the density and potential fluctuations showing strong turbulence in tokamaks [18]. We see similarities between the edge turbulence in tokamaks and that in Heliotron E and ATF [19]. However, it is not clear whether the origin of the edge turbulence is the same in both tokamaks and heliotron/torsatron or not. In this paper we will review our theoretical results of the g mode turbulence and compare some of them with the characteristics of fluctuations observed in Heliotron E.

Recently the theoretical model of electrostatic g mode is extended to study coupling to the resistive drift waves [20]. Although two branches corresponding to the g mode and the resistive drift waves appear in the linear dispersion, dependence of the g mode growth rate changes from $\gamma \propto \nu_e^{1/3}$ in the collisional limit to $\gamma \propto \nu_e$ in the weakly collisional regime, where ν_e is an electron collision frequency. This suggests that the turbulent diffusion coefficient depends on ν_e differently according to the collisionality regime [21]. The interesting result for the wavenumber-spectrum is that the dual cascade is seen in the collisional regime [21], which means normal cascade of density fluctuation $\tilde{n}$ and inverse cascade of electric potential fluctuation $\tilde{\phi}$. However, in the weakly collisional regime, it disappears since the tendency forcing the Boltzmann relation $\tilde{n}/n_0 \simeq e\tilde{\phi}/T_e$ becomes strong, where n_0 is a back-ground density and T_e is an electron temperature.

Another extension of the theoretical model for the g mode is to study coupling to the ion temperature gradient or η_i mode. Cordey et al. studied coupling between the g mode and the η_i mode in the Levitron experiment [22]. In heliotron/torsatron the unfavorable average curvature destabilizes both modes. In the weakly collisional regime the growth rate of η_i mode coupled to the g mode remains significant, even if the η_i parameter is close to -1 [23, 24]. This suggests that the ion thermal transport may become anomalous even for a peaked density profile in Heliotron E.

Recent studies of H mode in several tokamaks show the possibility that the poloidal velocity shear induced by the radial electric field suppresses the density and potential fluctuations in the edge region and improves the global confinement [25, 26]. In order to improve the confinement in heliotron/torsatron, we may exploit the same approach as in the L to H transition. According to the linear theory of g mode in the presence of the poloidal shear flow, the growth rate decreases substantially when the shear flow is chosen appropriately [27, 28]. When it is excessively strong, Kelvin-Helmholtz instabilities are excited [29].

When the nonlinear model equations including both the g mode and the resistive drift waves are solved, it is found that the poloidal shear flow or the radial electric field is generated by the convective nonlinearity [28, 30]. The self-generation of the poloidal shear flow is explained by the Reynolds stress [31]. The interesting result is that the self-generated poloidal shear flow affects the saturation level of the g mode turbulence; however, the externally imposed frozen poloidal shear flow does not [21, 28]. The similar property is found for the long wavelength drift waves [32]. At present there is no definite experimental data showing suppression of fluctuations in the edge region due to the poloidal shear flow or the radial electric field in Heliotron E. Although a symptom of H mode like discharge is seen in CHS [33] there is no clear change in the poloidal flow velocity.

In Sec. 2 linear and nonlinear theory of g mode are briefly summarized and in Sec. 3 extensions of g mode theory to include coupling to resistive drift waves and η_i modes, and to study poloidal velocity shear effect are discussed. In Sec. 4 the theoretical predictions for the g mode turbulence and the predicted anomalous transport are compared with experimental results of the fluctuations and the anomalous transport in Heliotron E. Concluding remarks are given in Sec. 5.

7.2.2 LINEAR AND NONLINEAR THEORY OF g MODE

Let us consider a large aspect ration torus with minor radius a and major radius R_0. We use toroidal coordinates r, θ, and ζ here. In the limit of $\varepsilon = a/R_0 \ll 1$, by applying the stellarator expansion to two-fluid equations and averaging them over the pitch length in the toroidal direction, we obtain reduced two-fluid equations in the normalized forms; the momentum balance equation

$$\left(\frac{\partial}{\partial t} + \mathbf{v}_\perp \cdot \nabla\right) \nabla_\perp^2 F - \alpha \nabla_\perp \cdot (\hat{\zeta} \times \nabla P_i \cdot \nabla \nabla_\perp F) =$$

$$\nabla_\| J_\| + \nabla(P_e + P_i) \times \nabla \Omega \cdot \hat{\zeta} + \mu_\perp \nabla_\perp^4 \phi . \tag{1}$$

Faraday's law combined with Ohm's law

$$\frac{\partial A}{\partial t} = -\nabla_\|(\phi - \alpha P_e) - \eta J_\| , \tag{2}$$

the electron pressure equation

$$\frac{1}{\beta}\left(\frac{\partial}{\partial t} + \mathbf{v}_E \cdot \nabla\right) P_e = -\mathbf{v}_{e\perp} \cdot \nabla \Omega - \nabla_\|(v_{\|i} - \alpha J_\|) + \chi_\perp \nabla_\perp^2 P_e , \tag{3}$$

the ion pressure equation

$$(\partial_t + \mathbf{v}_E \cdot \nabla) P_i = 0 , \tag{4}$$

the parallel momentum equation

$$\left(\frac{\partial}{\partial t} + \mathbf{v}_E \cdot \nabla\right) v_{\|i} = -\nabla_\|(P_e + P_i) , \tag{5}$$

where

$$F = \phi + \alpha P_i , \qquad \mathbf{v}_\perp = \hat{\zeta} \times \nabla F ,$$

$$\mathbf{v}_E = \hat{\zeta} \times \nabla \phi , \qquad \mathbf{v}_{e\perp} = \hat{\zeta} \times \nabla(\phi - \alpha P_e) ,$$

$$\nabla_\| = \frac{\partial}{\partial \zeta} + \nabla \Psi \times \hat{\zeta} \cdot \nabla , \qquad \Psi = A + \frac{1}{2} \overline{\nabla \Phi \times \nabla \int_0^{z'} \Phi \cdot \hat{\zeta} \, dz'} ,$$

$$J_\| = -\nabla_\perp^2 A , \qquad \Omega = 2r \cos\theta + \overline{|\nabla \Phi|^2} ,$$

$$\alpha = \frac{c}{a \omega_{pi}} , \qquad \beta = \frac{4\pi P_{e0}}{B_0^2} .$$

It is noted that the normalizations are given by

$$t \equiv (a/\varepsilon v_A)t , \qquad r \equiv ar , \qquad \zeta \equiv R_0 \zeta ,$$

$$\mathbf{v} \equiv \varepsilon v_A \mathbf{v} , \qquad \phi \equiv (\varepsilon a v_A B_0/c)\phi , \qquad F \equiv (\varepsilon a v_A)F , \qquad \Phi \equiv \varepsilon^{1/2} a B_0 \Phi ,$$

$$P \equiv \frac{\varepsilon B_0^2}{4\pi} P , \qquad A \equiv (\varepsilon a B_0) A , \qquad \eta \equiv (4\pi \varepsilon a v_A) \eta/c^2 ,$$

where $v_A = B_0/\sqrt{4\pi n_0 m_i}$ is the Alfvén velocity. Here the external helical magnetic field is $\mathbf{B}_h = \nabla \Phi$, and P_{e0} is a background electron pressure. The viscosity is denoted by $\mu_\perp$, the resistivity by η and the thermal conductivity by $\chi_\perp$. The ion plasma oscillation frequency is ω_{pi}. The bar in the expressions of Ψ and Ω denotes the averaging over the one pitch length in the $\hat{\zeta}$ direction [34]. The first term in Ω is related to the toroidal curvature and the second term in Ω to the averaged curvature due to the external helical magnetic field.

In the limit of $\beta = 0$ the parallel momentum equation (3) is decoupled from the other three equations. Furthermore under $\alpha = 0$, the reduced MHD equations for heliotron/torsatron are obtained as

$$\left(\frac{\partial}{\partial t} + \mathbf{v}_E \cdot \nabla\right) \nabla_\perp^2 \phi = \nabla_\| J_\| + \nabla P \times \nabla \Omega \cdot \hat{\zeta} , \qquad (6)$$

$$\frac{\partial A}{\partial t} = -\nabla_\| \phi - \eta J_\| , \qquad (7)$$

$$\left(\frac{\partial}{\partial t} + \mathbf{v}_E \cdot \nabla\right) P = 0 , \qquad (8)$$

where $P = P_e + P_i$. When $\eta = 0$, Eqs. (6)–(8) describe ideal MHD phenomena.

First we consider interchange modes to discuss the beta limit. In order to study the interchange mode stability the most useful condition is the Mercier criterion [6]. Figure 1 shows the stability diagram of Heliotron E in the plane $(\beta(0), \alpha^*)$, where α^* denotes the ratio between the toroidal field generated by additional toroidal coils and that generated by helical coils at the magnetic axis. Here the Mercier limits (continuous lines) for the resonant surface at $\iota = 0.75$ and $\iota = 1.0$ are shown, where ι is the rotational transform. $\beta(0)$ corresponding to the growth rate $\gamma = 0.01$ of the ideal (circles) or resistive (squares) mode is also plotted. The growth rate is normalized by the poloidal Alfvén transit time. For the resistive interchange modes the magnetic Reynolds number is assumed $S = 10^6$. The pressure profile is fixed at $P = P_0(1 - \Phi)^2$, where Φ is a toloidal flux function. Figure 1 shows that $\alpha^* < 0$ enhances the magnetic hill and the beta limit decreases. On the contrary $\alpha^* > 0$ improves beta limit moderately. The pressure gradient driven sawtooth oscillations related to the $\iota = 1$ resonant surface were observed in Heliotron E for $\beta(0) \gtrsim 1.5 - 2.0\%$ as shown in Fig. 2 [10]. When the pressure profile becomes more peaked, $\beta(0)$ corresponding to the appearance of the sawtooth decreases to $\beta(0) \sim 1\%$ and sometimes $(m, n) = (3, 2)$ mode excites the sawtooth [11]. By comparing Fig. 1 with Fig. 2 corresponding to the standard Heliotron E configuration with $\alpha^* = 0$, when the g mode with $(m, n) = (1, 1)$ has a substantial growth rate of $\gamma = 0.01$, the nonlinear evolution of the $(1, 1)$ mode seems consistent with the pressure gradient driven sawtooth in Heliotron E [12]. The nonlinear g mode produces the magnetic island as shown in Fig. 3, although the equilibrium plasma current is negligible. In Heliotron E the magnetic island overlap due to high-m g modes was discussed by Nakamura *et al.* [35]. The accompanied pressure profile deformation is also shown in Fig. 4. It is considered that the sawtooth collapse occurs when the pressure profile deformation is significant. It is noted that the magnetic island does not cover the whole plasma column at the collapse. The similar sawtooth oscillations were also observed in Heliotron DR [36].

Here we note that the Mercier criterion [6] becomes the Suydam criterion [7] in the cylindrical limit. In this case the growth rate of

Suydam mode, γ_s, is given by

$$\frac{R_0\gamma_s}{v_A} = 16r\frac{d\iota}{dr}\exp\left\{\frac{2}{u}\left[3\arg\Gamma\left(1+\frac{1}{2}iu\right)\right.\right.$$

$$\left.\left. - \arg\Gamma(1+iu) - \tan^{-1}(e^{-\pi u/2}) - \frac{3}{4}\pi\right]\right\}, \qquad (9)$$

where

$$u = \sqrt{\frac{16\pi}{B_\theta^2}\left(\frac{d}{dr}\log\iota\right)^{-2}\left(-\frac{dP_0}{dr}\right)\left(\frac{d\Omega}{dr}\right) - 1} \qquad (10)$$

and Γ is a gamma function. Here the poloidal magnetic field is given by $B_\theta = d\Psi_0/dr$, where Ψ_0 is the equilibrium poloidal flux function including both the external helical magnetic field component and the longitudinal plasma current component. Figure 5 shows γ_s as a function of $\beta(0)$. Here the growth rates of low m interchange modes excited at the $\iota = 1$ surface are also shown. It is seen that the growth rate becomes extremely small when $\beta(0)$ decrees toward the Suydam limit; however, the beta limit of low m modes is equal to the Suydam limit. This conjecture is supported by the behavior of the eigenfunction with $(m,n) = (1,1)$ that the radial mode width becomes extremely localized when $\beta(0)$ becomes close to the Suydam limit (see Fig. 6).

For the g modes the stability criterion was obtained by Glasser, Greene and Johnson,

$$D_R \equiv E + F + H^2 > 0, \qquad (11)$$

where expressions for E, F and H are referred in Ref. [37]. Figure 7 shows both the Mercier criterion $D_I > 0$ and the resistive stability criterion $D_R > 0$ for the standard configuration of Heliotron E with the pressure gradient $P = P_0(1-\Psi)^2$. It is seen that the edge plasma region of Heliotron E is vulnerable to the g modes.

From the reduced MHD model for heliotron/torsatron (6)–(8) the linearized equations are obtained for the g modes in the cylindrical approximation,

$$\frac{d^2\tilde{A}}{dx^2} = \frac{\gamma}{\eta}\tilde{A} - \frac{1}{\eta}msx\tilde{\phi}, \tag{12}$$

$$\frac{d^2\tilde{\phi}}{dx^2} = \frac{ms}{\gamma}\frac{d^2\tilde{A}}{dx^2} + \frac{m^2s^2}{\gamma^2}G\tilde{\phi}, \tag{13}$$

where $G = -\kappa\beta^*/s^2$, $\kappa = \frac{r}{P_0}\frac{dP_0}{dr}\Big|_{r=r_0}$, $\beta^* = \beta_p r \frac{d\Omega}{dr}\Big|_{r=r_0}$, $s = \frac{r}{q}\frac{dq}{dr}\Big|_{r=r_0}$, β_p is a poloidal beta and q is a safety factor given by $q = 1/\iota$. The eigenvalue in Eqs. (12) and (13) corresponding to the growth rate is given by

$$\gamma^\ell = \eta^{1/3}(ms)^{2/3}|h|^{2/3} \simeq \eta^{1/3}(ms)^{2/3}G^{2/3} \quad (G \ll 1), \tag{14}$$

where $h = (-1 + \sqrt{1-4G})/2$ [4]. The mode width is obtained from the eigenfunction as [4]

$$\Delta x^\ell = \eta^{1/3}|h|^{1/6}(ms)^{-1/3} \simeq \eta^{1/3}G^{1/6}(ms)^{-1/3} \quad (G \ll 1). \tag{15}$$

The expressions (14) and (15) show that the higher m mode has the larger growth rate and the more localized mode structure, where m is the poloidal mode number. These are consistent with Figs. 5 and 6.

Here we pay attention to the nonlinear theory of g modes. Our concern is in the saturation level of fluctuations and the anomalous transport driven by the g mode turbulence. The most popular approach to study these subjects is the mixing length theory [38] or the scale invariance [39]. The former depends essentially on the linear theory and the latter use the scale transformations for the original equations. Since the dissipative terms including the viscosity $\mu_\perp$ and the thermal conductivity $\chi_\perp$ (see Eqs. (1) and (3)) are neglected, the saturation mechanism becomes obscure in the scale invariance. However, the interesting point that the above both approaches give the same result for the g mode turbulence driven transport.

The mixing length theory gives the convective transport coefficient

$$D_0 = \eta \frac{\kappa\beta^*}{s^2} \tag{16}$$

by taking account of $\gamma^\ell(\Delta x^\ell)^2$ from (14) and (15) [44]. By assuming that collisional and collisionless stochastic transport coefficients induced by the magnetic surface destruction are $D_1 = \frac{v_{Te}^2}{\nu_e}\left(\frac{\delta B_r}{B}\right)^2$ and $D_2 = v_{Te} L_c \left(\frac{\delta B_r}{B}\right)^2$, the mixing length theory gives

$$D_1 = \eta^{4/3} \frac{v_{Te}^2}{\nu_e} \frac{\varepsilon^2}{q^2} (ms)^{2/3} |G|^{8/3} \qquad (17)$$

and

$$D_2 = \eta \frac{v_{Te}}{v_A} G^{3/2} , \qquad (18)$$

respectively [40], where v_{Te} is an electron thermal velocity and ν_e is an electron collision frequency. The magnetic fluctuating level is described by $\delta B_r/B$ and L_c is the correlation length of the fluctuating magnetic field lines. It is noted that the poloidal mode number appears in the transport coefficient D_1.

Next we consider the scale invariance for the correlation time and the correlation length which gives the scale transformations shown as [40]

$$\tau^s = G^{-1/2} , \qquad (19)$$

$$\Delta x^s = \eta^{1/2} G^{1/4} . \qquad (20)$$

These are different from $\tau^\ell = 1/\gamma^\ell$ given by (14) and Δx^ℓ given by (15); however, the convective transport coefficient given by $(\Delta x^s)^2/\tau^s$ become the same expression as (16). We note that the poloidal mode number m also has the scale transformation [41]

$$m = \eta^{-1/2} G^{-1/4} s^{-1} . \qquad (21)$$

If we substitute (21) into (14) and (15), then $\Delta x^\ell = \Delta x^s$ and $\tau^\ell(= 1/\gamma^\ell) = \tau^s$ are obtained. Since the mixing length theory assumes the maximum growth rate for τ^ℓ, the scale invariance (21) may describe the poloidal mode number corresponding to the maximum growth rate in the g mode turbulence.

By applying the scale invariance to the collisionless stochastic transport coefficient D_2 the result same as (18) is given. However, in the case of collisional stochastic transport coefficient D_1, we need to substitute

(21) into the expression given by the mixing length theory (17), which reproduces the scale invariance result. Thus the three transport coefficients (16), (17), (18) can be obtained by the significantly different two approaches, the mixing length theory and the scale invariance. Here we summarize scalings for the transport coefficients and the fluctuation levels driven by the g mode turbulence:

$$D_0 \sim n T_e^{-1/2} L_s^2 L_p^{-1} L_g^{-1}, \tag{22}$$

$$D_1 \sim n^2 T_e^{7/2} L_s^5 L_p^{-5/2} L_g^{-5/2}, \tag{23}$$

$$D_2 \sim n^2 T_e^{1/2} L_s^3 L_p^{-3/2} L_g^{-3/2}, \tag{24}$$

and

$$\frac{\tilde{P}}{P_0} \sim n^{1/2} T_e^{-1/2} L_s^{1/2} L_p^{-5/4} L_g^{-1/4}, \tag{25}$$

$$\frac{e\tilde{\phi}}{T_e} \sim n T_e^{-3/2} L_s L_p^{-1} L_g^{-1}, \tag{26}$$

$$\frac{\delta B_r}{B} \sim n^{3/2} T_e^{1/2} L_s^{5/2} L_p^{-5/4} L_g^{-5/4}, \tag{27}$$

$$\frac{\delta B_\theta}{B} \sim n^{3/2} T_e^{1/2} L_s^{3/2} L_p^{-5/4} L_g^{-5/4}, \tag{28}$$

where L_s, L_p and L_g are characteristic scale length of the shear, the pressure gradient and the magnetic curvature, respectively.

The transport coefficients D_0, D_1 and D_2 are compared with the electron thermal transport coefficient obtained from the power balance study of Heliotron E plasmas. Figure 10 shows both the theoretical values and the experimental result in the case of $T_e(x) = T_e(0)[(1-x^2)+\alpha]$ and $n(x) = n(0)[(1-x^2)+\alpha]$ with $T_e(0) = 500\,\text{eV}$, $n(0) = 1.0 \times 10^{14}\,\text{cm}^{-3}$ and $\alpha = 0.05$, where $x = r/a$. Here the magnetic field is $B_0 = 1\,\text{T}$. For a reference the neoclassical ripple transport coefficient [42] D_h is also plotted, which is negligible in this case. In the region, $0.5 \lesssim x \lesssim 0.8$, the three transport coefficients are lower than the experimental value with a factor of $(2 \sim 20)$. Also D_1 and D_2 have an opposite radial dependence compared with the experimental result increasing toward the plasma surface.

Carreras et al. studied the numerical coefficient required to reproduce the numerical results obtained by solving the reduced MHD equations (6), (7), (8) directly [43]. They confirmed that about a factor of 10 is necessary. Carreras et al. also improved the g mode turbulence theory by considering the fact that the radial correlation length of the magnetic fluctuation is not the same as the charateristic length of the velocity fluctuation [44].

One way to obtain the numerical factor of the transport coefficient based on an analytic theory is to employ the two-point remermalized turbulence theory [45]. The two-point correlation function of the pressure fluctuation is given approximately by

$$\langle \tilde{P}(1)\tilde{P}(2) \rangle \simeq \tau_{c\ell}(\mathbf{x}_-) S^0 , \qquad (29)$$

where $\tilde{P}(1)$ and $\tilde{P}(2)$ are fluctuations at $\mathbf{x}_1$ and $\mathbf{x}_2$, respectively, and $\mathbf{x}_- = \mathbf{x}_1 - \mathbf{x}_2$. S^0 is the value of fluctuation source term $S(1,2)$ in the case of $\mathbf{x}_1 \to 0$. Then $\langle \tilde{P}(1)\tilde{P}(2) \rangle$ depends on the relative coordinate $\mathbf{x}_-$ only through the clump life time $\tau_{c\ell}$. By applying Fourier transform to (29),

$$\langle \tilde{P}\tilde{P} \rangle_\mathbf{k} = S^0 \int k_{0x} dx_- \int dy_- \int dz_- e^{-ik_y y_- - ik_z z_-} \tau_{c\ell}(\mathbf{x}) , \qquad (30)$$

where $\mathbf{k} = (k_y, k_z)$ and k_{0x} is a representative wavenumber. Here integration (30) over k_z space yields the k_y-spectrum [46],

$$\langle \tilde{P}\tilde{P} \rangle_{k_y} \equiv \int \frac{dk_z}{2\pi} \langle \tilde{P}\tilde{P} \rangle_\mathbf{k} = \frac{2\pi S^0}{k_{0x}^2 D_0 k_{0y}} \left(\frac{k_y}{k_{0y}} \right) \left[1 - J_0 \left(\frac{k_y}{k_{0y}} \right) \right] , \qquad (31)$$

where D_0 is the convective diffusion coefficient and k_{0y} is also a representative wave number. The source term S^0 can be evaluated by comparing the energy change due to the convective motion induced by the curvature with the sink of energy corresponding to the ohmic dissipation,

$$S^0 = 2 \left(\frac{c}{B_0} \right)^2 \left(-\frac{dP_0}{dx} \right) \left(\frac{d\Omega}{dx} \right) \eta L_s^2 k_{0x}^2 \int \frac{d^2k}{(2\pi)^2} \langle \tilde{P}\tilde{P} \rangle_\mathbf{k} . \qquad (32)$$

By substituting (32) into (31) and integrating both side of the equation over k_y space, we obtain [46]

$$D_0 = 4\left(\frac{c}{B_0}\right)^2 \eta \left(-\frac{dP_0}{dx}\right)\left(\frac{d\Omega}{dx}\right) L_s^2 . \tag{33}$$

This expression is the same as (16) except the numerical factor 4. In order to explain the transport coefficient in the Heliotron E experiment shown in Fig. 10, the numerical factor 4 is a little small.

g Modes Coupled to Resistive Drift Waves and η_i Modes

Since the drift waves have basically electrostatic characteristics, the electromagnetic effect is unimportant. Let us consider the electrostatic limit by neglecting $\partial A/\partial t$ in Eq. (2). Furthermore we neglect the ion temperature and the parallel fluid velocity in Eqs. (1) $\sim$ (5), and assume constant electron temperature for simplicity. Then we obtain [21]

$$\left(\frac{\partial}{\partial t} + \hat{\zeta} \times \nabla\phi \cdot \nabla\right)\nabla_\perp^2 \phi = \frac{\omega_{ce}}{\nu_e}\nabla_\parallel^2(n-\phi) + \nabla n \times \nabla\Omega \cdot \hat{\zeta} + \mu_\perp \nabla_\perp^4 \phi \tag{34}$$

$$\left(\frac{\partial}{\partial t} + \hat{\zeta} \times \nabla\phi \cdot \nabla\right) = \frac{\omega_{ce}}{\nu_e}\nabla_\parallel^2(n-\phi) + \nabla(n-\phi)\times\nabla\Omega\cdot\hat{\zeta} + D_\perp \nabla_\perp^2 n , \tag{35}$$

where the following normalizations are taken into account; $t \equiv \omega_{ci}t$, $\mathbf{x} \equiv \mathbf{x}/\rho_s$, $\mathbf{v} \equiv \mathbf{v}/c_s$, $\phi \equiv e\phi/T_e$ and $n \equiv \tilde{n}/n_0$. Here $\rho_s = c_s/\omega_{ci}$, $c_s = \sqrt{T_e/m_i}$ is an ion acoustic velocity, ω_{ce} and ω_{ci} are the electron and ion cyclotron frequency, respectively. And $\nu_e = n_0 e^2\eta/m_e$ is an electron collision frequency. When the average curvature produced by the external helical magnetic field is neglected or $\nabla\Omega = 0$ in Eqs. (34) and (35), they describe resistive drift waves only.

By linearizing Eqs. (34) and (35) and neglecting the viscosity and diffusion terms, the second order differential equation is obtained. From a solution in a cylindrical plasma model the eigenvalue is shown as

$$\omega = \omega_g + i\frac{\pi}{4}\frac{\nu}{k_\parallel^2}\frac{\omega_g(\omega_{*e}-\omega_g)^2}{\omega_{*e}-\omega_g(m^2/r_0^2+2)} , \tag{36}$$

in the limit of $\nu \to 0$,[21] where $\nu = \nu_e k_\parallel/\omega_{ce}$, $\omega_{*e} = (m/r_0)(-dn_0/dr)|_{r=r_0}$ and $\omega_g = (m/r)(d\Omega/dr)|_{r=r_0}$. The prime denotes the derivative with respect to the radius. Real and imaginary parts of the eigenvalue (36) are compared with the numerical results obtained from the linearized equations of (34) and (35) for the $(m, n) = (1, 1)$ mode in the cylindrical Heliotron E model in Fig. 8. Here the growth rate ω_i proportional to ν in the weakly collisional regime becomes close to $\omega_i \propto \nu^{1/3}$ in the collisional limit, which agrees with the g mode growth rate in the resistive MHD model (14). It should be noted that the resistive drift wave branch is also included in the linear eigenmode equation for the cylindrical model. The eigenmode of the resistive drift wave shows a nonlocal and oscillatory behavior as shown in Ref. [20] and the eigenvalue is

$$\omega = \omega_{r0} + \frac{\sqrt{2}\pi\, r_0 (\rho\omega_{r0}\alpha_m^k)^2 \omega_{r0}^{1/2} J_m^2(\alpha_m^k r_0)}{\varepsilon |k_\parallel| \omega_{*e} J_{m-1}^2(\alpha_m^k)}(-1+i)\nu^{1/2} \qquad (37)$$

for the small ω_g case, where $\omega_{r0} \equiv \omega_{*e}/(1+(\rho\alpha_m^k)^2)$, $\rho = \rho_s/a$ and α_m^k is the k-th zero of the Bessel function J_m.

The nonlinear simulation solving Eqs. (34) and (35) reveals the role of the first terms in the RHS of Eqs. (33) and (34) in the spectrum cascade. When $(\omega_{ce}/\nu_e)(\rho_s^2/R^2)/(\kappa\rho_s)$ is small, a trend of a dual cascade, normal cascade of the density fluctuations and inverse cascade of the potential fluctuations, is seen in the wave-number spectra. For large $(\omega_{ce}/\nu_e)(\rho_s^2/R^2)/(\kappa\rho_s)$, the electrons become adiabatic and characteristic scale of the density fluctuations becomes comparable to that of the potential fluctuations.

Next we consider the coupling between the g mode and the η_i mode. The linear stability analysis of the η_i mode using Eqs. (1)-(5) are given in Ref. [23] and Ref. [24]. Figure 10 shows the η_i mode growth rate as a function of η_i, where $\eta_i = d\ell n T_i/d\ell n n$. When the collisionality is $\nu_e/\omega_{ce} = 5 \times 10^{-4}$ in the cylindrical Heliotron E model, the growth rate remains finite even for $\eta_i = -1$ which is the marginal value in the case of incompressible collisionless fluid model for the η_i mode. The destabilizing mechanism is the coupling to the g mode. However, in the collisionless limit, the marginal η_i value becomes -1 as shown in Fig. 9. The destabilizing tendency of the η_i mode in the presence of the unfavorable averaged magnetic curvature may produce difference between tokamak and heliotron/torsatron for a peaked density profile case where η_i becomes small. Recent pellet injection experiment in Heiotron E showed that the confinement improvement due to the peaked density profile is fairly mild [47].

In tokamaks there are several ways to improve the global confinement from the L mode scaling. One way is to produce a peaked density profile which may suppress the η_i mode in tokamaks [48]. The most promising way is to induce the L to H transition and obtain the H mode discharge [49]. The candidate to explain the L to H transition is the enhancement of the radial electric field which may increase the poloidal rotation velocity or produce the poloidal velocity shear in tokamaks. In heliotron/torsatron almost all results follow the LHD scaling $\tau_E \propto n^{0.6} P_h^{-0.5} B^{0.6}$ where P_h is a heating power [50]. The power degradation is similar to the L mode in tokamaks, although τ_E is enhanced with the increase of density n. It is recognized that confinement improvement is necessary to realize a compact heliotron/torsatron reactor.

Here we discuss the radial electric field effects on the g mode or possibility to suppress the g mode turbulence. In the five-fields equations (1)–(5), the three-fields equations (6)–(8) or the two-fields equations (34) and (35), the radial electric field can be included through the electrostatic potential $\phi(r,\theta,z) = \phi_0(r) + \tilde{\phi}(r,\theta,z)$ in the cylindrical coordinates (r,θ,z). $E_r(r) = -d\phi_0(r)/dr$ generates the poloidal rotation due to the $\mathbf{E} \times \mathbf{B}$ drift motion. The poloidal shear flow appears when the profile of $\phi_0(r)$ is chosen adequately. It is noted that the sign of E_r does not affect the g mode turbulence described by the reduced MHD equations (6)–(8), because under the transformation: $\phi(r,\theta,z) \to -\phi(r,\theta,z)$, $P(r,\theta,z) \to P(r,-\theta,-z)$ and $A(r,\theta,z) \to A(r,-\theta,-z)$, they are invariant [51]. However, when the diamagnetic drift effect or the FLR effect is included in the model equations, the above invariant transformation disappears. Therefore the sign of E_r may affect the linear and nonlinear properties described by Eqs. (1)–(5) or Eqs. (34) and (35).

From the nonlinear simulation of the turbulent behavior described by Eqs. (34) and (35), it is discovered that the radial electric field is generated by the convective nonlinearity [30], which does not occur in the case of the reduced MHD equation (6)–(8). This radial electric field usually generates the poloidal velocity shear in the edge region of the cylindrical plasma and changes the saturation level compared to the result without the self-generation of E_r. In the numerical simulation this is possible by suppressing the $(m,n) = (0,0)$ component intentionally. The poloidal flow generation process is recently described by using Reynolds stress [31]. At present there is no clear experimental evidence showing the self-generation of E_r by fluctuations in both tokamaks and heliotron/torsatron, although there is a possibility that it plays a role in the L to H transition.

Theoretically g modes can be stabilized by the poloidal velocity shear when $k_y v_0 > \gamma_g L_E/\Delta x$ and $\Delta x < L_E$ are satisfied [27], where $v_0 \tanh(x/L_E)$ is a velocity profile, $k_y = m/r_0$, γ_g and Δx are the linear growth rate and the radial mode width at $v_0 = 0$. The mode resonant surface is $r = r_0$ and $x = r - r_0$. When v_0 becomes too large, the well-known Kelvin-Helmoholtz instability is excited [29]. It is noted that the above poloidal shear flow destabilized the ideal interchange mode by introducing the driving term for the Kelvin-Helmholtz instability in heliotron/torsatron [52]. By considering these results the poloidal shear flow may not be always favorable to improve confinement in heliotron/torsatron. Only for an adequate choice of $E_r(r)$ favorable effects to suppress fluctuations are expected. In the heliotron/torsatron experiments, although there were several trials to generate a large radial electric field in the edge region, no clear result showing correlation between the poloidal flow and the fluctuation level or confinement improvement by the radial electric field is obtained.

An interesting result obtained by the nonlinear calculations based on Eqs. (34) and (35) including the externally imposed and frozen $E_r(r)$ is that a negative (positive) polarity, $E_r < 0$ ($E_r > 0$), the poloidal velocity shear suppresses (enhances) the fluctuation level in the growth phase; however, these effect practically disappear in the saturated state. If this is conclusive, E_r is not always effective in suppressing the fluctuations. Recently Carreras et al. showed the poloidal velocity v_p works to stabilize the nonlinear g modes when the second order derivative of $v_p(r)$ is large, in addition that the velocity shear or the first order derivative of $v_p(r)$ is not effective to stabilize them [28].

7.2.3 COMPARISON OF g MODE THEORY WITH HELIOTRON E RESULTS

As discussed in Sec. 2 fluctuations observed in the measurements of density, soft-X ray, electron cyclotron emission and poloidal magnetic field showing sawtooth oscillations are considered as manifestation of the low-m g modes. However, the magnetic fluctuations always show broad frequency spectra irrespective of the sawtooth oscillations (see Fig. 11) [15]. The magnitude of the magnetic fluctuation level usually depends on the heating power, and the fluctuation level is enhanced with the increase of the heating power. It is natural that the magnetic fluctuation with the broad frequency spectrum as shown in Fig. 11 comes from the high-m g modes destabilized in the magnetic hill region of Heliorton E.

Recently electron density fluctuation are measured by applying the laser phase contrast method in Heliotron E [16]. The density fluctuation in Fig. 12 shows clear correlation with the magnetic fluctuation. The phase velocity obtained from the power spectrum of density fluctuation, $S(k_\perp,\omega)$, seems consistent with the relation (36). This suggests that the density fluctuation is also induced by the high-m g modes. Since the density fluctuation have the broad frequency spectrum, it may correspond to the g mode turbulence.

Next problem is whether the predicted anomalous transport based on the g mode turbulence is consistent with the result by the power balance analysis. As discussed in Sec. 3 expressions obtained by the mixing length theory, the scale invariance or the renormalized turbulence theory seem to give a little smaller value than the experimental transport coefficient. Recently $K - \varepsilon$ model for the g mode turbulence is introduced by assuming propagation of the turbulence from the edge region to the inner region [53], which give a more plausible coincidence with experimental results. Also a new mode similar to the g mode is given by Itoh *et al.* [54] by assuming current diffusion due to the electron viscosity. This approach is also possible to explain the experimental results.

In ATF the radial profile of density fluctuation is shown to be consistent with the g mode theory [17]. Since ATF has a lower aspect ratio than Heliotron E, the finite beta effect easily expands the magnetic well region. It is expected that the confinement improves according to the increase of β, since the magnetic well stabilizes the g modes. This is demonstrated by the recent scaling law $\tau_E \propto \beta^{0.3}$ [55].

In comparison between the theory and the experiment, if we obtain information concerning the phase difference between the density fluctuation and the potential fluctuation, a more conclusive result may be given. The g mode based on the resistive MHD model has the phase difference of 90 degree, while the resistive drift wave may satisfy the Boltzmann relation $\tilde{n}/n_0 \simeq e\tilde{\phi}/T_e$.

7.2.4 CONCLUDING REMARKS

From the experimental results in Heliotron E, Heliotron-DR, ATF and CHS, it is shown that the anomalous transport governs the transport particularly in the edge region. Theoretically the g modes are unstable in the above all devices, since magnetic hill is unavoidable. From studies of linear and nonlinear g modes in heliotron/torsatron, the g mode turbulence seems a plausible candidate to explain the anomalous transport. In order to confirm this conjecture, we need more detailed comparisons between the turbulence theory and the fluctuation data. Recently the LHD scaling is established for heliotrons [50]. It is not easy to derive this scaling from the transport coefficients shown in Sec. 3 directly. Since the LHD scaling is close to the gyro-Bohm scaling [56], drift waves may play a role to explain it. Thus it is important to study trapped electron modes in heliotron/torsatron [57].

When we will try to elaborate the theoretical model of g mode turbulence, one important point is to study the toroidal effect particularly on the self-generation of radial electric field, since present results are limited in the cylindrical model.

Finally we note that comparison of fluctuation data in both tokamaks and heliotron/torsatron is important. Recently there are several tokamak results showing that the Bohm scaling is more relearnt than the gyro-Bohm scaling [58]. At present heliotron/torsatron results show the gyro-Bohm scaling. It is interesting to see whether this difference is related to the fluctuation properties or not.

The author acknowledges Dr. H. Sugama, Dr. M. Yagi and Dr. K. Watanabe for useful discussions.

Bibliography

[1] K. Uo, *et al.*, Proc. 9th IAEA Conf. (Baltimore, 1982) vol. 2, p. 209.

[2] J.F. Lyon, *et al.*, Fusion Technol. **10**, (1986) 179.

[3] K. Matuoka, *et al.*, Proc. 12th IAEA Conf. (Nice, 1988) vol. 2, p. 411.

[4] H.P. Furth, J. Killeen, and M.N. Rosenbluth, Phys. Fluids **6**, (1963) 459.

[5] M. Wakatani, IEEE Trans. Plasma Sci. **PS-9**, (1981) 243.

[6] C. Mercier, Nucl. Fusion **1**, (1960) 47.

[7] B.R. Suydam, Proc. Second United Nations International Conf. on the Peaceful Uses of Atomic Energy, Geneva, 1958 (Columbia University Press, New York, 1959) vol. 31, p. 157.

[8] H. Sugama and M. Wakatani, J. Phys. Soc. Jpn, **58**, (1989) 1128.

[9] N. Domingnez, J.N. Leboeuf, B.A. Carreras and V.E. Lynch, Nucl. Fusion **20**, (1989) 2079.

[10] J.H. Harris, *et al.*, Phys. Rev. Lett. **53**, (1984) 2242.

[11] H. Zushi, *et al.*, Nucl. Fusion **27**, (1987) 835.

[12] M. Wakatani, H. Shirai and M. Yamagiwa, Nucl. Fusion **24**, (1984) 1407.

[13] M. Wakatani, Proc. International School of Plasma Physics on 'Theory of Fusion Plasmas' (Varenna, 1987) p. 333.

[14] M. Wakatani et al., Proc. 13th IAEA Conf. (Washington DC, 1990) vol.2, p.567.

[15] H. Zushi, et al., Nucl. Fusion **28**, (1988) 433.

[16] K. Tanaka, et al., J. J. Appl. Phys. **31**, (1992) 2260.

[17] R.C. Isler, et al., Phys. Fluids B **4**, (1992) 2104.

[18] P.C. Liewer, Nucl. Fusion **25**, (1985) 543.

[19] C.P. Ritz et al., Proc. 13th IAEA Conf. (Washington DC, 1990) vol. 2, p. 589.

[20] H. Sugama, M. Wakatani and A. Hasegawa, Phys. Fluids **31**, (1988) 1601.

[21] M. Wakatani, K. Watanabe, H. Sugama and A. Hasegawa, Phys. Fluids B **4**, (1992) 1754.

[22] J.G. Cordey, E.M. Jones and D.F.H. Start, Nucl. Fusion **20**, (1980) 459.

[23] M. Yagi, M. Wakatani, H. Sugama, B.G. Hong, and W. Horton, J. Phys. Soc. Jpn. **58**, (1989) 4265.

[24] B.G. Hong, W. Horton, S. Hamaguchi, M. Wakatani, M. Yagi, and H. Sugama, Phys. Fluids **3**, (1991) 1638.

[25] J. Groebner, K.H. Burrel, and R.P. Seraydarian, Phys. Rev. Lett. **64**, (1990) 3015.

[26] K. Ida, et al., Phys. Rev. Lett. **65**, (1990) 1364.

[27] H. Sugama and M. Wakatani, Phys. Fluids B **3**, (1991) 1110.

[28] B.A. Carreras, V.E. Lynch, L. Garcia, and P.H. Diamond, "Resistive Pressure-Gradient-Driven Turbulence with Self-Consistent Flow Profile Evolution" ORNL/P-92/57995 (1992).

[29] T. Chiueh, P.W. Terry, P.H. Diamond, and J.E. Sedlak, Phys. Fluids **29**, (1986) 231.

[30] A. Hasegawa and M. Wakatani, Phys. Rev. Lett. **59**, (1987) 1581.

[31] P.H. Diamond and Y.B. Kim, Phys. Fluids B **3**, (1991) 1626.

[32] B.A. Carreras, K. Sidikman, P.H. Diamond, P.W. Terry, and L. Garcia, Phys. Fluids B **4**, (1992) 3115.

[33] K. Toi, *et al.*, Proc. 14th IAEA Conf. (Würzburg, 1992) paper H-1-2.

[34] H.R. Strauss, Plasma Physics **22**, (1980) 733.

[35] Y. Nakamura, M. Wakatani, C.C. Hegna, and A. Bhattachrjee, Phys. Fluids B **2**, (1990) 2528.

[36] N. Yanagi, *et al.*, Nucl. Fusion **32**, (1992) 1264.

[37] A.H. Glasser, J.M. Greene, and J.L. Johnson, Phys. Fluids **18**, (1975) 875.

[38] B.B. Kadomtsev, Plasma Turbulence (Academic Press, London, 1965).

[39] J.W. Connor and J.B. Taylor, Phys. Fluids **27**, (1984) 2676.

[40] M. Yagi, M. Wakatani, and K.C. Shaing, J. Phys. Soc. Jpn. **57**, (1988) 117.

[41] J.W. Connor, Nucl. Fusion **26**, (1986) 517.

[42] J.W. Connor and R.J. Hastie, Phys. Fluids **17**, (1974) 114.

[43] B.A. Carreras, L. Garcia, and P.H. Diamond, Phys. Fluids **30**, (1987) 1388.

[44] B.A. Carreras and P.H. Diamond, Phys. Fluids B **1**, (1989) 1011.

[45] P.H. Diamond, R.D. Hazeltine, Z.G. An, B.A. Carreras, and H.R. Hicks, Phys, Fluids **27**, (1984) 1449.

[46] H. Sugama and M. Wakatani, J. Phys. Soc. Jpn. **57**, (1988) 2010.

[47] T. Obiki, *et al.*, Proc. 14th IAEA Conf. (Würzburg, 1992) paper C-1-2.

[48] M. Greenwald, *et al.*, Proc. 10th IAEA Conf. (London, 1984) vol. 1, p. 45.

[49] F. Wagner, *et al.*, Phys. Rev. Lett. **49**, (1982) 1408.

[50] S. Sudo, *et al.*, Nucl. Fusion **30**, (1990) 11.

[51] H. Sugama and M. Wakatani, J. Phys. Soc. Jpn. **58**, (1989) 3859.

[52] K. Watanabe, H. Sugama and M. Wakatani, Nucl. Fusion **32**, (1992) 1647.

[53] H. Sugama, M. Okamoto, and M. Wakatani, Proc. 14th IAEA Conf. (Würzburg, 1992) paper D-4-20.

[54] K. Itoh, S.-I. Itoh, and A. Fukuyama, Phys. Rev. Lett. **69**, (1992) 1050.

[55] M. Murakami, *et al.*, Proc. 14th IAEA Conf. (Würzburg, 1992) paper C-1-1.

[56] F.W. Perkins, 'Issues in Tokamak/Stellarator Transport and Confinement Enhancement Mechanisms', PPPL-2708 (1990).

[57] N. Dominguez, B.A. Carreras, V.E. Lynch, and P.H. Diamond, Phys. Fluids B **4**, (1992) 2894.

[58] J.D. Callen, Phys. Fluids B **4**, (1992) 2142.

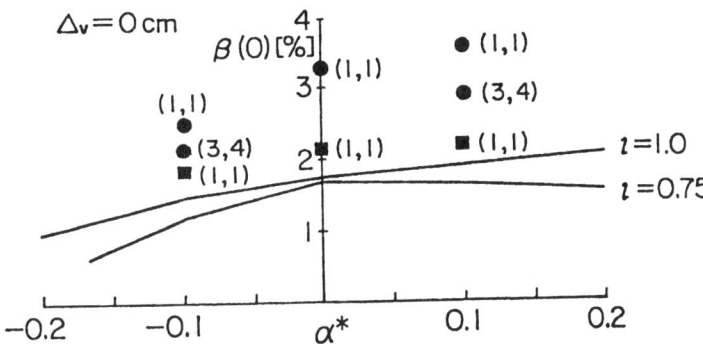

Fig. 1 Dependence of Mercier limit and $\beta(0)$ corresponding to growth rate of 0.01 normalized by poloidal Alfven transit time on the additional toroidal magnetic field in Heliotron E. $\alpha^* > 0$ reduces rotational transform and $\alpha^* < 0$ increases it. (Ref. 14)

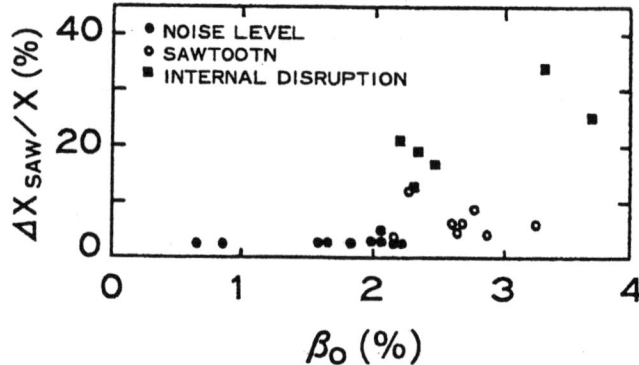

Fig. 2 fluctuation level of soft-X ray as a function of $\beta(0)$ in Heliotron E. Internal disruption is a large amplitute sawtooth here. (Ref. 10)

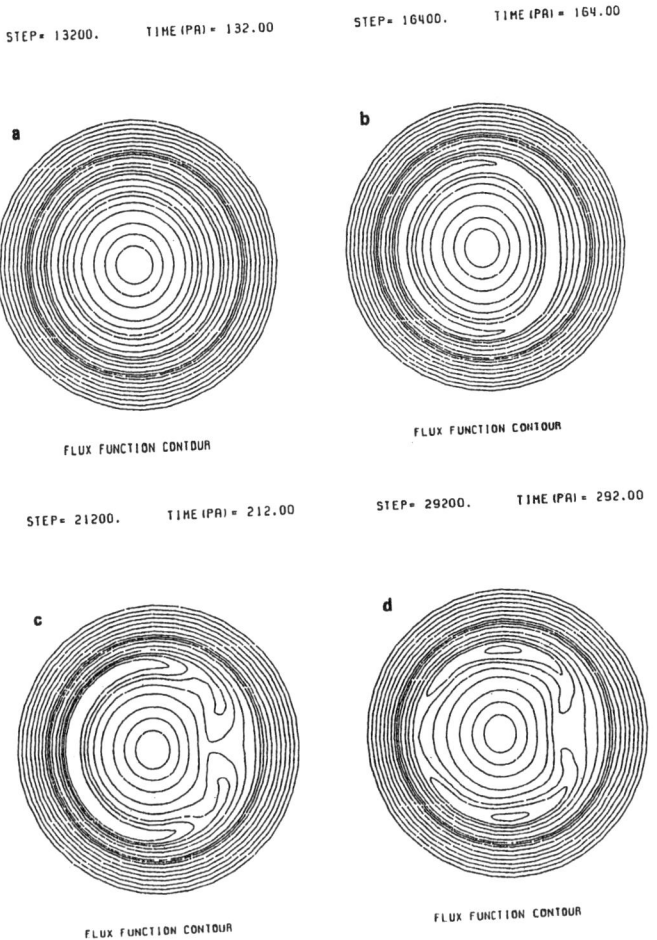

Fig. 3 Evolution of magnetic surfaces due to nonlinear g mode with $(m, n) = (1, 1)$ in Heliotron E model. (Ref. 12)

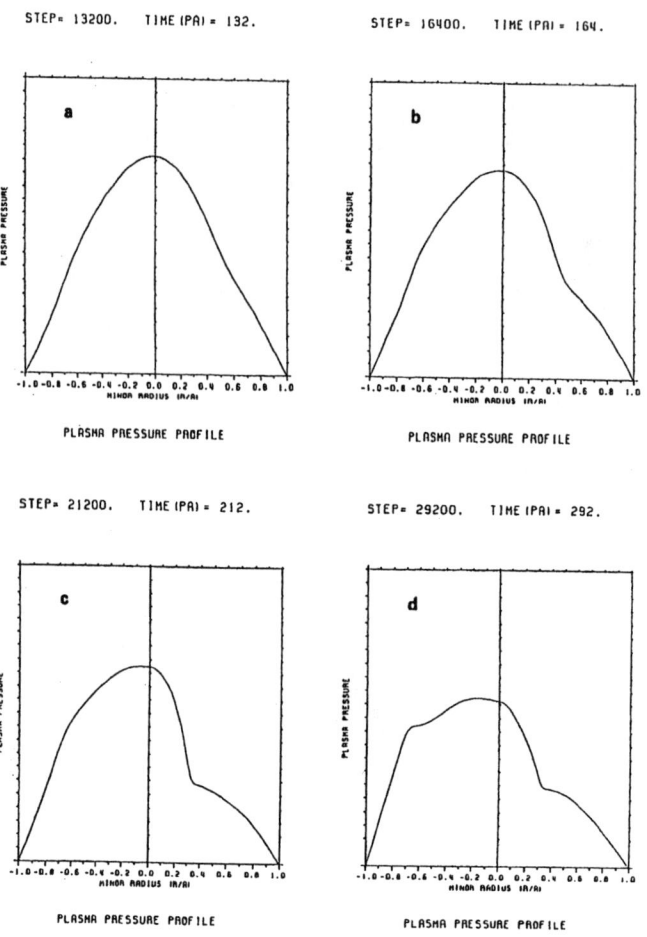

Fig. 4 Pressure profile changes due to nonlinear g mode with $(m, n) = (1, 1)$ corresponding to the case of Fig. 3. (Ref. 12)

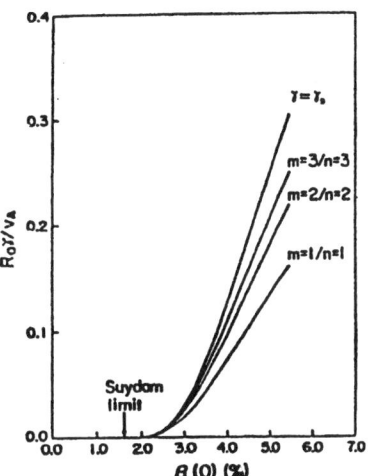

Fig. 5 Growth rate of Suydam mode and low m interchange modes as a function of $\beta(0)$ in Heliotron E model. (Ref. 8)

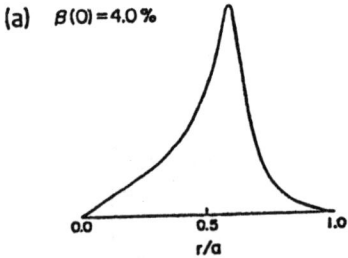
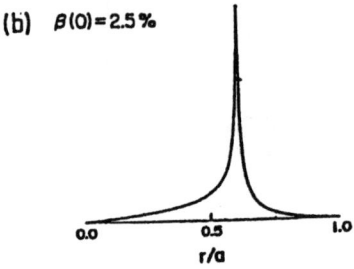

Fig. 6 Radial mode structures with $(m, n) = (1, 1)$ corresponding to Fig. 5. (Ref. 8)

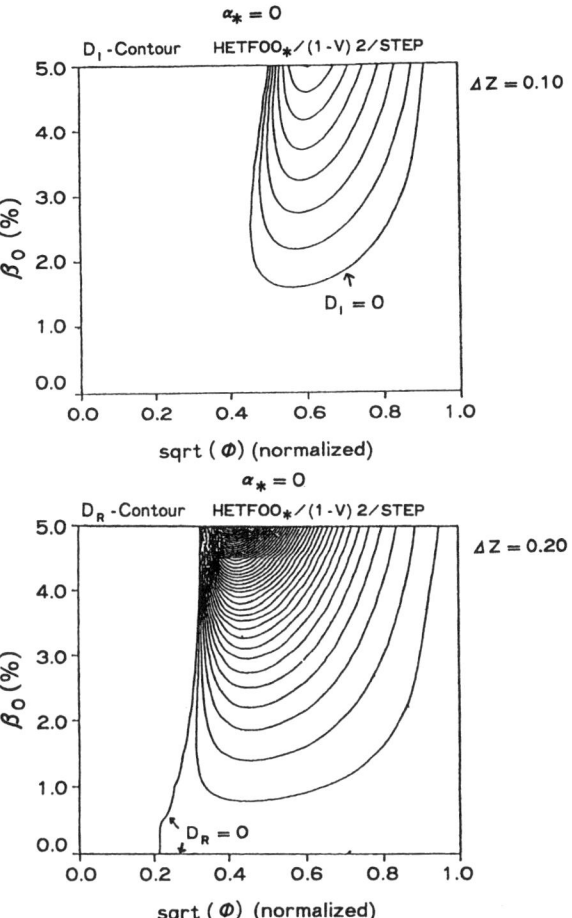

Fig. 7 Stability diagram given by $D_I > 0$ and $D_R > 0$ for Heliotron E in the plane $(\beta(0), \sqrt{\Phi})$, where $\sqrt{\Phi}$ denotes a radial variable. Pressure profile is assumed $P = P_0(1 - \Phi)^2$, where Φ is a toroidal flux function. (Ref. 35)

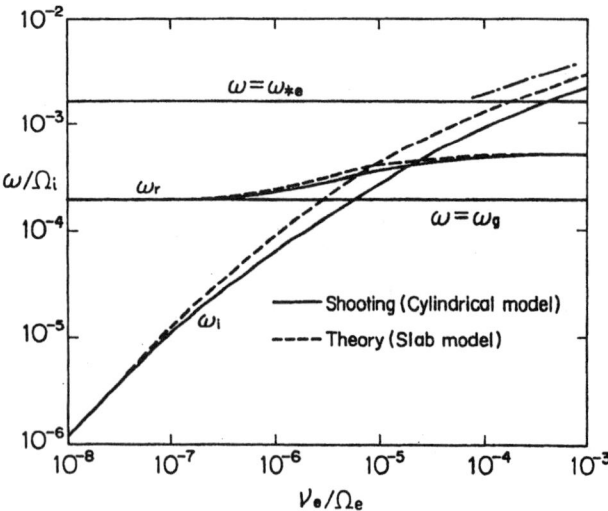

Fig. 8 Real and imaginary parts of the eigenvalue ω/Ω_i versus collision frequency ν_e/Ω_e for the $(m, n) = (1, 1)$ mode. The continuous lines with ω_r and ω_i are obtained by the shooting method and dotted lines are given by dispersion relation. The dash and dotted line is a reference line showing $\omega_i \propto \nu_e^{1/3}$. (Ref. 21)

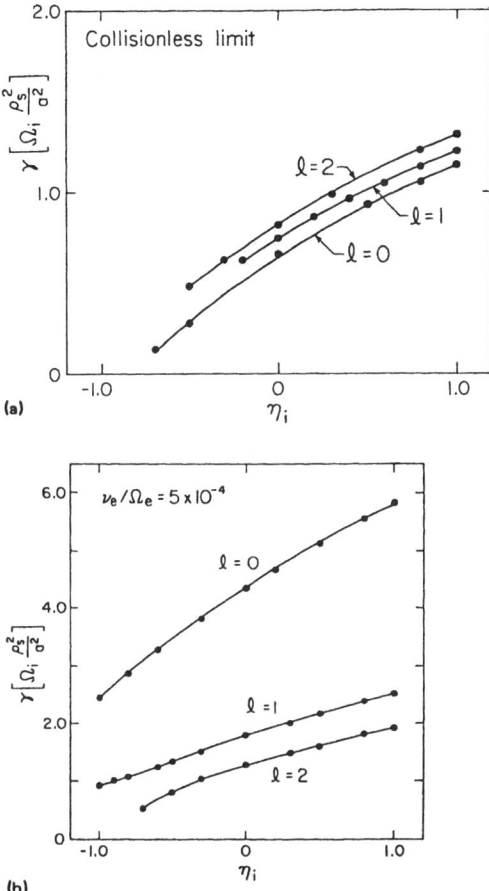

Fig. 9 Growth rate γ in unit of $\Omega_i(\rho_s^2/a^2)$ versus ion temperature gradient η_i for collisionless limit (a) and $\nu_e/\Omega_e = 5 \times 10^{-4}$ (b). (Ref. 24)

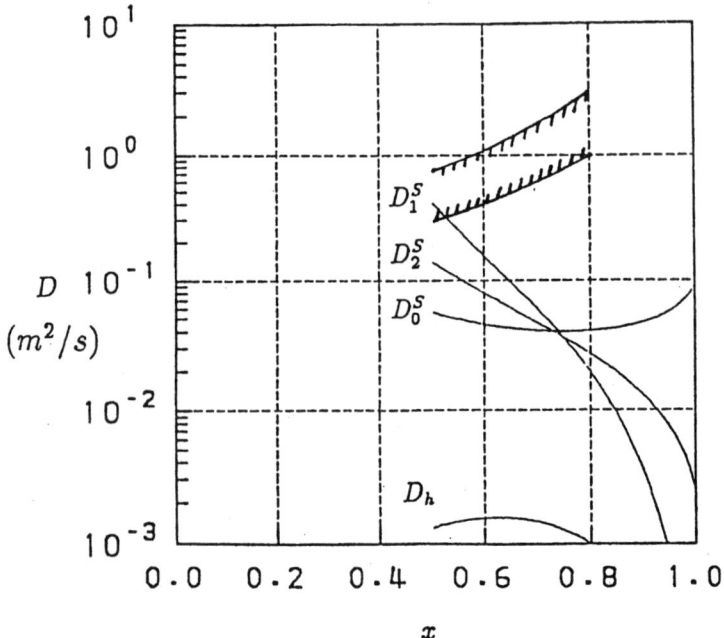

Fig. 10 Transport coefficient profiles in the case of $T_e(x) = T_e(0)[(1-x^2)+\alpha]$ and $n(x) = n(0)[(1-x^2)+\alpha]$ with $T_e(0) = 500eV$, $n(0) = 1.0 \times 10^{14} cm^{-3}$, $\alpha = 0.05$ and $B_0 = 1T$. Here $x = r/a$. Convective transport D_0^S, collisional stochastic transport D_1^S and collisionless stochastic transport D_2^S are plotted. For a reference neoclassical ripple transport D_h is also plotted. Experimental values in Heliotron E are also shown with the shaded region. (Ref. 40)

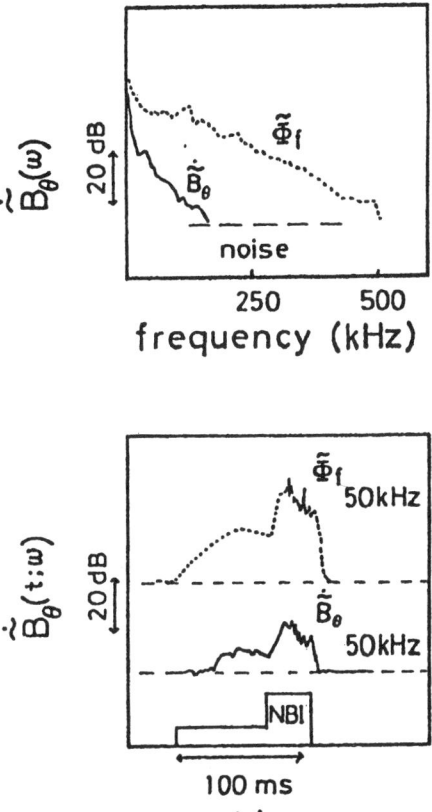

Fig. 11 Frequency spectrum and time evolution of $\tilde{B}_\theta$ and $\tilde{\Phi}_f$ in the typical NBI heated plasma, where $\tilde{B}_\theta$ is a fluctuating poloidal magnetic field and $\tilde{\Phi}_f$ is a fluctuating floating potential. (Ref. 15)

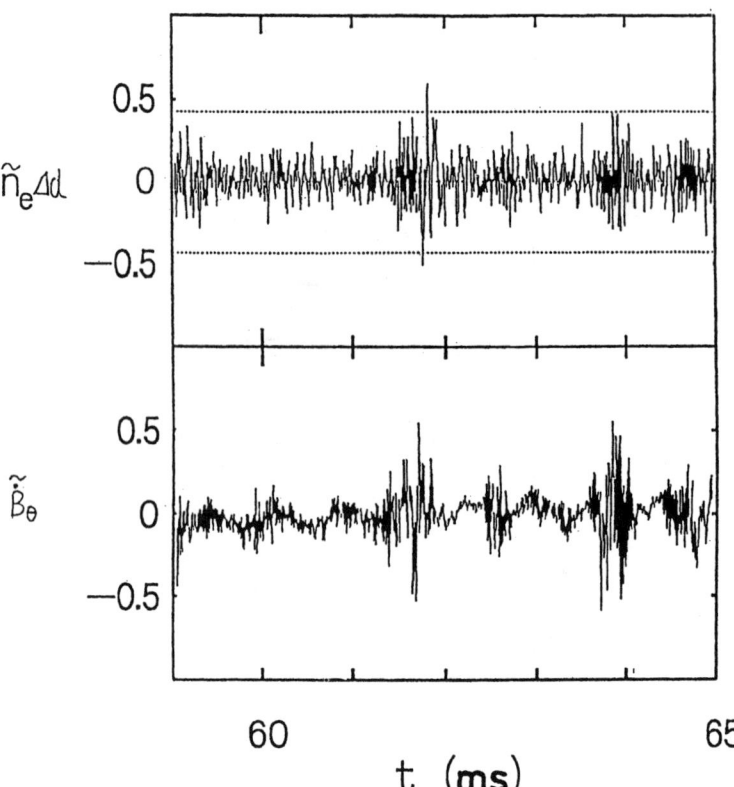

Fig. 12 Density fluctuation measured by laser phase contrast method, and magnetic fluctuation $\tilde{B}_\theta$ in the NBI heated plasma of Heliotron E. (Ref. 16)

7.3 POINT VORTEX DYNAMICS IN A MAGNETIZED PLASMA

Mitsuo Kono, Hideaki Shibahara
Research Institute for Applied Mechanics
Kyushu University
Kasuga, Fukuoka, 816, Japan
Kentaro Yabuki
Interdisciplinary Graduate School of Engineering Sciences
Kyushu University
Kasuga, Fukuoka, 816, Japan

Abstract. *A self-consistent theory describing vortex-wave dynamics in a magnetized plasma has been formulated based on the Hasegawa-Mima equation by extending the modulated point vortex model so as to include the vortex-wave interactions. The energy and enstrophy are shown to be conserved in contrast with the previous modulated point vortex description. Dynamical behaviors of vortices under the interaction with waves are numerically studied. A contour dynamics is also shown to be formulated for a magnetized plasma with the temperature gradient.*

7.3.1 INTRODUCTION

Dynamics of drift vortices is one of the key issues to get a full understanding of anomalous transport in a magnetically confined plasma as well as the nature of turbulence. Numerical works [1, 2, 3, 4, 5] based on the Hasegawa-Mima equation [6] revealed dynamical properties of the drift vortices which are found to be nicely recovered by the modulated point vortex model [7, 8, 9].

An advantage of introducing point vortices is to convert a nonlinear partial differential equation into a system of ordinary differential equations which are easier to solve. A crucial difference of the Hasegawa-Mima equation from Euler's equation is the existence of the drift term giving rise to dispersive waves, which requires that the vorticity attached to each point vortex is no longer constant but varies in space and time. This is in contrast with the point vortex description of Euler's equation characterized by constant strength vorticities. This modulated point vortex model was first introduced by Zabusky-McWilliams [7] based on the fact that the Hasegawa-Mima equation conserves the vorticity along the trajectory. They studied the configurations of the vortices corresponding to a stationary solution of the Hasegawa-Mima equation and the stability of the configurations. Then Kono and Horton [8] and Hobson [9] studied dynamical properties of the vortices based on the modulated point vortex model and showed excellent agreements with the results of numerical simulations based on the Hasegawa-Mima equation. It should be noted that Reznik [10] and Gryanik [11] also developed point vortex models for stationary atmospheric vortices. However the modulated point vortex model has a definite advantage to describe the dynamical behavior of the vortices.

In spite of the fact that the modulated point vortex model successfully describes many results of the numerical simulations, there is one problem that the energy and enstraphy are not conserved in the modulated point vortex model while they are originally conserved for the Hasegawa-Mima equation. Drift wave turbulence certainly has two main components, coherent structures on one side and background field on the other, whose interactions are not taken into account in the modulated point vortex model, resulting in the lack of the consevation laws. As a matter of fact the numerical simulations have shown that a vortex-pair emits wake fields though small when it changes its propagation direction. Reznik [12] formulated the vortex-wave interaction based on the model he developed previously. However his theory is a bit involved and hard to treat. We rather extend the modulated point vortex model so as to include the vortex-wave interaction. Furthermore we show how a contour dynamics can be formulated for a magnetized plasma with the temperature gradient as well as the density gradient.

In this paper we formulate a self-consistent theory of vortex-wave dynamics by taking account of vortex-wave interactions. In Sec. 2 we summarize vortex dynamics based on the modulated point vortex. In Sec. 3 we extend the modulated point vortex model to include the vortex-wave interactions and show that the energy and enstrophy are conserved. In Sec. 4 the self-consistent dynamics of the vortex-wave system is numerically studied. In Sec. 5 a contour dynamics is formulated for drift wave solitons which are excited when the temperature gradient is taken into account. Discussions are given in the last section.

7.3.2 MODULATED POINT VORTEX DYNAMICS

The modulated point vortex model for the Hasegawa-Mima equation in a slab geometry

$$\frac{\partial \Pi}{\partial t} + [\psi, \Pi] + v_* \frac{\partial \psi}{\partial y} = 0 \ , \tag{1}$$

where [,] denotes the Poisson bracket and $\Pi = \psi - \nabla^2 \psi$, was introduced by Zabusky and McWilliams [7] first and then later Kono and Horton [8] and Hobson [9] in the following form:

$$\frac{d\mathbf{r}_\alpha}{dt} = \hat{\mathbf{z}} \times \nabla \psi(\mathbf{r}_\alpha) \ , \tag{2}$$

$$\psi(\mathbf{r}) = \frac{1}{2\pi} \sum_\alpha (\kappa_{\alpha 0} + v_* x_\alpha) K_0(|\mathbf{r} - \mathbf{r}_\alpha|) \ , \tag{3}$$

where K_0 is the modified Bessel function of the second kind. We also assume v_* positive constant and $\kappa_{\alpha 0}$ constant.

The Hamiltonian is given by

$$H = \sum_\alpha \psi(\mathbf{r}_\alpha) \ , \tag{4}$$

from which we have

$$\frac{dx_\alpha}{dt} = -\frac{\partial H}{\partial y_\alpha} = \frac{1}{2\pi} \sum_\beta (\kappa_{\beta 0} + v_* x_\beta) \frac{y_\alpha - y_\beta}{r_{\alpha\beta}} K_1(r_{\alpha\beta}) \ , \tag{5}$$

$$\frac{dy_\alpha}{dt} = \frac{\partial H}{\partial x_\alpha} = -\frac{1}{2\pi} \sum_\beta (\kappa_{\beta 0} + v_* x_\beta) \frac{x_\alpha - x_\beta}{r_{\alpha\beta}} K_1(r_{\alpha\beta}) \ , \tag{6}$$

where $r_{\alpha\beta} = |\mathbf{r}_\alpha - \mathbf{r}_\beta|$. Since H is translationally invariant along the y-axis, the translational momentum in x is conserved:

$$P = \sum_\alpha (\kappa_{\alpha 0} + v_* x_\alpha)^2 \ . \tag{7}$$

For two vortices in addition to Eq. (7) the relative distance of the vortices is a constant of motion,

$$r_{12}^2 = (x_1 - x_2)^2 + (y_1 - y_2)^2 = r_0^2, \qquad (8)$$

which gives the solution as

$$\cos\theta = \gamma \frac{1 - \beta^2 sn^2(\omega t, k)}{1 - \alpha^2 sn^2(\omega t, k)}, \qquad (9)$$

where $x_1 - x_2 = r_0 \cos\theta$, $A = [\sqrt{2P} + (\kappa_{20} - \kappa_{10})]/v_* r_0$, and $B = [\sqrt{2P} - (\kappa_{20} - \kappa_{10})]/v_* r_0$. In Eq. (9) for $v_* r_0 + |\kappa_{10} - \kappa_{20}| > \sqrt{2P} > |v_* r_0 - |\kappa_{10} - \kappa_{20}||$ we have

$$\alpha^2 = \frac{1+B}{A+B}, \quad \beta^2 = A\alpha^2, \quad \gamma = -\text{sgn}\,(\kappa_{10} - \kappa_{20})$$

$$\omega = \sqrt{2(A+B)}\frac{v_* K_1(r_0)}{4\pi}, \quad k = \sqrt{\frac{(1+A)(1+B)}{2(A+B)}}, \qquad (10)$$

and for $\sqrt{2P} < |v_* r_0 - |\kappa_{10} - \kappa_{20}||$ we have similar expressions for $\alpha, \beta, \gamma, \omega$ and k. Equations for the center of gravity are given by

$$\frac{1}{2}(x_1 + x_2) = -\frac{\pi r_0}{v_* K_1(r_0)}\frac{d\theta}{dt} - \frac{\kappa_{10} + \kappa_{20}}{2 v_*}, \quad \frac{1}{2}\frac{d}{dt}(y_1 + y_2)$$

$$= \frac{K_1(r_0)}{4\pi}[\kappa_{10} - \kappa_{20} + v_* r_0 \cos\theta]\cos\theta,$$

whose explicit solutions are readily obtained by using Eq. (9).

A vortex-pair ($\kappa_{10} + \kappa_{20} = 0$) propagates in the y-direction without oscillation in the orbit for $\theta_0 (= \theta(t=0)) = n\pi$ (n : integer) and with oscillation for $\theta_0 \neq n\pi$ as is shown in Fig. 1, which has been numerically observed by Makino, Kamimura, and Taniuchi [1] based on the Hasegawa-Mima equation. In fact the frequency given by Eq. (10) shows the same type of v_*-dependence as those observed in numerical simulations by Makino et al. Recently Nycander and Isichenko [13] derived the equation for the center of gravity of a vortex-pair from the moment equations of the Hasegawa-Mima equation and obtained the frequency of the trajectory which is also well fitted to the results by Makino et al. A Non-propagating solutions like that in Fig. 1 are realized for such initial angles of the symmetry axis of the vortex-pair to the x-axis that the time-averaged velocity of the vortex-pair is zero.

Two like-signed vortices ($\kappa_{10} = \kappa_{20}$) are mutually trapped, rotating around the center of gravity. This mutual trapping leads to coarse-graining of the correlations over directions and may be considered a mechanism behind the fusion of vortices in the sense that a group of point vortices positioned sufficiently near one another act at large distances as a single vortex with the sum intensity, $\kappa_T \cong \sum_\alpha \kappa_\alpha$. A coalescence of like-signed vortices and a long-lived monopole numerically observed by Horton [14] may be interpreted by this mutual trapping process. The inverse cascade [4] of the energy associated by the conservation of enstrophy is also regarded in the vortex representation as a trapping as a kind of snowballing process.

Collision processes of two vortex-pairs are shown to recover those observed numerically [1, 3, 4] for elastic cases with zero [Fig. 2(a)] and non-zero impact parameters [Fig. 2(b)-(f)]. Since our point vortex model does not take into account effects of interaction with the wake fields, the inelastic collisions with an emission of wake fields observed by McWilliams and Zabusky [2] cannot be described by the vortex field component in Eq. (3) alone. However, the position dependence of the vorticity provides a variety of complicated behavior including an exchange scattering and a boomerang scattering, indicating that our point vortex system is likely to become turbulent when many vortices are involved. However the potential structure constructed from Eq. (3) is quite orderly: mutually trapped vortices behave as a single vortex though dynamics of constituent point vortices is very complicated. Therefore the complication of the dynamics of the point vortices is rather analogous to complicated behaviors of constituent particles in an ordinary gas or fluid dynamics and averaged properties may be of primary importance although the dynamical properties of the point vortex system are academically interesting since the chaotic behavior may be characterized by intermittent structures-clusters of vortices, in which local order is a preferred state because of the short-range interaction force between point vortices. The range of the interaction $\rho_s = c(m_i T_e)^{1/2}/eB$ is given by the parallel electron motion shielding the charge separation in the Euler vortex.

For a cylindrical plasma ($\mathbf{v}_* = (v_T^2/\Omega)\hat{\mathbf{z}} \times \nabla \ln n_0$), we have

$$\frac{d\mathbf{r}_\alpha}{dt} = \hat{\mathbf{z}} \times \nabla \psi(\mathbf{r}_\alpha), \qquad (11)$$

$$\psi(\mathbf{r}) = \frac{1}{2\pi} \sum_\alpha (\kappa_{\alpha 0} - \hat{v}_* r_\alpha^2) K_0(|\mathbf{r} - \mathbf{r}_\alpha|), \qquad (12)$$

where the background density is assumed to be a Gaussian. A vortex-pair revolves around the center of the cylinder while the constituent vortices rotate around each other, that is, the motion is characterized by two frequencies, implying the motion is mostly quasiperiodic. A single cluster consisting of many vortices with the same sign of the vorticity ($\kappa_{\alpha 0} = \kappa$) turns round the center of the cylinder clockwise, while the vortices rotate around the center of mass clockwise for $\kappa > 0$ and counterclockwise for $\kappa < 0$. When $|\kappa| > v_*(r_\alpha^2 - r_\alpha^2(t = 0))$, the vortices redistribute themselves to lattice points to form a circle. When $\kappa(> 0)$ decreases so as for some outer vortices in the cluster not to satisfy the above condition, those vortices rotate in the direction opposite to that of the core of the cluster and are left behind. Thus anticyclones ($\kappa < 0$) are long-lived while cyclones ($\kappa > 0$) are not as is shown in Fig. 3. This situation is just similar to the observations by Antipov et al. [15] Two clusters are almost independent on each other and their motion is a superposition of that of each cluster when κ is small compared with the relative distance between the clusters. However they coalesce into one when κ's are the same sign and become large.

There are two conserved quantities for the Hasegawa-Mima equation; one is the energy

$$E = \int d\mathbf{r} \left[\psi^2 + (\nabla \psi)^2\right] = \int d\mathbf{r} \psi \Pi, \qquad (13)$$

and the other is the enstrophy

$$W = \int d\mathbf{r} \left[(\nabla \psi)^2 + (\nabla^2 \psi)^2\right] = -\int d\mathbf{r} (\nabla^2 \psi) \Pi. \qquad (14)$$

In the point vortex description the energy and enstrophy are expressed as

$$E = \frac{1}{2\pi} \sum_{\alpha \neq \beta} (\kappa_{\alpha 0} + v_* x_\alpha)(\kappa_{\beta 0} + v_* x_\beta) K_0(r_{\alpha\beta}) , \qquad (15)$$

$$W = \frac{1}{8\pi} \sum_{\alpha \neq \beta} (\kappa_{\alpha 0} + v_* x_\alpha)(\kappa_{\beta 0} + v_* x_\beta)[K_0(r_{\alpha\beta}) + K_2(r_{\alpha\beta})] , \qquad (16)$$

where the self-energy has been subtracted since it diverges. Equations (15) and (16) are not conserved in general, since

$$\frac{dE}{dt} = -\frac{1}{2} \sum_\alpha v_* \frac{\partial}{\partial y_\alpha} \psi^2(\mathbf{r}_\alpha) , \quad \frac{dW}{dt} = -\frac{1}{2} \sum_\alpha v_* \frac{\partial}{\partial y_\alpha} [\nabla_\alpha \psi(\mathbf{r}_\alpha)]^2 .$$

The descrepancy comes from the fact that Eqs. (13) and (14) take account of contributions from both vortices and waves, while Eqs. (15) and (16) are based on vortices only. Therefore for the vortex-pairs corresponding to the stationary solutions of the Hasegawa-Mima equation, that is, for those propagating in straight in the y-direction, Eqs. (15) and (16) are certainly constants of motion since the wave are never excited. In general, however, changes in energy and enstrophy of the vortex system in the course of time evolution may occur whenever waves are involved in the fundamental processes of the vortices such as emission or absorption of waves by the vortices, which is the subject in the next section.

7.3.3 SELF-CONSISTENT DESCRIPTION OF VORTEX-WAVE DYNAMICS

For the Hasegawa-Mima equation we split the potential and vorticity into a vortex part and a continuous field part as

$$\psi = \psi_v + \psi_w , \quad \Pi = \Pi_v + \Pi_w , \qquad (17)$$

where subscripts v and w denote the vortex part and wave part, respectively. Since linear eigen-frequency of the drift wave is well separated from the turn-over frequency of the vortex, the above separation is always possible. Following Ref. 8, we introduce vortices as

$$\Pi_v = (1 - \nabla^2)\psi_v = \sum_\alpha \kappa_\alpha(t) V_\alpha(\mathbf{r} - \mathbf{r}_\alpha(t)), \qquad (18)$$

where $V_\alpha(\mathbf{r} - \mathbf{r}_\alpha(t))$ is a localized function centered at $\mathbf{r} = \mathbf{r}_\alpha(t)$ and $\mathbf{r}_\alpha(t)$ is determined by the characteristics of Eq. (1)

$$\frac{d\mathbf{r}_\alpha(t)}{dt} = \hat{\mathbf{z}} \times \nabla(\psi_v + \psi_w)|_{\mathbf{r}=\mathbf{r}_\alpha}(t). \qquad (19)$$

Substituting Eq. (17) into Eq. (1), we obtain

$$\sum_\alpha \frac{d\kappa_\alpha}{dt} V_\alpha(\mathbf{r} - \mathbf{r}_\alpha) - \sum_\alpha \kappa_\alpha \left\{ \frac{d\mathbf{r}_\alpha}{dt} - \hat{\mathbf{z}} \times \frac{\partial \psi(\mathbf{r},t)}{\partial \mathbf{r}} \right\} \cdot \frac{\partial}{\partial \mathbf{r}} V_\alpha(\mathbf{r} - \mathbf{r}_\alpha)$$
$$+ \frac{\partial \Pi_w}{\partial t} + [\psi, \Pi_w] + v_* \frac{\partial \psi}{\partial y} = 0. \qquad (20)$$

Since $V_\alpha(\mathbf{r}-\mathbf{r}_\alpha)$ is a function localized around $\mathbf{r} = \mathbf{r}_\alpha(t)$, we may replace $\mathbf{r}$ appeared in the coefficients of $V_\alpha(\mathbf{r} - \mathbf{r}_\alpha)$ by $\mathbf{r}_\alpha(t)$. Then the second term of the left-hand side of Eq. (20) vanishes according to Eq. (19). The remaining is

$$\frac{\partial \Pi_w}{\partial t} + [\psi, \Pi_w] + v_* \frac{\partial \psi}{\partial y} + \sum_\alpha \frac{d\kappa_\alpha}{dt} V_\alpha(\mathbf{r} - \mathbf{r}_\alpha) = 0. \qquad (21)$$

Multiplying Eq. (21) by $V_\beta(\mathbf{r} - \mathbf{r}_\beta(t))$ and integrating with respect to $\mathbf{r}$, we have

$$\frac{d}{dt}(\kappa_\beta - v_* x_\beta + \Pi_w(\mathbf{r}_\beta, t)) = 0, \qquad (22)$$

which gives for a constant diamagnetic drift velocity v_*

$$\kappa_\alpha(t) = \kappa_{\alpha 0} + v_* x_\alpha(t) - \Pi_w(\mathbf{r}_\alpha(t), t). \qquad (23)$$

This shows that the vorticity changes under the influence of the wave fields, which is the reaction of emission or absorption of the wave fields by the vortex.

The vortex part of the potential is obtained through Eq. (18) as

$$\psi_v(\mathbf{r}, t) = \int d\mathbf{r}' G(\mathbf{r} - \mathbf{r}')\Pi_v(\mathbf{r}') = \sum_\alpha \psi_\alpha(\mathbf{r} - \mathbf{r}_\alpha), \quad (24)$$

where the Green funcction $G(\mathbf{r})$ is defined by

$$(1 - \nabla^2)G(\mathbf{r}) = \delta(\mathbf{r}), \quad (25)$$

whose solution is given by

$$G(\mathbf{r}) = \frac{1}{2\pi} K_0(|\mathbf{r}|). \quad (26)$$

Now we examine conserved quantities for our system. In our present description the energy and enstrophy are carried by both vortices and waves, while in the previous description they are carried by only vortices. First we examine the energy conservation. The time derivative of E is given as

$$\frac{dE}{dt} = 2\int d\mathbf{r}\psi \frac{\partial \Pi_w}{\partial t} + 2\sum_\alpha \frac{d\kappa_\alpha}{dt}\psi(\mathbf{r}_\alpha, t) - 2\sum_\alpha \kappa_\alpha \int d\mathbf{r}\psi(\mathbf{r}, t)\frac{d\mathbf{r}_\alpha}{dt}\cdot\frac{\partial}{\partial \mathbf{r}}V_\alpha(\mathbf{r} - \mathbf{r}_\alpha)$$

$$= 2\int d\mathbf{r}\psi \frac{\partial \Pi_w}{\partial t} + 2\sum_\alpha \frac{d\kappa_\alpha}{dt}\psi(\mathbf{r}_\alpha, t) + 2\sum_\alpha [\psi(\mathbf{r}_\alpha, t), \psi(\mathbf{r}_\alpha, t)]$$

$$= 2\int d\mathbf{r}\psi \frac{\partial \Pi_w}{\partial t} + 2\sum_\alpha \frac{d\kappa_\alpha}{dt}\psi(\mathbf{r}_\alpha, t). \quad (27)$$

On the other hand, multiplying Eq. (21) by ψ and integrating over the space gives

$$\int d\mathbf{r}\psi \left\{\frac{\partial \Pi_w}{\partial t} + [\psi, \Pi_w]\right\} + \sum_\alpha \frac{d\kappa_\alpha}{dt}\psi(\mathbf{r}_\alpha, t) = 0. \quad (28)$$

Noting an identity

$$\int d\mathbf{r}\varphi[\psi, \phi] = \int d\mathbf{r}\phi[\varphi, \psi], \quad (29)$$

which holds for a periodic boundary condition, Eq. (27) reduces to

$$\int d\mathbf{r}\psi \frac{\partial \Pi_w}{\partial t} + \sum_\alpha \frac{d\kappa_\alpha}{dt}\psi(\mathbf{r}_\alpha, t) = 0. \quad (30)$$

Thus the energy is conserved.

In a similar way the enstrophy is also conserved. Equation (14) is rewritten as

$$W = \int d\mathbf{r} \Pi^2 - E , \qquad (31)$$

from which we have

$$\frac{dW}{dt} = 2\int d\mathbf{r}\Pi \frac{\partial \Pi_w}{\partial t} + 2\sum_\alpha \frac{d\kappa_\alpha}{dt} \Pi(\mathbf{r}_\alpha, t) + 2\sum_\alpha \kappa_\alpha [\psi(\mathbf{r}_\alpha, t), \Pi(\mathbf{r}_\alpha, t)] . \qquad (32)$$

Multiplying Eq. (21) by Π and integrating the resultant equation over $\mathbf{r}$, we have

$$\int d\mathbf{r} \left\{ \Pi \frac{\partial \Pi_w}{\partial t} + \Pi[\psi, \Pi_w] + v_*\Pi \frac{\partial \psi}{\partial y} \right\} + \sum_\alpha \frac{d\kappa_\alpha}{dt} \Pi(\mathbf{r}_\alpha, t) = 0 . \qquad (33)$$

Noting that

$$\int d\mathbf{r} \Pi \frac{\partial \psi}{\partial y} = \frac{1}{2}\int d\mathbf{r} \frac{\partial}{\partial y} \left\{ \psi^2 + (\boldsymbol{\nabla}\psi)^2 \right\} = 0 , \qquad (34)$$

and

$$\int d\mathbf{r} \Pi[\psi, \Pi_w] = \int d\mathbf{r} \left\{ \Pi[\psi, \Pi] - \Pi[\psi, \Pi_v] \right\}$$

$$= \int d\mathbf{r} \Pi_v[\psi, \Pi] = \sum_\alpha \kappa_\alpha [\psi(\mathbf{r}_\alpha, t), \Pi(\mathbf{r}, t)] . \qquad (35)$$

Equation (32) reduces to

$$\int d\mathbf{r}\Pi \frac{\partial \Pi_w}{\partial t} + \sum_\alpha \left\{ \frac{d\kappa_\alpha}{dt} \Pi(\mathbf{r}_\alpha, t) + \kappa_\alpha[\psi(\mathbf{r}_\alpha, t), \Pi(\mathbf{r}_\alpha, t)] \right\} = 0 , \qquad (36)$$

which implies that

$$\frac{dW}{dt} = 0 . \qquad (37)$$

7.3.4 SELF-CONSISTENT DYNAMICS OF VORTEX-WAVE SYSTEM

In this section we study the self-consistent dynamics of vortex-wave system governed by Eqs. (19), (21), (23), and (24). It is noted that Eq. (21) is free from singularity since the motion associated with the vortex cores is subtracted because of Eq. (23), that is, Eq. (21) is symbolically rewritten with the aid of Eq. (23) as

$$\left\{\frac{\partial \Pi_w}{\partial t} + [\psi_w, \Pi_w] + v_* \frac{\partial \psi_w}{\partial y} + [\psi_v, \Pi_w - v_* x]\right\}_{\overline{D}} = 0, \quad (38)$$

where $\overline{D}$ implies that the vortex trajectory is subtracted from the whole space. Here we need some approximations to Eq. (38). First we may omit the wave nonlinearity $[\psi_w, \Pi_w]$, since this vector nonlinearity produces vortical motions which are already subtracted as vortices and therefore the wave is assumed small. Because of the same reason, we may approximate the last term of Eq. (38) as

$$[\psi_v, \Pi_w - v_* x] \simeq v_* \frac{\partial \psi_v}{\partial y}. \quad (39)$$

Furthermore the vortex part of the potential is approximated as

$$\psi_v(\mathbf{r}, t) = \sum_\alpha \kappa_\alpha \int d\mathbf{r}' G(\mathbf{r} - \mathbf{r}') V_\alpha(\mathbf{r}' - \mathbf{r}_\alpha)$$

$$\simeq \sum_\alpha \kappa_\alpha \int d\mathbf{r}' [G(\mathbf{r} - \mathbf{r}_\alpha) - \mathbf{r}'\mathbf{r}' : \nabla\nabla G(\mathbf{r} - \mathbf{r}_\alpha)] V_\alpha(\mathbf{r}')$$

$$\simeq \sum_\alpha \frac{\kappa_\alpha}{2\pi} K_0(|\mathbf{r} - \mathbf{r}_\alpha|), \quad (40)$$

Thus our basic system of equations for the vortex-wave interaction are

$$\frac{d\mathbf{r}_\alpha}{dt} = \hat{\mathbf{z}} \times \nabla \psi(\mathbf{r}_\alpha, t), \quad (41)$$

$$\frac{\partial \Pi_w}{\partial t} + v_* \frac{\partial \psi_w}{\partial y} + v_* \frac{\partial \psi_v}{\partial y} = 0, \quad (42)$$

$$\psi(\mathbf{r}, t) = \psi_v(\mathbf{r}, t) + \psi_w(\mathbf{r}, t). \quad (43)$$

Numerical simulations are performed for various initial configurations of the vortices. In solving Eq. (42) the contributions from the vortex core are subtracted. When a small wave exists initially, a monopole is driven to propagate in contrast with the simple modulated vortex model in which a monopole cannot propagate since there is no field acting on it. A vortex-pair emits wake fields behind it, and the wake, in turn, affects the motion of the vortex-pair depending on the difference between the average velocity of the vortex-pair and the drift velocity. Vortex-pairs tilted initially with small angles paropagate faster than the drift waves and are not affected appreciably. However the others deviate far to the x-direction and then the position dependence of κ dominates the motion. Those travelling in the direction opposite to drift waves are certainly not affected. Because of the same reason collision processes with zero impact parameter are not affected greatly while those with non-zero impact parameters are inevitably under the influence of the wake fields.

7.3.5 CONTOUR DYNAMICS OF TEMPERATURE GRADIENT DRIVEN MODES

The density gradient has been taken into account so far. In this section we study if localized structures are also possible in the system with the temperature gradient as well. The equation for a cylindrical plasma with the temperature plasma is given by [16, 17]

$$\frac{\partial \Pi}{\partial t} + [\psi, \Pi] + \mathbf{v}_* \cdot \nabla \left(\psi - \frac{A}{2}\psi^2\right) = 0 \ . \tag{44}$$

If the nonlinearity and dispersion term are assumed to be small, Eq. (49) may be further simplified to give

$$\frac{\partial \psi}{\partial t} + [\psi, \Pi] + \mathbf{v}_* \cdot \nabla \left(\psi - \frac{A}{2}\psi^2 + \nabla^2 \psi\right) = 0 \ . \tag{45}$$

Now we look for a solution expressed as

$$\psi(\mathbf{r}, t) = \sum_\alpha \kappa_\alpha F_\alpha(|\mathbf{r} - \mathbf{r}_\alpha(t)|) \ , \tag{46}$$

where F_α is a function localized at $\mathbf{r} = \mathbf{r}_\alpha$. Substituting Eq. (46) into Eq. (45), we have

$$\sum_\alpha \frac{d\kappa_\alpha}{dt} F_\alpha + \sum_\alpha \kappa_\alpha \frac{\mathbf{r}-\mathbf{r}_\alpha}{|\mathbf{r}-\mathbf{r}_\alpha|} F'_\alpha \left\{ -\frac{d\mathbf{r}}{dt} - \sum_{\beta\neq\alpha} \kappa_\beta \frac{\hat{\mathbf{z}} \times (\mathbf{r}-\mathbf{r}_\beta)}{|\mathbf{r}-\mathbf{r}_\beta|} F'''_\beta \right.$$

$$\left. +\hat{v}_* \hat{\mathbf{z}} \times \mathbf{r} \left[1 - A\kappa_\alpha F_\alpha + \frac{F'''_\alpha}{F'_\alpha} - A \sum_{\beta\neq\alpha} \kappa_\beta F_\beta \right] \right\} = 0 . \quad (47)$$

There are several ways to determine the localized function F_α. One is to choose so as the dispersion and the nonlinearity to balance

$$A\kappa_\alpha F_\alpha + \frac{F'''_\alpha}{F'_\alpha} = 0 , \quad (48)$$

which gives

$$F_\alpha = \frac{12}{A\kappa_\alpha} \frac{1}{|\mathbf{r}-\mathbf{r}_\alpha|^2} . \quad (49)$$

In this case Eq. (47) gives

$$\frac{d\kappa_\alpha}{dt} = 0 , \quad (50)$$

$$\frac{d\mathbf{r}_\alpha}{dt} = \frac{288}{A} \sum_{\beta\neq\alpha} \frac{\hat{\mathbf{z}} \times (\mathbf{r}_\alpha-\mathbf{r}_\beta)}{|\mathbf{r}_\alpha-\mathbf{r}_\beta|^6} + \hat{v}_* \hat{\mathbf{z}} \times \mathbf{r}_\alpha \left[1 - 12\sum_{\beta\neq} \alpha \frac{1}{|\mathbf{r}_\alpha-\mathbf{r}_\beta|^2} \right] . \quad (51)$$

In this case vortices are all equal, which can be seen from Eqs. (46) and (49). More preferable choice is to have vortices with two signs and various sizes. If we put $F'''_\alpha/F'_\alpha = q_\alpha^2 = $ const, then we have $F_\alpha = e^{-q_\alpha|\mathbf{r}-\mathbf{r}_\alpha|}$. Thus Eq. (47) yields Eq. (50) and

$$\frac{d\mathbf{r}_\alpha}{dt} = \sum_{\beta\neq\alpha} \kappa_\beta q_\beta^3 \frac{\hat{\mathbf{z}} \times (\mathbf{r}_\alpha-\mathbf{r}_\beta)}{|\mathbf{r}_\alpha-\mathbf{r}_\beta|} e^{-q_\beta|\mathbf{r}_\alpha-\mathbf{r}_\beta|} + \hat{v}_\alpha \hat{\mathbf{z}} \times \mathbf{r}_\alpha \left[1 - A_\alpha \sum_{\beta\neq\alpha} \kappa_\beta e^{-q_\beta|\mathbf{r}_\alpha-\mathbf{r}_\beta|} \right] , \quad (52)$$

where $v_\alpha = (1 - A\kappa_\alpha + q_\alpha^2)\hat{v}_*$ and $A_\alpha = A/(1 - A\kappa_\alpha + q_\alpha^2)$.

Equations (51) and (52) describe the formation of vortices whose dynamical behaviors are similar to those described by Eqs. (11) and (12), though we have not studied detail yet.

7.3.6 DISCUSSION

In this paper we have formulated self-consistent vortex-wave dynamics in a magnetized plasma based on Hasegawa-Mima equation by extending the modulated point vortex model so as to include the vortex-wave interactions. Dynamical behaviors of vortices under the interaction with waves are numerically studied. A contour dynamics is also formulated for a plasma with the temperature gradient and now under study.

Bibliography

[1] M. Makino, T. Kamimura and T.Taniuchi, J. Phys. Soc. Jpn. **50**, 980 (1981).

[2] J.C. McWilliams and N.J. Zabusky, Geophys. Astrophys. Dyn. **19**, 207 (1982).

[3] J.D. Meiss and W. Horton, Phys. Fluids **26**, 990 (1983).

[4] M. Kono and E. Miyashita, Phys. Fluids **31**, 326 (1988).

[5] V. Naulin, K.H. Spatschek and A. Hasegawa, Phys. Fluids B **4**, 2672 (1992).

[6] A. Hasegawa and K. Mima, Phys. Rev. Lett. **39**, 205 (1977).

[7] N.J. Zabusky and J.C. McWilliams, Phys. Fluids **25**, 2175 (1982).

[8] M. Kono and W. Horton, Phys. Fluids B **3**, 3255 (1991).

[9] D.D. Hobson, Phys. Fluids A **3**, 3027 (1991).

[10] G.M. Reznik, Oceanology **26**, 119 (1986).

[11] V.M. Gryanik, Oceanology **26**, 126 (1986).

[12] G.M. Reznik, J. Fluid Mech. **240**, 405 (1992).

[13] J. Nycander and M.B. Isichenko, Phys. Fluids B **2**, 2042 (1990).

[14] W. Horton, Phys. Fluids B **1**, 524 (1989).

[15] S.V. Antipov, M.V. Nezlin, V.K. Rodionov, A.Yu. Rylov, E.N. Snezhkin, A.S. Trubnikov, and A.V. Khutoretskii, Sov. J. Plasma Phys. **14**, 648 (1988).

[16] W. Horton, D.-I. Choi, and W.M. Tang, Phys. Fluids **24**, 1077 (1981).

[17] H. Tasso, Phys. Lett. **96A**, 33 (1983).

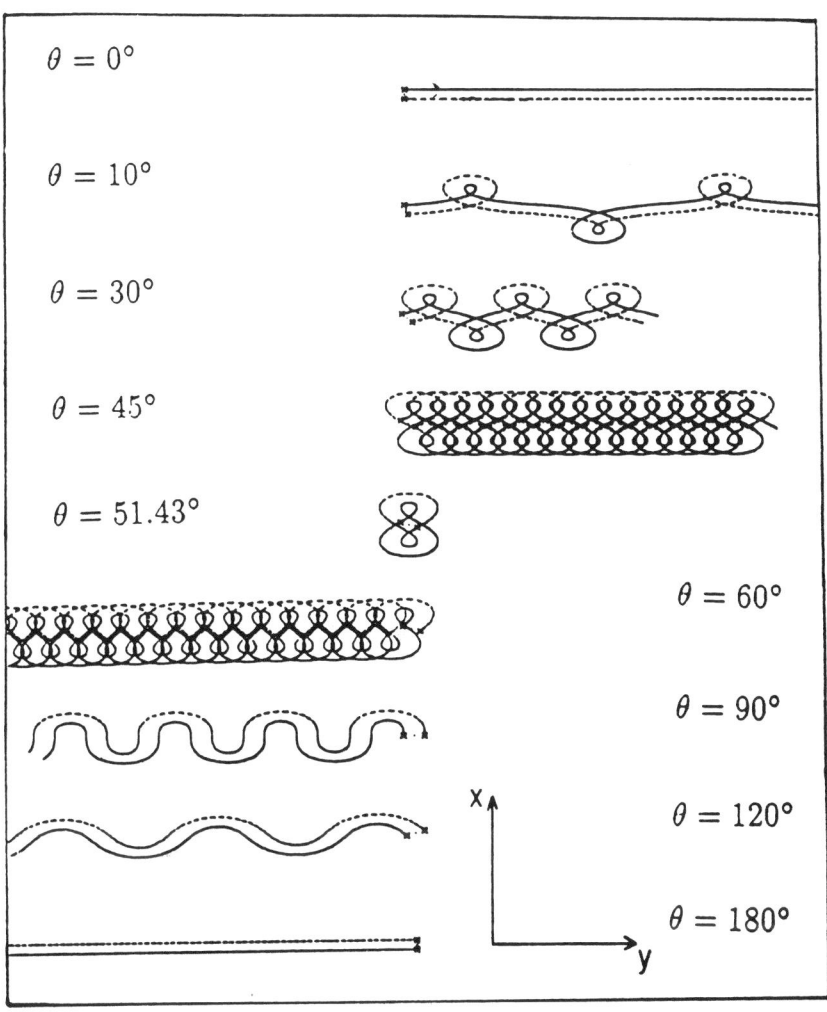

Fig. 1 Trajectories of a vortex-pair for various initial tilted angle.

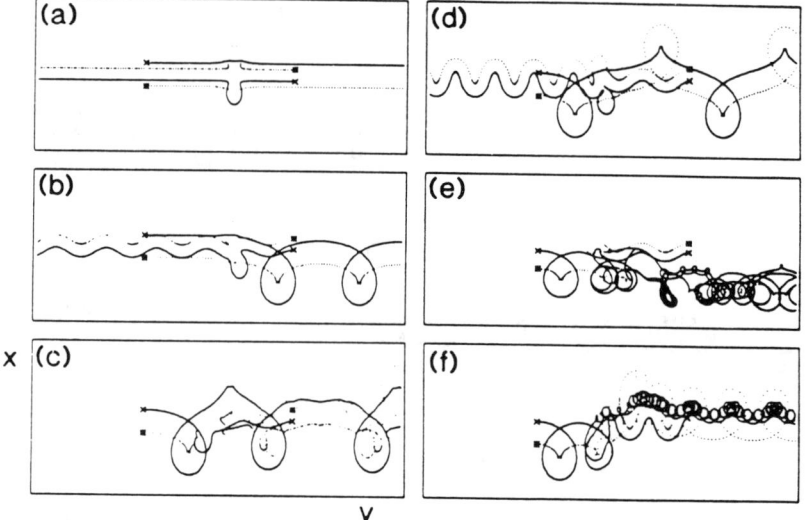

Fig. 2 Head-on collision of vortex-pairs for (a) zero and (b)–(f) non-zero impact parameters.

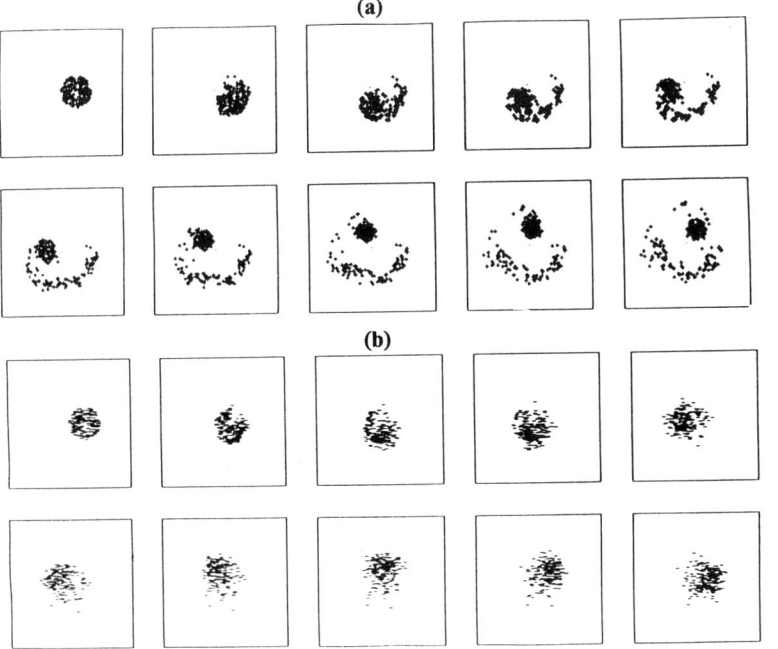

Fig. 3 Time evolution of (a) a cyclone ($\kappa > 0$) and (b) an anticyclone ($\kappa < 0$).

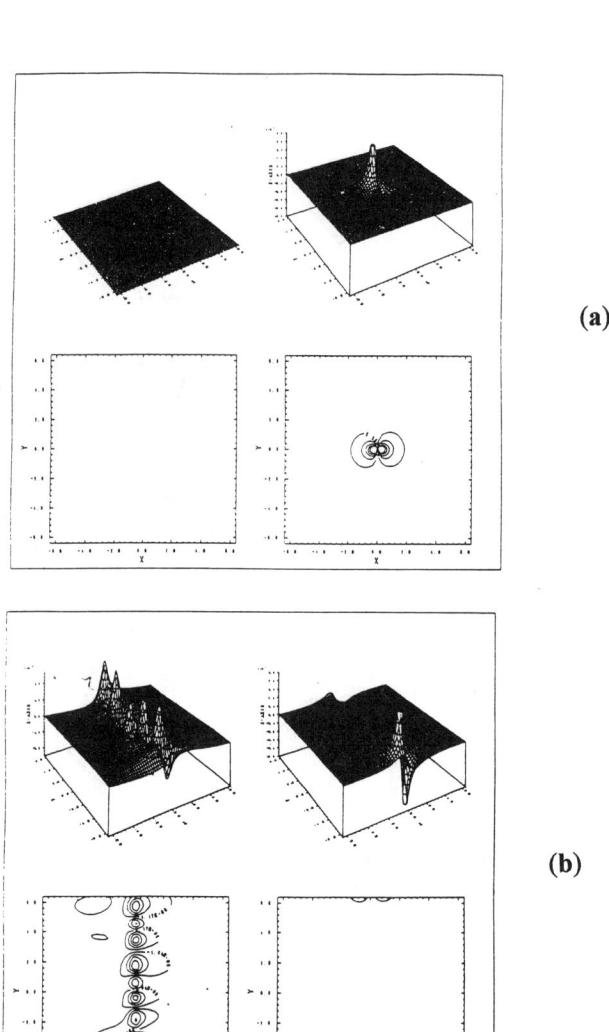

Fig. 4 Vortex-wave system at (a) the initial moment and (b) the time when the vortex-pair reaches the boundary.

7.4 COMMENTS ON SINGLE-HELICITY VERSUS MULTI-HELICITY SIMULATIONS

H. Sugama[1], A. D. Beklemishev[2], and W. Horton
Institute for Fusion Studies, The University of Texas, Austin, Texas 78712

In the sheared magnetic field, several important instabilities appear as modes radially localized in the vicinity of their resonant surfaces (or rational surfaces). There are two types of nonlinear simulations for plasma turbulence consisting of such unstable modes: one is 2D single-helicity simulation in which only modes having the same helicity and therefore the same resonant surface are included, and the other 3D multi-helicity simulation considering also different helicity modes. Naturally, 2D simulations are easier to do because of less computer memory and time. However, it is important to clarify the implications of the single-helicity simulation results for the real 3D plasma turbulence and the resultant anomalous transport. Thus, our main concern here is in the relation between the anomalous transport in the single-helicity case and that in the multi-helicity case. In the 3D multi-helicity simulation, compared to the 2D case, we have one additional parameter, namely, the density of radial distribution of unstable modes (or states). We call it the 'density of states'. The key issue in this short comment is to examine the effects of the density states on the turbulence and transport.

Calculating the density flux from electrostatic fluctuations, we easily obtain the following general expression for the diffusion coefficient:

$$D(r) = \frac{cL_n}{rB\bar{n}} \sum_{m,n} \langle \delta\phi_{mn}(r)\delta n_{-m-n}(r)\rangle \ . \tag{1}$$

---

[1] Permanent address: National Institute for Fusion Science, Nagoya 464-01, Japan
[2] Permanent address: Kurchatov Institute of Atomic Energy, Moscow 123182, Russia

Thus the transport coefficient is a function of the fluctuation spectrum. In the conventional mixing length theory [1], the structure of the spectrum is not taken into account and only the total fluctuation amplitude is estimated. Therefore, there appears no dependence of the transport coefficient on the density of states. It means that, even if the density of states increases, each mode shares the energy with one another such that the total energy is kept unchanged. This assumption is justified when the modes with different helicities overlap each other so that there are enough interactions among them.

On the other hand, it is assumed in the Beklemishev-Horton model [2, 3] that only the interactions among the modes with the same helicity are strong while the different helicity modes interact so weakly that they behave as independent modes. The energy redistribution occurs among the modes belonging to each independent subset (characterized by the same helicity) such that the total energy in the subset does not change. Then the transport coefficient is proportional to the number of independent subsets per unit radial length, i.e., the density of states. The transport coefficient in the multi-helicity case is given by

$$D_{MH}(r) = F(r)\overline{D}_{SH} \qquad (2)$$

where $\overline{D}_{SH}$ denotes the radially averaged transport transport coefficient in the single-helicity case and $F(r)$ the density of states factor defined in terms of the mode width Δr_m and the radial distribution density f_m as

$$F(r) = \sum_m f_m \Delta r_m . \qquad (3)$$

This model is considered to be valid when the radial localization of the modes is so strong that there is less overlapping of different helicity modes. This effect of the density of states is attractive because it predicts the enhancement of the anomalous transport in the tokamak edge region which cannot be explained by the conventional mixing length models such as the gyro-Bohm diffusivity.

In order to investigate the density-of-states effects, we have done 2D and 3D simulations of the resistive g turbulence in the sheared slab configuration. We used the reduced MHD equations in the electrostatic approximation:

$$\frac{\rho_m c}{B_0}\left(\frac{\partial}{\partial t} - \mu_\perp \nabla_\perp^2 + \frac{c}{B_0}\hat{z} \times \nabla\phi \cdot \nabla\right)\nabla_\perp^2 \phi = -\frac{B_0}{c\eta}\nabla_\parallel^2 \phi - \Omega'\frac{\partial \tilde{p}}{\partial y} \quad (4)$$

$$\left(\frac{\partial}{\partial t} - \chi_\perp \nabla_\perp^2 + \chi_\parallel |\nabla_\parallel| + \frac{c}{B_0}\hat{z} \times \nabla\phi \cdot \nabla\right)\tilde{p} = \frac{c}{B_0}P_0'\frac{\partial \phi}{\partial y} \quad (5)$$

where ϕ is the electrostatic potential, $\tilde{p}$ the pressure fluctuation, ρ_m the average mass density, η the resistivity, $\chi_\perp$ the perpendicular pressure diffusivity, $\mu_\perp$ the kinematic viscosity, $P_0' \equiv dP_0/dx$ (<0) the background pressure gradient and $\Omega' \equiv d\Omega/dx$ (>0) the average curvature of the magnetic field line. The gradient along the static sheared magnetic field line is given by $\nabla_\parallel = \frac{\partial}{\partial z} + \frac{x}{L_s}\frac{\partial}{\partial t}$. Here the background pressure gradient is fixed and the Landau damping type of parallel diffusivity $\chi_\parallel$ is included in the pressure equation to control the radial localization of the modes. We employed the following units for normalization of the variables:

$$\begin{aligned}
&[t] = (-P_0'\Omega'/\rho_m)^{-1/2} &&[x] = [y] = cL_s\eta^{1/2}(-\rho_m P_0'\Omega')^{1/4}/B_0 \\
&[z] = L_s &&[\chi_\perp] = [\mu_\perp] = [x]^2/[t] = c^2\eta(-P_0')\Omega' L_s^2/B_0^2 \\
&[\phi] = c\eta(-P_0')\Omega' L_s^2/B_0 &&[\tilde{p}] = cL_s\eta^{1/2}\rho_m^{1/4}(-P_0')^{5/4}\Omega'^{1/4}/B_0 \\
&[\chi_\parallel] = [z]/[t] = L_s(-P_0'\Omega'/\rho_m)^{1/2}
\end{aligned} \quad (6)$$

Preliminarily, we had two cases of multi-helicity simulations. In these simulations, we included all the modes with the poloidal and toroidal mode numbers (m, n) satisfying $m \leq 50$, $n \leq 70$ and $4/3 < n/m < 3/2$ (the total number of the modes was 184). The parallel wavenumber and the rotational transform are given as $k_\parallel = (m\iota - n)/L_z$ and $\iota = (L_z/L_y)x + 17/12$ in our units where L_y and L_z are the period lengths in the y and z directions, respectively, and we put the origin of x at $\iota = 17/12$. In Case I, we used $L_y = 300 \times 2\pi$, $L_z = (2/3) \times 2\pi$, $\chi_\perp = \mu_\perp = 0.25$ and $\chi_\parallel = 3.0$ while in Case II only the value of L_z was changed as $L_z = (4/3) \times 2\pi$ so that the rational surface density in Case II is twice higher than in Case I. The simulation results are found in Figs. 1–3. The contours of potential ϕ at the saturated states in Case I and II are shown in Fig. 1. Figures 2 and 3 show the convective flux and the transport coefficients in Case I and II, respectively.

In the both cases, turbulent energy tends to concentrate on the modes with lower mode numbers and the profile of the turbulent transport has peaks at the corresponding low mode number rational surfaces. Other modes, which are located around the peaks within the width of the low wavenumber modes, have small amplitudes since their energy is transferred into the peaked low wavenumber modes by the inverse cascade process. In the higher rational surface density case, i.e., in Case II, the energy concentrates dominantly on the lowest wavenumber mode $(m, n) = (5, 7)$. We found that the peak values of transport in the both multi-helicity cases are roughly the same as each other and they are also the same as in the single-helicity case. These results show that the interactions between neighboring modes with different helicities promote the energy inverse cascade and reduce the number of the rational surfaces contributing to transport. Thus, the total transport is not simply proportional to the unstable mode rational surface density and the density of state theory is not applicable in our preliminary simulation cases. However, our results does not support the conventional model either since, instead of the energy equipartition between the different helicity modes, we find the energy inverse cascade or coalescence of vortices yielding the complicated transport profile. Up to this point, our results is too preliminary to derive clear conclusion on the density-of-states effects and further investigation on this subject is required.

Bibliography

[1] J.W. Connor, to be published in Phys. Fluids B **5** (1993).

[2] A.D. Beklemishev and W. Horton, Phys. Fluids B **4**, 200 (1992).

[3] A.D. Beklemishev and W. Horton, Phys. Fluids B **4**, 2176 (1992).

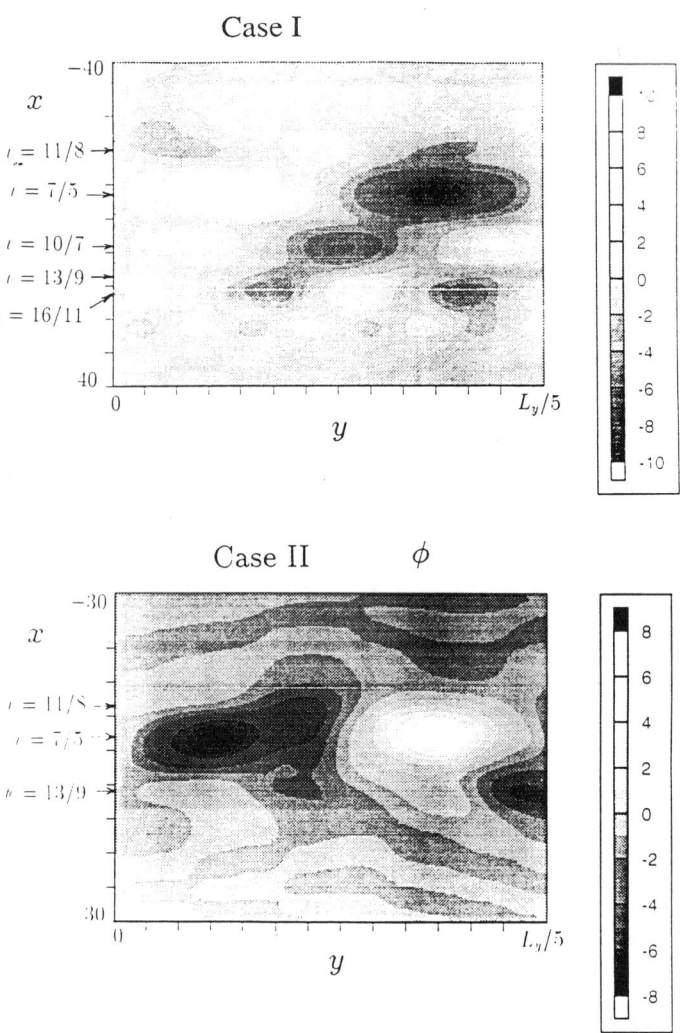

Figure 1. The contours of potential ϕ in the saturated states. The rational surface density in Case II (bottom) is twice higher than in Case I (top).

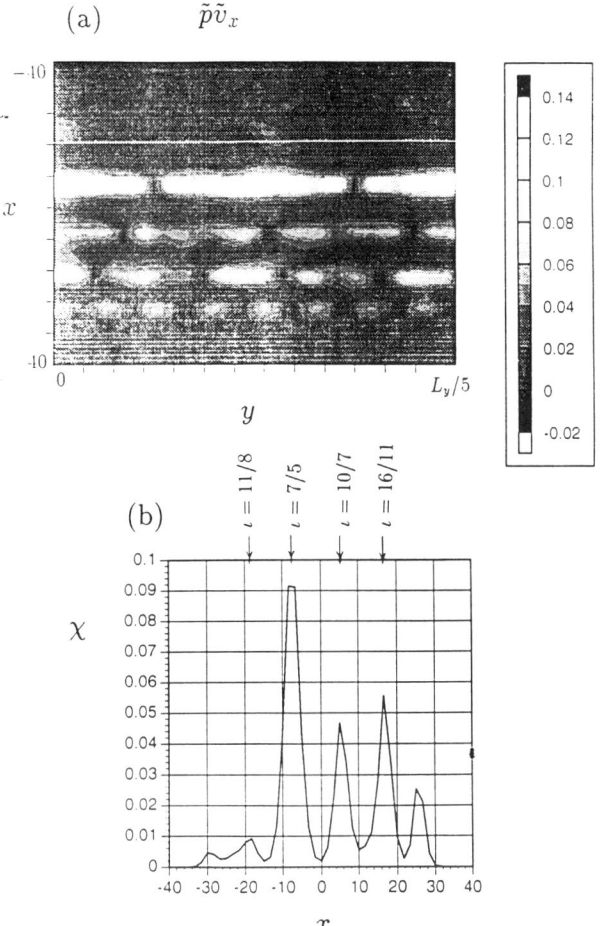

Figure 2. The convective flux and the transport coefficient at the saturated state in Case I. (a) The contour of the convective flux $\widetilde{p}\widetilde{v}_x$ in the (x,y) plane at $z=0$ and $t=2000$. (b) The radial profile of the transport coefficient defined by $\chi \equiv \frac{1}{L_y L_z} \int_0^{L_y} \int_0^{L_z} dy\, dz\, \widetilde{p\widetilde{v}_x}/\left(-\frac{dP_0}{dx}\right)$ at $t=2000$.

586 Single-Helicity Versus Multi-Helicity

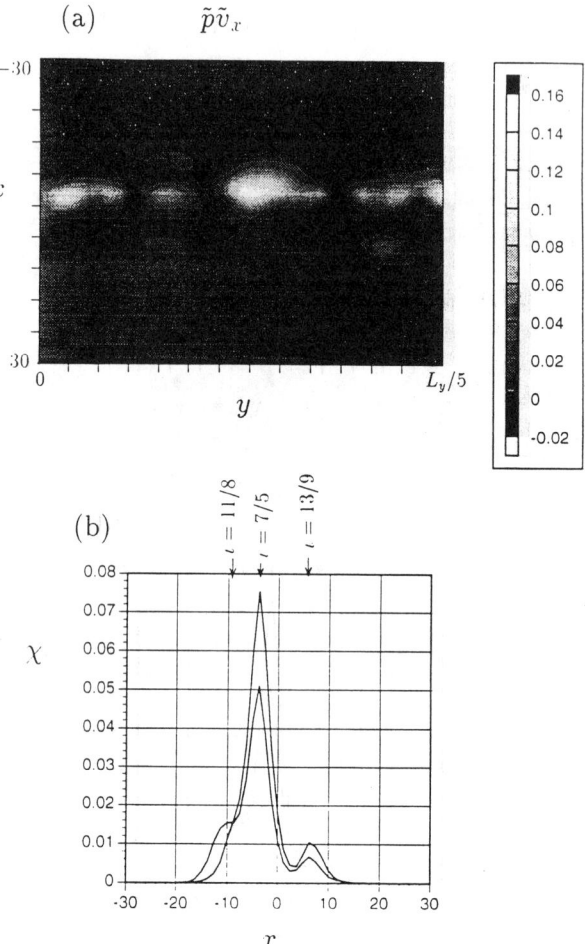

Figure 3. The convective flux and the transport coefficient in Case II where the rational surface density is twice higher than in Case I (Fig. 2). (a) The contour of the convective flux $\widetilde{pv}_x$ in the (x, y) plane at $z = 0$ and $t = 3000$. (b) The radial profile of the transport coefficient defined by $\chi \equiv \frac{1}{L_y L_z} \int_0^{L_y} \int_0^{L_z} dy dz\, \widetilde{pv}_x / \left(-\frac{dP_0}{dx}\right)$ at $t = 2000, 3000$. The transport coefficient at $t = 3000$ shows more peaked profile around the lowest mode number rational surface $\iota = 7/5$ than at $t = 2000$.

Subject Index

adiabatic electron density, 147, 349
ALCATOR-C ion diamagnetic drift direction, 174
anisotropic η_i, 137, 142
anomalous diffusivity D, 518
anomalous edge transport, 526
anomalous ion heat transport, 22
anomalous particle and electron energy pinches, 444
anomaly factor, 116
ATF, 526

ballooning mode representation, 310, 372, 393
   3D nonlinear, 315
beam contribution, 279
beam power scans, 43
beam-heated plasmas, 25
Beklemishev-Horton model, 580
Bohm transport, 39

charge-exchange recombination spectroscopy (CXRS), 26
circulating electrons, 445
Columbia Linear Machine (CLM) transport studies, 135
compact helical system (CHS), 188, 526
confinement degradation, 170
connection machine, 213
continuous eigenvalue spectrum, 428, 429, 431, 434
contour dynamics of temperature gradient driven modes, 559, 572
coupling between the g mode and the η_i mode, 528
critical η_c, 430
cross-field energy flux, 428, 432
current profile experiment, 414

density of states factor, 580
diagnostics, 27
DIII-D off-axis ECH experiment, 300, 444
dissipative trapped electron mode, 145

DKES code, 191
Dominguez and Waltz model, 98
drift vortices, 559
drift wave turbulence (DWT) models, 191

$E \times B$ drifts, 168
$E \times B$ rotation, 138
ECH plasmas in Heliotron E, 519
edge turbulence, 412, 527
eikonal representation, 393
electromagnetic effect, 408
electron cyclotron heated (ECH) plasmas, 444
electron particle flux Γ_e, 373
electron thermal transport, 189
   η_i mode stability, 100
   η_i mode turbulence in JT-60, 87
experimental evidence for ion temperature gradient driven turbulence in tokamaks, 165

fast-ion diffusion, 46
finite $k_\parallel$ effect, 408
finite-β stabilization of micro-instabilities, 213
fluctuation levels driven by the g mode turbulence, 535
'flute'-like mode, 145
free streaming ion effect, 428

gas puffing, 39
GIOTA code, 190
global energy confinement time, 35
global linear radial mode width, 264
g-mode turbulence, 193
Gyro-Landau fluid (GFL) models, 310
gyro-reduced Bohm (GRB) difffusivity, 189, 325, 519
gyrofluid equations, 345
gyrokinetic equation, 429
gyrokinetic initial value code, 277
gyrokinetic integral equation, 492

gyrokinetic particle simulations, 213, 224, 227

Hamaguchi and Horton model, 89
Hamiltonian dynamics, 561
Hasegawa-Mima equation, 559
helically trapped electron mode, 192
Heliotron E plasmas, 509, 526, 535, 541
historical survey of χ_i, 24
homogeneous turbulence, 509
hot beam ions effects, 405
hot-ion H-mode, 25

ideal ballooning mode, 409
image modes, 326
improved ohmic confinement regime (IOC), 23
impurities, 279
impurity drift mode, 470, 486
impurity effect, 406, 486
impurity modes, 469
inverse cascade, 582
inward pinch of impurities, 213
ion feature, 165
ion heat diffusivity as a function of local ion temperature, 45
ion mode, 165
$\nabla T_{i\parallel}$ production
  by D.C. ion heating, 137
  by transit time RF heating, 140
T_i/T_e effect, 405
ion transport in Ohmic plasmas, 29
isotopic stabilization effect, 494
ITG instability, 135
ITG ion thermal transport, 146

JET experiments, 302, 473
JET pellet enhanced performance, 392
JIPP T-IIU ion diamagnetic, 174
JT-60 discharges, 433
JT-60 L-mode ion transport analysis, 98

K-ϵ anomalous transport model, 509, 515
Kelvin-Helmholtz instability, 540

kinetic and fluid model equations for η_i modes, 101
kinetic toroidal analysis, 371
Krook relaxation collision operator, 280

L-mode plasmas, 39
Landau damping model, 311, 344
Large Helical Device (LHD), 188
Large Helical Device (LHD) transport simulations, 196
LHD scaling, 509, 521
Linsker effect, 314
low shear limit, 397

marginal stability ϵ_{T_c}, 258
marginal stability hypothesis, 296
marginal stability of experimental profiles, 276
measured χ_i^{tot} versus heating power, 44
measurements, 19
Mercier criterion, 526, 532
microturbulence propagating in the ion diamagnetic drift, 165, 168
mixed plasmas, 495
mixing length formula, 372, 411, 533
modeling of anomalous ion thermal transport, 403
moderate shear limit, 394
modulated point vortex model, 559
Monte-Carlo code AURORA, 196
multi-helicity, 579
multiple helicity (m,n,r) representation, 311
multiplier on neoclassical χ_i, 37

neoclassical effect on η_i mode, 108
neoclassical magnetic pumping, 320
neoclassical MHD equations, 109
neoclassical models of thermal conductivities, 190
neoclassical multiplier, 49
neutron emission, 29, 33, 34
nondimensional scans, 39
nonlinear FLR phase-mixing, 344
nonlinear gyrofluid model, 344
nonlinear simulation of η_i mode turbulence, 88

Subject Index 589

nonlinear theory of g mode, 529
nonquasilinear (NQL) simulation, 93

parallel and transverse ion temperature gradient, 136
parallel velocity shear, 91
parallel wavelength, 139
parametric variations, 19
particle diffusion, 147
peaked $n_e(r)$, 168
peaked T_i profile, 137
pellet injection experiments, 22, 118
perpendicular shear flow, 359
perpendicular velocity shear, 344
pinch fluxes, 296
plasma edge, 469
point vortices, 559
poloidal shear flow effect, 526
PROCTR-MOD code, 194
profile consistency, 296

quasilinear (QL) effect on slab η_i modes, 90
quasilinear (QL) simulation, 93
quasilinear particle and energy fluxes, 372, 375, 449
quasilinear transport equations, 448

radial correlation length L_r, 391
radial correlation length of the magnetic fluctuation, 536
radial electric field, 170
radial electric field effects on the g mode, 539
radial mode width, 411
radial modes, 319, 327
radial profile problem, 403
radially elongated turbulence, 392
rational surface model
  2D single helicity, 232
  3D multiple helicity, 235
reactive drive wave model (RDWM), 295
relaxation model, 260, 413
resistive g turbulence, 580
resistive drift waves and η_i modes, 537, 538
resistive interchange, 193

resistive interchange instabilities in Heliotron E, 509, 526
resistive stability criterion $D_R >$, 532
resonance broadening, 452
Reynolds stress, 528
ripple transport, 190

sawtooth type relaxation oscillation, 527
scale invariance, 533
self-organized relaxed state, 255
shear velocity flows, 172
sheared E×B rotation, 319
sheared slab geometry, 491
sheared velocity flows, 345
shearless slab approximation, 488
simulation algorithms utilizing massively parallel architecture, 213
simulation issues, 351
single-helicity simulation, 579
slab branch of ITG mode, 137
SNAP, 28
stabilizing factors, 403
strong ballooning mode, 431
strong coupling approximation, 395
subcritical turbulence, 325
supershot, 25
Suydam criterion, 526
Suydam mode, 532

temperature dependence of χ_i, 49
temperature profile relaxation, 147
temperature profiles for Ohmic, 31
temperature relaxation, 255
temperatures in the Ohmic, 32
test of η_i and trapped electron model for JT-60 L-mode, 105
Texas Experimental Tokamak (TEXT), 165
TFTR confinement experiments, 19, 371
TFTR current ramp experiment, 392
TFTR data sets, 277
TFTR density and power scans, 33
TFTR L-mode shot, 372, 375
TFTR shot, supershot, 374
TFTR supershot discharges, 412
thermal conduction, 147

thermal diffusivity χ_i^{Exp} in TEXT, 116
toroidal η_i mode turbulence, 99
toroidal ITG mode, 143
toroidal kinetic code, 372
toroidal magnetic configuration, 492
toroidal microinstability code, 371
toroidal particle code (TPC), 256
transport analysis, 27
transport of the turbulent energy, 513
transport simulations for the Large Helical Device (LHD), 196
transport studies in the Columbia Linear Machine (CLM), 136
trapped electron effect, 101
trapped electron mode, 192
turbulent energy production, 510, 515
turbulent pinch mechanism, 445
turbulent resonance broadening, 447

two-dimensional toroidal fluid model, 297, 393
Two-Scale Direct-Interaction Approximation (TSDIA), 511
TWODQ double integral, 430

variation of χ_i with temperature, 38
variation of thermal diffusivity with density, 40
various stabilizing factors in the core region, 404
viscous dissipation rate, 510
VMEC code, 194
vortex dynamics, 559
vortex-pair, 562
vortex-wave dynamics, 566
weak turbulence, 360
Z-mode discharge, 487

AUTHOR INDEX

A

Amano, T., 188
Azumi, M., 60, 87

B

Barnes, C. W., 19
Beer, M. A., 344
Beklemishev, A. D., 579
Bell, M. G., 19
Bell, R. E., 19
Brower, D. L., 165
Bush, C. E., 19

C

Chen, J., 135
Cummings, J. C., 213

D

Dominguez, R. R., 444
Dong, J. Q., 486
Dorland, W., 344

E

Ernst, D., 19

F

Fredrickson, E. D., 19

G

Gray, M. G., 255
Greaves, R. G., 135
Grek, B., 19

H

Hahm, T. S., 344
Hammett, G. W., 344
Hill, K. W., 19
Hirayama, T., 60
Horton, W., 255, 403, 486, 579

J

Janos, A., 19
Jassby, D. L., 19
Johnson, D., 19

K

Kerbel, G. D., 310
Kim, J. Y., 255, 403
Kishimoto, Y., 255
Koide, Y., 60
Kono, M., 559
Kotschenreuther, M., 255, 276

L

LeBrun, M. J., 255
Lee, W. W., 213

M

Mansfield, D. K., 19
Migliuolo, S., 469

N

Nordman, H., 295

O

Okamoto, M., 509
Ouroua, A., 116
Owens, D. K., 19

P

Park, H., 19
Parker, S. E., 213

R

Ramsey, A. T., 19
Rewoldt, G., 371
Romanelli, F., 391

S

Santoro, R. A., 213
Schivell, J., 19
Scott, S. D., 19
Sen, A. K., 135
Shibahara, H., 559
Shirai, H., 60
Song, B., 135
Stratton, B. C., 19
Su, X. N., 486
Sugama, H., 509, 579
Sydora, R. D., 224

Synakowski, E. J., 19

T

Tajima, T., 255
Tang, W. M., 371
TEXT Group, 116
Thompson, M., 19

W

Wakatani, M., 509, 526
Waltz, R. E., 310
Watanabe, K. Y., 188
Weiland, J., 295
Wong, V., 255
Wootton, A. J., 116

Y

Yabuki, K., 559
Yagi, M., 87
Yamada, H., 188
Yamagishi, T., 428
Yamazaki, K., 188

Z

Zarnstorff, M. C., 19

AIP Conference Proceedings

	L.C. Number	ISBN
No. 197 Drops and Bubbles (Monterey, CA, 1988)	89-46360	0-88318-392-7
No. 198 Astrophysics in Antarctica (Newark, DE, 1989)	89-46421	0-88318-398-6
No. 199 Surface Conditioning of Vacuum Systems (Los Angeles, CA, 1989)	89-82542	0-88318-756-6
No. 200 High T_c Superconducting Thin Films: Processing, Characterization, and Applications (Boston, MA, 1989)	90-80006	0-88318-759-0
No. 201 QED Structure Functions (Ann Arbor, MI, 1989)	90-80229	0-88318-671-3
No. 202 NASA Workshop on Physics From a Lunar Base (Stanford, CA, 1989)	90-55073	0-88318-646-2
No. 203 Particle Astrophysics: The NASA Cosmic Ray Program for the 1990s and Beyond (Greenbelt, MD, 1989)	90-55077	0-88318-763-9
No. 204 Aspects of Electron-Molecule Scattering and Photoionization (New Haven, CT, 1989)	90-55175	0-88318-764-7
No. 205 The Physics of Electronic and Atomic Collisions (XVI International Conference) (New York, NY, 1989)	90-53183	0-88318-390-0
No. 206 Atomic Processes in Plasmas (Gaithersburg, MD, 1989)	90-55265	0-88318-769-8
No. 207 Astrophysics from the Moon (Annapolis, MD, 1990)	90-55582	0-88318-770-1
No. 208 Current Topics in Shock Waves (Bethlehem, PA, 1989)	90-55617	0-88318-776-0
No. 209 Computing for High Luminosity and High Intensity Facilities (Santa Fe, NM, 1990)	90-55634	0-88318-786-8
No. 210 Production and Neutralization of Negative Ions and Beams (Brookhaven, NY, 1990)	90-55316	0-88318-786-8
No. 211 High-Energy Astrophysics in the 21st Century (Taos, NM, 1989)	90-55644	0-88318-803-1
No. 212 Accelerator Instrumentation (Brookhaven, NY, 1989)	90-55838	0-88318-645-4
No. 213 Frontiers in Condensed Matter Theory (New York, NY, 1989)	90-6421	0-88318-771-X 0-88318-772-8 (pbk.)

No. 214	Beam Dynamics Issues of High-Luminosity Asymmetric Collider Rings (Berkeley, CA, 1990)	90-55857	0-88318-767-1
No. 215	X-Ray and Inner-Shell Processes (Knoxville, TN, 1990)	90-84700	0-88318-790-6
No. 216	Spectral Line Shapes, Vol. 6 (Austin, TX, 1990)	90-06278	0-88318-791-4
No. 217	Space Nuclear Power Systems (Albuquerque, NM, 1991)	90-56220	0-88318-838-4
No. 218	Positron Beams for Solids and Surfaces (London, Canada, 1990)	90-56407	0-88318-842-2
No. 219	Superconductivity and Its Applications (Buffalo, NY, 1990)	91-55020	0-88318-835-X
No. 220	High Energy Gamma-Ray Astronomy (Ann Arbor, MI, 1990)	91-70876	0-88318-812-0
No. 221	Particle Production Near Threshold (Nashville, IN, 1990)	91-55134	0-88318-829-5
No. 222	After the First Three Minutes (College Park, MD, 1990)	91-55214	0-88318-828-7
No. 223	Polarized Collider Workshop (University Park, PA, 1990)	91-71303	0-88318-826-0
No. 224	LAMPF Workshop on (π, K) Physics (Los Alamos, NM, 1990)	91-71304	0-88318-825-2
No. 225	Half Collision Resonance Phenomena in Molecules (Caracas, Venezuela, 1990)	91-55210	0-88318-840-6
No. 226	The Living Cell in Four Dimensions (Gif sur Yvette, France, 1990)	91-55209	0-88318-794-9
No. 227	Advanced Processing and Characterization Technologies (Clearwater, FL, 1991)	91-55194	0-88318-910-0
No. 228	Anomalous Nuclear Effects in Deuterium/Solid Systems (Provo, UT, 1990)	91-55245	0-88318-833-3
No. 229	Accelerator Instrumentation (Batavia, IL, 1990)	91-55347	0-88318-832-1
No. 230	Nonlinear Dynamics and Particle Acceleration (Tsukuba, Japan, 1990)	91-55348	0-88318-824-4
No. 231	Boron-Rich Solids (Albuquerque, NM, 1990)	91-53024	0-88318-793-4
No. 232	Gamma-Ray Line Astrophysics (Paris-Saclay, France, 1990)	91-55492	0-88318-875-9

No.	Title		
No. 233	Atomic Physics 12 (Ann Arbor, MI, 1990)	91-55595	088318-811-2
No. 234	Amorphous Silicon Materials and Solar Cells (Denver, CO, 1991)	91-55575	088318-831-7
No. 235	Physics and Chemistry of MCT and Novel IR Detector Materials (San Francisco, CA, 1990)	91-55493	0-88318-931-3
No. 236	Vacuum Design of Synchrotron Light Sources (Argonne, IL, 1990)	91-55527	0-88318-873-2
No. 237	Kent M. Terwilliger Memorial Symposium (Ann Arbor, MI, 1989)	91-55576	0-88318-788-4
No. 238	Capture Gamma-Ray Spectroscopy (Pacific Grove, CA, 1990)	91-57923	0-88318-830-9
No. 239	Advances in Biomolecular Simulations (Obernai, France, 1991)	91-58106	0-88318-940-2
No. 240	Joint Soviet-American Workshop on the Physics of Semiconductor Lasers (Leningrad, USSR, 1991)	91-58537	0-88318-936-4
No. 241	Scanned Probe Microscopy (Santa Barbara, CA, 1991)	91-76758	0-88318-816-3
No. 242	Strong, Weak, and Electromagnetic Interactions in Nuclei, Atoms, and Astrophysics: A Workshop in Honor of Stewart D. Bloom's Retirement (Livermore, CA, 1991)	91-76876	0-88318-943-7
No. 243	Intersections Between Particle and Nuclear Physics (Tucson, AZ, 1991)	91-77580	0-88318-950-X
No. 244	Radio Frequency Power in Plasmas (Charleston, SC, 1991)	91-77853	0-88318-937-2
No. 245	Basic Space Science (Bangalore, India, 1991)	91-78379	0-88318-951-8
No. 246	Space Nuclear Power Systems (Albuquerque, NM, 1992)	91-58793	1-56396-027-3 1-56396-026-5 (pbk.)
No. 247	Global Warming: Physics and Facts (Washington, DC, 1991)	91-78423	0-88318-932-1
No. 248	Computer-Aided Statistical Physics (Taipei, Taiwan, 1991)	91-78378	0-88318-942-9
No. 249	The Physics of Particle Accelerators (Upton, NY, 1989, 1990)	92-52843	0-88318-789-2

No. 250	Towards a Unified Picture of Nuclear Dynamics (Nikko, Japan, 1991)	92-70143	0-88318-951-8
No. 251	Superconductivity and its Applications (Buffalo, NY, 1991)	92-52726	1-56396-016-8
No. 252	Accelerator Instrumentation (Newport News, VA, 1991)	92-70356	0-88318-934-8
No. 253	High-Brightness Beams for Advanced Accelerator Applications (College Park, MD, 1991)	92-52705	0-88318-947-X
No. 254	Testing the AGN Paradigm (College Park, MD, 1991)	92-52780	1-56396-009-5
No. 255	Advanced Beam Dynamics Workshop on Effects of Errors in Accelerators, Their Diagnosis and Corrections (Corpus Christi, TX, 1991)	92-52842	1-56396-006-0
No. 256	Slow Dynamics in Condensed Matter (Fukuoka, Japan, 1991)	92-53120	0-88318-938-0
No. 257	Atomic Processes in Plasmas (Portland, ME, 1991)	91-08105	0-88318-939-9
No. 258	Synchrotron Radiation and Dynamic Phenomena (Grenoble, France, 1991)	92-53790	1-56396-008-7
No. 259	Future Directions in Nuclear Physics with 4π Gamma Detection Systems of the New Generation (Strasbourg, France, 1991)	92-53222	0-88318-952-6
No. 260	Computational Quantum Physics (Nashville, TN, 1991)	92-71777	0-88318-933-X
No. 261	Rare and Exclusive B&K Decays and Novel Flavor Factories (Santa Monica, CA, 1991)	92-71873	1-56396-055-9
No. 262	Molecular Electronics—Science and Technology (St. Thomas, Virgin Islands, 1991)	92-72210	1-56396-041-9
No. 263	Stress-Induced Phenomena in Metallization: First International Workshop (Ithaca, NY, 1991)	92-72292	1-56396-082-6
No. 264	Particle Acceleration in Cosmic Plasmas (Newark, DE, 1991)	92-73316	0-88318-948-8
No. 265	Gamma-Ray Bursts (Huntsville, AL, 1991)	92-73456	1-56396-018-4
No. 266	Group Theory in Physics (Cocoyoc, Morelos, Mexico, 1991)	92-73457	1-56396-101-6
No. 267	Electromechanical Coupling of the Solar Atmosphere (Capri, Italy, 1991)	92-82717	1-56396-110-5

No. 268	Photovoltaic Advanced Research & Development Project (Denver, CO, 1992)	92-74159	1-56396-056-7
No. 269	CEBAF 1992 Summer Workshop (Newport News, VA, 1992)	92-75403	1-56396-067-2
No. 270	Time Reversal—The Arthur Rich Memorial Symposium (Ann Arbor, MI, 1991)	92-83852	1-56396-105-9
No. 271	Tenth Symposium Space Nuclear Power and Propulsion (Vols. I–III) (Albuquerque, NM, 1993)	92-75162	1-56396-137-7 (set)
No. 272	Proceedings of the XXVI International Conference on High Energy Physics (Vols. I and II) (Dallas, TX, 1992)	93-70412	1-56396-127-X (set)
No. 273	Superconductivity and Its Applications (Buffalo, NY, 1992)	93-70502	1-56396-189-X
No. 274	VIth International Conference on the Physics of Highly Charged Ions (Manhattan, KS, 1992)	93-70577	1-56396-102-4
No. 275	Atomic Physics 13 (Munich, Germany, 1992)	93-70826	1-56396-057-5
No. 276	Very High Energy Cosmic-Ray Interactions: VIIth International Symposium (Ann Arbor, MI, 1992)	93-71342	1-56396-038-9
No. 277	The World at Risk: Natural Hazards and Climate Change (Cambridge, MA, 1992)	93-71333	1-56396-066-4
No. 278	Back to the Galaxy (College Park, MD, 1992)	93-71543	1-56396-227-6
No. 279	Advanced Accelerator Concepts (Port Jefferson, NY, 1992)	93-71773	1-56396-191-1
No. 280	Compton Gamma-Ray Observatory (St. Louis, MO, 1992)	93-71830	1-56396-104-0
No. 281	Accelerator Instrumentation Fourth Annual Workshop (Berkeley, CA, 1992)	93-072110	1-56396-190-3
No. 282	Quantum 1/f Noise & Other Low Frequency Fluctuations in Electronic Devices (St. Louis, MO, 1992)	93-072366	1-56396-252-7